MICHAEL MENAKER

Editor

Biochronometry

PROCEEDINGS

OF A

SYMPOSIUM

Friday Harbor

Washington

September 1969

NATIONAL ACADEMY

OF SCIENCES

Washington, D.C.

1971

This symposium was supported by the
National Aeronautics and Space Administration
(Contract NSR 09–012–077)

ISBN 0–309–01866–8

Available from:

Printing and Publishing Office
National Academy of Sciences
2101 Constitution Avenue
Washington, D.C. 20418

Library of Congress Catalog Card Number 70–612078

Printed in the United States of America

Foreword

In 1729 the French astronomer Jean Jacques de Mairan recorded in the *Histoire de l'Académie Royale des Sciences* the observation of periodic movements of leaves in total darkness—apparently the first scientific evidence of the existence of an endogenous periodicity. Indeed, these observations were subsequently confirmed at least twice during the eighteenth century. By the early twentieth century the case for endogenous periodicities—especially those approximating 24 hours (later to be designated as *circadian*)—was well documented, even though far from generally accepted. The recognition of the phenomenon of entrainment of the endogenous periodicity into a daily periodicity, together with the role therein of the *Zeitgeber*, has had a tortuous history fraught with argument. But perhaps even more difficult has been the development of the concept of endogenous (circadian) periodicities as the bases of chronometric functions.

The Symposium on Biological Clocks in June 1960 at Cold Spring Harbor, organized under the chairmanship of C. S. Pittendrigh, contributed extensively to a crystallization of the concept of a biochronometry based on endogenous (circadian) periodicities. This concept of chronometry based on circadian function, naturally entrained into daily periodicities, was strongly reinforced at the highly successful Summer School on Circa-

dian Clocks at Feldafing in Bavaria in September 1964, organized by
Jürgen Aschoff.

At its meeting of January 12–13, 1968, the Executive Committee of the
Division of Biology and Agriculture of the National Research Council
reached the conclusion that so much significant progress in biochronometry had occurred since the summer school at Feldafing that it was highly
desirable that arrangements be made for an international symposium that
would convene many of the leading investigators from throughout the
world. Through kind offices of Professor Robert Fernald, Director of the
Friday Habor Laboratories, arrangements were made with the University
of Washington to convene the Symposium at Friday Harbor in June 1969.
Michael Menaker, Department of Zoology, University of Texas, Austin,
kindly agreed to serve as chairman of the organizing committee. It is clear
from the papers presented in this volume that our knowledge and understanding of biochronometry have increased substantially since 1964 and
that this interesting and important field has attracted to it a corps of able
young investigators.

DONALD S. FARNER
Chairman
Division of Biology and Agriculture

Preface

In the published record of this Symposium and in the many hours of informal discussion that occurred while it was in session, one senses the beginning of an important change. The study of circadian rhythmicity has been unified primarily by the remarkable degree of formal similarity among rhythms at very different levels of biological organization. In terms of its input–output relations, especially its response to light and temperature signals, the circadian "black box" has remarkably consistent, if complex properties. It is still very much an open question whether or not these similarities reflect a common time-keeping mechanism at the cellular level; it seems likely that they do reflect, to some degree, common adaptive significance but that they do not reflect a common mechanism at the physiological and biochemical level in multicellular organisms.

In contrast to the Cold Spring Harbor and Feldafing symposia, this volume contains many papers that analyze circadian systems at the physiological level and that lead one to expect further rapid progress. As this level of analysis is pursued, major differences among groups of organisms seem certain to appear. Vigorous growth of this field gives promise of greatly increased understanding but carries with it the risk of fragmentation. It will become ever more difficult to maintain a sense of unity among workers studying vertebrate brains, the leaves of higher plants, single invertebrate neurons, and fungal growth patterns. But the necessary effort must be made, for there is an essential meeting ground in the overriding interest that all of us have in the temporal organization of living systems. On that ground each of us has much to learn from those who study systems with mechanisms that may differ widely from those operating in his particular experimental subject.

We are grateful to Dorothy Farner, who ably handled the local arrangements, to the staff of the Friday Harbor Station, and to Celeste Cromack, who helped with typing and general organization of the manuscripts. The editor is especially grateful to Russell Stevens of the staff of the National Research Council, who provided invaluable assistance in organizing the symposium and preparing this volume and to Linda Moore who shared importantly in the task of scientific editing.

ORGANIZING COMMITTEE FOR THE SYMPOSIUM
ON BIOCHRONOMETRY
Michael Menaker, *Chairman*
Donald S. Farner
Eberhard Gwinner
J. Woodland Hastings
Frank B. Salisbury
Felix Strumwasser

Contents

II. PHOTOPERIODIC TIME MEASUREMENTS

III. PHOTORECEPTION IN RHYTHMS AND PHOTOPERIODISM

IV. CIRCANNUAL RHYTHMS AND PHOTOPERIODIC CONTROL

V. NEURAL AND ENDOCRINE CONTROLS

VI. CELLULAR AND BIOCHEMICAL MECHANISMS

I

SYSTEMS AND MODELS

JÜRGEN ASCHOFF   PETER RIEGER
URSULA GERECKE   URSULA v. SAINT PAUL
ARMIN KURECK   RÜTGER WEVER
HERMANN POHL

Interdependent Parameters of Circadian Activity Rhythms in Birds and Man

Gross motor activity is probably the function most often measured in studies of circadian rhythms. Activity is easily recorded, and there are distinct, unambiguous points in the daily cycle (e.g., onset of activity) from which the free-running period can be computed. In many species, the whole circadian period (τ) can be divided into two sections—activity-time (a—wakefulness, measured from onset of activity in the subjective morning to end of activity in the subjective evening), and rest-time (ρ—sleep, measured from end to following new onset of activity). At least in man, the study of sleep as a special physiological state is older than the study of the sleep-wakefulness cycle; 8 hr of sleep have been postulated as normally necessary for functional reasons and independent of any considerations of rhythmicity. We will investigate here the question whether the relationship between activity-time and sleep-time is the same in free-running rhythms as in entrained rhythms.

The durations of a and ρ vary, within limits, from cycle to cycle. It may be asked whether, under constant conditions, the variability of the whole period τ is simply a summation of the variabilities of a and ρ or whether τ is more or less stable than its two components. This question leads to problems of precision in general: Is it correct, as has often been suggested, that onset of activity always provides a more precise measure of τ than

3

does end of activity? And is precision of τ, measured from whatever phase point, the same throughout the whole range of realized free-running periods, or does precision of τ depend on the τ value itself?

These problems will be discussed on the basis of data from finches that were caged singly in soundproof boxes under constant conditions and different intensities of illumination. Data on humans come from experiments in which the subjects lived for several weeks in complete isolation in an underground bunker (Wever, 1969b). Results from experiments in which the subjects showed internal desynchronization (Aschoff, Gerecke, and Wever, 1967) have been excluded from computation.

FREE-RUNNING ACTIVITY RHYTHMS

In Figure 1, original records of activity of five chaffinches are reproduced. Bird I was kept in 5 lux, all others in 0.5 lux. In all cases, onset of activity as well as end of activity can be determined, although sometimes the decision is difficult (e.g., onset in bird V). The bird with the shortest period (I) has very precise onsets and long solid blocks of activity; it ends its activity-time with about 2 hr of scattered activity. Birds II and III show less marked onsets, and their solid blocks of high activity are shorter. In bird IV, activity is more equally distributed over the whole activity-time; onsets and ends of activity have nearly the same variability. The bird with the longest period (V) has the least well defined onsets of activity and ends of activity that are as sharply marked as are onsets in bird I. The record of bird V shows a short solid block of high activity at the end instead of at the beginning of activity-time.

These observations suggest several questions. In order to specify them, two typical activity records are redrawn in Figure 2. The two records differ in the following ways:

• The rhythm with the shorter period (upper record) has a longer activity-time and a shorter rest-time than the rhythm with the longer period (lower record); the ratio of activity-time to rest-time is 3.47 in the first case and 1.38 in the second case.

• Mean values for the period τ can be computed from onsets of activity (τ_{onset}) or from ends of activity (τ_{end}). In the upper as well as in the lower record, these two τ values differ only slightly, indicating that the $a : \rho$ ratio is not changing during either of the two records. There are, however, large differences in standard deviation between the two records. In the upper record, the standard deviation for τ_{end} is twice as large as the standard de-

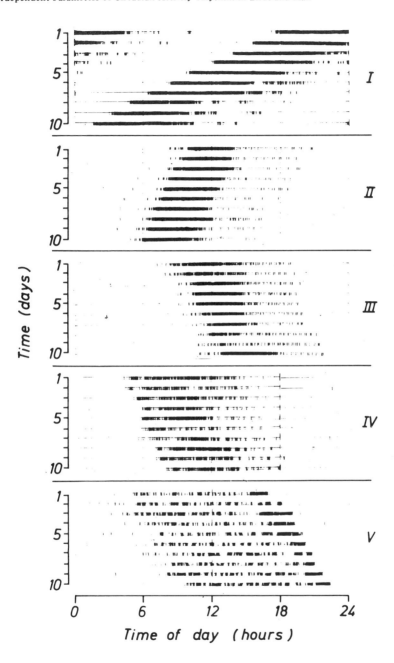

FIGURE 1   Activity records of chaffinches (*Fringilla coelebs* L.) kept in constant conditions. Intensity of illumination: 5 lux for I, or 0.5 lux for II–V.

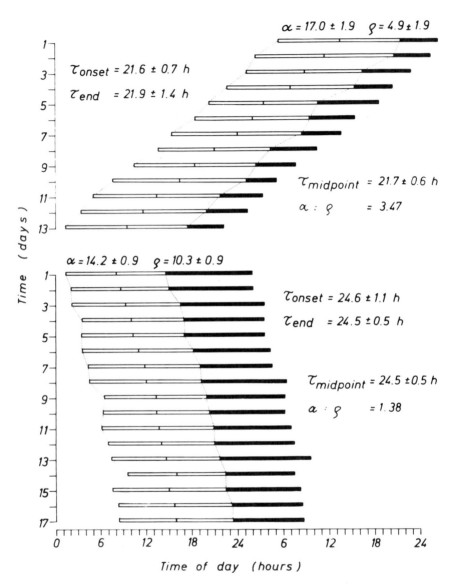

FIGURE 2   Schematic drawings of original activity records of two chaffinches kept in 5 lux constant illumination. White bars: activity-time a. Black bars: rest-time ρ. Vertical lines: midpoints of activity-time.

viation for τ_{onset}. In contrast, the lower record shows a smaller standard deviation for τ_{end} than for τ_{onset} (note the thin dotted lines drawn through onsets and ends of activity).

● Standard deviations for a and for ρ are half as large in the lower record as in the upper record.

● Midpoints of activity-times, indicated in Figure 2 by black marks, can be used for the computation of τ. The standard deviation for these values is of the same order of magnitude as the standard deviation for τ_{end} in the lower record and is smaller than the standard deviation of either τ_{onset} or τ_{end} in the upper record.

These differences in the properties of the rhythm, indicated in Figure 2, are related to differences in the free-running period. None of them is self-evident. The question arises whether they can be generalized and, if so, whether they can be explained as consequences of a special type of oscillation. This paper attempts to show the significance of the following features of and interactions among rhythm parameters:

1. Dependence of the $a : \rho$ ratio on the period τ.
2. Systematic correlations between a and either the following or the preceding ρ.
3. Precision of a, ρ, and τ as a function of τ.

During the discussion of 1, it will become clear that possible models and their implications must be considered. From these theoretical considerations, predictions follow concerning 2 and 3.

ACTIVITY-TIME a AND REST-TIME ρ

As mentioned in the introduction, a sleep-time of about 8 hr is considered "normal" in man. It has also been suggested that the degree of fatigue produced during activity-time demands a certain sleep-time for recovery. If this is correct, one would expect a longer than average activity-time to be followed by a correspondingly longer than average sleep-time. At least under constant conditions where no *Zeitgeber* forces the subject to wake up or to rise at a given time, such a positive correlation between a and ρ should become evident. As Figure 3 illustrates, the opposite effect has been observed in man. With an increase of a from 14 to 19 hr, ρ decreases from 11 to 6 hr. A similar negative correlation has been observed in birds (Figure 4). There is, however, a difference between the two sets of data.

FIGURE 3   Rest-time as a function of activity-
time in 28 human subjects, each kept in isolation in an underground bunker. Each point
represents the mean of at least 7 periods. Connected points are from the same subject, kept
under two different conditions.

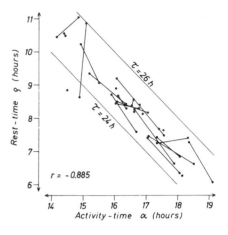

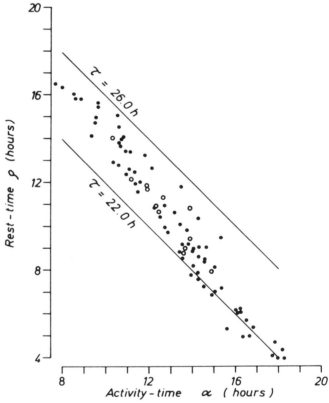

FIGURE 4   Rest-time as a function of activity-time in 12 chaffinches
kept in constant conditions with different intensities of illumination.
Solid circle: mean value from at least 7 periods of one bird for one condition. Open circle: mean value of one bird for all conditions.

In man, a and ρ always add up to the same mean τ value of about 25 hr. In the birds, the combination of a short a with a long ρ comes close to 26 hr, while a long a and a short ρ add up to 22 hr. In other words, in these birds, the $a : \rho$ ratio is a function of τ (or τ a function of the $a : \rho$ ratio!). The correlation is demonstrated in Figure 5. According to this diagram, a free-running period of 24 hr has an $a : \rho$ ratio of about 1.2. The ratio decreases slightly with an increase in τ, and increases markedly when τ decreases from 24 to 22 hr. It could be argued that these changes in the $a : \rho$ ratio are consequences of different intensities of illumination to which the birds had been exposed, and that the correlation between $a : \rho$ ratio and τ is an indirect one, both parameters depending directly on light intensity. However, the same dependence of the $a : \rho$ ratio on τ is shown by birds that have been kept continuously in DD and in which τ has changed spontaneously (inset, Figure 5; Eskin, 1969).

According to Figure 3, the $a : \rho$ ratio does not seem to depend on τ in man. While the bird data were gathered in experiments in which only the

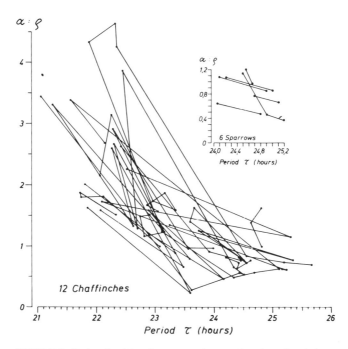

FIGURE 5   Ratio of activity-time to rest-time as a function of period τ in 12 chaffinches kept in constant conditions with different intensities of illumination. Lines connect data from the same bird. Inset: similar data from 6 sparrows (*Passer domesticus* L.) kept in constant darkness and showing spontaneous changes of period τ (after Eskin, 1969).

FIGURE 6   Ratio of activity-time to rest-time
as a function of period τ in 10 human subjects
living isolated in an underground bunker under
two different conditions (with and without a
10-cps electric field). Each point represents the
mean of at least seven periods under one condition. Lines connect data from the same subject.

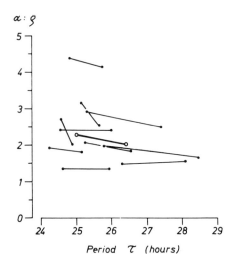

intensity of (continuous) illumination was varied, all other conditions remaining the same, the data collected for Figure 3 come from experiments
in which the subjects have been kept in a variety of different conditions,
which may have had quite different effects on τ as well as on other parameters (cf. Wever, 1969a). It is possible that opposing effects on the $a : \rho$
ratio may have obscured the picture in Figure 3. We have therefore re-
analyzed these results, using only data from 10 subjects who lived in constant illumination all the time, but for part of the time with and part
without a 10-cps electric field. As Wever has demonstrated (1968), under
the influence of such a field the circadian period is shortened by 1.33 hr
on the average. Results are summarized in Figure 6. Concurrent with the
shortening of the period, the $a : \rho$ ratio increases in 7 out of 10 subjects
and remains the same in 3 subjects. For the whole group, the increase in
the $a : \rho$ ratio is statistically significant at the 0.01 level (Wilcoxon test).
To summarize: In man as well as in birds, a and ρ are negatively correlated
with each other, and $a : \rho$ ratio is a function of period τ. The relevance of
these findings will be discussed in the following two sections.

THEORETICAL CONSIDERATIONS

The negative correlation between a and ρ may be surprising to a physiologist who thinks in terms of fatigue and recovery processes. The same correlation may seem trivial to someone whose interest concentrates on

rhythmic phenomena and to whom a and ρ are only sections of and have to add up to the period τ. That the problem may not be trivial is suggested in Figure 7. In the upper half of the diagrams, two models of an activity rhythm are drawn. Both models include one circadian oscillator, represented by a sinusoidal curve. Model A assumes that activity is "released" once during each period. The "releaser" may be arbitrarily chosen to be the maximum of the oscillation (arrows in Figure 7). Consequently, only onset of activity is determined by the circadian oscillator; duration of a and of ρ depend on other circumstances. Therefore, the standard deviation of τ_{onset} is always less than standard deviation of τ_{end}. Results such as those shown in the lower half of Figure 2 would be difficult to explain on the

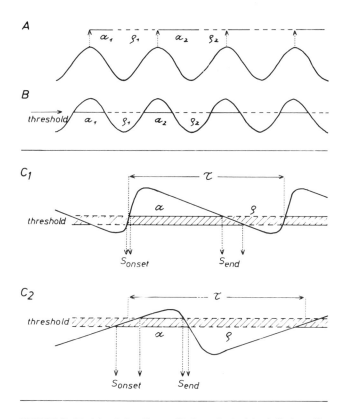

FIGURE 7   Models of circadian oscillations. A: Activity is "released" once during each period. B: Oscillator passes through a threshold, resulting in activity when it is above the threshold. C: Extended versions of model B. a: Activity-time. ρ: Rest-time. τ: Period. S: Standard deviation of onset and end of activity.

basis of such a model. Furthermore, if a is long, ρ necessarily is short, and vice versa, that is, model A demands a strong negative correlation between a and ρ. But this correlation is restricted to a and the following ρ; there is no correlation between a and the preceding ρ.

If it could be demonstrated that a is correlated with the following as well as with the preceding ρ, model A would not be applicable. Model B, however, demands such a two-sided correlation. This model assumes that the circadian oscillator passes through a threshold twice during each period (Wever, 1960; Aschoff and Wever, 1962a). The organism is supposed to be active as long as the oscillation is above the threshold and to rest when the oscillation is below the threshold. It is further assumed that threshold and mean level of oscillation can change their relative positions. When the threshold is lowered or the level raised, a becomes longer and ρ becomes shorter (and vice versa). If those changes of level or threshold can occur at any phase of the rhythm, a negative correlation is to be expected between a and the preceding as well as the following ρ.

Some other predictions can be made if one takes into account that the circadian oscillation is probably never sinusoidal. There is good reason to assume that the shape of the oscillation depends on τ as do so many other parameters. In the mathematical model developed by Wever (1965, 1966), the oscillation is skewed to the left (positive skewness) if the period is short and skewed to the right (negative skewness) if the period is long. The mathematical model also shows that, in accordance with the findings illustrated in Figures 2 and 5, the short period has a large $a : \rho$ ratio and the long period a small $a : \rho$ ratio. In order to match all these properties, model B is redrawn in two versions, C_1 and C_2, in the lower half of Figure 7. The positively skewed oscillation C_1 is drawn as a short period with a high level of oscillation, relative to the position of the threshold. The negatively skewed oscillation C_2 is drawn as a long period with a relative low level. It is further assumed that the threshold is subject to "noise." This is indicated by drawing the threshold not as a line but as a band (the same could be done with regard to level).

From the diagrams C_1 and C_2 it is evident that small fluctuations of the threshold (between upper and lower limits of the band) have quite different effects on onset and end of activity in the two versions. If activity is recorded during several periods, a mean value for the period can be derived from either onsets of activity (τ_{onset}) or from ends of activity (τ_{end}). In oscillation C_1, the standard deviation for τ_{onset} will be small (S_{onset} in Figure 7), while the standard deviation for τ_{end} will be large (S_{end} in Figure 7). The opposite holds true for oscillation C_2. Consequently, in oscillation C_1 the negative correlation between a and ρ is stronger when measured between a and the following ρ, whereas in oscillation C_2 the negative correlation is stronger between a and the preceding ρ.

In model C, the standard deviations for τ_{onset} and for τ_{end} are partly determined by the steepness with which the oscillation crosses the threshold band. This steepness is not only a function of skewness but also of amplitude of oscillation. With an increase of amplitude, both S_{onset} and S_{end} become smaller. Amplitude in self-sustained oscillations can be expected to depend on τ; it does so, at least, in the mathematical model (Wever, 1966). That is, changes in τ produce changes of standard deviation for τ_{onset} and for τ_{end} in opposite directions because of skewness, but in the same direction because of amplitude. According to the mathematical model, amplitude is largest somewhere toward the middle of the range of realizable τ values. If the model is applicable to the circadian oscillator, increases of standard deviation may be expected toward the extremes of τ values.

If model B has any relevance at all, and if the shape of the oscillation is a function of τ as indicated in the more specific versions C_1 and C_2, then the following predictions can be made:

1. Duration of a will be negatively correlated with duration of the preceding as well as of the following ρ.

2. In rhythms with short periods, the negative correlation will be stronger between a and the following ρ; long periods will show a stronger negative correlation between a and the preceding ρ.

3. The standard deviation of τ will be smaller for τ_{onset} than for τ_{end} in rhythms with short periods, the same for onset and for end at intermediate τ values, and smaller for τ_{end} than for τ_{onset} when the period is long.

4. Standard deviation of τ will be smaller than the summed standard deviations of a and ρ.

5. When drawn as functions of τ, the standard deviations for a, ρ, and τ will pass through minima.

SHAPE OF OSCILLATION

Since practically nothing is known about the real nature of circadian oscillators, it seems premature to discuss problems of shape. However, if one assumes that some of the functions measured can reflect in one way or another general characteristics of the basic oscillation, it may not be too speculative to take "form of function" as an indication of "form of oscillation," at least by way of trial. On this basis, the changes in pattern shown in Figure 1 are already informative and are in agreement with tendencies illustrated by models C_1 and C_2 in Figure 7. The five records reproduced in Figure 1 demonstrate that peak activity occurs early in a short period (record I) and that it occurs progressively later with an increase in period, eventually concentrating at the end of activity-time in the record that has the longest period (record V). Record I may be called "skewed to the left" (see model C_1), and record V may be called "skewed to the right" (see model C_2 in Figure 7).

Comparable to changes in activity distribution (Figure 1) are changes in those patterns characterized by two peaks of activity. Usually, in a short period the first peak is larger than the second peak, producing a bigeminus-type of pattern (Aschoff, 1957), whereas in a long period the second peak is larger than the first peak, producing an alternans-type. Systematic changes from bigeminus-type to alternans-type of pattern, occurring as a function of duration of activity-time and hence of period, have been documented elsewhere (see Aschoff, 1966a, Figure 5).

The indirect hints about form mentioned so far are limited in their usefulness by the fact that activity gives information only for parts of the period. Functions such as oxygen consumption and body temperature, which can be measured throughout the whole period, should be more instructive. In order to describe the shape of the curve of those functions, a form factor has been introduced (Wever, 1968). The form factor is given by the ratio of the interval during which the function descends to its minimum to the interval during which it ascends to its maximum (form factor = duration of descending slope ÷ duration of ascending slope). A sinusoidal curve has a form factor of 1.0; a left (positively) skewed curve has a form factor of more than 1.0 (oscillation C_1 in Figure 7); a right (negatively) skewed curve a form factor of less than 1.0 (oscillation C_2 in Figure 7). In Figure 8, form factors of chaffinches and human subjects are drawn as a

FIGURE 8 Form factor of circadian oscillation as a function of period τ. Data from continuous measurements of oxygen consumption in 7 chaffinches and of rectal temperature in nine isolated human subjects all kept in constant conditions. Upper margin: schematic oscillations, one with a form factor of 2.0, skewed to the left, and one with a form factor of 0.8, skewed a little to the right.

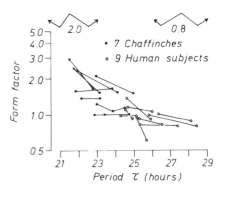

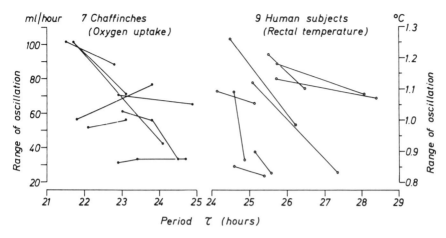

FIGURE 9   Range of oscillation ("amplitude") as a function of τ. Left: from continuous measurements of oxygen consumption in seven chaffinches kept in constant illumination. Right: from continuous measurements of rectal temperature in nine human subjects, kept underground in isolation.

function of period τ. Oxygen consumption has been measured in finches kept in constant dim illumination (Pohl, 1970) and rectal temperature in men kept in isolation underground (Wever, 1968). The form factor decreases from greater than 2.0 to less than 1.0, concurrent with an increase of τ from 21.5 to 28.5 hr. At the upper margin of Figure 8, schematic curves with a form factor of 2.0 and 0.8 respectively are drawn at the extreme ends of the abscissa (the two sets of data could have been drawn in two separate figures without loss of information relevant to this discussion).

Even less information is available on changes of amplitude than on changes in skewness. The striking variations in amount of activity that have been described in connection with changes in period length (Aschoff and Wever, 1962b, Figure 1; Aschoff, 1966b, Figure 1) may be taken as rather doubtful indirect evidence for changes in amplitude. Complications (discussed in Aschoff and Wever, 1962b) make it difficult to infer amplitude from amount of activity. Again, oxygen consumption and rectal temperature are probably more reliable indicators. In Figure 9, the range of oscillation is given as a function of period τ. The right half of the figure shows rectal temperature data from human subjects; in all nine subjects, range of oscillation decreases with an increase of period. In the case of the birds, measurements of oxygen consumption give conflicting results. In most birds, range of oscillation increases with decreasing period length, even to less than 23 hr; in three birds, however, range of oscillation becomes smaller when the period shortens to less than 23 hr. These results may indicate that range of oscillation, in general, increases with decreasing periods, but can also reach a maximum and decline thereafter when the period is shortened below a certain limit. (*A priori*, it is not necessary that the τ value at which amplitude reaches its maximum should be the same in birds as it is in man.)

In summary, statements about amplitude are still vague and only of little value for a critical discussion of models. The findings concerning skewness are highly significant and support the hypothesis illustrated by oscillations C_1 and C_2 in Figure 7. The conformity between model and experimental data is reason enough to test further predictions made by this model.

CHANGING CORRELATIONS BETWEEN a AND ρ

Data points in Figures 3 and 4 that describe negative correlations between a and ρ are mean values for longer time spans, each point being the aver-

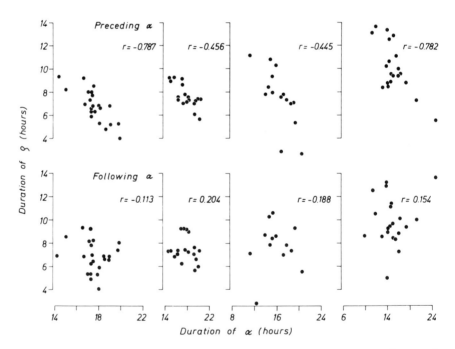

FIGURE 10   Correlation between rest-time ρ and preceding (upper row) or following (lower row) activity-time. Data from four human subjects kept underground in isolation; r = coefficient of correlation.

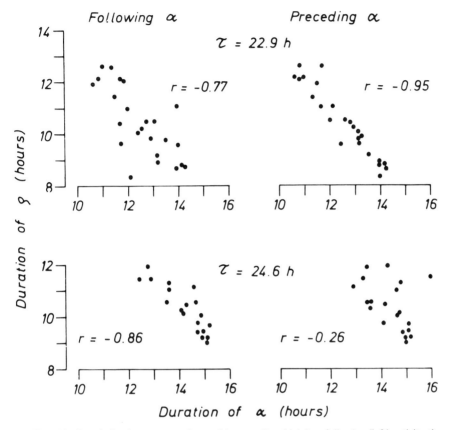

FIGURE 11 Correlation between rest-time and its preceding (right) or following (left) activity-time in two chaffinches kept in constant conditions. τ: mean period; r: coefficient of correlation.

age of at least seven periods. These two diagrams, therefore, give no information on the correlation between adjacent sections of each period and do not take into account the sequence of a and ρ. This has been done in Figure 10 for some of the human data. The upper row shows the dependence of ρ on the preceding a, the lower row the dependence of ρ on the following a. The correlations are clearly negative in the upper row, but close to zero in the lower row. These results, which are significant in a sample of 10 subjects ($p < 0.01$; Wever, 1969b), do not preclude a model such as model A in Figure 7.

Two typical sets of chaffinch data are drawn in Figure 11. In all four diagrams, the correlation between ρ and a is negative. The two upper diagrams come from records of a bird with a short period (22.9 hr). Here, the negative correlation between ρ and the preceding a is stronger (r = –0.95) than that between ρ and the following a (r = –0.77). The situation is reversed in the lower two diagrams, based on data from a bird with a long period (24.6 hr). Now the negative correlation is stronger between ρ and the following a (r = –0.86) than between ρ and the preceding a (r = –0.26). Altogether, seven birds have been tested under three intensities of constant illumination. The results are summarized in Figure 12. In all but one case, ρ and a are negatively correlated with each other. When averaged, the negative correlation is stronger between ρ and preceding a (upper diagram, mean r = –0.772) than between ρ and following a (lower diagram, mean r = –0.575). With increasing τ values, the correlation coefficient between ρ and following a changes from about –0.4 to –0.75, while the correlation coefficient between ρ and preceding a shows, if anything, a tendency to change in the opposite direction.

We can conclude, then, that in finches the negative correlation between ρ and a depends on τ in a predictable manner when skewness of the circadian oscillation depends on τ as illustrated in Figure 7. In man, no such dependence has yet been demonstrated.

VARIABILITY OF a, ρ, AND τ

As Figure 1 illustrates, duration of a is quite variable; Figure 13 shows that variability of ρ is of the same order of magnitude. Each point in these diagrams represents a standard deviation (S.D.) around a mean value computed from at least seven consecutive measurements. Standard deviations of a and ρ are drawn as a function of the standard deviation of their combined value τ. The values come from eight of the same chaffinches from which the data of the foregoing figures were gathered. Results from measurements in only two intensities of illumination are shown. In 5 lux, the eight birds have a mean period of 22.8 hr; all standard deviations of a and of ρ (with one exception) are greater than the standard deviations of τ. In 0.5 lux, the birds have a mean period of 24.0 hr; standard deviations of a and ρ are closer to standard deviations of τ, but the mean is still greater than the mean standard deviation of τ. That is, in all cases, standard deviation of τ is smaller than the summated standard deviations of a and ρ.

From the two diagrams in Figure 13, it is also clear that all three parameters (a, ρ, and τ) have smaller standard deviations when the period is 24 hr than when it is 22.8 hr (0.67 versus 0.8 for τ; 0.9 versus 1.3 for a and for ρ). In contrast to birds, the variability of τ in man is similar to or even a little greater than the variability of either a or ρ. It is nonetheless

FIGURE 12 Coefficient of correlation between rest-time and its preceding or its following activity-time a, drawn as a function of period τ. Data from seven chaffinches kept in constant conditions with three intensities of illumination.

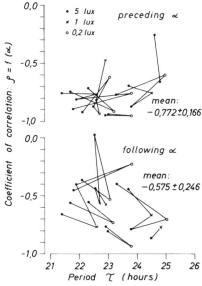

FIGURE 13 Standard deviation of activity-time (open circles) and of rest-time (closed circles) as a function of standard deviation of period τ. Data from eight chaffinches from repeated trials in constant conditions with two different intensities of illumination. Mean period of all birds in 5.0 lux is 22.8 hr (upper diagram); in 0.5 lux it is 24 hr (lower diagram).

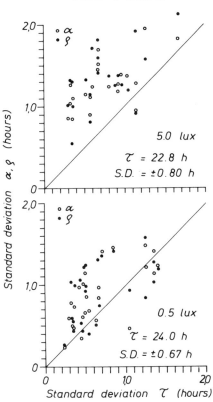

possible, to demonstrate that variability of τ is not just an additive effect of variabilities in a and ρ. To test this, one must take the square root of the sum of squares (in accordance with the law of error propagation), instead of simply adding the standard deviations. In Figure 14, this factor is drawn as a function of τ_{onset} (left diagram) and as a function of τ_{end} (right diagram). The diagram on the left shows that in 53 out of 64 cases the standard deviation of τ_{onset} is smaller than the combined standard deviations of a and ρ. Variability of τ_{onset} cannot be explained as being composed randomly of the variabilities of a and ρ. In contrast to this, standard deviation of τ_{end} (right diagram) is, in general, of the same order of magnitude as the combined standard deviations of a and ρ. In other words, the period τ, as measured between consecutive onsets of activity, is better "stabilized" than is activity-time or rest-time. This statement does not exclude the possibility that stabilization, i.e., precision, may be itself a function of τ, a proposition that will be discussed in the last section.

PROBLEMS OF "NOISE" AND "PRECISION"

In this paper, the noise of rhythm parameters is expressed by standard deviation. (This notation is not used for statistical tests, but only as a descriptive quantity.) As explained previously, standard deviation of τ de-

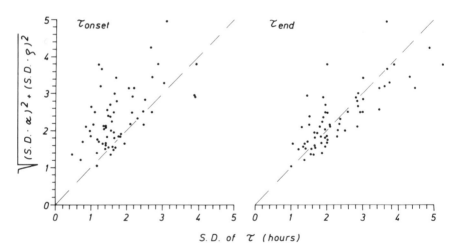

FIGURE 14   Square root of the sum of squares of standard deviation of activity-time a and of rest-time ρ, drawn as a function of standard deviation of period τ. Data from 39 human subjects most of whom lived under at least two different conditions. Altogether there are 64 separate values ("man-conditions"). Standard deviation of period is computed from onsets of activity (left) and from ends of activity (right).

pends partly on fluctuations of threshold (or level). A second and main source of noise is given by irregularities in length of the basic oscillation. In the graphic scheme presented in Figure 7, this would mean that oscillator C fluctuates in a horizontal plane, in contrast to vertical fluctuations of the threshold. For reasons of brevity, in the following discussion the two sources of noise will be called "vertical noise" and "horizontal noise."

With both sources of noise, amount of noise depends on skewness of the oscillation and hence on period length (see version 1 and 2 of model C in Figure 7; see also Figure 8). With an increase in τ, it is expected that standard deviation of τ_{onset} would increase and standard deviation of τ_{end} decrease. A simple expression of this relationship is given by the ratio of the two standard deviations. If model C is valid, the quotient

$$\frac{\text{S.D. } (\tau_{end})}{\text{S.D. } (\tau_{onset})}$$

should become progressively smaller as τ becomes longer. As shown in Figure 15, this seems to be the case in both men and birds. In the human experiments, the quotient changes in accordance with the hypothesis in 6 out of 10 subjects. In only one subject is the change in strong contrast to expectation. If this one case is excluded, the trend of the remaining data is significant ($p < .05$) (Wever, 1969b). In birds, the same general trend can be seen. The data have been averaged according to four classes of τ (open circles); the difference in ordinate values between the two extreme classes is significant ($p < .01$). (Since the ordinate values in Figure 15 are related to both onset and end of activity, the τ values for the abscissa have been computed from midpoints of activity-times.)

The changes of noise, shown in Figure 15, depend on changes of skewness that have opposite effects on τ_{onset} and on τ_{end}. In contrast, the effects of "horizontal noise" on τ are the same, regardless of the phase point used to measure τ. Therefore, a dependence of "horizontal noise" on τ should look alike for all possible τ measurements. According to the mathematical model, "horizontal noise" results in a minimum standard deviation when τ is near the middle of its range (see Wever, this volume, Figure 7). For the bird data, this is demonstrated in Figure 16. Minimal values of standard deviation are indicated around 24 hr for τ_{onset}, at or above 25 hr for τ_{end}, and possibly between these two values for $\tau_{midpoint}$. Two further phenomena warrant discussion:

1. With decreasing τ, standard deviation of τ_{end} increases from its minimal value more than does standard deviation of τ_{onset}. This can be ex-

plained by the fact that skewness depends on τ in such a way that, with decreasing period length, it augments standard deviation of τ_{end} but diminishes standard deviation of τ_{onset}. In other words, when the period is shortened, two sources of noise increase together for τ_{end} but partly cancel each other for τ_{onset}.

2. As already shown in Figure 2, standard deviation of $\tau_{midpoint}$ does not reach a magnitude intermediate between the standard deviations of τ_{onset} and τ_{end}; it is close to or even smaller than the standard deviation of τ_{onset}. This phenomenon, which also is indicated in Figure 15, can be explained on the basis of model C. In this model, vertical movements of the threshold influence onset and end of activity in opposite directions, resulting in only small shifts of midpoint. Thus, increasing "vertical noise" increases standard deviation of τ_{onset} and of τ_{end} but does little or nothing—depending on skewness—to standard deviation of $\tau_{midpoint}$. (The same argument favors the idea of midpoint being, in general, a less erroneous measure for phase; see Aschoff and Wever, 1962a; Aschoff, 1969.)

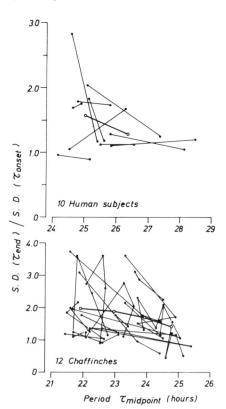

FIGURE 15   Ratio of standard deviation of τ_{end} (period computed from ends of activity) to standard deviation of τ_{onset} (period computed from onsets of activity), drawn as a function of period τ. Chaffinches kept in constant conditions, human subjects kept in isolation in an underground bunker.

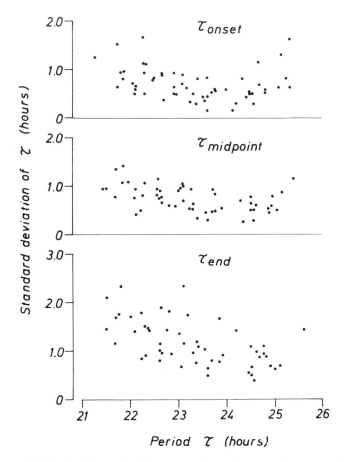

FIGURE 16   Standard deviation of period τ as a function of τ, computed from onset, midpoint, and end of activity-time. Data from 12 chaffinches kept in constant conditions with different intensities of illumination.

The use of standard deviation as a measure of noise is advantageous insofar as it gives an estimate of amount of noise in absolute units. On the other hand, the same standard deviation describes different relative noise levels in long and short periods. This ambiguity is corrected by the coefficient of variability (standard deviation divided by the mean value). The reciprocal of coefficient of variability (i.e., mean value divided by standard deviation) has been suggested as an expression for "precision" (Wever, this volume). According to Figure 17, precision of the activity rhythm of birds depends on τ. Two regression lines have been computed for all data

on the left side and on the right side, respectively, of a line drawn arbitrarily at a τ of 24 hr. There would have been little difference in the results
if the line had been moved to the right or to the left by half an hour. In
any case, highest precision of the rhythm occurs around 24 hr.

Since the $a:\rho$ ratio is correlated with τ (see Figures 5 and 6), precision
of a and ρ may depend on their mean values as does precision of τ on its
own value. The data shown in Figure 18 support this idea. In activity-time
as well as in rest-time, precision increases with increasing duration of a and
ρ, and seems to decline again after a certain maximum has been reached.
There is great intraindividual and interindividual variability in the data,
and nothing can be said about statistical significance. However, a tendency
for higher precision somewhere between very long and very short a and ρ
values is indicated. When checked directly on the basis of standard deviations, a minimum of noise is evident when a is about 13 hr and ρ about

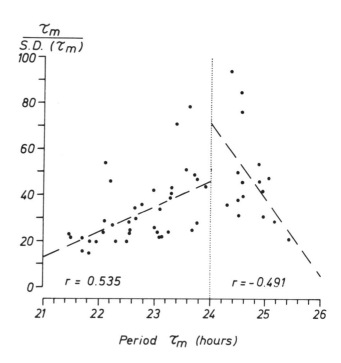

FIGURE 17 Precision of period τ (computed from midpoint of activity-
time) as a function of τ. Data from 12 chaffinches kept in constant conditions with different intensities of illumination; r: coefficient of correlation for all data on the right and the left side, respectively, of vertical line
arbitrarily drawn at τ = 24 hr.

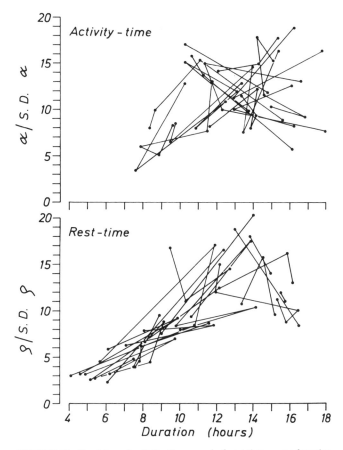

FIGURE 18 Precision of activity-time a and of rest-time ρ as a function of duration of a and ρ. Data from 12 chaffinches kept in constant conditions with different intensities of illumination.

11 hr. Least variable, therefore, is an $a : \rho$ ratio of 1.2, corresponding to a τ of 24 hr (see Figure 5).

All these results, indicating maximal precision of a, ρ, and τ for periods around 24 hr, confirm tentative results published 10 years ago (Aschoff, 1960, Figure 10).

CONCLUDING REMARKS

In free-running activity rhythms of birds and men, the following parameters have been shown to depend on the period τ:

Ratio of activity-time a to rest-time ρ,

Degree of negative correlation between a and ρ, which, in addition, depends on the sequence of the two sections,

Standard deviation of τ_{onset} and of τ_{end}, and

Precision of τ, measured as mean value divided by standard deviation.

It is also demonstrated that variability of τ is less than could be expected from "summed" variabilities of a and ρ, and that midpoint of activity-time can be less variable than either onset or end of activity. All these characteristics are not only compatible with, but demanded by, a model that assumes (a) that the circadian oscillator passes through a threshold twice during each period, (b) that the organism is active as long as the oscillation is above the threshold, and (c) that shape, amplitude, and level of oscillation depend on period τ.

It is hard to believe that the nice coincidence between experimental results and predictions from the model should be due to chance. The predictions, on the other hand, are based on a "one-oscillator-model." Unfortunately, it is becoming more and more unlikely that such a model sufficiently explains all circadian phenomena. It then becomes important to ask, Which of the "more-than-one-oscillator" models presently under discussion best predicts or describes the kind of data we have presented?

REFERENCES

Aschoff, J. 1957. Aktivitätsmuster der Tagesperiodik. Naturwiss. 44:361–367.
Aschoff, J. 1960. Exogenous and endogenous components in circadian rhythms. Cold Spring Harbor Symp. Quant. Biol. 25:11–28.
Aschoff, J. 1966a. Circadian activity pattern with two peaks. Ecology 47:657–662.
Aschoff, J. 1966b. Circadian activity rhythms in chaffinches (*Fringilla coelebs*) under constant conditions. Jap. J. Physiol. 16:363–370.
Aschoff, J. 1969. Phasenlage der Tagesperiodik in Abhängigkeit von Jahreszeit und Breitengrad. Oecologia (Berl.) 3:125–165.
Aschoff, J., U. Gerecke, and R. Wever. 1967. Desynchronization of human circadian rhythms. Jap. J. Physiol. 17:450–457.
Aschoff, J., and R. Wever. 1962a. Beginn und Ende der täglichen Aktivität freilebender Vögel. J. Ornithol. 103:1–27.
Aschoff, J., and R. Wever. 1962b. Aktivitätsmenge und $a:\rho$-Verhältnis als Messgrössen der Tagesperiodik. Z. Vgl. Physiol. 46:88–101.
Eskin, A. 1969. The sparrow clock: Behavior of the free running rhythm and entrainment analysis. Ph.D. Thesis. University of Texas at Austin.
Pohl, H. 1970. Zur Wirkung des Lichtes auf die circadiane Periodik des Stoffwechsels und der Aktivität beim Buchfinken (*Fringilla coelebs* L.). Z. Vgl. Physiol. 66:141–163.
Wever, R. 1960. Possibilities of phase-control, demonstrated by an electronic model. Cold Spring Harbor Symp. Quant. Biol. 25:197–206.

Wever, R. 1965. A mathematical model for circadian rhythms, p. 47 to 63. *In* J. Aschoff [ed.], Circadian clocks. North-Holland Publ. Co., Amsterdam.

Wever, R. 1966. Ein mathematisches Modell für die circadiane Periodik. Z. Angew. Math. Mech. 46:148–157.

Wever, R. 1968. Gesetzmässigkeiten der circadianen Periodik des Menschen, geprüft an der Wirkung eines schwachen elektrischen Wechselfeldes. Pfluegers Arch. 302:97–122.

Wever, R. 1969a. Autonome circadiane Periodik des Menschen unter dem Einfluss verschiedener Beleuchtungs-Bedingungen. Pfluegers Arch. 306:71–91.

Wever, R. 1969b. Untersuchungen zur circadianen Periodik des Menschen, mit besonderer Berücksichtigung des Einflusses schwacher elektrischer Wechselfelder. Bundesminist. Wiss. Forsch., Forschungsber. W 69–31, Weltraumforsch. 212 p.

DISCUSSION

LOBBAN: I am a bit staggered by all this; it is not at all what I expected. I would like Prof. Aschoff to comment on the studies that people such as Lewis have done in the Arctic and that various medical officers have done in the Antarctic. There in the military camps you have communities that are the equivalent of the free-running ones of your experiments. The men live in huts that can be completely closed off, and they sleep when they wish. Although sometimes the outside conditions may determine whether or not they stay indoors, quite often the people are completely free to choose their own ways of living. The conditions are such that certain people choose to work at odd times of the day because generators are working better, or something like that. The only thing they do all together—and that not all the time—is to take their meals together. And under these conditions—that is, a large number of different schedules—the mean sleep-time, ρ, works out to be about 7.9 hr.

ASCHOFF: That's fine; there is no reason for the mean ρ not to be 7.9 hr. But

I think that if you would check this day after day, you would see a negative correlation between a and ρ. I don't say this would be true for all the subjects all the time, but I believe in many you would find it.

LOBBAN: I would be glad if you would look at some data that I can furnish from my own experience. Once when I finished an experiment in the high Arctic, I went out for a walk by myself. It was a long walk; it lasted 15 days. And during that time I went about 100 miles and saw a great deal of the country. I was eating, sleeping, walking entirely as I wished, apart from one occasion when there was a bad blizzard and I lay up for 28 hr. I worked out the mean sleeping time for myself, and it came out to 7.8 hr.

ASCHOFF: We had more than 100 subjects in the bunkers, and their mean rest-time was only a little longer (8.4 hr). Nevertheless, we have this strong negative correlation between consecutive a and ρ.

LOBBAN: I should like to add one more thing. We have a little data on macaque monkeys. There, too, the onset of activity is much more consistent than end of activity, where you find much greater scatter.

ASCHOFF: This has been demonstrated very often. The next question to ask would be whether you can see in these monkeys whether the precision depends on τ.

STRUMWASSER: I have just one comment. It worries me that the standard deviation of τ_{end} is so small at 24 hr; I immediately think of laboratory artifacts. Your subjects were all free-running?

ASCHOFF: Yes, they were all free-running, I don't believe we are getting disturbance from the outside. I believe the effect is coming from within the system; the system is most precise at about $\tau = 24$ hr. That is my interpretation of the data.

STRUMWASSER: If it were 25 hr, or some other number, I wouldn't worry about it. But since it is 24, can one exclude the possibility that you are getting some synchronization with something in the laboratory? Your technique is detecting the point of minimum noise in the system, and minimum noise would exist around $\tau = 24$ hr if there is a *Zeitgeber*, even if it is weak.

ASCHOFF: I'm sure the subjects were all perfectly free-running, but let us just check the possibility of synchronization. The range of minimal error is from 23.5 hr to 24.5 hr. And in Figure 17, the dividing line can be moved half an hour left or right, and the slopes on either side of the line would still differ in the same way. There is not one point at exactly $\tau = 24$ hr. I see no reason to think that those animals that are close to $\tau = 24$,

but nevertheless free-running, are less disturbed than those a little farther away from $\tau = 24$. They are all exposed to the same noise, in the same amounts.

STRUMWASSER: This is one kind of experiment that can be done in space, where you are pretty sure there is no 24-hr *Zeitgeber*, to see if the crossover point is still around $\tau = 24$.

ASCHOFF: I still see no real possibility of explaining the better precision by disturbance.

PITTENDRIGH: It may be that you have got either complete or partial entrainment to laboratory "noise," and even with partial entrainment the variance of τ will go down.

ASCHOFF: Excuse me; it goes up, because of relative coordination, which is what happens if you are not entrained by crossing a powerful *Zeitgeber*. And since we have 20 days of data, of course the noise will increase.

PITTENDRIGH: That is about the strongest argument I've seen.

SOLLBERGER: You studied the scatter of midpoints compared to the scatter of endpoints. Now quite mathematically, when you define the midpoint by the two endpoints, it has to have less error than the sum of that of the endpoints, though not necessarily the mean value of the two.

ASCHOFF: You are correct. However, I did not compare the scatter of midpoints to the scatter of the sum of the two endpoints, but to the individual scatters of onset and of end. In one case, the scatter of the midpoints is between that of onset and that of end, which you might expect, but in the other it is even less than that of onset. Now that tells you something about where the noise in the system is coming from and is indicative for model C. If a large amount of the total scatter is due to changes in relative position between oscillation and threshold, i.e., due to "noise" of threshold, then you have to expect less scatter for midpoint than for either onset or end. The reason is that, in this case, onset and end of activity move in opposite directions, with midpoint remaining more or less unchanged! Up to now, I could not find any better explanation for the fact that sometimes standard deviation of midpoint is less than that of onset as well as that of end of activity.

SOLLBERGER: My other point is that when you correlated τ divided by standard deviation with τ, you correlated τ with itself, with some disturbance from the standard deviation. So you would expect a correlation. I would prefer that you just correlate SD with τ; your modifications of this add no new information and are just confusing. Your observation about range is quite interesting, since for small groups range is about as powerful a measure of scatter as the standard deviation.

ASCHOFF: Would you prefer to compare standard deviation to a coefficient of variation?

SOLLBERGER: The coefficient of variation is an antique thing, rarely used nowadays and useful only when you have incommensurable units that you can't compare any other way. Otherwise it is just confusing.

THOMAS J. CROWLEY
FRANZ HALBERG
DANIEL F. KRIPKE
G. VERNON PEGRAM

Individual Variation in Circadian Rhythms of Sleep, EEG, Temperature, and Activity among Monkeys: Implications for Regulatory Mechanisms

Rhythms can be defined in terms of five basic characteristics:

Frequency, the number of rhythmic fluctuations per unit of time;

Level, C_o, the mean value above and below which fluctuations occur (such as the approximately $37°C$ mean rectal temperature of a man);

Amplitude, C, a measure of the extent of change (such as the peak value minus the level when a mathematical function is used to approximate the rhythm);

Phase, the temporal relation of the rhythm to some reference point (dividing the 24-hr clock into 360 degrees, we use the term "external acrophase," ϕ, to indicate the time in degrees of arc from the middle of the dark span to the peak of that cosine curve most nearly approximating a given rhythm); and

Wave form, the shape of the periodic change (Halberg, 1969a).

In all five of these basic characteristics, biological rhythms are highly variable and seldom conform to an ideal. Although biological rhythms are reliable enough to be recognized as periodic by inferential statistical means (Halberg, 1969a), their characteristics vary from subject to subject and

Individual Variation in Circadian Rhythms of Sleep, EEG, Temperature, and
Activity among Monkeys: Implications for Regulatory Mechanisms

31

from day to day in the same subject. This study examines intersubject variation of three of these basic characteristics in several physiological variables in a group of monkeys. Many simian functions vary rhythmically (Halberg, 1969b), each function having its own typical characteristics. For instance, in monkeys, as in many species, the circadian periodic high in the body temperature rhythm is markedly out of phase with sleep.

Although the several rhythms of a given individual stand in various phase relationships to one another, these phase relationships tend to be similar in all members of the same species. For example, if under certain environmental conditions some rats sleep during one part of a 24-hr span and their temperature peaks during another part, other rats similarly standardized will probably behave likewise. Within a certain range the phase relation of sleep and body temperature rhythms appears to be phylogenetically determined. For a group of synchronized subjects there will be a mean time of occurrence for the peaks of temperature and activity; but there also will be a scatter of individual acrophases around each mean.

This scatter prompts several questions: If the acrophase of a given variable in an animal is advanced in relation to the interanimal mean acrophase for that function, will the acrophase of another variable be similarly advanced in that animal? If, for example, the peak of a monkey's activity on a given lighting regimen precedes that of his fellows on the same regimen, will his body temperature peak also come earlier? Conversely, if one function is delayed in certain subjects, will another function also be delayed in those subjects? In general, is there a correlation between the phases of two circadian functions measured in each member of a group of subjects?

If the acrophases of temperature and activity, for instance, were not correlated—if they did not advance together in some individuals and delay together in others—the functions could probably be assumed to be physiologically unrelated. Conversely, if the rhythms were found to be phase-locked, the functions would appear either to be interdependent or subject to some common regulation.

The above example concerns only phase, but it is clear that similar correlations may exist between the amplitudes of two circadian functions measured in a group of subjects or between the circadian levels, frequencies, or wave forms of those functions. We have previously determined the mean acrophase, amplitude, level, and wave form of circadian changes in the sleep, electroencephalogram (EEG), motor activity, and brain temperature of a group of rhesus monkeys (Crowley et al., manuscript in preparation). We now examine these data for any parameter correlations between each pair of variables.

MATERIALS AND METHODS

The details of our methods are reported elsewhere (Kripke *et al.*, 1968a; Crowley *et al.*, manuscript in preparation). We used 24 male *Macaca mulatta* monkeys implanted with EEG electrodes over the occipital dura, thermistors in the parietal white matter, electromyographic (EMG) electrodes in the temporalis muscle, and electrodes at each orbital margin for recording ocular movements (providing an electro-oculogram, or EOG). A cannula directed into the third ventricle of each animal remained sealed during this experiment. It was used later in biochemical studies, which are beyond the scope of this report.

After surgery the animals lived together in a large room illuminated by artificial light from 0800 to 2000 Mountain Standard Time (*LD 12*:12). Daily at about 0930 the room was cleaned and the animals, loosely restrained by metal neck-rings, were fed.

At least one month after surgery each animal was transferred from the communal quarters into an individual recording chamber (Figure 1), where the previous lighting and feeding schedules were maintained. Data reported here include only those obtained during the last 24 hr of each 48- to 72-hr recording session.

Brain temperature was measured every 5 min through a bridge circuit and digital voltmeter. The EEG, EMG, and EOG were continuously recorded on magnetic tape and polygraph paper. Approximately 50 km of polygraph records were obtained and were divided into 20-sec sections. Each section was visually classified into one of four stages—Stage Awake, 1–2, 3–4, or REM (rapid eye movements)—following criteria slightly modified from those of Dement and Kleitman (1957), Reite *et al.* (1965), and Rechtschaffen and Kales (1968). Stage Awake was characterized by a high-frequency, low-voltage EEG pattern with considerable myogram activity and frequent eye movements. In Stage 1–2 sleep, general EEG slowing occurred with increased voltage, while the EOG and EMG became quiescent. In the deeper Stage 3–4 sleep, eyes and muscles remained quiescent, and at least 25 percent of the record was taken up by very high-voltage, slow delta waves (over 75 μV amplitude, 1–3 Hz). In Stage REM ("rapid eye movement" sleep in which human beings do most of their dreaming) the EEG reverted to a low-voltage, high-frequency pattern while the eyes rapidly darted about, the animal remaining behaviorally in deep sleep with very little EMG activity.

The analogue magnetic tape records of EEG were electronically filtered into the five clinically used frequency bands (Figures 2 and 3) by a device previously described (Kripke *et al.*, 1968b). The approximate band widths

Individual Variation in Circadian Rhythms of Sleep, EEG, Temperature, and
Activity among Monkeys: Implications for Regulatory Mechanisms

33

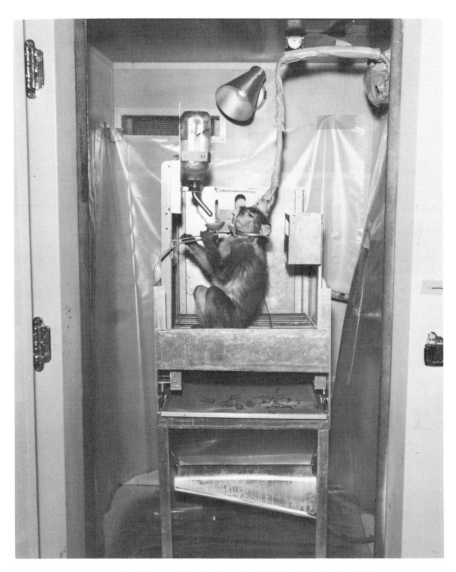

FIGURE 1   Rhesus monkey in recording chamber. Surgically implanted electrical connector mates
to cable and to implanted sensors for recording brain temperature, electroencephalogram, electro-
myogram, and electrooculogram. Each animal was studied for 2 to 3 days in this booth after spending at least one postoperative month in a room with light from 0800 to 2000 hr, alternating with
darkness.

AWAKE

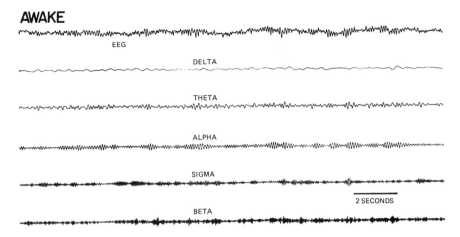

FIGURE 2 Frequency analysis of waking EEG. Original tracing from human subject at top of figure. Below, five systems of electronic filters have separated the EEG into five frequency bands, eliminating all activity outside each band's approximate frequency range: delta 1–3 Hz, theta 4–7, alpha 8–13, sigma 13–16, and beta 18–22.

were delta 1–3 Hz, theta 4–7, alpha 8–13, sigma 13–16, and beta 18–22. After rectification each band was integrated every 5 min for 24 hr to obtain a 24-hr plot of the amplitude, or amount of activity, in each band for each animal. The integrals of the various EEG bands were then recombined to produce a plot of total EEG abundance. EMG and EOG were similarly averaged over 5-min intervals. Characteristic artifacts appeared in the EEG when the subjects moved. The times at which such artifacts occurred were also coded as a rough measure of gross motion.

We thus obtained time-series on 15 partially overlapping variables: abundance of total EEG; abundance in the frequency bands delta, theta, alpha, sigma, and beta; brain temperature; gross motion and two other activity variables, EMG and EOG; and the occurrence of the sleep stages Awake, 1–2, 3–4, REM, and NREM (non-REM sleep, or Stages 1–2 plus 3–4). For technical reasons one or another of these variables was not documented on some animals, but the remaining data are regarded as sufficient for the analyses; 17 monkeys contributed data on all variables.

Sleep research has tended to focus more closely on the ultradian organization of sleep than on its circadian organization. Simian sleep, like human sleep, has a clear ultradian rhythm. Stage Awake is followed by Stage 1–2, followed in turn by Stage 3–4, followed by Stage REM. This ultradian cycle reportedly lasts on the average about 50 min in the monkey (Weitzman *et al.*, 1965); it recurs continuously throughout the night. Several authors have also noted a circadian modulation of this ultradian

Individual Variation in Circadian Rhythms of Sleep, EEG, Temperature, and
Activity among Monkeys: Implications for Regulatory Mechanisms

35

STAGE 4

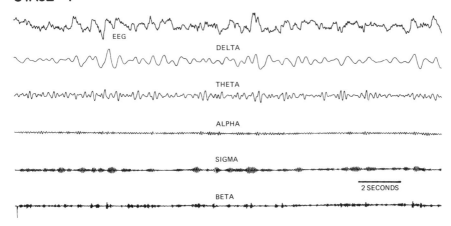

FIGURE 3   Frequency analysis of Stage 4 EEG. Same subject as in Figure 2. The eye discerns
alpha band prominence during waking (Figure 2) and delta and theta prominence during Stage 4
sleep (this figure). Subsequent electronic analysis quantifies the differences.

cycle, with relatively more Stage 3–4 in the early cycles of the dark span
and relatively more Stage REM in the later cycles (Kripke *et al.*, 1968b;
Williams, Agnew, and Webb, 1964). In the present study such circadian
changes were assessed by fitting 24-hr cosine curves to the data.

The individual time series were first expressed as a percent of the series
mean. These plots of relative values for each animal were then averaged
for the group of animals as a whole at each time-point, to obtain a mean
24-hr curve for each function. To each of these means a 24-hr cosine curve
was fitted by the least squares method, facilitating comparisons of phase
and amplitude in the irregular curves of the raw data. From such a fit, an
acrophase, ϕ, as well as an amplitude, C, and the standard error (SE) of
this C, were obtained. The confidence interval of the ϕ was estimated from
the SE/C ratio (Halberg, Tong, and Johnson, 1967). The circadian frequency was assumed to be one cycle per 24 hr, since synchronization of
monkey circadian rhythms to a 24-hr cycle of light and darkness has been
observed by Simpson and Galbraith (1905), Migeon *et al.* (1955), Erikson
(1960), and Winget, Card, and Hetherington (1968), among others, as
reviewed by Halberg (1969b). The cosines fitted to the mean curves were
then compared with respect to acrophase and amplitude (Figure 4).

In addition to analysis of the mean time series on each variable, individual time series on each variable from each animal were also analyzed by
the least squares fit of a cosine curve with a 24-hr period. This estimate of
the circadian rhythm of a given variable in each animal was paired succes-

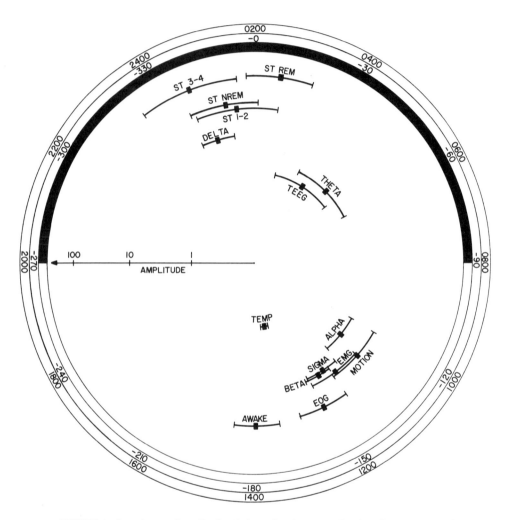

FIGURE 4   Acrophase and amplitude relations of cosine curves fitted to the mean curves for sleep, EEG, temperature, and motor activity of all monkeys studied. The outer circular scale represents a 24-hr clock. Another scale shows corresponding angular degrees with the phase reference or zero degrees at the middle of the daily dark span (0200 hr). Darkness is indicated as a heavy black arc. Negative degrees indicate clockwise rotation. The acrophase (peak of cosine curve best fitting each variable) is a point determined by its position on the circle in relation to the phase reference. The logarithmic distance of each point from the center of the circle represents the amplitude as percent of the mean level. A 95 percent confidence arc is also shown for each acrophase.

Individual Variation in Circadian Rhythms of Sleep, EEG, Temperature, and
Activity among Monkeys: Implications for Regulatory Mechanisms

37

sively with the cosine estimate of every other variable in that animal. Thus, possible correlations of acrophases, of levels, and of amplitudes were studied in the group of animals as a whole. Table 1 lists, as an example, the relations among acrophases, levels and amplitudes of the 24-hr cosine curves fitted to Stage Awake and Stage NREM. (For brevity many animals are excluded from the list.) Similar correlations were computed for every other pair of variables.

Since the EEG recording equipment was not calibrated in absolute units and amplification varied somewhat from animal to animal, the levels of delta, theta, alpha, sigma, and beta were determined as proportions of the total EEG abundance for each animal. Gross motion levels were not calculated because of similar amplification differences. The amplitude of the circadian modulation of EEG was expressed as a percentage of the mean level.

RESULTS

Correlating acrophases as in Table 1—each of the 15 variables eventually being paired against all of the others—results in 105 separate correlation

TABLE 1   Examples of Correlation Determinations: Acrophases, Levels, and Amplitudes of Stage Awake and Stage NREM Sleep in Rhesus Monkeys

Animal Number	Acrophase in Degrees of Arc from Mid-Dark Reference[a] to Peak of Fitted Cosine Curve		Level, Measured during 24-hr, Mean Number of Min per 5-Min Span Spent in Each Stage		Amplitude, Peak Value of Fitted 24-hr Cosine Minus Mean Level, in Min per 5-Min Span	
	Awake	NREM	Awake	NREM	Awake	NREM
1	−209	−25	2.7	1.9	2.4	1.9
2	−212	−32	2.6	2.1	2.3	1.8
3	−208	−22	3.1	1.5	2.1	1.6
4	−205	−16	2.7	2.0	1.3	0.9
⋮	⋮	⋮	⋮	⋮	⋮	⋮
32	−208	−23	2.4	2.2	2.4	1.9
33	−202	−20	2.4	2.3	2.3	1.9
34	−194	−8	2.1	2.6	1.7	1.4
n	34		34		34	
r	0.98		−0.96		0.96	
$p<$	0.01		0.01		0.01	

[a]Compare Figure 4.

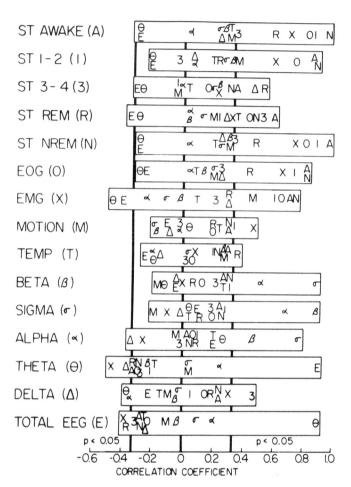

FIGURE 5 Correlations among circadian acrophases of simian sleep,
EEG, temperature, and activity. Variables investigated listed at left, with
abbreviations in parentheses. Correlation coefficients (*r*) on abscissa. For
example, at far right in bar associated with Stage Awake, N is an abbrevi-
ation for Stage non-REM sleep. The position of the N on the r scale corresponds to a correlation coefficient of 0.98 between Stage Awake and
Stage non-REM. Vertical bars are drawn at r = −.33, 0, and +.33. Values
outside the limit (−.33, +.33) are all statistically significant (p < .05), as
are a few of those just within these limits.

Individual Variation in Circadian Rhythms of Sleep, EEG, Temperature, and
Activity among Monkeys: Implications for Regulatory Mechanisms

39

coefficients for phase relations, graphically presented in Figure 5. Correlations of level and of amplitude similarly appear in Figures 6 and 7, respectively.*

Cosine fitting to time series, although a rough approximation, is a standardization allowing comparisons of acrophase and amplitude among irregularly shaped curves. The fit of the 24-hr cosine curve was generally very good. Fewer than 5 percent of the 422 separate time series approximated by the fit of a 24-hr cosine in this study failed to meet our empirical criterion for "goodness of fit" (that the standard error of the amplitude divided by the amplitude should be less than 0.35).

ACROPHASE

When there is a high positive correlation between the acrophases of two functions, those animals with an early acrophase in one variable tend to be early in the other variable as well, and those animals that show a late ϕ in one tend to exhibit a late ϕ in the other. A high negative correlation occurs when animals having a relatively early ϕ in one variable show a relatively late ϕ in the other. A low correlation occurs when the timing of the acrophase of one function tends to be unrelated to that of the other.

The extent of data manipulation in this study requires that the validity of the method be tested to rule out systematic errors of data reduction. For instance, the acrophases of variables known to be interdependent should correlate highly. In Figure 5 Stage Awake correlates significantly with each of the sleep stages, bearing out this prediction and supporting the methodology.

There is a significant phase correlation of Stage 3–4 with the circadian modulation in the delta band. Except for this, none of the frequency bands correlates positively with any function except other frequency bands. Stage 3–4 is defined by the presence of high-voltage delta activity, so it is not surprising that the acrophases correlate. But this correlation of Stage 3–4, read from polygraph paper by a technician, and delta, determined from magnetic tape by a machine, represents an especially important agreement between two separate approaches to related data, since one of the procedures is relatively free from observer bias.

Certain other correlations in Figure 5 also result from interdependence of variables. Stage 1–2 makes up most of Stage NREM, with Stage 3–4 con-

*Original correlation matrices are available as NAPS document 01023 from ASIS National Auxiliary Publications Service, c/o CCM Information Sciences Incorporated, 22 West 34th Street, New York, New York 10001. Remit $1 for microfiche or $3 for photocopies.

tributing a smaller part; the correlations reflect this relationship. The indices of activity—EOG, EMG, and gross motion—tend to correlate with one another and with Stage Awake. The detection of these predictable correlations lends some credence to more unexpected results.

One such unexpected result is the statistically significant positive correlation of Stage 3–4 and Stage REM. Each occupies a relatively small part of the sleep span. Conceivably, these two ϕ's could be independent—an early REM ϕ being associated in some animals with an early Stage 3–4 ϕ, and in others with a late Stage 3–4 ϕ. However, the high correlation indicates instead that animals with an early massing of Stage REM tend also to have an early massing of Stage 3–4, while those with later REM also have later Stage 3–4. The timing of Stage 3–4 is more closely related to the timing of REM than to that of Stage 1–2.

Since the mammalian temperature acrophase is somewhat reproducible, it is often used as an internal phase-reference for other circadian rhythms within an organism (Halberg, 1959, 1969a, 1969b; Halberg *et al.*, 1969). Surprisingly, in this study the timing of temperature had the least number of statistically significant correlations and the narrowest range of correlation coefficients. Of the 15 measured variables, brain temperature is least likely to advance or delay its acrophase in unison with the others, a finding to be discussed further below.

The circadian ϕ of brain temperature does correlate significantly with the Stage REM ϕ. This point deserves further study in the ultradian as well as the circadian ranges.

The ϕ's for the three bands of higher frequency EEG—alpha, sigma, and beta—are very highly correlated among themselves, but not with the ϕ's of any other functions. Abundance in these frequencies is prominent during wakefulness, but the precise circadian timing of each band is more related to the corresponding timing of the other high-frequency EEG bands than to the timing of Stage Awake. Alpha, sigma, and beta thus appear to be subject to some common regulation of circadian timing separate from such other events as sleep stage and motor activity. This circadian temporal relationship, so apparent in these "microscopic" computer analyses (Halberg, 1969a), is lost in the gross "clinical" record of EEG, where spans of alpha predominance simply alternate with spans of beta or sigma predominance.

Except for the previously noted interdependence of delta and Stage 3–4, the circadian rhythms in the EEG frequency bands do not correlate in acrophase with rhythms of activity, sleep, or temperature. Like brain temperature, the intersubject variation of EEG circadian timing seems to remain relatively independent of the intersubject variation in the cycle of sleep and activity.

Individual Variation in Circadian Rhythms of Sleep, EEG, Temperature, and
Activity among Monkeys: Implications for Regulatory Mechanisms

41

The acrophases of theta and total EEG correlate very highly; the ϕ's of
these functions correlate negatively with those of several others. An early
ϕ in total EEG or theta appears to be associated with a late ϕ in these other
variables, while late ϕ's in the former pair are associated with early ϕ's in
the latter group.

LEVELS

There is a high positive correlation between the levels of two functions
when those animals having a high 24-hr level for one variable tend to have
a high 24-hr level for the other, and those low in one tend to be low in the
other. A high negative correlation occurs when subjects high in the level of
one function tend to be low in the other.

Again, correlations of certain obviously interrelated functions serve to
check our method. Clearly, those animals spending the most time awake
during the 24-hr day must spend the least time asleep. Figure 6 does show
a statistically significant negative correlation between the levels of Stage
Awake and each of the sleep stages. Stage NREM, including all sleep time
except the small amount of Stage REM, is almost the inverse of Stage
Awake, and Awake and NREM have a very strong negative correlation.
The previously noted relationships of Stage NREM, Stage 1–2, and Stage
3–4 are again demonstrated in Figure 6, as are the expected relationships
of Stage Awake, EOG, and EMG.

The proportion delta/total EEG does not significantly correlate in level
with Stage 3–4. Although Stage 3–4 is defined by the presence of high
amplitude delta waves, lower amplitude delta activity is always measurable
in the filtered EEG. This continuous low amplitude delta apparently contributes more to the mean 24-hr level than do the bursts of high amplitude
delta during Stage 3–4. However, the timing of these bursts apparently
weighs considerably in determining the circadian phase of delta, for the
acrophases of delta and Stage 3–4 do correlate significantly (Figure 5).

The mean level of brain temperature does not correlate significantly
with the level of any other function. This circumstance is even more per-
vasive in level than in acrophase, where at least Stage REM correlates significantly with temperature.

The level of Stage REM does not correlate with the level of Stage 3–4,
although they do correlate in acrophase. Stage REM does correlate significantly with beta. There is a similar, though statistically less significant, correlation between Stage 1–2 and theta.

The amounts of beta and sigma in the total 24-hr EEG are highly correlated. They also correlate with respect to acrophase (Figure 5). Similarly,

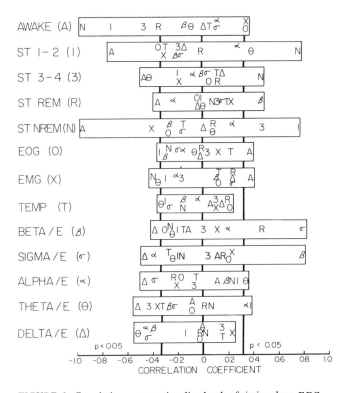

FIGURE 6 Correlations among circadian levels of simian sleep, EEG, temperature, and activity. Notation corresponds to that in Figure 5. Before computing the correlation matrix, original data on the five EEG frequency bands were expressed as a proportion of the total EEG abundance.

theta and alpha, weakly correlated with respect to acrophase, correlate significantly with respect to level. The circadian rhythm of delta remains independent. In acrophase and level, it negatively correlates with alpha and theta, and in level with sigma and beta as well.

AMPLITUDE

As defined herein and earlier (Halberg, 1959), the amplitude of a rhythm is the peak minus the mean value of the fitted cosine curve. In the amplitude calculations certain *a priori* predictions again serve to check the methodology. An animal awake during a great part of the day but seldom awake at night would have a high amplitude in Stage Awake. (If wakeful-

dividual Variation in Circadian Rhythms of Sleep, EEG, Temperature, and
ctivity among Monkeys: Implications for Regulatory Mechanisms

43

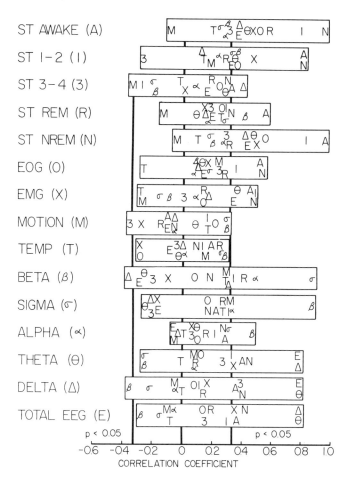

FIGURE 7 Correlations among circadian amplitudes of simian sleep,
EEG, temperature, and activity. The notation corresponds to that of
Figures 5 and 6.

ness were equally distributed throughout the day and night, the fitted circadian "curve" for Stage Awake would be flat.) Obviously, this same animal
would have a high amplitude for Stage NREM, which includes most of the
sleep time. The amplitudes should thus be correlated. Indeed, in Figure 7,
Stage Awake significantly correlates with all the sleep stages, and especially with Stage NREM. Stage NREM again makes the predictably high
correlation with Stage 1–2.

The negative correlation of Stage 1–2 and Stage 3–4 just barely attains statistical significance. High amplitudes of Stage 1–2 result when the animal spends little of the light span and much of the dark span in Stage 1–2. Since Stage 3–4 is almost entirely confined to the dark span, animals with a large amount of nighttime Stage 1–2 would have less time for Stage 3–4; hence the negative correlation. The positive correlations with Stage 1–2 indicate that those animals with high amplitudes of 1–2 will also have high amplitudes of beta, sigma, theta, EOG, and EMG.

Stage 3–4 and delta have a statistically significant correlation in amplitude, as in acrophase. The amount and timing of the bursts of delta activity must determine the height and timing of the fitted circadian wave in delta abundance. But as noted earlier, the low-voltage delta activity that continues day and night appears to be more important in determining the mean 24-hr level of delta.

A number of other predictable correlations appear in Figure 7. Animals concentrating sleep in the dark span and wakefulness in the light span have a high amplitude in wakefulness, and the correlations demonstrate that these animals also tend to have high amplitudes in EMG and EOG, in Stage NREM, and in the EEG frequencies of sleep, delta and theta. These correlations could be predicted *a priori* because of the interdependence of sleep, wakefulness, EEG, and motor activity.

In amplitude, brain temperature again fails to correlate significantly with any other function.

In addition to its correlation with Stage Awake noted above, the circadian amplitude of REM also correlates significantly with the circadian amplitude of beta. Thus, in both level and amplitude, REM and beta are related.

Both EOG and EMG correlate significantly in amplitude with Stage Awake, with Awake's near-mirror-image, Stage NREM, and with Stage 1–2, which constitutes most of Stage NREM. EMG also correlates in amplitude with theta and total EEG abundance. Motion has no statistically significant correlations. EEG artifact, our parameter of motion, may be so much influenced by uncontrolled factors of electrical conduction that in individual cases it is an unreliable measure, even though it usefully describes the group of animals taken together.

As in acrophase and level, beta and sigma amplitudes correlate highly. The lower frequencies, delta and theta, also have a high positive correlation with one another and with total EEG abundance, which consists largely of delta and theta. The higher frequencies tend to correlate negatively with the lower, so that a subject with a high circadian amplitude in beta would be likely to have a high amplitude in sigma and low amplitudes in delta

Individual Variation in Circadian Rhythms of Sleep, EEG, Temperature, and
Activity among Monkeys: Implications for Regulatory Mechanisms

45

and theta. Alpha correlates more with the higher frequencies in its circadian amplitude modulation.

DISCUSSION

This project extends earlier temporal maps of the frequency-analyzed EEG done on mice (Harner, 1961; Harner and Halberg, 1958), on healthy men (Frank *et al.*, 1966), and on diseased man (Engel, Halberg, and Gully, 1952; Engel *et al.*, 1954, 1952; Halberg *et al.*, 1952, 1955). We have made frequent measurements of EEG and other variables in a group of monkeys in order to examine relationships among the variables. Several factors could produce relationships in the data:

Interdependence. Some of our variables are clearly not independent of the others. An animal that is asleep obviously cannot be awake, so that the amount and timing of his wakefulness are not independent of his sleep. We have used these interdependent measures as a check on our method of data analysis. The method does appear valid, yielding the expected correlations.

Common regulation of responses to the light cycle. The light cycle is a major synchronizer of circadian rhythms (Halberg 1969a, 1969b). However, rhythms are not perfectly synchronized even among animals exposed to the same light–dark cycle. This variability may reflect some different biological characteristics of the animals. If the rhythms of several indices vary together across the group of animals, the indices may be subject to common regulation by something that is slightly different within each animal. This information could be useful in illuminating the processes that synchronize biological rhythms.

Common response to uncontrolled secondary synchronizers. Two variables might both be affected by such uncontrolled changes in the environment as variations in the time or duration of chamber inspection, small changes in feeding time, differences in barometric pressure, or changes in ambient electromagnetic field. But unless the uncontrolled secondary synchronizer changes all the measured variables together, those that do change together would appear to be commonly regulated apart from the others, and again this information could help elucidate regulatory processes.

Statistical artifact. In a matrix of 100 correlations, about 5 can be expected to reach the 0.05 level of significance on the basis of chance alone. Many of the correlations we report are far beyond this level, and we find many more than 5 significant correlations per 100 in our data, but some of these correlations may still be mere statistical artifacts. Although we can *a priori* recognize interdependence of the functions, we cannot distinguish in our data correlations arising from variations in the biological response to the primary synchronizer, variations in exposure to secondary synchronizers, or statistical artifacts. However, when a consistent pattern of correlations between two functions appears in acrophase, level, and amplitude, the likelihood of statistical artifact decreases.

Table 2 reviews certain patterns that emerge from the data in Figures 5, 6, and 7. The table shows that motor activity, especially as measured by EMG and EOG, correlates in acrophase, level, and amplitude with Stage Awake. Stage Awake is defined in part by EMG and EOG activity, and the method of analysis clearly detects the relationship. We feel that this example and the others noted above demonstrate the validity of our approach, involving first a cosine fit and next a correlation analysis—a procedure resembling an approach by Künkel (1966) combining correlations with prior variance spectral analysis.

The table shows the relative independence of the circadian curve of brain temperature from sleep, EEG, and motor activity curves. Simian brain temperature may vary over shorter spans in association with activity and sleep (Hayward, Smith, and Stewart, 1966; Reite and Pegram, 1968), but its circadian curve appears to be largely unrelated to them. A considerable degree of independence of the circadian rhythms in temperature and activity was emphasized by Jürgensen's finding of a persistent temperature

TABLE 2 Correlation Patterns

Functions	Acrophase	Level	Amplitude
Motor activity–Stage Awake	+	+	+
Temperature–Stage REM	+	0	0
Temperature–any function other than Stage REM	0	0	0
Beta–sigma	+	+	+
Beta or sigma–alpha	+	–	+
Beta or sigma–delta	0	–	–
Delta–theta	–	–	+
Delta–alpha	–	–	0
Theta–total EEG	+	Not available	+
Stage REM–beta	0	+	+
Stage 1–2–theta	0	+	+
Stage 3–4–REM	+	0	0

Positive and negative correlations of $p < 0.05$ indicated by + and –, respectively. No statistical significance is represented by 0. For details of these and other correlations, see Figures 5–7.

Individual Variation in Circadian Rhythms of Sleep, EEG, Temperature, and
Activity among Monkeys: Implications for Regulatory Mechanisms

47

rhythm in continually recumbent men (Jürgensen, 1873); this independence has also been commented upon by Aschoff in this symposium and
elsewhere (Aschoff, 1967). Our data, derived differently in a correlation
analysis, support these earlier findings and interpretations.

For some purposes it is useful to determine the phase of one biological
rhythm in relation to another within the organism rather than in relation
to an external event such as the lighting cycle. The circadian rhythm of
body temperature has been suggested as one such "internal phase reference"
with which the phases of other rhythms may be compared (Halberg *et al.*,
1969; Halberg, 1969b). Our data, taken over only one day, do not bear on
the question of which biological rhythm is more stably reproducible from
day to day in each subject—a question highly relevant to the choice of an
internal phase reference. But the data do show that intersubject variations
of temperature tend to be independent of intersubject variations in sleep,
EEG, and motor activity, suggesting that changes in the circadian rhythms
of the latter functions may not be reflected by changes in the temperature
rhythm, and vice versa.

Stage REM does correlate significantly in acrophase with brain temperature. The circadian curve of Stage REM thus may be related to the timing
of the circadian curve of brain temperature, even though their acrophases
are almost in antiphase. There is some biochemical evidence that Stage
REM and brain temperature may be subject to common regulation. Injections of norepinephrine into the third ventricle or into the hypothalamus
of monkeys produce a fall in body temperature (Feldberg, Hellon, and
Lotti, 1967; Crowley, 1967; Myers and Sharpe, 1968). Ventricular injections of norepinephrine suppress Stage REM in cats (Rakic, Radulovacki,
and Beleslin, 1968), and Crowley and Smith noted the same effect in monkeys (unpublished). Monkeys undergoing norepinephrine depletion with
alphamethylparatyrosine spend less time in Stage REM (Weitzman *et al.*,
1968–1969; Crowley *et al.*, 1968). Brain norepinephrine itself shows a
circadian rhythm in cats and rats (Reis, Weinbrum, and Corvelli, 1968;
Scheving *et al.*, 1968). Thus CNS norepinephrine may participate in regulating both Stage REM and temperature, contributing to their common
circadian timing.

The curves of circadian modulation in beta and sigma are highly correlated in acrophase, level, and amplitude (Table 2). It seems very unlikely
that these consistently high correlations are statistical artifacts, and the
electronic filter performance ensured that their measurement was virtually
independent. The finding strongly suggests common regulation for circadian modulation in these bands.

Theta and total EEG have high acrophase and amplitude correlations

(Table 2) and overlapping confidence arcs for acrophase (Figure 4). This correlation is unavailable for level, since for level determinations all frequency bands were expressed as percent of total EEG. The circadian modulation in the two variables appears similar even though theta constitutes only about 30 percent of the total EEG abundance in our data.

Theta and alpha are not highly correlated in any parameter, but they have a consistent tendency toward positive correlations, which suggests (but does not prove) commonality in their regulation. Theta clearly lacks this kind of correlation with beta and sigma. However, alpha tends to correlate positively with beta and sigma in acrophase and amplitude, though not in level. Thus, alpha seems to occupy an uncertain middle position between the higher frequencies and theta. Delta generally does not correlate with the other functions.

Encephalographers are accustomed to seeing relatively brief polygraph tracings in which the various frequencies alternately appear, predominate, and disappear in relation to the functional state of the subject. For example, very little low-frequency activity is apparent in the polygraph record of the waking EEG. Alpha becomes very prominent, especially in human beings but also in some monkeys, during quiet waking periods when the eyes are shut. Beta briefly predominates during early drowsiness. "Sleep spindles" of the sigma frequency appear during the middle sleep stages, while delta occupies much of the Stage 3–4 record. This alternation of predominant frequencies gave rise to the concept of independently functioning neural generators, each responsible for activity in one or more of the frequency bands (*cf.* Johnson *et al.*, 1969).

But when the EEG is filtered into the five component frequencies, each proves to be continuously active, and the activity shows not only brief, state-dependent changes but also a prominent circadian component. In our circadian analysis beta and sigma appear to be commonly regulated, although they are perhaps separate in short-term regulation. Circadian changes in delta and theta are probably each regulated independently, while alpha occupies an uncertain position. The data fit the view that at least three, and perhaps as many as five, separate circadian "governors" regulate the frequency-band generators, of which there may be as many as five. The lack of parsimony in this view suggests that the discrete generator hypothesis may eventually prove inadequate to explain the complexities of ultradian, circadian, and perhaps even infradian rhythms in the EEG.

Stage REM correlates highly with beta in both level and amplitude, though not in acrophase. Beta peaks in the light span. When more beta appears during darkness the circadian curve becomes flatter and its amplitude lower. This would be the case if beta were simply appearing in increased quantity during Stage REM—that is, animals with more Stage REM would

Individual Variation in Circadian Rhythms of Sleep, EEG, Temperature, and
Activity among Monkeys: Implications for Regulatory Mechanisms

49

show decreased amplitude of beta. Instead, animals with more Stage REM apparently have more beta with a higher beta circadian amplitude. The increased beta is thus not simply beta occurrence during REM periods. Studies on humans (Johnson et al., 1969) do not show any unique rise in beta during Stage REM. It appears that animals generating more beta, and especially more in the day with less at night, simply have a greater tendency to produce Stage REM at night.

Stage 1-2 correlates in the same fashion with theta. But theta is definitely a characteristic of Stage 1-2 sleep, and the ϕ's for theta and Stage 1-2 are much closer than those for Stage REM and beta. Unlike the correlation between beta and REM, then, this may be a simple dependent relationship in which animals having more Stage 1-2 at night have extra theta as part of the extra time spent in Stage 1-2.

Stage 3-4 and Stage REM correlate in acrophase. Stage REM is usually viewed as a unitary state, uniquely apart from Stage 1-2 and Stage 3-4 and regulated separately from them. But in certain respects Stage REM and Stage 3-4 are known to act as a unit apart from Stage 1-2. Stage 3-4 and Stage REM are both reduced in depressed psychiatric patients, while Stage 1-2 persists (Mendels and Hawkins, 1967). Alterations in brain 5-hydroxytryptamine (5HT) metabolism have been reported to alter the amounts of both Stage 3-4 (Weitzman et al., 1968; Crowley et al., 1968) and Stage REM (Hartmann, 1967; Oswald et al., 1966). Our observation of a circadian temporal relationship in these two variables is further evidence that they may be subject to some common regulation, such as a brain 5HT rhythm (Albrecht et al., 1956). Since only the acrophases, and neither the levels nor amplitudes, of Stage REM and Stage 3-4 significantly correlate, this single correlation may be only a statistical artifact. Alternatively, further study might demonstrate that other rhythms are also independent in level and amplitude but correlated in timing.

Finally, delta and theta correlate in amplitude with one another and with Stage NREM, Stage Awake, and total EEG (not shown in Table 2). These two slow frequencies predominate during sleep, and animals with more nocturnal sleep and more diurnal wakefulness have more delta and theta at night. Since these frequencies make up most of the total EEG, its amplitude rises with theirs. These amplitude correlations seem to reflect the extent to which subjects concentrated sleep in the dark span.

SUMMARY AND CONCLUSIONS

Among 34 male *Macaca mulatta* monkeys on an *LD 12*:12 schedule, circadian rhythms appeared in each of 15 variables related to sleep, EEG, temperature, and motor activity. Statistical procedures assessed the variables for evidence of common regulation in these aspects of their circadian rhythms: acrophase (timing), amplitude (extent of change), and level (24-hr mean value). Patterns appearing in the data suggested that the circadian rhythms of certain variables are regulated in common.

The circadian modulation of activity in the beta and sigma frequency bands of the EEG was correlated with statistical significance in acrophase, level, and amplitude. The delta frequency band appeared to be under circadian regulation distinct from that of the other bands. The circadian rhythm of Stage REM (rapid eye movement) sleep was like that of beta activity in level and amplitude. The data indicate that Stage REM may share some common regulation of circadian timing with both Stage 3–4 sleep and with temperature. Generally, however, the circadian rhythm of temperature appeared to bear little relation to the circadian rhythms of motor activity, EEG, or sleep.

ACKNOWLEDGMENTS

The efforts of many people made this work possible. O. F. Lewis, J. O'Donaghue, and J. Kechle assisted at surgery and maintained the animals. C. Clark, C. Merritt, and K. Schweikert developed and maintained the recording assembly. C. Cook interpreted polygraph records. D. Hillman, J. Patterson, G. Jordan, R. Butler, and H. Lerum developed computer programs and supervised their application. D. Stilson and E. A. Johnson made helpful suggestions on statistical matters. M. K. Parker assisted in preparing the manuscript.

Animals used in this research were handled in accordance with the Principles of Laboratory Animal Care published by the National Academy of Science–National Research Council. Further reproduction of this article is authorized as needed to meet the requirements of the U.S. Government.

This study was supported by ARL Project 6892, Holloman Air Force Base, by the U.S. Public Health Service (CA-5-K6-GM 13981), by NASA (NAS 2-2738 and NGR-24-005-006), and by the U.S. Air Force (F 29608-69-0011). Doctors Crowley and Kripke were on active duty with the U.S. Air Force during part of this project.

REFERENCES

Albrecht, P., M. B. Visscher, J. J. Bittner, and F. Halberg. 1956. Daily changes in 5-hydroxytryptamine in mouse brain. Proc. Soc. Exp. Biol. Med. 92:703–706.

Individual Variation in Circadian Rhythms of Sleep, EEG, Temperature, and
Activity among Monkeys: Implications for Regulatory Mechanisms

51

Aschoff, J. 1967. Adaptive cycles: their significance for defining environmental
hazards. Int. J. Biometeorol. 11:255–278.

Crowley, T. 1967. Norepinephrine and serotonin effects on sleep, alertness, and temperature regulation. *In* Behavioral problems in aerospace medicine. NATO Advisory
Group for Aerospace Research and Development, Paris.

Crowley, T., E. Smith, O. Lewis, and G. V. Pegram. 1968. The biogenic amines and
sleep in the monkey—a preliminary report. Paper presented at the annual meeting,
Association for the Psychophysiological Study of Sleep.

Dement, W., and N. Kleitman. 1957. Cyclic variation in EEG during sleep and their
relation to eye movements, body motility, and dreaming. Electroencephalogr. Clin.
Neurophysiol. 9:673–690.

Engel, R., F. Halberg, and R. J. Gully. 1952. The diurnal rhythm in EEG discharge and
in circulating eosinophils in certain types of epilepsy. Electroencephalogr. Clin.
Neurophysiol. 4:115–116.

Engel, R., F. Halberg, F. Y. Tichy, and R. Dow. 1954. Electrocerebral activity and
epileptic attacks at various blood sugar levels (with a case report). Acta Neuroveg.
9:147–167.

Engel, R., F. Halberg, M. Ziegler, and I. McQuarrie. 1952. Observations on two children
with diabetes mellitus and epilepsy. J.-Lancet 72:242–248.

Erickson, L. 1960. Diurnal temperature variation in the rhesus monkey under normal
and experimental conditions. Nature (Lond.) 186:83–84.

Feldberg, W., R. F. Hellon, and V. J. Lotti. 1967. Temperature effects produced in
dogs and monkeys by injections of monoamines and related substances into the
third ventricle. J. Physiol. (Lond.) 191:501–515.

Frank, G., F. Halberg, R. Harner, J. Matthews, E. Johnson, H. Gravum, and V. Andrus.
1966. Circadian periodicity, adrenal corticosteroids, and the EEG of normal man.
J. Psychiatr. Res. 4:73–86.

Halberg, F. 1959. Physiologic 24-hour periodicity: general and procedural consideration with reference to the adrenal cycle. Z. Vitam.-Horm. Fermentforsch. 10:225–
296.

Halberg, F. 1969a. Chronobiology. Annu. Rev. Physiol. 31:675–725.

Halberg, F. 1969b. Circadian system of non-human primates: summary of a symposium
in 1968 and of earlier work, p. 106 to 1260. *In* F. H. Rohles [ed.], Circadian
rhythms in non-human primates. Bibl. Primatol. No. 9. Karger, Basel.

Halberg, F., R. Engel, E. Halberg, and R. J. Gully. 1952. Diurnal variations in amount
of electroencephalographic paroxysmal discharge and diurnal eosinophil rhythm of
epileptics on days with clinical seizures. Fed. Proc. Fed. Am. Soc. Exp. Biol. 11:63.

Halberg, F., R. J. Gully, R. Engel, and M. R. Ziegler. 1955. P. 130 to 135. *In* R. A.
Good and E. S. Platou [ed.], Essays on pediatrics in honor of Dr. Irvine McQuarrie.
Further studies on coexisting diabetes and epilepsy. P. 130 to 135. *In*: Essays on
Pediatrics, in honor of Dr. Irvine McQuarrie. R. A. Good and E. S. Platou [eds.]
Lancet Publ., Minneapolis, Minn.

Halberg, F., J. Reinhardt, F. Bartter, C. Delea, R. Gordon, A. Reinberg, J. Ghata,
H. Hoffman, M. Halhuber, R. Gunther, E. Knapp, J. C. Pena, and M. Garcia Sainz.
1969. Agreement in endpoints from circadian rhythmometry on healthy human
beings living on different continents. Experientia 25:107–112.

Halberg, F., Y. L. Tong, and E. A. Johnson. 1967. Circadian system phase—an aspect
of temporal morphology: procedures and illustrative examples, p. 20 to 48. *In*:
H. von Mayersbach [ed.], Cellular aspects of biorhythms: Symposium on biorhythms.
Proc. Int. Congr. Anat. Springer Verlag, New York.

Harner, R. N. 1961. Electrocorticography and frequency analysis in mice; circadian
periodicity in electrocerebral activity. Electroencephalogr. Clin. Neurophysiol.
13:752–761.

Harner, R. N., and F. Halberg. 1958. Electroencephalographic differences in D2 mice at times of high and low susceptibility to audiogenic convulsions. Physiologist 1:34–35.

Hartmann, E. 1967. Some studies on the biochemistry of dreaming sleep. Excerpta Med. Found. Int. Congr. Ser. 150:3100–3102.

Hayward, J. N., E. Smith, and D. Stewart. 1966. Temperature gradients between arterial blood and brain in the monkey. Proc. Soc. Exp. Biol. Med. 121:547–551.

Johnson, L., A. Lubin, P. Naitoh, C. Nute, and M. Austin. 1969. Spectral analysis of the EEG of dominant and non-dominant alpha subjects during waking and sleep. Electroencephalogr. Clin. Neurophysiol. 26:361–370.

Jürgensen, T. 1873. Die Körperwärme des gesunden Menschen. Leibzig.

Kripke, D., C. Clark, and J. A. Merrit. 1968. A system for automated sleep analysis and physiological data reduction. Air Force Tech. Rep. ARL–TR–68–12. Aeromedical Research Laboratory, Holloman Air Force Base, New Mexico.

Kripke, D., T. Crowley, G. V. Pegram, and F. Halberg. 1968a. Circadian rhythmic modulation of Berger-region frequencies in electroencephalograms from *Macaca mulatta.* Rass. Neurol. Veg. 22:519–525.

Kripke, D., M. Reite, G. V. Pegram, L. Stephens, and O. Lewis. 1968b. Nocturnal sleep in rhesus monkeys. Electroencephalogr. Clin. Neurophysiol. 24:582–586.

Künkel, H. 1965. Die periodik der paroxysmalen Dysrhythmie in Elektroenzephalogramm. Habililationsschrift. Freie Universität Berlin.

Mendels, J., and D. Hawkins. 1967. Sleep and depression: a controlled EEG study. Arch. Gen. Psychiatry 16:344–353.

Migeon, C. J., A. B. French, L. T. Samuels, and J. Z. Bowers. 1955. Plasma 17-hydroxy-corticosteroid levels and leucocyte values in the rhesus monkey, including normal variation and the effect of ACTH. Am. J. Physiol. 182:462–468.

Myers, R., and L. Sharpe. 1968. Intracerebral injections and perfusions in the conscious monkey, p. 449 to 465. *In:* H. Vagtborg [ed.], Use of subhuman primates in drug evaluation. Univ. Texas Press, Austin.

Nelson, W., and F. Halberg. 1966. Phase relations of circadian rhythms: animals, p. 586 to 596. *In:* P. Altman and D. Dittmer [eds.], Environmental biology. Committee on Biological Handbooks, Federation of American Societies for Experimental Biology, Bethesda, Maryland.

Oswald, I., G. W. Ashcroft, R. J. Berger, D. Eccleston, J. I. Evans, and V. R. Thacore. 1966. Some experiments in the chemistry of normal sleep. Br. J. Psychiatry 112:391–404.

Rakic, M., M. Radulovacki, and D. Beleslin. 1968. The effect of adrenalin on different states of sleep. Experientia 24:243.

Rechtschaffen, A., and A. Kales. 1968. A manual of standardized terminology, techniques, and scoring system for sleep stages of human subjects. U.S. Dep. of Health, Education, and Welfare, Bethesda, Maryland.

Reis, D., M. Weinbrum, and A. Corvelli. 1968. A circadian rhythm of norepinephrine regionally in cat brain: its relationship to environmental lighting and to regional and diurnal variations in brain serotonin. J. Pharmacol. Exp. Ther. 164:135–145.

Reite, M., and G. V. Pegram. 1968. Cortical temperature during paradoxical sleep in the monkey. Electroencephalogr. Clin. Neurophysiol. 25:36–41.

Reite, M., J. Rhodes, E. Kavan, and W. R. Adey. 1965. Normal sleep patterns in macaque monkey. Arch. Neurol. (Chic.) 12:133–144.

Scheving, L., W. Harrison, P. Gordon, and J. Pauly. 1968. Daily fluctuations (circadian and ultradian) in biogenic amines of the rat brain. Am. J. Physiol. 214:166–174.

Simpson, S., and J. Galbraith. 1905. Observations on the normal temperature of the monkey and its diurnal variation, and on the effect of change in the daily routine on this variation. Trans. R. Soc. Edinb. 45:65–104.

Individual Variation in Circadian Rhythms of Sleep, EEG, Temperature, and
Activity among Monkeys: Implications for Regulatory Mechanisms

53

Weitzman, E., D. Kripke, C. Pollack, and J. Domingues. 1965. Cyclic activity in sleep
of *Macaca mulatta*. Arch. Neurol. (Chic.) 12:463–467.
Weitzman, E., P. McGregor, C. Moore, and J. Jacoby. 1968–1969. The effect of
α-methylparatyrosine on sleep patterns of the monkey. Psychophysiology 5:210.
Weitzman, E., M. Rappaport, P. McGregor, and J. Jacoby. 1968. Sleep patterns of the
monkey and brain serotonin concentration: effect of p-chlorophenylalanine. Science
160:1361–1365.
Williams, R., H. Agnew, and W. Webb. 1964. Sleep patterns in young adults: an EEG
study. Electroencephalogr. Clin. Neurophysiol. 17:376–381.
Winget, C. M., D. H. Card, and N. W. Hetherington. 1968. Circadian oscillations of
deep-body temperature and heart rate in a primate (*Cebus albafrons*). Aerosp. Med.
39:350–353.

DISCUSSION

DAVIS: Dr. Crowley, you have presented a very intriguing discussion of a number of areas involving techniques with which many of us are unfamiliar. I would like to commend you for a lack of complex terminology about the cycles themselves.

It occurred to me, as you were talking, that in a free-running situation, the positive phase correlations would tend to get the rhythms further out of synchrony, while the negative correlations would tend to keep the animal nearer a 24-hr rhythmicity. Would you care to comment on that?

CROWLEY: I don't think our data would answer that question. We looked only at 24 hr, and I can't say that the correlated advances or delays would be the same from day to day. We can only say that they vary together in individual animals on a given day, and can say nothing about what they do over a longer span of time.

JUSTICE: I was struck by your remarks about the lack of correlation of other variables with temperature. Could it be that many of your variables simply reflect the noise in the system and therefore would not be expected to correlate with a stable measure such as temperature?

CROWLEY: Temperature has the least relation, over one 24-hr period, to these other functions. I hesitate at the term "noise" because noise implies randomness. I think the high correlations we've found indicate that we are not dealing with randomness, but either with some common biological response to a known synchronizer or with a common susceptibility in certain functions to the effect of some secondary synchronizers we may not have been able to control. Essentially what we did was to correlate, for each animal, the advance or delay in phase in terms of the group mean for one particular function, against that same measure for another function.

HALBERG: The assumption—yet to be proved—underlying our approach is that in comparing animals for one day we are simulating the behavior of any one animal over many days. This is the reason for Dr. Crowley's reluctance to accept the fluctuation in many of the variables as "noise," and the rationale for his endeavor to see whether the correlations represent coupling among functions. He very cautiously—and properly, I believe—stated that we cannot say anything about couplings until we look at the behavior of the system under other conditions. Dr. Crowley has presented a method, results, and a new set of what you may wish to call oscillatory circuits that have been teased out of the coordinated system, in addition to the metabolic, adrenal, pituitary, and hypothalamic ones previously described. So far we can present these only as an anatomy in time, as Dr. Menaker so nicely put it at the outset of this meeting. We are now exploring by phase-shifting in these phenomena to see what this phase-coupling really means.

ARNOLD ESKIN

Some Properties of the System Controlling the Circadian Activity Rhythm of Sparrows

FREE-RUNNING RHYTHM

Since little is known about the behavior of the free-running period (FRP) of an individual organism under particular environmental conditions, it is often not clear what measurements of FRP, widely reported in the literature, mean in terms of the state of the underlying circadian system. The studies reported here were performed to define the behavior of the FRP of the house sparrow, *Passer domesticus*.

The existence of intra-individual variation in the FRP values of reptiles, insects, birds, and mammals has led to the conclusion that the FRP of an organism is labile (Aschoff, 1963; DeCoursey and DeCoursey, 1964; Palmer, 1964; Roberts, 1960). The value of an individual's FRP has been observed to change spontaneously (Palmer, 1964) and after various experimental treatments (Pittendrigh, 1967), a phenomenon Pittendrigh (1960, 1968) has termed "after-effects." After-effects were observed in the FRP's of cockroaches after exposure to LL (Roberts, 1960), but the values exhibited by individual cockroaches following many such treatments were significantly different from the values of other individuals treated in the same way. This result suggests that the range of FRP values that an individual (or species) may exhibit is an expression of genetic influence (Pittendrigh, 1960; Roberts, 1960). However, it has also been asserted (Palmer, 1964)

that the spontaneous changes observed in the FRP rule out the interpretation that the FRP is under genetic control.

Gradual changes in the FRP have been noted in a few organisms.
Changes occurred over about 20 days in lizards (Hoffmann, 1960), 50
days in Arctic ground squirrels (Swade and Pittendrigh, 1967), 100 days in
bats (Rawson, 1960), and 200 days in mice (Pittendrigh, 1968). A number
of explanatory hypotheses were suggested by these investigators: a spontaneous physiological change in the case of the lizards (although FRP
changes were gradual), a change from the winter to the summer condition
in the case of the bats, and changes related to the gradual decay of after-
effects in mice.

In none of the experiments in which either after-effects or spontaneous,
or gradual changes of the FRP were observed has it been established
whether these were transient or steady-state effects. Before these phenomena—and thus specific FRP values—can be adequately interpreted, this
question must be resolved.

METHODS

The results here reported were obtained from the activity records of more
than 60 house sparrows. Birds were housed singly in cages enclosed in light-
tight boxes. In each cage were two perches attached to microswitches wired
to an Esterline Angus event recorder. Each day's activity record was removed from the recorder and pasted beneath the preceding day's record on
a poster board. Remotely controlled fluorescent lamps were mounted in
each box, providing light intensities from 200–300 lux on the cage floor.
There was a constant background random noise of 92 dB (re 0.0002 μ Bar)
throughout all experiments. The birds were taken from the field and entrained for about 2 months by photoperiods ranging from LD 2:22 to
LD 15:9 before being placed into DD.

To study the characteristics of the FRP changes, FRP's were calculated
by regression analysis over 10 to 16 day segments of the activity data
(Eskin, 1969). Curves depicting free-run change as a function of length of
time in DD were then constructed for each bird by plotting each value of
the FRP calculated versus the last day in DD used in that calculation.

RESULTS

FRP Changes

When birds were placed in DD they exhibited various patterns of FRP
change (Figure 1). The changes or transients of the FRP lasted for at least

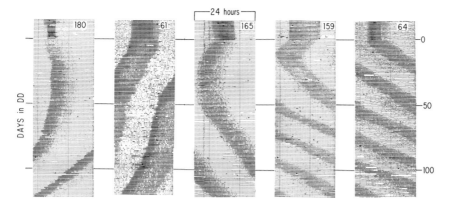

FIGURE 1   Patterns of FRP change. The activity records of five different birds are shown. The last day that a light cycle was presented to the birds before they were placed into DD was 0. Prior to being placed in DD, the birds were entrained for at least one month.

2 months. About half the birds had initial FRP's shorter, and about half longer, than 24 hr. FRP's of about 95 percent of the birds eventually became longer than 24 hr. The transition from less than 24 hr to more than 24 hr was abrupt in some cases and gradual in others.

Short term changes or oscillations in the value of the FRP were observed in the activity records of many birds (Figure 2). The periods of these oscillations ranged from 4 to 6 days. Such oscillations of the FRP occurred at various times throughout the records and in some cases continued to occur for about 2 months.

Examination of the FRP profiles obtained from the activity data (Figure 3) reveals several other general features of the FRP change. After the first few weeks of DD, the FRP usually began to increase. A "knee" or temporary decrease in the rate of change of the FRP often occurred between days 30 and 50 of DD. In the raw activity data, these "knees" of the FRP profiles appeared as spontaneous changes of the FRP. This decrease in the rate of change of the FRP generally lasted about two weeks, after which the FRP began to increase again. Around day 80 of DD a maximum or peak FRP value was reached. After the maximum FRP value was reached, the FRP decreased for a few weeks and then remained relatively constant for the next several months.

In order to compare specific values of the FRP following different treatments, two measures of the FRP were operationally defined (Eskin, 1969). The earliest measure of the FRP that could be obtained reliably was the value of the FRP on day 10 of DD (FRP-10). The mean FRP-10 value of

59 birds was 24.08 ± .33 hr (1 SD). To have a measure of the FRP after it ceased to change, the maximum or peak value of the FRP during the first 120 days of DD (FRP-PK) was selected. Though this is an arbitrary definition for FRP-PK, chosen mainly for convenience, the FRP-PK value is a good approximation of the FRP values during the time that the FRP is relatively stable, that is, after the peak. The mean FRP-PK value of 49 birds was 24.87 ± .54 hr (1 SD). To determine if FRP-10 and FRP-PK values were related to each other, correlation analysis was performed between these values for individual birds that were previously entrained by similar photoperiods. A significant, though weak, correlation (r = .57) existed between an individual's FRP-10 and FRP-PK values (Figure 4A).

Effects of Pretreatment on the FRP

Photoperiod The results just presented were obtained from birds entrained by cycles from *LD 2:22* to *LD 15:9*. The effect of photoperiod on the FRP was examined by grouping the FRP's of these birds on the basis

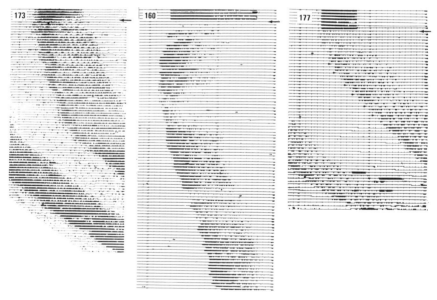

FIGURE 2 Oscillations in the value of the FRP. The arrows on the right sides of these records indicate the first day of DD. Examine the onsets of activity and note the FRP oscillations in the early part of the records of 173 and 177 and throughout the record of 160. The arrow points within the record of 177 indicate cycles that had long periods relative to the other cycles.

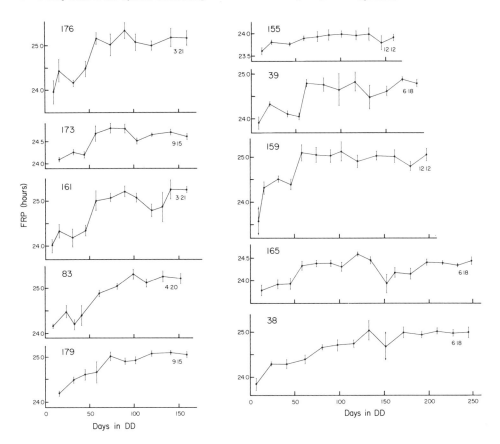

FIGURE 3 FRP changes as a function of time in DD. Each point represents a FRP value calculated over a segment of data at a particular time. Each curve is from a different bird. The light cycle to which each bird was exposed before being placed into DD is shown on the right side of each graph. The vertical lines on the curves represent 95% confidence intervals.

of the photoperiod of the pretreatment entrainment (Table 1). The photoperiod of the pretreatment had a slight effect on FRP-10 values. There were no significant differences between any of the mean FRP-10 values for the groups previously entrained by photoperiods LD 2:22 to LD 9:15. However, the mean FRP-10 for the group entrained by 12–15-hr photoperiods was significantly shorter than the mean FRP-10's of the 2–6-hr photoperiod groups. Due to the variance of the FRP-PK values and the size of the groups, the effect of photoperiod on FRP-PK cannot be established with certainty, although there were no significant differences in mean FRP-PK values between any of the groups.

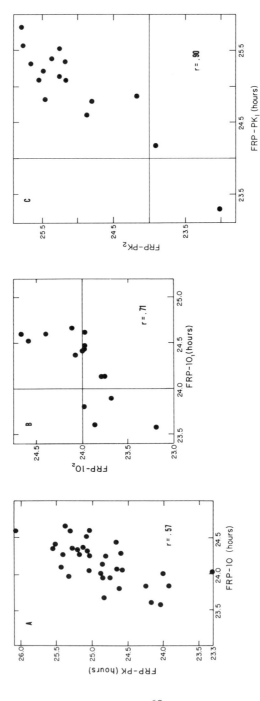

FIGURE 4 Correlations between different FRP values. A: The points represent FRP-10 vs. FRP-PK values from the first free-run for individual birds. B: The points represent FRP-10$_1$ (from the first free-run) vs. FRP-10$_2$ (from the second free-run) values of individual birds. C: The points represent FRP-PK$_1$ (from the first free-run) vs. FRP-PK$_2$ (from the second free-run) values of individual birds. The r values are correlation coefficients.

TABLE 1   F R P versus Photoperiod

Previous Photoperiod (T = 24 hr)	Number of Birds per Group	FRP-10[a] (hr)	Number of Birds per Group	FRP-PK[a] (hr)
2 hr	5	24.08 ± .23	5	24.78 ± .43
3 hr	8	24.07 ± .20	7	24.68 ± .70
4 hr	15	24.15 ± .19	12	24.99 ± .29
2–4 hr	28	24.12 ± .10	24	24.86 ± .22
6 hr	18	24.12 ± .15	12	24.98 ± .37
9 hr	6	24.10 ± .31	–	–
12–15 hr	7	23.70 ± .33	–	–
9–15 hr	13	23.88 ± .18	5	24.65 ± .63

[a]The plus and minus values are confidence intervals (95%).

Repeatability   To investigate the repeatability of the F R P changes and of specific F R P values, a group of birds was entrained a second time by the same light cycle that had been used for entrainment prior to the first free-run. Representative results from this experiment are shown in Figure 5. For the first month, the F R P's of the second free-run were generally shorter than those of the first free-run. Although the F R P-10's of the second free-run were shorter (.20 hr on the average) than those of the first free-run, a medium-strength correlation (r = .71) existed between the two FRP-10 values of individual birds (Figure 4B). After about two months of DD, the F R P's of the second free-run reached values similar to those of the first free-run. The F R P-PK's of individual birds were similar (.08-hr difference on the average) for the two free-runs.

Period of the cycle   To investigate the influence of the period of the light cycle on the F R P, a group of birds was first entrained by cycles with 24-hr periods (T = 24 hr) and then allowed to free-run. After the first free-run, these birds were entrained by cycles with periods different from 24 hr (6 hr of light per cycle) and placed into DD for a second time. Representative results from this experiment are shown in Figure 6. After T = 28 hr entrainment, the birds initially had F R P's longer than after T = 24 hr entrainment. The F R P's then decreased for the next 30–50 days, when they began to increase in the usual manner. Eventually, F R P-PK's similar to those following T = 24 hr entrainment were reached. Following T = 22 hr entrainment, the F R P's were initially shorter than after T = 24 hr entrainment but rapidly became similar to those after T = 24 hr. "Knees" in the F R P curves were very prominent following T = 22 hr entrainment. Following T = 20 hr entrainment, the early F R P values were about the same or

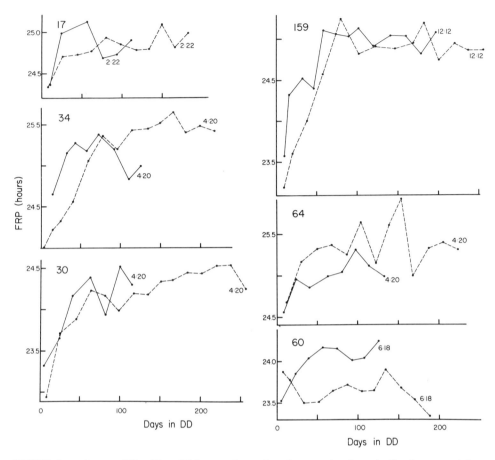

FIGURE 5 FRP repeatability. The solid lines are drawn through FRP values from the first free-run and the dashed lines are drawn through FRP values from the second free-run.

longer than after T = 24 hr entrainment. In none of the experiments did the period of the entraining cycle have an effect on the FRP-PK values. Since FRP-PK is independent of T, the results were combined with those from the repeatability study to further examine FRP-PK repeatability. Correlation analysis was performed between individual FRP-PK values; a strong correlation (r = .90) existed between FRP-PK values from the first and second free-runs (Figure 4C). The overall mean difference in FRP-PK values for the two free-runs was 0.02 hr (mean absolute difference was .26 hr). Different treatments affect the early or transient FRP values but do not appear to affect the FRP-PK values.

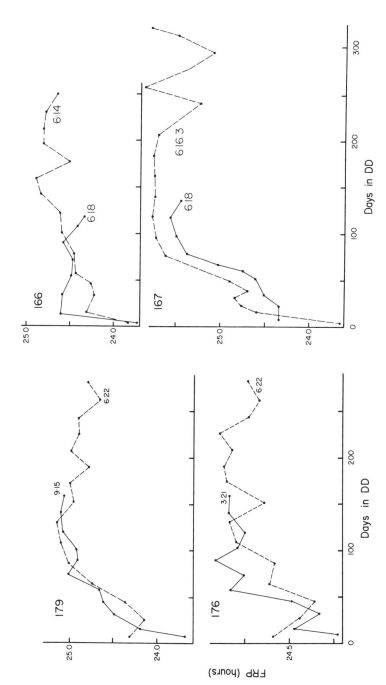

FIGURE 6 The effect of the period of the light cycle on the F R P. The solid lines are drawn through F R P values from the first free-run and after T = 24 hr entrainment. The dashed lines are drawn through F R P values from the second free-run and after T = 28 hr entrainment for 179 and 176, T = 20 hr entrainment for 166, and T = 22.3 hr entrainment for 167.

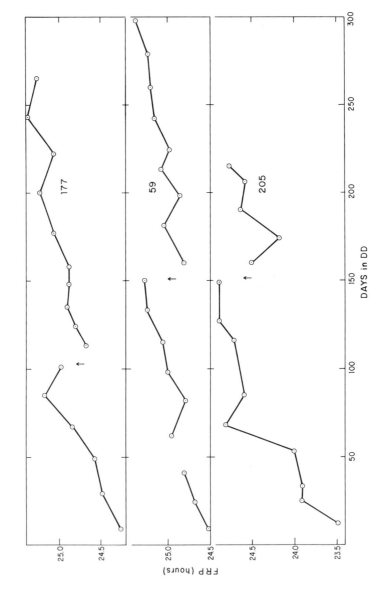

FIGURE 7 Effects of single light pulses on the F R P. The arrows indicate the day that the birds received a single pulse of fluorescent light. The durations of the pulses were 6 hr for 177 and 59 and 24 hr for 205. The onset of the pulse coincided with the onset of activity of 205 and was 40 min after the onsets of 177 and 59.

Light pulses   During the course of a phase response curve study, it was discovered that single pulses of light as short as 15 min in duration often produced significant FRP changes. Since the measurement of a phase shift is ambiguous when the light pulse produces long-lasting FRP changes, it was necessary to determine when (relative to number of days in DD) birds could be pulsed without accompanying FRP changes. The fewest FRP changes were observed when birds were pulsed around day 200 of DD, that is, after the FRP had remained relatively constant for 50 to 100 days. Pulsing before or just after FRP-PK usually produced long-lasting FRP changes.

Three birds that had large FRP changes when pulsed were observed for a long period of time after the pulse to determine the duration of the effect (Figure 7). The FRP's of these birds decreased after the pulse, but after about 75 days FRP values similar to those before the pulse were reached. Thus, under certain conditions a single pulse of light can effectively reset the FRP change and, presumably, the underlying process by at least 75 days.

FRP Change of Peromyscus

The circadian rhythm of running wheel activity of *Peromyscus spp.* was examined for comparison with the FRP changes of *P. domesticus*. The activity data from the mice were treated in the manner previously described for the birds. The FRP change of the mice appeared to last longer than that of the sparrows (Figure 8). The FRP's of several mice were still decreasing after 300 days of DD. In addition, the rate of change of the FRP of the mice (about 0.2 min/day) was considerably less than that of the sparrows (about 0.9 min/day).

DISCUSSION

FRP Change

The FRP of the sparrow is a function of the number of days in DD at least for the first 3 or 4 months. The general manner in which the FRP changes is predictable. Most of the FRP changes that appeared as spontaneous changes in the activity data can be associated with general features of the rhythm, for example, the abrupt changes in FRP from less than 24 hr to greater than 24 hr, the "knees," and the decrease in the FRP after the peak or maximum FRP value is reached.

The FRP results reported here are not peculiar to *P. domesticus*, as is demonstrated by data from *Peromyscus*. In addition, gradual FRP changes can be seen in activity data in the literature from lizards, rats, mice, bats,

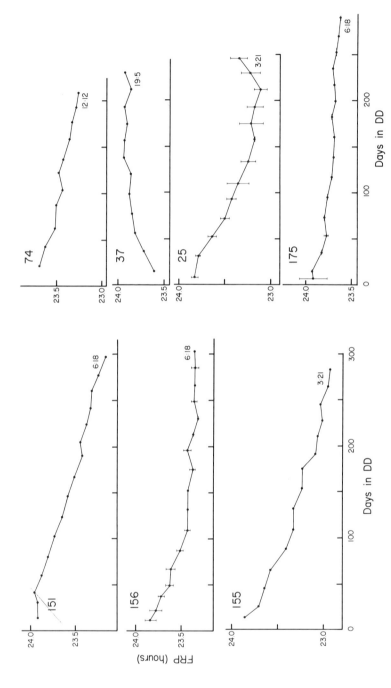

FIGURE 8 FRP changes of *Peromyscus*. The vertical lines represent 95% confidence intervals. In some cases the confidence intervals were too small for separation on this scale. These confidence intervals are represented by single horizontal lines.

66

birds, flying squirrels, and humans (Hoffmann, 1960; Richter, 1965, 1968; Pittendrigh, 1968; Aschoff, 1968; Aschoff *et al.*, 1962; DeCoursey, 1959; DeCoursey and DeCoursey, 1964; Aschoff and Wever, 1962). Therefore, at least in complex organisms, a gradual, long-term change in the FRP after an organism is placed into constant conditions appears to be a general property of the system responsible for overt circadian rhythms.

In many reported experiments, rhythms were not recorded over a long enough time for the FRP changes to become evident. In these experiments and those in which it was evident that the FRP was changing, a transient value of the FRP was probably measured. Change in FRP has been taken into consideration in only one experiment (Rawson, 1960), and in this case it was assumed that the change was linear. In another experiment (Hoffmann, 1960), the existence of gradual FRP changes was discussed; however, single values of the FRP were used for individual organisms in this experiment and it was not clear what these values represented. In a few other experiments (Enright, 1966a; Lohmann, 1967), attempts were made to establish that the FRP was not changing at the time of measurement, but the period of time and the manner in which the records were examined were certainly not adequate to make such a determination.

The manner in which nearly all experiments dealing with effects of changes of environmental conditions (temperature, light intensity, and so forth) on the FRP were performed raises questions concerning the interpretation of the results from these experiments. Many of these experiments doubtless measured experimental change superimposed on intrinsic change. In addition, the results from pulsing birds with light suggest that different results may be obtained depending on when experiments are performed in relation to the course of the FRP change.

In repeatability studies or studies to determine whether the FRP represents an individual characteristic, care must be taken that similar measures of FRP are compared, since different treatments can affect the time course of the change in the FRP. In the repeatability studies reported here, different results would have been obtained if FRP's had been compared at different arbitrary times. Perhaps this explains why large variations in FRP values for one individual have been observed when FRP's were compared following different treatments (DeCoursey, 1959; Hoffmann, 1960; Roberts, 1960).

After-Effects

It is not clear from Pittendrigh's (1960) use of the term "after-effects" whether he intends it to apply to transient or steady-state changes of the FRP, or to both. Any environmental variable that can change the FRP during its application will produce after-effects in the FRP, for when the organism is returned to prior conditions a certain number of transient cycles will always precede the attainment of steady-state. In the sparrow, entrainment by light cycles produced effects on the FRP that were similar to effects that previously had been labeled "after-effects." These effects decayed after 3 to 4 months, at which time an FRP-PK value was reached that was similar to the FRP-PK value before entrainment. Single light pulses affected the FRP of the sparrow, but these effects also were transient. To my knowledge, no results have been published that adequately demonstrate a permanent modification of the FRP by previous environmental conditions, though in a few cases FRP modifications have been referred to as steady-state effects (Lohmann, 1967) or permanent changes (DeCoursey, 1964). Since spontaneous changes and after-effects apparently represent transient changes of the FRP and can be considered predictable features of the individual rhythm, it is incorrect to cite these changes and effects as evidence for FRP lability. The results reported here are consistent with the idea that the FRP-PK value represents an individual characteristic.

Conclusion

Though 120 to 200 days is a relatively long time for the FRP to continue to change, only 120 to 200 cycles of the oscillating system will have occurred during this time—compare this to the time it takes the heart pacemakers of most organisms to complete a similar number of cycles. The number of cycles that have occurred, not length of time in DD, appears to be the important factor in the FRP change of the sparrow; that is, the FRP change is neither a "DD effect" *per se* nor an aging effect. The evidence for this is that a single brief pulse of light can set the FRP change back for many days. It seems unlikely that this pulse of light could be reversing the effects of 100 days of DD. In addition, FRP changes in dim LL (0.1 lux Panelescent light) are similar in most respects to those that occurred in DD (unpublished observations). Re-entrainment that can completely reset the FRP change can hardly be reversing the aging process.

From the FRP results reported here, it seems clear that more information is contained in the response of the biological clock of an organism to constant conditions than is revealed by a single FRP value. To specify what these results indicate about the nature of the mechanism underlying the rhythm, I believe it would first be necessary to fit them with a mathematical model. It would be useful to know if the behavior of Richie's (1966) coupled oscillator model or Winfree's (1967) and Pavlidis's (1969) models, which employ populations of coupled oscillators, could fit the sparrow FRP results.

ENTRAINMENT

ENTRAINING AGENTS

Several types of stimuli other than light cycles are capable of entraining the sparrow activity rhythm. It has been shown that bird song cycles as well as mutual interactions between birds are effective, though weak, entraining agents (Gwinner, 1966; Menaker and Eskin, 1966). Mechanical noise cycles can also entrain birds (Lohmann and Enright, 1967), and, at least for sparrows, white noise cycles are as effective entraining agents as are bird song cycles (Eskin and Menaker, unpublished). Temperature cycles of very large amplitude (around 32°C) can also entrain sparrows (Figure 9)

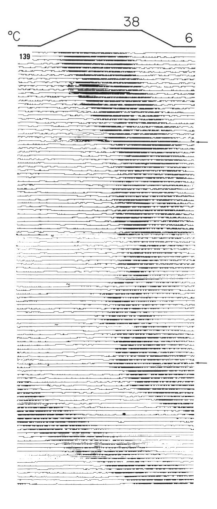

FIGURE 9 Entrainment by temperature cycles. The bird whose activity record is shown in this figure was housed in a cage within a light-tight and relatively soundproof B.O.D.-type incubator. A background white noise (92 dB) was provided within the incubator; the bird was in DD throughout the experiment. The temperature cycle drawn on the top of the record was presented to the bird on the days between the arrows on the right side of the record. For the days before and after the temperature cycle, the temperature was constant at 25 ± 1°C.

(Eskin and Richie, unpublished). The phase angles of the birds entrained by the temperature cycles varied; in some birds, most of the activity was confined to the warm part of the cycle, while in others most of the activity time was confined to the cold part of the cycle.

PHASE ANGLE TRANSIENTS

When the sparrows were first exposed to a light cycle, their phase angles went through an interval of transients. The duration of the transients was to some degree a function of the photoperiod of the cycle. The type of transients (advancing or delaying) that occurred depended on the phase at which the light cycle was first presented. For LD 6:18 to LD 9:15 cycles, the transients lasted about one month (Figure 10); they persisted much longer for photoperiods shorter than LD 6:18 (Figure 11) and persisted a slightly shorter time for photoperiods longer than LD 9:18 (Figure 10). If the light cycle began during the interval from 9 hr before to 3 hr after the onset of activity, the initial transients were in the advancing direction. At other phases, delaying initial transients occurred (Figure 10). Overshoots and undershoots usually followed the initial transients and preceded the establishment of steady-state phase angles. A relatively stable phase angle was usually reached during the second month of entrainment for LD 6:18 and cycles of longer photoperiod.

ENTRAINMENT MODEL STUDY

The entrainment model to be examined was developed mainly by Pittendrigh (1965, 1966, 1967; Pittendrigh and Minis, 1965). He has tested the model in a variety of ways and found that it works very well for *Drosophila pseudoobscura*. The model has been applied to circadian rhythms of other organisms (Aschoff, 1965a; Enright, 1965, 1966b) but has not been tested for any organism other than *Drosophila*. Since there are a number of differences between the *Drosophila* and sparrow systems, a test of the model utilizing the sparrow as the experimental organism appeared to be a good test of its general applicability. The entrainment model specifies that during each steady-state entrainment cycle a phase shift, $\Delta\phi$, occurs. This phase shift is equal to the difference between the FRP of the rhythm and the period of the light cycle; that is, $\Delta\phi = FRP-T$. In other words, the phase shift produced by the light pulse during each cycle "corrects" the period of the rhythm so that it matches the period of the light cycle. In this manner, the light cycle controls the period of the rhythm. For phase control, the model specifies that the steady-state phase

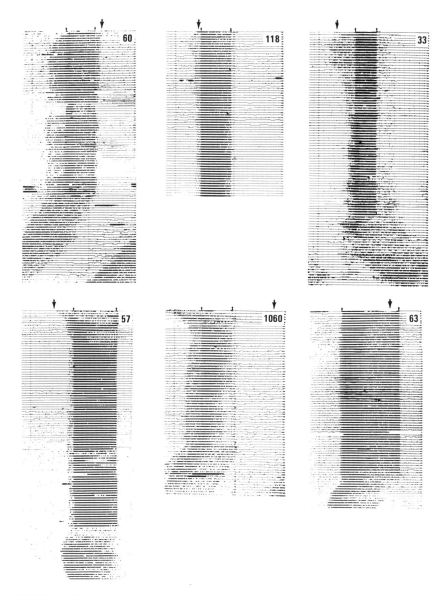

FIGURE 10   Phase-angle transients. The small arrow heads on the top of the records indicate the time that the light was on. The single large arrow head on the top of the records indicates the time at which the onset of activity occurred on the day before the light cycle began. The photoperiods were *LD 4*:20 for 33; *LD 6*:18 for 60, 118, and 1080; *LD 9*:15 for 57, and *LD 12*:12 for 63.

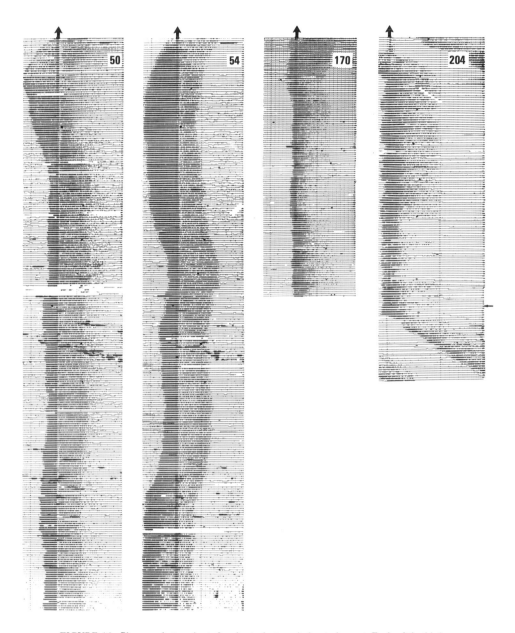

FIGURE 11   Phase-angle transients for short photoperiod entrainment. Each of the birds was entrained by *LD 0.5:23.5*. The base of the arrow on top of each record spans the time that the light
was on each day. The records for 50 and 54 are 9 months in length. Note the phase angle oscillations that occurred through the record of 204.

angle during entrainment can be predicted from the phase response curve. The phase angle during entrainment defines that point in the cycle at which a single pulse of light administered while the organism is free-running will produce a phase shift equal to the difference: FRP–T. If one has measured a response curve for a pulse of a certain duration, the results of several kinds of entrainment experiments can be predicted by applying the model. Conversely, one can test the model by doing entrainment experiments and using the model to derive response curve information. The test of the model is, then, the comparison of the measured response curve and the derived one.

To test the model, a phase response curve for 6-hr pulses of light, the range of entrainment, and the dependence of the phase angle of the rhythm on the period of the cycle were measured. The data points of Figure 12 represent results from presenting 6-hr pulses of light to individ-

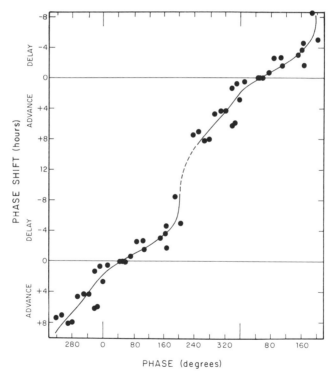

FIGURE 12   Phase response curve. The data of this figure were duplicated on both the phase and phase shift axes. The points represent the phase shifts of individual birds pulsed with 6 hr of light.

ual birds that had been in DD for 5 to 7 months. The data have been duplicated along the ordinate and an "eye fit" line drawn through the data
points. The reference points were the onset of activity and the onset of
the light pulse. Zero phase was taken as the onset of activity. (For the
method of phase angle and phase shift measurement, see Eskin, 1969.) The
6-hr response curve of *P. domesticus* is similar in most respects to the 4-hr
response curve of *D. pseudoobscura* (Pittendrigh, 1960).

The dependence of the phase angle of the rhythm on the period of the
light cycle is shown in Figure 13. Each point represents the mean phase
angle of a group of birds entrained by a light cycle of a particular period.
The light portion of the cycle was always 6 hr regardless of the period
length. Positive phase angles indicate that the onset of activity preceded
the onset of light and negative phase angles indicate that the onset of light
preceded the onset of activity. The phase angle of the rhythm increased
or became more positive as the period of the cycle increased. This positive
correlation is similar to that found for all other organisms thus far investigated (Aschoff, 1965b).

The range of entrainment of the sparrow is also represented in Figure 13.
Only the groups of birds in which at least 75 percent entrained are plotted.
Only 25 percent of the birds exposed to $T = 15.8$ hr entrained, whereas
80 percent of the birds exposed to $T = 17.8$ hr entrained. Therefore, the
lower limit of the range of entrainment is between $T = 17.8$ hr and $T = 15.8$
hr. By the same criterion, the upper limit of the range of entrainment was
found to be between $T = 28.0$ hr and $T = 28.7$ hr.

If the model is correct and the mean FRP for each group of birds entrained by the different period cycles is assumed to be 25 hr (actual mean
FRP-PK for sparrows was 24.9 hr), phase response curve information can
be derived from the $\phi \lambda$ vs. T relation (Figure 13) in the following manner:
For $T = 28$ hr, the model and entrainment experiment results predict that a
phase shift equal to -3 hr (FRP $-T = 25-28$) would occur if a bird were
pulsed at a phase of $175°$. Phase shifts derived in this manner using the
model (open squares) are compared to the actual phase shifts measured
(closed circles) in Figure 14. The phase shift values derived from the entrainment experiment agree very well with the phase shifts measured from
free-running birds. This agreement between derived and measured phase
shifts represents a confirmation of the model.

The range of entrainment for light cycles with different periods can be
predicted from the maximum phase shifts that occur in the region of the
response curve where the slope of the response curve is less than 2; that is,
$T_{limit} = FPR -(\Delta\phi)_{max}$ (Pittendrigh, 1966; Pavlidis, 1967). In the vicinity

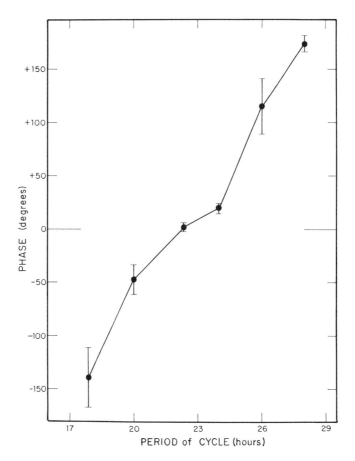

FIGURE 13 Phase-angle *vs.* period of the cycle. The points represent mean phase angles of groups of birds that were entrained by light cycles with different periods. The vertical lines represent 95% confidence intervals.

of phase 180–240° the slope of the response curve is greater than 2 (Figure 12); thus stable entrainment is not possible in this region. Since the largest advance phase shift was 8.2 hr, and if a mean FRP of 25 hr is assumed, T_{limit} = 25.0 – 8.2 = 16.8 hr for the lower limit of entrainment. The lower measured limit of entrainment, between T = 15.8 hr and T = 17.8 hr, agrees very well with this prediction. The upper measured limit, between T = 28.0 hr and T = 28.7 hr, is reasonably close to that predicted, T = 29.7 hr. This latter comparison is considered to be a reasonable

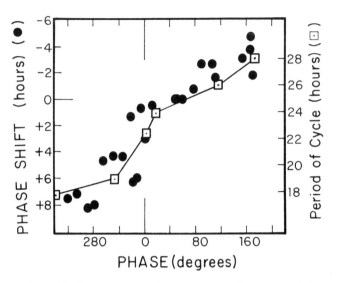

FIGURE 14   Entrainment model test. The closed circles represent phase shifts obtained from pulsing free-running birds with light. The open squares are phase shift values derived from the phase angle *vs.* period of the cycle experiment. On the right ordinate are plotted the periods of the cycles that correspond to the phase shift values on the left ordinate for **FRP** = 25hr.

agreement since the prediction was based on the value of a single phase shift obtained from an individual bird, and it was found that the reproducibility of such phase shifts was about ± 1 hr (Eskin, 1969).

CONCLUSIONS

Predictions made using Pittendrigh's model agree fairly well with results of two entrainment experiments, namely, the range of entrainment and $\phi\overline{\lambda}$ *vs.* T experiments. In addition, predictions made using Pittendrigh's model are consistent with results from other kinds of entrainment experiments ($\phi\overline{\lambda}$ *vs.* FRP and $\phi\overline{\lambda}$ *vs.* photoperiod) performed on *P. domesticus* (Eskin, 1969).

   These agreements between predictions from the model and experimental results represent the first confirmation of the model for an organism other than *Drosophila*. That the model works well in both these cases is surprising when the differences between the two organisms and the two processes being assayed are considered. This fact indicates that the model may apply not only to different organisms but also to different levels of organization. The entrainment results combined with the FRP results from *P. domesticus* should prove useful in further limiting the class of models applicable to circadian rhythms.

ACKNOWLEDGMENTS

The work reported here forms the basis of a Ph.D thesis completed at the University of Texas at Austin. I am extremely grateful to M. Menaker for his support throughout all phases of this work. Some of the data reported here were obtained from experiments performed in collaboration with C. G. Richie. I would like to thank Dr. Richie for allowing me to use these data and for many valuable discussions concerning most aspects of this work. This work was supported in part by an NSF Grant (GB 3806) to M. Menaker, a USPHS Training Grant to the Department of Zoology, and a PHS Predoctoral Fellowship (I-FI-GM-32,240-01).

REFERENCES

Aschoff, J. 1963. Comparative physiology: diurnal rhythms. Annu. Rev. Physiol. 25:581–600.

Aschoff, J. 1965a. Response curves in circadian periodicity, p. 95 to 111. *In* J. Aschoff [ed.], Circadian clocks. North-Holland Publ. Co., Amsterdam.

Aschoff, J. 1965b. The phase-angle difference in circadian periodicity, p. 262 to 276. *In* J. Aschoff [ed.], Circadian clocks. North-Holland Publ. Co., Amsterdam.

Aschoff, J. 1968. Human circadian rhythms in activity, body temperature, and other functions, p. 159 to 173. *In* A. H. Brown and F. G. Favorite [ed.], Life sciences and space research. North-Holland Publ. Co., Amsterdam.

Aschoff, J., I. Diehl, U. Gerecke, and R. Wever. 1962. Aktivitätsperiodik von Buchfinken (*Fringella coelebs*) unter Konstanten Bedingungen. Z. Vgl. Physiol. 45:605–617.

Aschoff, J., and R. Wever. 1962. Über Phasenbeziehungen zwischen biologischer Tagesperiodik und Zeitgeberperiodik. Z. Vgl. Physiol. 46:115–128.

DeCoursey, G., and P. J. DeCoursey. 1964. Adaptive aspects of activity rhythms in bats. Biol. Bull. 126:14–27.

DeCoursey, P. J. 1959. Daily activity rhythms in the flying squirrel, *Glaucomys volans*. Ph.D. Thesis. University of Wisconsin, Madison.

DeCoursey, P. J. 1964. Function of a light response rhythm in hamsters. J. Cell. Comp. Physiol. 63:189–196.

Enright, J. T. 1965. Synchronization and ranges of entrainment, p. 112 to 124. *In* J. Aschoff [ed.], Circadian clocks. North-Holland Publ. Co., Amsterdam.

Enright, J. T. 1966a. Temperature and the free-running circadian rhythm of the house finch. Comp. Biochem. Physiol. 18:463–475.

Enright, J. T. 1966b. Influences of seasonal factors on the activity onset of the house finch. Ecology 47:662–666.

Eskin, A. 1969. The sparrow clock: behavior of the free running rhythm and entrainment analysis. Ph.D. Thesis. University of Texas at Austin.

Gwinner, E. 1966. Entrainment of a circadian rhythm in birds by species-specific song cycles. Experientia 22:765.

Hoffmann, K. 1960. Versuche zur Analyse der Tagesperiodik I. Der Einfluss der Lichtintensität. Z. Vgl. Physiol. 43:544–566.

Lohmann, M. 1967. Phase-dependent changes of circadian frequency after light steps. Nature (Lond.) 213:196–197.

Lohmann, M., and J. T. Enright. 1967. The influence of mechanical noise on the activity rhythms of finches. Comp. Biochem. Physiol. 22:289–296.

Menaker, M., and A. Eskin. 1966. Entrainment of circadian rhythms by sound in *Passer domesticus*. Science 154:1579–1581.

Palmer, J. D. 1964. Comparative studies in avian persistent rhythms: spontaneous change in period length. Comp. Biochem. Physiol. 12:273–282.

Pavlidis, T. 1967. A mathematical model for the light affected system in the *Drosophila* eclosion rhythm. Bull. Math. Biophys. 29:291–310.

Pavlidis, T. 1969. Populations of interacting oscillators and circadian rhythms. J. Theor. Biol. 22:418–436.

Pittendrigh, C. S. 1960. Circadian rhythms and the circadian organization of living systems. Cold Spring Harbor Symp. Quant. Biol. 25:159–182.

Pittendrigh, C. S. 1965. On the mechanism of the entrainment of a circadian rhythm by light cycles, p. 227 to 297. *In* J. Aschoff [ed.], Circadian clocks. North-Holland Publ. Co., Amsterdam.

Pittendrigh, C. S. 1966. The circadian oscillation in *Drosophila pseudoobscura* pupae: a model for the photoperiodic clock. Z. Pflanzenphysiol. 54:275–307.

Pittendrigh, C. S. 1967. Circadian systems. I. The driving oscillation and its assay in *Drosophila pseudoobscura*. Proc. Nat. Acad. Sci. U.S. 58:1762–1767.

Pittendrigh, C. S. 1968. Circadian rhythms, space research and manned space flight, p. 122 to 134. *In* A. H. Brown and F. G. Favorite [ed.], Life sciences and space research. North-Holland Publ. Co., Amsterdam.

Pittendrigh, C. S., and D. H. Minis. 1964. The entrainment of circadian oscillations by light and their role as photoperiodic clocks. Am. Nat. 98:261–294.

Rawson, K. S. 1960. Effects of tissue temperature on mammalian activity rhythms. Cold Spring Harbor Symp. Quant. Biol. 25:105–113.

Richie, C. G. 1966. A mathematical model for the biological clock of *Passer domesticus*. Ph.D. Thesis. University of Texas at Austin.

Richter, C. P. 1965. Biological clocks in medicine and psychiatry. C. C. Thomas, Springfield, Illinois.

Richter, C. P. 1968. Inherent twenty-four hour and lunar clocks of a primate—the squirrel monkey. Commun. Behav. Biol., Part A. 1:305–332.

Roberts, S. K. 1960. Circadian activity rhythms in cockroaches. I. The free-running rhythm in steady state. J. Cell. Comp. Physiol. 55:99–110.

Swade, R. H., and C. S. Pittendrigh. 1967. Circadian locomotor rhythms of rodents in the Arctic. Am. Nat. 101:431–466.

Winfree, A. T. 1967. Biological rhythms and the behavior of populations of coupled oscillators. J. Theor. Biol. 16:15–42.

DISCUSSION

PITTENDRIGH: I have a few comments to make other than saying how much I enjoyed the presentation. I am frankly surprised that the model has held up so well for these sparrows. We have had for some time a similar study in hamsters (not published) where again, at least to a first approximation, the model behaves very well—so it works for nonflying organisms too! I am also delighted to see new examples of extremely long runs of data. I am interested in the curious fact that in sparrows FRP is not constant but changes more or less systematically with time. A few years ago we found that in hamsters also FRP increases as a function of time, and more so in LL than in DD. After a year or so the period begins to assume quite ridiculous proportions. What I don't understand from your discussion is why the model works so well with FRP-PK values, defined as the longest value of FRP. And would you comment on what it is that you are observing when you observe FRP-10?

ESKIN: I chose the FRP-PK value because it is related to the level part of the curve, and as such is a good representation of steady-state FRP. FRP-PK differs by less than 5 or 10 min from the mean value of the steady-state part of the curve.

PITTENDRIGH: But FRP is not constant. And when you release the rhythm from entrainment, the free-running period is very different from what you are calling FRP-PK, which is the FRP you use in the model manipulation.

ESKIN: I think that the FRP-PK measures reflect the inherent natural period of the oscillator underlying the rhythm. FRP-10 values are transient FRP values determined in part by the entrainment conditions and in part by the natural period of the driving oscillator. Evidence for this can be seen in Figure 4A where it was shown that FRP-10 and FRP-PK values are correlated.

PITTENDRIGH: The point I was heading toward is that FRP exhibits after-

effects, or changes as a function of the preceding entrainment cycle. Now, are these after-effects a function of the driving oscillator of the system? And if so, does the model work well only with the FRP-PK values?

ESKIN: A distinction cannot be made with my data between using FRP-10 or FRP-PK values. If FRP-10 values are used instead of FRP-PK values to test the model, the derived phase shift values of Figure 14 would be shifted upward one hour. The derived phase shift curve based on FRP-10 values fits the measured phase shift values nearly as well as the derived phase shift curve based on FRP-PK values.

ENRIGHT: I would like to compliment Dr. Eskin on the large quantity of data he has shown us here. It was originally my intention to point out the remarkable similarity between his response curves for sparrows and the response curve for the house finch that I published in the Feldafing Symposium, but I probably shouldn't even mention it because in that case I had only about 60 days of data from only one bird. There are, however, remarkable similarities in the amplitude, the direction, and the point of turnover in the curves for the house sparrow and the house finch. Mine was a derived response curve, and there were far too many assumptions involved and far too few data to justify my conclusions. Only large amounts of data such as you have presented can yield a reliable response curve.

The other comment I have is about the use of the word "transient." It strikes me that we are using that one word to cover such a broad variety of phenomena that perhaps we should begin to look for a different set of terms. First there are transients that

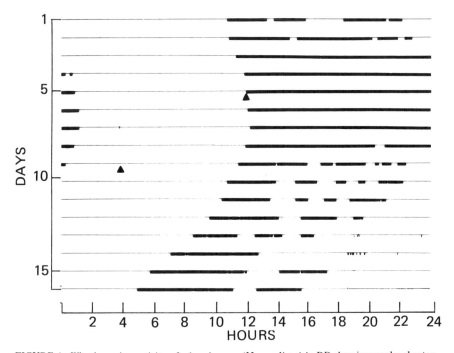

FIGURE 1   Wheel-running activity of a hazelmouse (*Muscardinus*) in DD showing regular shortening of FRP. Prior to start of recording, the animal was in DD. ▲ indicates feeding in dark.

arise as a result of a single stimulus
given to a free-running organism, and
which occur before it reachieves a
steady-state free-run. Then there is a
second class of phenomena called
transients, which occur when one re-
entrains an organism after a phase shift
of the light cycle. Now we have a third
class of transients—very long persisting, slow changes in the free-running
period. We may well have three very
different biological processes going on
here, so perhaps we should look for
words to differentiate them descrip-
tively as well.

COMMENT BY DECOURSEY: I wish to
congratulate Dr. Eskin on his meticu-
lous experiments, and to thank him
for answering some perplexing ques-

tions about free-running rhythms. In
three individuals of European hazelmouse (*Muscardinus*), we noticed
striking changes in the FRP (Figure 1).
The changes of FRP that you found
in *Peromyscus* were on the order of
0.2 min per day, but in *Muscardinus*
regularly increasing changes in FRP
were observed, reaching a maximum
in one individual of about 20 min/day.
The data available suggest that a spectrum exists, ranging from the relatively
labile *Muscardinus*, through *Passer* and
Peromyscus, and finally to such animals as *Glaucomys*, which show little
or no change in FRP even in long term
experiments. It would be of great interest to have comparable data for
other species.

ARTHUR T. WINFREE

Corkscrews and Singularities in Fruitflies: Resetting Behavior of the Circadian Eclosion Rhythm

It can be shown plainly that rotation is the primary locomotion (*primum mobile*). . . . Moreover rotatory motion is prior to rectilinear motion because it is more simple and complete.

—ARISTOTLE
Physics, VIII, 9

The experiments to be described here were undertaken to define the phase-resetting behavior of the circadian rhythm of pupal eclosion in populations of the fruitfly *Drosophila pseudoobscura*, a subject made familiar by the work of Pittendrigh (1954, 1960, 1965, 1966, 1967), Engelmann (1966, 1967), Chandrashekaran (1967, 1969), Honegger (1967), Pavlidis (1967, 1968), Skopik and Pittendrigh (1967), and Zimmerman (1968, 1969). In these experiments, I have asked how and why the resetting response depends on the *duration* of a standard perturbation as well as on the time at which it is given. These questions bear directly on the underlying dynamics of the timing process. The experiments were designed to identify the sets of stimuli that reset the circadian rhythm to the same phase.

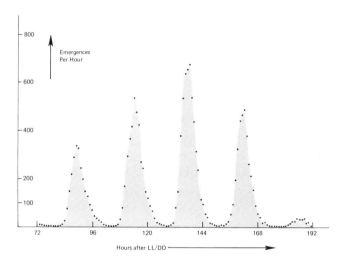

FIGURE 1   Daily emergence peaks in DD control populations (pooled). Resetting alters phase, but not period, of this rhythm.

METHODS

Young pupae previously reared in LL are transferred to DD. This change induces circadian rhythmicity. Then at some time (T) after the LL/DD transfer they are exposed to a standard dim blue light[†] for a duration (S). The experiment tested 136 combinations of S and T, with S varying from 15 to 120 sec and T varying from 0 to 24 hr. The daily bursts of adult flies are monitored 4–8 days later. The arithmetic mean of emergence time in each of the discrete, narrow peaks of emergence is computed as in Figure 1; the interval from the end of the light pulse to mean emergence time is called centroid time, θ (Figure 2).

There is nothing novel in the operations involved in these phase-resetting measurements. Our methods are virtually identical to those of Engelmann (1969). Many investigators have published phase-resetting experiments in which the stimulus duration was fixed and only T was varied; i.e., the stimulus was given at different phases of the rhythm (Bruce, 1960; DeCoursey, 1960; Pittendrigh, 1960; Pittendrigh and Minis, 1964; Engelmann, 1969;

[†] Light is from a 500-w, 3200°K incandescent bulb, passed through copper sulfate solution and a Wratten 47 filter to final intensity about 10 μw per cm^2, 31 in. from the filament.

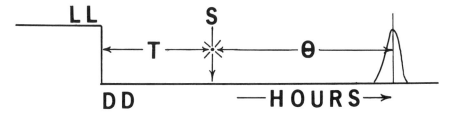

FIGURE 2   Format of resetting experiments: T, S, and θ defined. Only one of the three to four emergence peaks is shown.

Minis, 1965; Feldman, 1967; Sargent and Briggs, 1967). In other investigations, T was fixed, and only stimulus intensity or duration was varied (Wilkins, 1962; Engelmann, 1969). Pittendrigh has varied both parameters simultaneously, but using light pulses a thousand times more energetic than in the present experiment (Ottesen, 1965). It turns out that resetting is qualitatively different following weak and strong light stimuli.

RESULTS

The results of these experiments are plotted in three dimensions, since three variables are involved. The stimulus coordinates, T and S, are the independent variables, and the measurement of emergence time, θ, is the dependent variable.

Figure 3 is a photograph of such a three-dimensional graph, in which the first third of the data collected on *Drosophila pseudoobscura* is represented. Other data gathered are consistent with the major features of this graph. The S-axis represents duration of the stimulus, from 15 to 120 sec. The T-axis represents the time at which the stimulus is given, from 0 to 24 hr after the LL/DD transfer. Each vertical wire represents a single experiment in which a population of pupae received a specific (T, S) stimulus, and the buttons on the wires represent the centroid times, θ, of emergence peaks after transients. This graph extends upward along the θ axis for roughly 2½ days of emergence, although the more complete data cover 4 days.

Notice the helicoidal feature of the graph; the data spiral up around a vertical rotation axis. There are no centroid data shown at this rotation axis because resetting is erratic following this stimulus (T^*, S^*), and flies emerge not in discrete peaks but at all hours of the day, as discussed below.

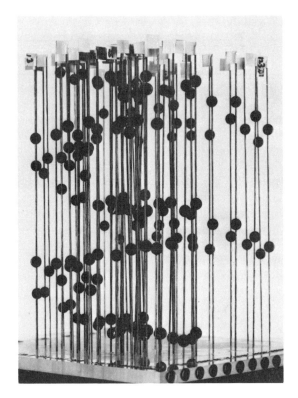

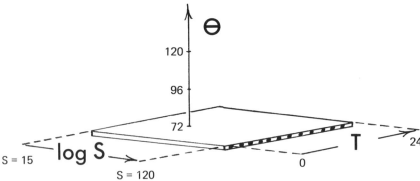

FIGURE 3   Three-dimensional graph of measured emergence centroids, θ, vs. the stimulus variables, T and S. The graph is oriented in the sketch below to emphasize spiral structure. From Lectures on Mathematics in the Life Sciences, p. 121.

THE RESETTING SURFACE

It would be easier to grasp the dependence of the phase resetting on the stimulus parameters T and S if we could fit a smooth surface, θ (T, S), to the complete cloud of 468 centroid data points, in the way we usually fit a smooth curve to data on two-dimensional graphs. A digital computer was used to find a smooth surface, called the Resetting Surface, that approximately fits this cloud of data points. (The exact mathematical form used is described in Winfree, 1970.)

This surface is perhaps best described as a vertical corkscrew linking together tilted planes. The tilted planes result from sets of stimuli having little effect on the phase of the rhythm.

Such a conformation is difficult to visualize. We can depict this surface in two dimensions by giving the curves in the T X S plane along which θ is constant, for different values of θ. That is, we can construct a θ-contour map of the resetting surface, θ being the height of the surface at each point above the T X S plane. The contours describe the sets of stimuli that result in emergence after the same interval, θ; i.e., the stimuli that reset the rhythm to the same phase.

Figure 4 shows the contour map computed from my data; the duration axis S runs from 0 to 120 sec exposure, and the T axis from 0 to 12 hr after the LL/DD transfer. This region of the map contains the corkscrew-shaped surface; to the right, from 12 to 24 hr (not shown in Figure 4), is the relatively flat upslope segment joining onto the next corkscrew in the next cycle along the T-axis.

Each contour line, θ, represents the time after the stimulus at which emergence peaks occur, plus multiples of 24 hr. The limit of resolution of the emergence time measurements is ± 2.0 hr (root-mean-square error), so although every (T, S) point is on some θ contour, contours are indicated only at 2-hr intervals.

DISCUSSION

That a single contour map can adequately represent a multilayer surface is intriguing. This result is an expression of the isoperiodism[†] of the rhythm: After three days of transients the intervals between centroids in the reset rhythms are distributed as closely around 24.0 hr as in the unperturbed

[†]I can measure constancy of period only to ± 2 percent. In other circadian systems it has been reported that the period can be changed by as much as 1 percent by prior entrainment to a light cycle (Pittendrigh and Bruce, 1957; Harker, 1964).

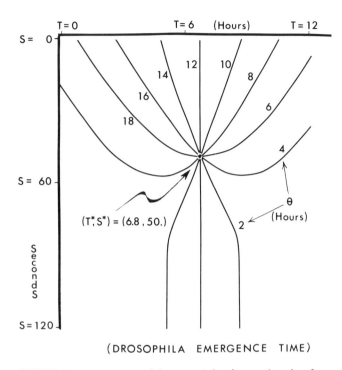

(DROSOPHILA EMERGENCE TIME)

FIGURE 4 θ–contour map of the computed corkscrew-shaped surface
fitting the measured emergence times $\theta(T,S)$.

controls. The apparent bilateral symmetry of the contour map and the
(coincidental?) appearance of $\theta = 12$ and $\theta = 24$ along the mirror axis are
curious observations I cannot explain.

Figure 4 reveals a curious implication of the corkscrew shape of the resetting surface. A corkscrew surface has a singularity, a central axis along
which the slope is infinite. The θ-contour lines of the corkscrew surface
converge on this central axis, which apparently represents a critical stimulus time, $T^* = 6.8$ hr, and duration, $S^* = 50$ sec.

Taken literally,[†] this feature of the resetting surface seems to mean that
there is an isolated perturbation following which there is either no circadian rhythm of emergence in the steady-state, or one of unpredictable
phase. Any minute error in the effective T or S coordinates of a stimulus

---

†The "noise" in my measurements precludes asking whether the contours "really" converge to a mathematical *point* in the T $\times$ S plane; strictly, we can claim convergence
only to a "small" region.

close to the singularity radically alters the resulting phase of the reset rhythm, i.e., puts the clock on a very different θ-contour, as Figure 4 shows.

The tendency toward arrhythmicity following stimuli near (T^*, S^*) is shown in Figure 5. The bottom panel presents five representative emergence records following stimuli *not* close to 50 sec at 6.8 hr. The horizontal axis is time in hours after the stimulus; each x represents the emergence of one fly at that hour. These rhythms are perfectly normal, although phase-shifted. The top panel shows five emergence records following stim-

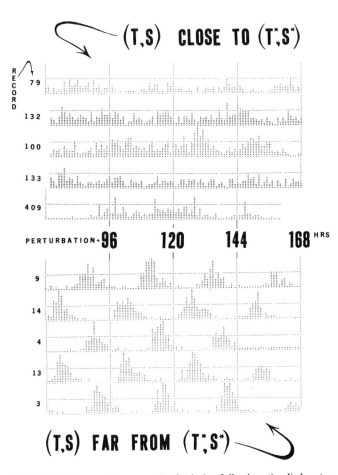

FIGURE 5 Representative records of eclosion following stimuli close to and stimuli far from (T*,S*). Abscissa is in hours after stimulus and each x indicates one fly. Record "3" is a DD control (S = 0). From Lectures on Mathematics in the Life Sciences, p. 132–3.

uli close to 50 sec at 6.8 hr. It is only in this region of the stimulus plane, near the rotation axis, that we find emergence occurring at every hour of the day, rather than in discrete daily pulses.

Now note again the resetting surface of Figure 4. If no perturbation is given, then when T is an hour greater, the time to the next emergence is an hour less. Thus, as T varies over 24 hr, so does the corresponding θ, and therefore all θ-contours must touch the S = 0 axis. The bottom three contours of Figure 4 curve back around to the S = 0 axis, as the θ = 4 contour is beginning to do.

All the contours originate along the upper border of the diagram, and all terminate at the singularity, (T*, S*). Along any horizontal line (S = constant) above S*, all θ-contours are present. Below S*, some are absent (e.g., θ = 10, 12, 14), and the other contours cross each S level twice as they curve back up to the singularity (e.g., θ = 2, 4).

This means that, using a sufficiently short light pulse, one can reset the rhythm to *any* phase by a proper choice of T, which may be interpreted as the phase at which the stimulus is given. The resetting curve, θ (T), therefore, resembles Figure 6a. This is called Type 1 resetting because the average slope in this representation is 1.

In contrast, we can reset only to some limited range of θ's if the stimulus duration exceeds the critical S*, regardless of choice of T. Each possible θ can be achieved with either of two different T's. The resetting curve, θ (T), looks like Figure 6b. This is called Type 0 resetting because the average slope is 0.

This qualitative change in the resetting behavior at a critical pulse duration requires some explanation, as does the corkscrew appearance of the resetting surface with its rotation axis identifying an isolated stimulus that results in unpredictable phase resetting.

AN INTERPRETATION OF CLOCK DYNAMICS

I will now briefly sketch a hypothesis that seems to me sufficient to encompass the main features of known resetting data using single discrete pulses of any perturbing agent.[†] I make no claims of deductive rigor, but rather argue heuristically, by plausible inference. Further, I have avoided unnecessary specificity; few physical assumptions are necessary—none of them new or surprising—but these few seem to be very necessary.

It seems safe to assume of the *Drosophila* eclosion rhythm that the underlying clock, or A-oscillator of Pittendrigh and Bruce (1957), is always

---

[†]This approach was originally suggested by Fitzhugh's (1961) treatment of neural pacemakers and their response to ionic perturbations.

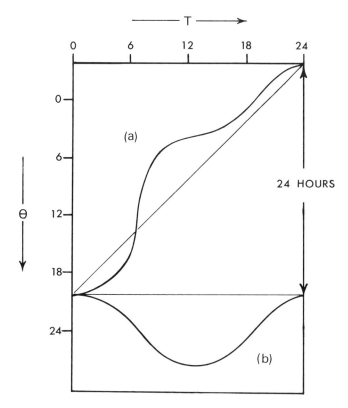

FIGURE 6   Resetting curves $\theta(T,S)$, at fixed stimulus duration (S = constant). Ordinate is θ minus multiples of 24 hr; abscissa is T in hours. (a) Type 1, with S = 45 sec; (b) Type 0, with S = 120 sec, peak-to-peak amplitude = 8 hr.

in some state or condition that can, in principle, be adequately identified by some small number of measurements, e.g., of two or three flux rates in biochemical reaction pathways. We can visualize the state of the clock at any instant as a point in the space formed by using these measurable variables as perpendicular coordinates. If, for example, no more than three independent measurements are needed to identify the state of the clock, then we can visualize that state at any instant unambiguously as a point in three-dimensional Euclidean space.[†]

[†]The timing process may, of course, involve many more than three physical variables and its state description thus require more dimensions (e.g., the "circadian oscillator" may actually be a community of autonomous cellular oscillators) (Winfree, 1967). If the oscillation depends on time-lags or a spatially distributed reaction, then the state space may have infinitely many dimensions. Moreover, this state-space representation

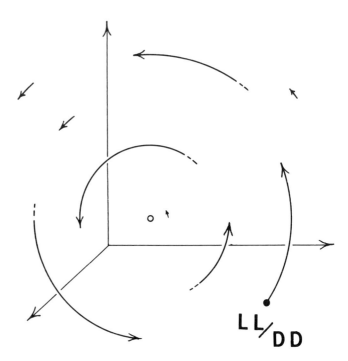

FIGURE 7   Oscillatory trajectories (DD free-run) through the states of a hypothetical clock process, indicated in a space of three coordinate variables. The initial state is the condition of the clock immediately after prolonged LL.

If the circadian timing process is autonomous in the dark at constant temperature, then with each state of the clock there is associated a spontaneous rate of change, indicated by a vector. These vectors describe the motion of the clock along smooth trajectories through its state space as in Figure 7. This manner of representing a complex continuous process, much used by mathematical physicists, chemists, and biologists (e.g., Lewontin and White, 1960; Fitzhugh, 1961; Higgins, 1967; Pavlidis, 1967b, 1968; Lewontin, 1968; Rosen, 1971) allows us to reduce simple hypotheses to abstract geometrical propositions. We can then derive their implications for phase-resetting behavior by using geometric intuition in

---

would be a gross distortion of reality if the clock process is more nearly an epigenetic logical automaton, as Kauffman (1969) has proposed for the mitotic cycle, or an expression of some spatial rotatory motion, as, for example, of an mRNA polymerase around a circular episome (Ehret and Trucco, 1967). In either case the topology of our state space is unrealistic, in the first case for its continuity, and in the second for its connectedness, which should be cylindrical rather than Euclidean.

place of the exact differential equations of a mechanism about which we still know very little.

The flow defined by this geometric description of the clock's dynamics in a fixed environment must somehow circulate in closed or nearly closed loops as in Figure 7, since the process is oscillatory and returns near to a given state at 24-hr intervals.

The three-dimensional space of Figure 7 is not the space in which the resetting surface was described. Here, the coordinates represent the variables, as yet not measured, of an unknown mechanism. Only emergence time, θ, can be measured: we must deduce the relation between the observable θ and the hypothetical state of the clock.

We could do this as follows: Let the state of the clock be represented by a dot in the state space. As the clock is released into DD from any initial state at time zero, this dot follows the flow field in loops; this movement somehow regulates the daily emergence activity. From time zero we can measure the time θ of the daily emergence bursts. With each point in the state space we can associate this θ value (0–24 hr, plus arbitrary multiples of 24 hr, assuming approximate isoperiodism), which describes the emergence rhythm that would result were the clock started from that state at time zero and left to free-run in DD (Figure 8).

It is convenient to think of the set of states having the same θ value as a surface in this state space. I will call such surfaces isochrons. They can be analytically computed only in the simplest models; computer simulation has been used in other cases. Since in each daily cycle the clock passes through each θ value once, each isochron cuts the cycle in only one place. These isochron surfaces therefore must converge somewhere, possibly along a curve from which they extend radially, filling up the state space in pinwheel fashion (Figure 9).

Thus it seems a necessary consequence of this hypothesis that in the center of the rotating flow field there exist states not on any isochron. These states must be phaseless, in the sense that taking any such state as the initial condition, the subsequent emergence rhythm, if it exists, has indeterminate phase. This formulation provides a basis for an interpretation of the behavior observed following the critical stimulus, (T^*, S^*), at the rotation axis of the resetting corkscrew.

CONSEQUENCES OF THE HYPOTHESIS

At this point we are ready to consider whether this hypothesis has any testable implications. We interpret a phase-resetting experiment as follows:

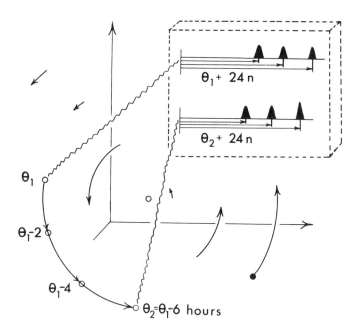

FIGURE 8   To each state in Figure 7 there corresponds a θ value. This is
the centroid time of emergence peaks, measured modulo 24 hr from time
zero, the clock having started from the given state as initial condition at
time zero.

A stimulus is a change of something in the environment, something that
affects the working of the clock, e.g., light. In geometrical terms, the
stimulus modifies the dynamical flow, i.e., alters the trajectories as, for
example, in Figure 10. The longer the clock is held in the perturbing environment, the further its state moves from the normal trajectory.
The clock starts its DD free-run after the LL/DD transition, and follows
oscillatory dynamics. Exposure to light after T hours causes it to follow a
modified trajectory for S seconds, so that the clock is left on a new isochron when DD dynamics are resumed. The state at the end of the light
pulse is the initial condition for the dark free-run that follows (Figure 11).
Knowing how the clock moves from one isochron to the next during perturbation is equivalent to predicting the θ values in the reset rhythm. Of
course, we don't know this without knowing how the clock works and
how light affects it, but a few consequences are independent of the details
of the mechanism.
   For example, I have computed the resetting surface for a variety of
models fitting the postulates we made about continuity, smoothness, etc.,

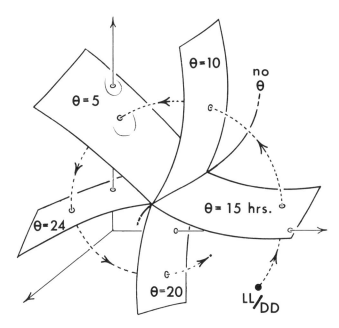

FIGURE 9 The same state space (with one DD trajectory shown as a
dotted curve) illustrating "isochron" surfaces joining states with the same
θ values. Because each isochron cuts each loop trajectory only once, there
must be some states with no θ assignable.

of dynamical flows and inhibition of oscillation in LL. Almost every resetting surface so generated resembles a corkscrew linking tilted planes,
and its θ-contour map converges to a point (critical stimulus) at the corkscrew axis.[†]

Resetting curves, θ (T), are cross-sections through the resetting surface,
θ (T, S), constructed with S held constant. They represent Type 1 resetting
between the S = 0 line and the corkscrew axis, and Type 0 at greater S. It
is easy to see why by considering the geometry of our dynamical state
space. As long as S is short (e.g., consider a zero-duration pulse), the clock
can be moved to any isochron by appropriate choice of T. In fact, the set
of states to which the clock can be moved by a given stimulus duration is
just a somewhat distorted image of the DD loop trajectory that started at
the LL/DD initial state (Figure 12). This ring of potential new states
crosses every isochron because, like the original unperturbed dark cycle,

---

[†]Qualitatively different resetting behavior can emerge from certain exotic models involving a minimum of 3 interacting variables.

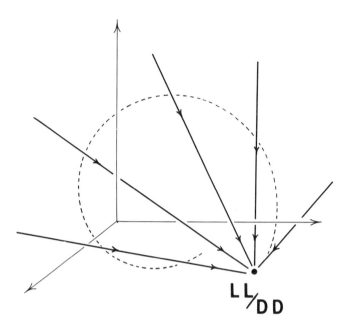

FIGURE 10 The same state space, with trajectories shown as they
might be modified by a perturbing agent, e.g., light, that suppresses the
oscillation. Prolonged LL may send the clock to the state from which it
resumes DD free-run as in Figure 7. The dotted curve represents a DD
trajectory.

it encircles the place where the isochrons meet. Thus as T is varied over a
full cycle using a short S, the θ of the reset rhythm likewise varies over a
full cycle in the opposite direction. This movement results in Type 1 resetting like that seen in the measured θ-contour map for *D. pseudoobscura*
when S < 50 sec (Figure 6a).

During a more prolonged perturbation, the same modified trajectories
are followed, but further. If the modified dynamics are sufficiently different from the unperturbed dynamics—e.g., if during perturbation it is
nonoscillatory continuous decay to a unique state as in Figure 10—then
the resulting states may lie along a ring that does not encircle the confluence of isochron surfaces (Figure 13). Thus as T is varied over a full cycle,
θ probably (but not necessarily) varies over a limited range of values, and
necessarily crosses each value twice—once as θ is increasing with T and
once as it is decreasing. This is Type 0 resetting, like that of the resetting
surface measured with S > 50 sec (Figure 6a).

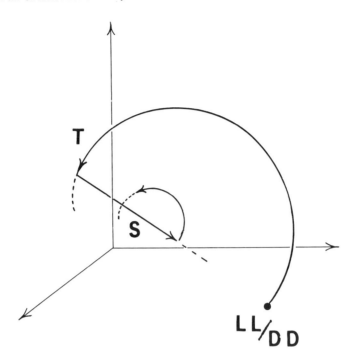

FIGURE 11   T is measured from the LL/DD transition along the initial
DD trajectory, and from that state S is measured along a modified, "perturbed" trajectory. At the end of S, the clock follows a new DD
trajectory.

Notice in Figure 13 that if Type 0 resetting is possible, it seems likely
that some modified, "perturbed" trajectory passes through the confluence
of isochron surfaces. If so, then a stimulus given at a critical time, T*,
which drives the clock along the modified trajectory for a critical duration,
S*, will send the clock to a state following which the phase of the emergence rhythm is unpredictable. Although not yet demonstrable, it may be
that there is only one such state of the fruitfly clock, a time-invariant,
stationary state or "equilibrium" of the clock reaction.

If the clock is indeed periodic in DD, then the stimulus either carries a
ring of initial states into one that still encircles the confluence of isochrons,
and the dependence of resetting on T therefore has Type 1 (or other non-
zero integer) topology, or else it does not and resetting is Type 0.

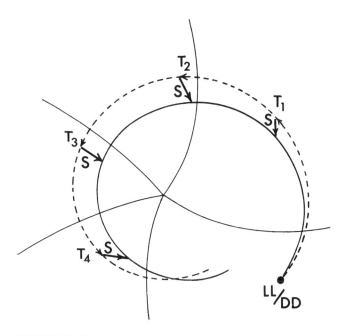

FIGURE 12   The same state space (with one DD trajectory shown as a dotted curve), showing the set of states to which the clock may be driven by a brief exposure to light, S, from any moment, T, on that DD trajectory. The fine lines indicate isochrons, in cross-section. This is Type 1 resetting.

I cannot visualize any continuous dynamical flow in the plane (i.e., any mechanism with only two degrees of freedom) that permits non-zero types other than Type 1. Other types, e.g., Type−1 with θ varying from 0 to 24 through a full cycle as T scans the cycle in the same direction, seem possible but only if the clock process has at least three degrees of freedom. Following the rubric of Occam's razor, let us assume until it is proven otherwise that the fruitfly clock has no more than two important degrees of freedom.

EMPIRICAL GENERALIZATIONS ABOUT CIRCADIAN RHYTHM RESETTING

Accordingly, if this way of thinking about the clock process can be generalized to other circadian systems, three properties are to be expected of data from resetting experiments. The first is the limitation of resetting

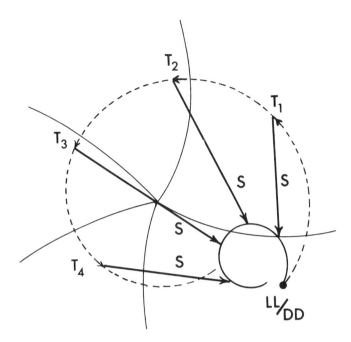

FIGURE 13 The analog of Figure 12, with a more prolonged interlude
of altered dynamics (larger S). This is Type 0 resetting.

topologies to Type 0 and Type 1; all published resetting data employing
perturbations by discrete stimuli do have this property (Table 1 and Figures 14 and 15 in Winfree, 1970a).

Second, it should be possible to change Type 0 resetting to Type 1 by
using a briefer perturbation of the same kind, but never by using a longer
one. This also is consistent with published resetting measurements on
circadian rhythms, as far as I know.

Third, to such rules about the topology of the dependence of θ on T,
some rules about its shape can be added by closer analysis of the geometry
of this kind of dynamics. For example, the amplitude[†] of a resetting curve
should rarely much exceed 6 hr (or 12 hr, peak-to-peak), and as S approaches S*, this limit should be approached and the curve should become
sawtooth in shape. This limits the range of admissible shapes in a surprising
way. The results of the many careful studies of resetting behavior of circadian rhythms now in the literature do seem to illustrate these restrictions
on shape, at least to a respectable first approximation.

[†]Measured as described in Winfree (1970b).

From these three generalizations alone it is possible to infer the existence of a corkscrew-shaped resetting surface and of an isolated critical stimulus that results in unpredictable phase resetting or abolishes the rhythm (Winfree, 1970a). The convergence of these predictions with the predictions of the dynamical theory above made it seem worthwhile to measure the full resetting surface of some single circadian system. It was in order to check these predictions that I chose the *Drosophila* system, to measure the resetting surface. And it does seem to follow the same pattern.

AMPLITUDE RESETTING (?)

Assuming the geometrical dynamics outlined above to be a reasonable interpretation of the timing process, we should then also like to know whether perturbation alters the clock oscillation in any respect other than advancing or delaying the phase of the rhythm.

Stimuli near (T^*, S^*) broaden the emergence peaks. This effect could be interpreted as a result of variegated phase-resetting of rhythms in a heterogeneous population. Pupae whose clocks were slightly precocious or retarded, or whose sensitivity to the light was somewhat more or less acute than the norm, would receive effective stimuli differing slightly from (T^*, S^*), and would therefore have markedly diversified θ values.

But are the phase-scattered clocks running normally otherwise, or could the observed broadening of peaks be due also to a resetting of amplitude, for example? One plausible interpretation of the convergence of θ-contours could be that (T^*, S^*) reduces the amplitude of an oscillation to zero, so that the clock loses control over the timing of emergence behavior, and "phase" becomes indeterminate.

One way to put this question experimentally would be to ask whether resetting behavior is altered by a first perturbation sending the clock toward the convergence of isochrons, and, if so, how. If the first stimulus simply phase-shifts without having other important effects, then the resetting curves measured any time subsequently will be normal, except for a displacement along the T-axis corresponding to the phase shift (the resetting surface is rigid).

If the clock is a limit-cycle process with an unstable equilibrium (as proposed by Pavlidis, 1967b, among others), and perturbation displaces it from the limit-cycle (as it must to produce a corkscrew-shaped resetting surface), then during recovery to the limit-cycle, the resetting curves will be abnormal. After full recovery they will again be normal except for a phase shift (the resetting surface is elastic). The interval during which resetting curve shape remains abnormal after perturbation is a measure of the rate of recovery to the limit-cycle.

If the character of the oscillation is lastingly altered by weak perturbations, then the deformation persists (the resetting surface is plastic). Biochemical reaction schemes that might suffer a lasting change of vigor or amplitude of oscillation have been proposed by Lotka (1920) and by Goodwin (1963).

Thus far, the results of measuring the duration of distortion seem somewhat surprising. After weak perturbation, especially one near the singularity (T*, S*), the resetting curves are altered and stay altered with no obvious tendency to return to normal for at least 48 hr. That is, there is so far no evidence either of persistence in a fixed cycle of clock states nor of rapid recovery to the limit-cycle mode of oscillation. On the contrary, the specific distortion of resetting curves measured with weak perturbations (a flattening) is consistent with the hypothesis that, given at time 0<T<16 when they have some effect, weak perturbations move the clock's state closer to the phaseless locus, reducing its range of oscillation, and that in the dark the clock continues to run at this reduced amplitude,† as we could call it. On this interpretation, the singular stimulus, (T*, S*), may be viewed as resetting the amplitude of the oscillator to zero (manuscript in preparation).

This distortion of the phase-resetting behavior by a weak perturbation is compactly illustrated in the following experiment:

If pupae are transferred from LL to DD and their resetting curves measured by application of 120-sec pulses of light at varying times T, the emergence rhythm in each case is normal but shifted in phase. That is, eclosion peaks are of normal width, but the position of the centroids, θ, is shifted, varying over a range of eight hours depending on when the resetting pulse is given (Figure 6b).

If, on the other hand, a stimulus (T*, S*) is applied after the LL/DD transition but before the 120-sec light pulse is given, the range of reset phases is reduced almost to zero. Regardless of when the 120-sec pulse is given, normal-looking emergence peaks appear at a single characteristic time after that second pulse, making the resetting curve, θ(T), almost a straight line.

In Figure 14 are illustrated the pooled, nearly identical results of 12 such experiments, in which the second pulse was given 6, 12, 18, or 24 hr

---

†This "amplitude," of course, has nothing to do with the amplitude of the eclosion rhythm itself, which merely registers the number of pupae used in the experiment.

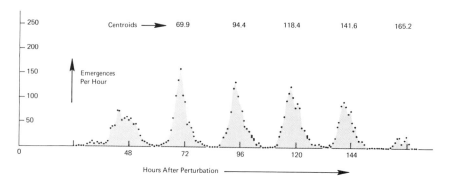

FIGURE 14   Populations previously given (T*,S*) are given 120 sec of light 6, 12, 18, or 24 hr later. Essentially identical emergence rhythms result in the 12 experiments here pooled (separated in Figure 12 of Winfree, 1970a). Time is measured from the 120-sec pulse.

after three different approximations of (T*, S*). The pooled controls, which received only critical perturbation, (T*, S*), show nearly continuous, arrhythmic emergence.

THE SINGULAR STATE

The result reported above is consistent with the supposition, also supported by other evidence (Winfree, 1970a; 1971), that S* seconds of light, applied at T*, drives the clock along a modified trajectory to a state in the confluence of isochrons, possibly much like the time-invariant, stationary state and that it stays there until the second pulse reinitiates the oscillation by driving the clock some distance further along the same trajectory (Figure 15). At the end of this second stimulus, no matter when it is applied, the clocks are all left on the same isochron, and consequently emergences occur at the same time after the second pulse.

If this is so, then the question arises whether the nonoscillating state reached by giving (T*, S*) after an LL/DD transition might be the same as the nonoscillatory condition suggested by Zimmerman (1969) as underlying the arrythmicity of emergence in dark-reared populations of pupae. An obvious test of this hypothesis is to subject dark-reared pupae to light pulses and see if the resulting emergence rhythms are the same as those of pupae given the (T*, S*) stimulus following an LL/DD transition (Figure 15). As Figure 16 shows,[†] the rhythms of the two populations are essen-

[†]In each of seven independent experiments, 100 to 200 pupae received 5 to 120 sec of the standard blue light. In three more experiments of comparable size, 15 min of bright white light was given. The results were indistinguishable.

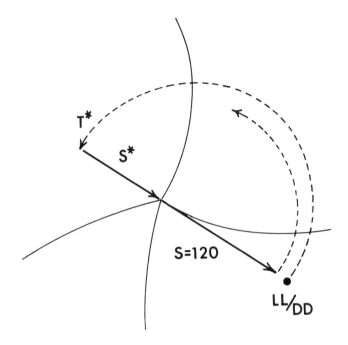

FIGURE 15 An interpretation of the experiment of Figure 14, analogous to Figure 11. The DD trajectory is shown to time T*, followed by light-modified dynamics for duration S*, leaving the clock with no θ value, until a later stimulus (S = 120 sec) reinitiates oscillations.

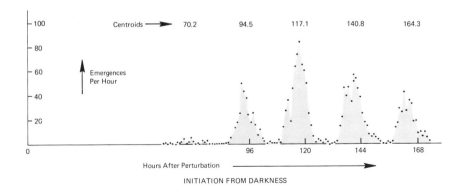

FIGURE 16 Ten dark-reared populations were exposed to light at an arbitrary time, labeled zero on the abscissa. All 10 emergence rhythms (here pooled) are essentially identical to those obtained in Figure 14 by reinitiating after (T*,S*).

tially identical. A similar experiment performed on a different strain of
D. pseudoobscura (Honegger, 1967) gave the same results.

This experiment at least failed to find any difference between the condition of the clock as it first appears in developing fruitflies and its condition following induction of oscillations plus application of the critical
perturbation. Perhaps the clock can be returned to its aboriginal arrhythmic condition by an application of the same sort of stimulus that can be
used to reset the clock when it is given for any other duration or at any
other time; this condition may be a steady-state of the clock, and if so
must necessarily lie at a meeting of isochrons.

SUMMARY OF THE HYPOTHESIS

The point of the foregoing is not that this fact or that is incontroverti-
bly established about fruitfly clocks, but only that thinking in terms of
geometrical dynamics allows a diversity of familiar questions to be asked
in an integrated framework of concepts, and suggests experimental approaches by which to answer them.

The view presented above assumes that clock dynamics is autonomous
and derives from continuous interaction among at least two continuously
variable quantities, and that the geometric description of this interaction
is reasonably simple and smooth when portrayed in a Euclidean state
space. It assumes that the timing process is approximately oscillatory in
some standard fixed environment (e.g., dark) and that in the presence of
the perturbing agent (e.g., light) the dynamical flow is substantially altered
in such a way that some trajectory now passes through the confluence of
isochrons. Isochrons are a bookkeeping trick for associating the *phases* of
an observed isoperiodic rhythm with the *states* of a hypothetical timing
process causing that rhythm.

This interpretation of the clock process is compatible with a diversity
of physical models, e.g., Higgins' biochemical oscillators (1967), the
Lotka-Volterra predator–prey interaction [which M. Rosenzweig (personal
communication) has suggested as a chemical–kinetic model for circadian
clocks], and the simple pendulum. A case more pertinent to our interests
is the model of Pavlidis for the *D. pseudoobscura* clock (1967b).

The hypothesis outlined in this paper is not, of course, all-inclusive.
Alternative interpretations of clock behavior have been quite useful in
some applications, but seem to falter[†] when confronted with the mea-

---

[†] Added in proof: A conspicuous exception has appeared in the simple and plausible
"time-lag" model of A. Johnsson and H. G. Karlsson, First European Biophysics Congress, Baden, Austria, 14–17 Sept. 1971.

sured corkscrew shape of the resetting surface and the inferred existence and accessibility of an isolated singular state in the fruitfly clock.

To consider one class of such alternatives: it seems quite reasonable to think of the clock as being always somewhere on a fixed cycle and to imagine that a light pulse alters its rate of progress through that cycle, either continuously, or only at the "on" or "off," or both. An example is the relaxation–oscillator model (Pittendrigh and Bruce, 1957; Bunning, 1960, 1964; Roberts, 1962; Pavlidis 1967a), according to which some substance accumulates (and/or is destroyed at a light-sensitive rate) until a threshold is crossed and the process starts over.

A second example is Swade's (1969) "velocity response-curve" model for mammalian activity rhythms.

A third example is the simplest version of the well-known chronon model of Ehret and Trucco (1967), according to which phase corresponds to the position of a transcribing polymerase on a very long polycistronic replicon (chronon). The effect of light would presumably be to alter its rate of progress along the chronon.

In more general terms, these three models represent the "simple clock" hypothesis, described by Campbell (1964), according to which:

there is a single variable on which all the interesting properties of the system depend; that any [pupa] can be assigned a time on the clock and that the expected behavior of any two clocks which read the same time will be the same regardless of their histories.[†]

But if such a "simple clock" interpretation of the circadian process were literally accurate, the resetting surface would not have a corkscrew shape. The θ-contours could neither converge to a point nor descend, run horizontal, and turn back up on the T $\times$ S plane. In the measured resetting surface of *D. pseudoobscura*, they do both. These properties lead me to favor the alternative hypothesis of continuous interplay among two or more variable factors, as suggested above.

ACHILLES' HEELS

In closing, let me follow the injunction of Alfred North Whitehead to "seek simplicity and distrust it" in doing science. I would like to draw your attention to several facts that might be found—although they have not yet been found, as far as I know—that could irretrievably undermine this approach to circadian clocks:

- Resetting curve shapes very much unlike sections through a screw surface could be demonstrated. It is suggestive, and in no way anticipated

[†]We have substituted "pupa" for "cell" in Campbell's analysis of the mitotic cycle.

by these geometrical arguments, that most of the published Type 1 circadian resetting maps have a segment whose slope approaches zero.

• A case of Type −1 resetting topology could be found. It would suffice to exhibit this response for any kind of perturbation, not necessarily visible light, since the physical nature of the stimulus in no way enters into the formulation I have presented. This would subvert the two-variable simplification and compel us to revert to a three-variable analysis, as in the state-space diagrams above.

• It could be shown that there is not really any Type 0 resetting; i.e., that all candidates are really extreme forms of Type 1, involving a near discontinuity in the dependence of θ on T (Winfree, 1970).

• It could be found that some circadian system has Type 0 resetting but no singularity is approachable using any briefer stimulus.

• A circadian system could be found to have several apparent singular states or even a continuum of them; for example, it could be shown that a state of indeterminate phase can be produced from every T, given the right choice of S (as is the case with a large class of relaxation-oscillators).

In more general terms, different kinds of possible clock processes are characterized by different topological properties of the measured resetting surface. The shape of the contours of equivalent stimuli (θ-contours), and the way they are put together in the contour map can serve as strong discriminating criteria between alternative classes of models.

ACKNOWLEDGMENTS

This work was done while enjoying the support of an NSF Regular Fellowship at Princeton University. The experimental parts were supported by grants NAS r-223 and NONr-1858 (28) to C. S. Pittendrigh. The first drafts of the manuscript were written during a Theoretical Biology Colloquium sponsored by NASA and AIBS, and improved by S. Kauffman, R. Levins, and my wife, Trisha.

I am grateful to the many investigators of circadian rhythms who shared with me their unpublished resetting data, and especially to K. Brinkmann, W. Englemann, T. Pavlidis, and C. S. Pittendrigh for their stimulating response to my "Puzzles and Paradoxes" mimeo of 1967, in which these thoughts were first circulated.

Figures 3 and 5 are reprinted with permission of the publisher, the American Mathematical Society, from Lectures on Mathematics in the Life Sciences, Copyright © 1970, Vol. II, pp. 121, 132–3.

REFERENCES

Bruce, V., F. Weight, and C. S. Pittendrigh. 1960. Resetting the sporulation rhythm in *Pilobolus* with short light flashes of high intensity. Science 131:728–730.

Bunning, E. 1960. Opening address: Biological clocks. Cold Spring Harbor Symp. Quant. Biol. 25:1–9.

Bunning, E. 1964. The physiological clock. Academic Press, New York. 69–70, 88.

Campbell, A. 1964. The theoretical basis of synchronization by shifts in environmental conditions, p. 469 to 484. *In* E. Zeuthen [ed.], Synchrony in cell division and growth. Interscience Division, John Wiley & Sons, New York.

Chandrashekaran, M. K. 1967. Studies on phase-shifts in endogenous rhythms. Z. Vgl. Physiol. 56:154–162.

Chandrashekaran, M. K., and W. Loher. 1969. The effect of light intensity on the circadian rhythms of eclosion in *Drosophila pseudoobscura*. Z. Vgl. Physiol. 62:337–347.

DeCoursey, P. 1960. Daily light sensitivity rhythm in a rodent. Science 131:33–35.

Ehret, C. F., and E. Trucco. 1967. Molecular models for the circadian clock: 1. The chronon concept. J. Theor. Biol. 15:240–262.

Engelmann, W. 1966. Effect of light and dark pulses on the emergence rhythm of *Drosophila pseudoobscura*. Experientia 22:1–5.

Engelmann, W. 1969. Phase-shifting in eclosion in *Drosophila pseudoobscura* as a function of the energy of the light pulse. Z. Vgl. Physiol. 64:111–117.

Engelmann, W., and H. W. Honegger. 1967. Versuche zur Phasenverschiebung endogener Rhythmen: Blütenblattbewegung von *Kalanchoe blossfeldiana*. Z. Naturforsch. B. 22:200–204.

Feldman, J. 1967. Lengthening the period of a biological clock in *Euglena* by Cycloheximide, an inhibitor or protein synthesis. Proc. Nat. Acad. Sci. U.S. 57:1080–1087.

Fitzhugh, R. 1961. Impulses and physiological states in theoretical models of nerve membrane. Biophys. J. 1:445–466.

Goodwin, B. C. 1963. Temporal organization in cells. Academic Press, New York.

Harker, J. 1964. The physiology of diurnal rhythms. Cambridge Univ. Press, London.

Hastings, W. 1964. The role of light in persistent daily rhythms, p. 333 to 340. *In* A. C. Giese [ed.], Photophysiology. Vol. 1. Academic Press, New York.

Higgins, J. 1967. The theory of oscillating reactions. Ind. Eng. Chem. 59:18–62.

Honegger, H. W. 1967. Zur Analyse der Wirkung von Lichtpulsen auf das Schlupfen von *Drosophila pseudoobscura*. Z. Vgl. Physiol. 57:244–262.

Kauffman, S. 1969. Metabolic stability and epigenesis in randomly constructed genetic nets. J. Theor. Biol. 22:437–467.

Lewontin, R. 1968. The evolution of complex genetic systems, p. 62 to 87. *In* M. Gerstenhaber [ed.], Lectures on mathematics in the life sciences. Vol. 1. American Mathematical Society, Providence, Rhode Island.

Lewontin, R., and M. White. 1967. Interaction between inversion polymorphism of two chromosome pairs in the grasshopper, *Moraba scurra*. Evolution 14:116–131.

Lotka, A. J. 1920. Undamped oscillations derived from the law of mass action. J. Am. Chem. Soc. 42:1595–1598.

Minis, D. 1965. Parallel peculiarities in the entrainment of a circadian rhythm and photoperiodic induction in the pink boll worm (*Pectinophora gossypiella*), p. 333 to 343. *In* J. Aschoff [ed.], Circadian clocks. North-Holland Publ. Co., Amsterdam.

Ottesen, E. 1965. Analytical studies on a model for the entrainment of circadian systems, unpublished A.B. thesis, Biology Department, Princeton University.

Pavlidis, T. 1967a. A mathematical model for the light affected system in the Drosophila eclosion rhythm. Bull. Math. Biophys. 29:291–310.

Pavlidis, T. 1967b. A model for circadian clocks. Bull. Math. Biophys. 29:781–791.

Pavlidis, T. 1968. Studies on biological clocks: A model for the circadian rhythms of nocturnal organisms, p. 88 to 112. *In* M. Gerstenhaber [ed.], Lectures on mathematics in the life sciences. Vol. 1. American Mathematical Society, Providence, Rhode Island.

Pittendrigh, C. S. 1954. On temperature independence in the clock system controlling emergence time in Drosophila. Proc. Natl. Acad. Sci. U.S. 40:1018–1029.

Pittendrigh, C. S. 1960. Circadian rhythms and the circadian organization of living systems. Cold Spring Harbor Symp. Quant. Biol. 25:159–184.

Pittendrigh, C. S. 1965. On the mechanism of the entrainment of a circadian rhythm by light cycles, p. 277 to 297. *In* J. Aschoff [ed.], Circadian clocks. North-Holland Publ. Co., Amsterdam.

Pittendrigh, C. S. 1966. The circadian oscillation in *Drosophila pseudoobscura* pupae: A model for the photoperiodic clock. Z. Pflanzenphysiol. 54:275–307.

Pittendrigh, C. S. 1967. Circadian systems. 1. The driving oscillation and its assay in *Drosophila pseudoobscura*. Proc. Natl. Acad. Sci. U.S. 58:1762–1767.

Pittendrigh, C. S., and V. G. Bruce. 1957. An oscillator model for biological clocks, p. 75 to 109. *In* D. Rudnick [ed.], Rhythmic and synthetic processes in growth. Princeton Univ. Press, Princeton, New Jersey.

Pittendrigh, C. S., and D. H. Minis. 1964. The entrainment of circadian oscillations by light and their role as photoperiodic clocks. Am. Nat. 98:261–294.

Roberts, S. 1962. Circadian activity rhythms in cockroaches—entrainment and phase-shifting. J. Cell. Comp. Physiol. 59:175–186.

Rosen, R. 1971. Dynamical system theory in biology, Vol 1. Wiley-Interscience, New York.

Sargent, M. L., and W. R. Briggs. 1967. The effects of light on a circadian rhythm of conidiation in *Neurospora*. Plant Physiol. 42:1504–1510.

Skopik, S. D., and C. S. Pittendrigh. 1967. Circadian systems. II. The oscillation in the individual *Drosophila* pupa; its independence of developmental age. Proc. Natl. Acad. Sci. U.S. 58:1862–1869.

Swade, R. H. 1969. Circadian rhythms in fluctuating light cycles: Toward a new model of entrainment. J. Theor. Biol. 24:227–239.

Wilkins, M. B. 1962. Effects of temperature changes on phase and period of the rhythm. Proc. R. Soc., Ser. B. 156:220–241.

Winfree, A. T. 1967. Biological rhythms and the behavior of populations of coupled oscillators. J. Theor. Biol. 16:15–42.

Winfree, A. T. 1970a. The temporal morphology of a biological clock, p. 109. *In* M. Gerstenhaber [ed.], Lectures on mathematics in the life sciences. Vol. 2. American Mathematical Society, Providence, Rhode Island.

Winfree, A. T. 1970b. An integrated view of the resetting of a circadian clock. J. Theor. Biol. 27:327–374.

Winfree, A. T. 1971. The investigation of oscillatory processes by perturbation experiments, Parts I and II. K. Pye, B. Chance [ed.], Biochemical oscillators. Academic Press, N.Y.

Zimmerman, W. F. 1969. On the absence of circadian rhythmicity in *Drosophila pseudoobscura* pupae. Biol. Bull. 136:494–500.

Zimmerman, W. F., C. S. Pittendrigh, and T. Pavlidis. 1968. Temperature compensation of the circadian oscillation in *Drosophila pseudoobscura* and its entrainment by temperature cycles. J. Insect Physiol. 14:669–684.

DISCUSSION

WEVER: You have shown that you can stop the *Drosophila* clock with a very specific push. But will it start again spontaneously? Each oscillation of second order, regardless of its special properties, has at least one point of singularity, but a very thorough understanding of the oscillation is necessary to locate this point. The question is whether the equilibrium at this point is stable or unstable. Statements concerning the properties of the oscillation can only be derived from analysis of its behavior at the equilibrium point—persistence or restarting. What I am asking is, can you decide from your experiments whether the *Drosophila* clock is self-sustained, and, if it is, whether it is self-exciting?

WINFREE: Those are certainly most important questions, which we would all like to answer. I can suggest one way of getting at the question of self-excitation. First allow me the liberty of rewording the question: I tried to show that you can almost stop the clock, that is, that you can severely diminish its rhythmicity. (One never does anything exactly with living organisms.) And I believe that the stability properties of an oscillation may be derived from analysis of its behavior near, but not exactly at, equilibrium. Now suppose we give a perturbation that almost stops the clock. If we then separated the pupae into aliquots and gave each one another perturbation, at different times, and measured resetting behavior, we could tell whether the clocks stayed (almost) stopped until the

second stimulus or started spontaneously before the second stimulus. If the clocks all stayed (almost) stopped, we would get nearly identical and normally distinct emergence rhythms in every case. And this is what happened when the experiment was done. At the other extreme, if the normal oscillation had recovered before the second stimulus, then the resulting peaks should have been broad and/or their timing dependent on *when* the second stimulus was given.

ASCHOFF: As I understood Dr. Wever, he suggested that the behavior you describe is typical for a self-exciting system. But I don't think so; you said that the clock stops and then doesn't start again until disturbed. That is not self-exciting.

WINFREE: Yes, that is how I see it, too. But to be very exact, the behavior I described is typical for both non-self-exciting systems and systems that are slowly self-exciting. I cannot strictly discriminate between the two types because I don't have the infinite time required by the mathematical description; I have only nine days of emergence rhythms. It is not possible in that time to discriminate between a stable system and one which slowly spirals out. You can only tell that if it self-excites, it does so very slowly.

WEVER: You have described the system as a single clock; but it is really a population clock. And it may be that for a population clock the terms "self-exciting" and "self-sustained" are not appropriate.

WINFREE: They might be adequate

descriptions for the clock of the individual fly; the question is, what can we infer from population experiments?

WEVER: Perhaps each individual oscillator is self-sustained, but they are not interacting, not in mutual entrainment. And when the individual free-running periods are not equal, the population oscillation damps out.

WINFREE: Data Dr. Pittendrigh has gathered, though, indicate that the free-running periods are quite similar, and the population rhythm free-runs quite well for nine or ten days.

BRUCE: I have an entirely different question, about the experimental details of the resetting signals. Am I correct that you present a range of signals that are below threshold to above threshold in the sense that they do or do not cause a phase shift?

WINFREE: I don't find a threshold effect, but a continuously graded influence of the signal.

BRUCE: I was going to ask what you thought about Zimmerman's action spectrum experiments, which he interprets in terms of differential effects of the stimulus on different pupae. The resetting effect varies from one to another pupa because in this range of signal strengths, different pupae effectively get different light pulses. I understood you to be working in the same range.

WINFREE: My first response is really a quibble: I am using a different strain of flies, and that may conceivably make a difference. I think my results are different from Zimmerman's. He inferred an age-dependence of resetting. Though the flies that emerged in the later peaks of the four days of eclosion were younger when they were

given the perturbation than those that emerged earlier, we see no difference in the shape of their resetting surface. Thus we see no age effect in my experiments.

BRUCE: I was thinking of the generation of arrhythmicity by a weak signal given at a specific time. We might expect some differential effects on different individuals here. The time of your singular stimulus is such that a strong stimulus given then will reset phase by a large number of hours, while a weak stimulus given then will reset them not at all. If the pupae are receiving effectively different stimuli, you might expect to reset them to many different phases, and the resulting arrhythmicity could be interpreted as a scattering of phases.

WINFREE: That is certainly very true. (T^*, S^*) probably does diversely rephase the rhythms of pupae of any given age. But I tried something that seems to discriminate a population of merely phase-scattered clocks from a population of nearly static, nonfunctioning clocks. The experiment is too involved to describe in a moment (Winfree, 1970a). The essential point is that an artificially constructed population of phase-scattered clocks and a population previously exposed to (T^*, S^*) gave very different results in this experiment.

EHRET: Wever and Bruce touched on the principal points I had to make, but perhaps I can focus more sharply upon them. In working with a population phenomenon, it is most important to be aware of the focal-plane of your question—whether you are dealing with the hyperset, the set, or some subset—whether with a population of organisms, of cells, of macromolecules, or whatever. In that respect your "isochron" surfaces are quite ingenious and could be very helpful. But they remind me of Kepler's wheels; your *fiats* and *caveats* may not extend from the population level downward and may not define the boundary conditions for molecular clocks.

WINFREE: I agree 100 percent. Without separate proof of homogeneity, observations of population behavior may tell us nothing about the dynamics of any lower level.

THEODOSIOS PAVLIDIS

Mathematical Models of Circadian Rhythms: Their Usefulness and Their Limitations

During the last five years a series of models for circadian rhythms has been developed at Princeton, starting with Pittendrigh's empirical model for the effects of light on the *Drosophila pseudoobscura* eclosion rhythm (Pittendrigh and Minis, 1964; Pittendrigh, 1966). That model has since been generalized to other circadian systems and expanded to account for temperature effects as well as light effects (Ottesen, 1965; Pavlidis, 1967a, 1967b, 1968; Pavlidis, Zimmerman, and Osborn, 1968; Zimmerman, Pittendrigh, and Pavlidis, 1968; Pavlidis, 1969a, 1969b; Pavlidis and Kauzmann, 1969). Of the large variety of models that has been proposed—the various forms of the Princeton models and those of other investigators—some are quite elementary mathematically, while others are very complicated. Some are purely phenomenological or abstract in their formulation, while others involve very specific suggestions about the physical structure of the system. Since considerable effort has gone into the development of such models, it is reasonable to ask what purpose they serve in the study of circadian rhythms and whether we understand such rhythms better now than before.

To evaluate the proposed models, it is necessary to sketch what one may expect from a mathematical model of a physical system, in general. Such a model may serve one or more of the following functions:

- Facilitation of the bookkeeping. One would like to keep in mind only a small number of "laws" instead of a large collection of experimental facts. Kepler's three laws of motion of the planets illustrate this function. Aschoff's rule is an example from the area of circadian rhythms.
- Guidance of experimentation. A model can make predictions about the behavior of the physical system and suggest experiments to prove or disprove them.
- Understanding of the nature of the physical system. This is the most important function and, strictly speaking, one that can never be fully achieved. However, a model can point out the importance of various factors for the system, as in the Hodgkin-Huxley model of the neural membrane. Although some specific features of their model were arbitrarily chosen, it goes far in describing what may actually happen in the membrane during the generation and transmission of a neural pulse.

It should be emphasized that a model may perform the first two functions extremely well and yet say nothing about the structure of the system. This situation may not be at all objectionable as long as one realizes the limitation. However, this point seems to have been overlooked in the design of many early models for circadian rhythms. For example, many aspects of circadian rhythms are typical of any nonlinear self-sustained oscillator. Hence any model based on nonlinear self-sustained oscillation can perform very well the bookkeeping function, but the success of such a model is not a reflection of the significance of its specific structure.

In order to avoid the pitfall of attributing to specific characteristics success that is really due to generic characteristics, it is very important not to overspecify the model. Because this point has had careful attention in the Princeton models, the models have not only facilitated the "bookkeeping" of facts about circadian rhythms, but have also pointed out simple constraints required for the structure of the system responsible for the overt behavior. Models that have been overspecified in terms either of differential equations or of electronic analogs (Barlow, 1962; Wever, 1965) cannot define such constraints.

USES OF DESCRIPTIVE MODELS

A good descriptive model discourages the undertaking of "trivial" experiments. An advertisement appeared in the *New York Times* (1969) criticizing biologists for continuing to experiment on live animals when the same results could be obtained from computer simulation. Although the tone of the advertisement was rather naïve, it certainly contained an element of truth. In circadian rhythm study such a "trivial" experiment is entrainment by periodic light stimuli. Any nonlinear oscillator (and most investigators are persuaded that circadian rhythms are of that nature) that is sensitive to light in the way described by Aschoff's rule is expected to be entrainable. Even certain apparently complex and surprising results are readily predictable from the properties of nonlinear oscillators in general (Pavlidis, 1969a).

Models can guide research positively as well. Mathematical models (Wever, 1965; Pavlidis, 1967b, 1968) suggesting a singularity in the system responsible for circadian rhythmicity prompted experiments to determine whether such a singularity indeed exists (Zimmerman, 1969; Winfree, this volume).

Secondly, a good descriptive model could indeed provide some insight into the nature of the system. Essential to all the Princeton models is the assumption that light drives the system toward a sequence of states that is also traversed during the absence of light between Circadian Time 4 and 12 (approx.). This is illustrated in Figure 1. Some evidence of light adaptation was also found (Pavlidis, 1967b, 1969a), which is not surprising. Although one could conceive of altogether different classes of models that might also simulate the results of various behavioral experiments, the simplicity of this assumption and the wide range of its simulations is reason to believe that there is some physical significance in it.

Mathematical reasoning could be used to draw conclusions about the nature of the real system. This, however, requires certain caution which is not always taken by life scientists and, on occasion, claims have been made about the nature of circadian rhythms, the validity of which can be seriously challenged by anyone familiar with the classical texts on nonlinear oscillations.

UNDERSTANDING THE SYSTEM

The failures of a model, rather than its successes, provide the most important information for proceeding toward an understanding of the system. There is at this point a great temptation to dismiss some of the discrepancies as the result of interference from systems outside the scope of the main study, as was done in some cases when experimental data differed from those predicted by the Princeton models. One investigator has proposed that his model (designed on the basis of data mainly from mammals) failed to simulate the large phase shifts observed in the *Drosophila pseudoobscura* eclosion rhythm after the application of light pulses of very short

FIGURE 1 Phase plane portrait of the limit cycle of a model for circadian clocks (solid line). The trajectories under light exposure are shown by broken lines; x and y can be any two variables. Any topological equivalent of the above could also serve as model. S_0 is a state of equilibrium. (Numbers indicate approximate circadian time.)

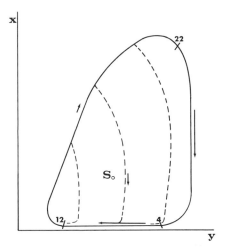

duration because the latter rhythm was of an essentially different nature from the others (Wever, 1965). Such explanations may, of course, be valid, but one has to prove them instead of taking them for granted in order to excuse the inadequacies of a given model.

Another "easy way out" is to modify the model in order to assimilate new evidence as it becomes available. Probably the most notorious example of this strategy is the development of the geocentric model of the planetary system. Before making trivial modifications in a model (i.e., changing numerical values of the parameters by adding extra terms in the equations, etc.), one should first ask whether a more fundamental change is in order.

For example, deviations from steady-state rhythmicity, in the form of spontaneous changes of free-running period and similar changes after perturbation by an external stimulus, have been observed in some organisms (Roberts, 1962; Richie and Womack, 1966; Eskin, 1969). Such changes could be dismissed as due to a "noisy biological environment" or by claiming that one deals with a system that varies with time. Existing models could be modified to simulate these phenomena, or one could conclude that second-order models oversimplify complex systems. On the other hand, high-order oscillators or interacting populations of simpler oscillators present very complex oscillatory patterns (Moser, 1966; Jenks, 1968) and, therefore, one can observe changes in the frequency of oscillations after a certain disturbance moves the state of the system from the region of attraction of one limit cycle to that of another. Some studies have already been made in this direction (Richie and Womack, 1966; Pavlidis, 1969b), and it has also been shown that interacting oscillators may also offer a way of frequency reduction. Going from a lower-order model to a higher-order one obviously offers many more degrees of freedom; it is not surprising that the latter model can simulate additional classes of experiments. What we do not know yet is whether there is any other type of system that could simulate the multifrequency aspects of circadian rhythms. However, it should be pointed out that it will be almost impossible to devise an experiment that would decide against a higher-order system in favor of a simpler one, since a complex system can always behave like a simple one.

In spite of extensive modeling we have made hardly any progress toward understanding the nature of the physical system responsible for circadian rhythms. All we have are generic models, difficult to disprove,* which successfully simulate the overt behavior. The Princeton model, for example, could be disproved by the discovery of a phase response curve of the form shown in Figure 2, but it could be argued that such a curve will have no survival value for the organism involved and that a search for it would be in vain.

The disappointing performance of mathematical models proposed to date is, I believe, due to the limited amount of information provided by the behavioral experiments on which most of them have been based. This is particularly true of behavioral experiments involving features having a survival value. It seems that the circadian clock contains a sturdy mechanism that cannot be sufficiently disturbed by light or temperature changes to reveal much information about its structure. What the mathematical models have told us is that many behavioral experiments indicate little more than that the clock is a well-designed mechanism.

For this reason a new approach to circadian systems is necessary. We proposed a quantitative biochemical model (Pavlidis and Kauzmann, 1969) that simulates all the behavioral data the earlier Princeton models do and, in addition, has a specific physical structure. The model can now be tested on the basis of its biochemical predictions:

● The existence of enzymes having the property that the concentration of their active form is a decreasing function of temperature,

● The "dominance" of various reactions during different parts of the cycle. An agent that slows down the clock between CT 12 and 18 should speed it up between CT 18 and 26(2). Verifying this prediction will not necessarily prove the model, but finding agents that do not obey this law

---

*This is not true, of course, for overspecified models, which can be easily disproved. But the only information resulting from disproof is that the overspecification was in error.

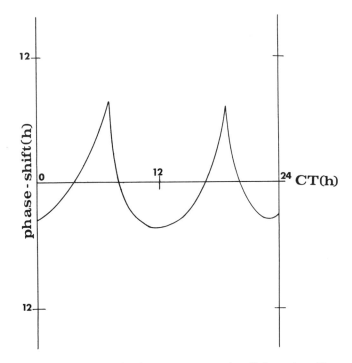

FIGURE 2 Outline of a phase response curve that, if observed, could disprove the type of model shown in Figure 1.

will disprove it. In general, selective effects on various parts of the cycle should provide us with a great amount of information about the clock.

• Existence of reactions slow enough to produce the 24-hr period or of certain frequency reduction mechanisms; for example, a chemical system with subsystems sharing the same enzyme pool, which will have a slow oscillation as well as interactions that simulate spontaneous changes in the free-running period.

Earlier biochemical models that suggested specific physical structures did not attempt to simulate the overt behavior and therefore could not be tested (Ehret and Barlow, 1960; Goodwin, 1963: Ehret and Trucco, 1967; Chance, Pye, and Higgins, 1967).

REFERENCES

Barlow, J. S. 1962. A phase-comparator model for the diurnal rhythm of emergence in Drosophila. Ann. N.Y. Acad. Sci. 98:788–805.

Chance, B., K. E. Pye, and J. Higgins. 1967. Waveform generation by enzymatic oscillators. IEEE Spectrum 4:79–86.

Ehret, C. F., and J. S. Barlow. 1960. Toward a realistic model of a biological period-measuring mechanism. Cold Spring Harbor Symp. Quant Biol. 25:217–220.

Ehret, C. F., and E. J. Trucco. 1967. Molecular models for the circadian clock: I. The chronon concept. J. Theor. Biol. 15:240–262.

Eskin, A. 1969. The sparrow clock: Behavior of the free running rhythm and entrainment analysis. Ph.D. Thesis. University of Texas at Austin.

Goodwin, B. 1963. Temporal organization in cells. Academic Press, New York.

Jenks, R. D. 1968. Quadratic differential systems for interactive population models. J. Differ. Equations 5:497–514.

Moser, J. 1966. On the theory of quasiperiodic motions. SIAM (Soc. Ind. Appl. Math.) 8:145–172.

New York Times. 1969. News Summary Section. May 18.

Ottesen, E. 1965. Analytical studies on a model for the entrainment of circadian rhythms. A.B. Thesis. Princeton University.

Pavlidis, T. 1967a. A mathematical model for the light affected system in the *Drosophila* eclosion rhythm. Bull. Math. Biophys. 29:291–310.

Pavlidis, T. 1967b. A model for circadian clocks. Bull. Math. Biophys. 29:781–791.

Pavlidis, T. 1968. Studies on biological clocks: A model for the circadian rhythms of nocturnal organisms, p. 88 to 112. *In* M. Gerstenhaber [ed.], Some mathematical problems in biology. American Math. Society, Providence, R.I., Publisher.

Pavlidis, T. 1969a. An explanation of the oscillatory free-runs in circadian rhythms. Am. Nat. 103:31–42.

Pavlidis, T. 1969b. Populations of interacting oscillators and circadian rhythms. J. Theor. Biol. 22:418–436.

Pavlidis, T., and W. Kauzmann. 1969. Toward a quantitative model for circadian oscillators. Arch. Biochem. Biophys. 132:338–348.

Pavlidis, T., W. F. Zimmerman, and T. Osborn. 1968. A mathematical model for the temperature effects on circadian rhythms. J. Theor. Biol. 18:210–221.

Pittendrigh, C. S. 1966. The circadian oscillation in *Drosophila pseudoobscura* pupae: A model for the photoperiodic clock. Z. Pflanzenphysiol. 54:275–307.

Pittendrigh, C. S. 1967. Circadian systems. 1. The driving oscillation and its assay in *Drosophila pseudoobscura*. Proc. Nat. Acad. Sci. 58:1762–1767.

Pittendrigh, C. S., and V. Bruce. 1959. Daily rhythms on coupled oscillator systems and their relation to thermoperiodism and photoperiodism, p. 475 to 505. *In* R. Withrow [ed.], Photoperiodism and related phenomena in plants and animals. American Association for the Advancement of Science, Washington, D.C.

Pittendrigh, C. S., and D. H. Minis. 1964. The entrainment of circadian oscillations by lights and their role as photoperiodic clocks. Am. Nat. 98:261–294.

Richie, C. G., and B. F. Womack. 1966. A mathematical model for the biological clock of *Passer domesticus*. Univ. Texas, Laboratory for Electronics, Tech. Rep. No. 28.

Roberts, S. K. F. 1962. Circadian activity rhythms in cockroaches: II. Entrainment and phase shifting. J. Cell. Comp. Physiol. 59:175–186.

Wever, R. 1965. A mathematical model for circadian rhythms, p. 47 to 63. *In* J. Aschoff [ed.], Circadian clocks. North-Holland Publ. Co., Amsterdam.

Zimmerman, W. F. 1969. On the absence of circadian rhythmicity in *Drosophila pseudoobscura* pupae. Biol. Bull. 136:494–500.

Zimmerman, W. F., C. S. Pittendrigh, and T. Pavlidis. 1968. Temperature compensation of the circadian oscillation cycles in *Drosophila pseudoobscura* and its entrainment by temperature. J. Insect Physiol. 14:669–684.

RÜTGER WEVER

Influence of Electric Fields on Some Parameters of Circadian Rhythms in Man

It will be useful at the outset to illustrate how raw data are presented in this paper. Figure 1 is a typical example of a free-running circadian rhythm in man. The bars represent the activity rhythm, black indicating activity and white indicating rest. Maxima and minima of rectal temperature are indicated by the triangles. Each successive period is drawn beneath the preceding one. The figure illustrates the stability of the rhythm; after an interval of transients subsequent to transfer from an entraining LD cycle to constant conditions, a free-running period (τ) of 25.3 hr is maintained. In the course of the experiment, the rhythm shifts with respect to local time for nearly 2 days. This shows that the rhythm is autonomous.

In the past five years, we have examined more than 100 human subjects under constant conditions; with only one exception, the rhythms of these subjects have proved to be autonomous (Wever, 1969b). The mean autonomous or free-running period for 102 subjects is 25.05 hr, averaged over all the different conditions tested, with the remarkably small interindividual standard deviation of ± 0.63 hr. This mean value of τ (tau) proved to be independent of the sex and age of the subjects. On the other hand, τ deviates significantly from the period values of all known or suspected *Zeitgebers* in the environment, including the apparent period of the moon (Wever, 1969b).

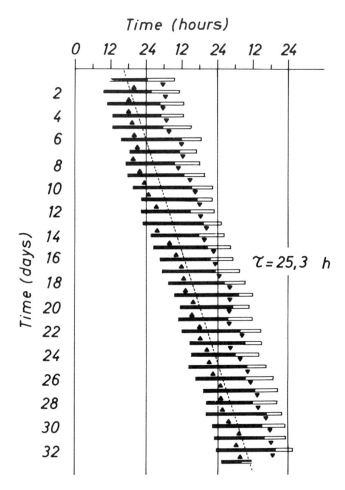

FIGURE 1 Free-running rhythm in a human subject: constant conditions (230 lux) during the total experiment. Abscissa: local time; ordinate: successive periods. Black bars represent activity and white bars represent rest; ▲ indicates a maximum value of the rectal temperature rhythm and ▼ indicates a minimum value.

In most of these studies, the experimental conditions have been changed (up to three times) to test their effect on certain rhythm parameters. It is of interest to know, on the one hand, which environmental factors are able to control the rhythm (either to influence the rhythm when constant, or to entrain the rhythm when periodically changing), and, on the other hand, which parameters of circadian rhythms are influenced by external factors. This paper deals with the second question only. Correlations in the changes of different parameters will be compared with corresponding correlations that have been observed frequently in animal experiments and formalized into certain generalizations (Aschoff, 1960, 1964). Because these rules are predictable, at least qualitatively, from a special mathematical model (Wever, 1965, 1966), the problem can be reduced to the question whether this model is applicable to human circadian rhythms (Wever, 1968b).

To answer experimentally the question posed above, we need an external stimulus that can influence human circadian rhythms and that fulfills the following conditions: It must not be perceptible consciously, and it must be effective during rest time as well as during activity time. Light as the controlling stimulus violates both these conditions. As a consequence, the human free-running rhythm under the influence of varying light intensities does not change its behavior according to any rule (Wever, 1968b, 1969a). In some light-active mammals also, not only the correlation between light intensity and period value, but the internal correlation between different rhythm parameters, is irregular. One theoretical reason for this irregular behavior, as predicted from the mathematical model, is the difference in the light perception during activity time (when the subject has eyes open) and rest time (when he has eyes closed) (Wever, 1969a). As a consequence, self-controlled environmental changes may influence the rhythm (Aschoff *et al.*, 1968).

On the other hand, a weak a-c electric field, with a frequency of 10 cps, influences human circadian rhythms and fulfills the two conditions mentioned above. Figure 2 demonstrates the influence of such a field on the circadian period: with the field in operation, the period is shorter than without it. (The illustrated results come from the first experiment of this type, where the subject happened to have an unusually long period.) "Without" means that not only is the artificial field switched off, but the natural electric and magnetic fields are eliminated by means of special shielding (Wever, 1967). Neither this subject nor any of the other subjects examined under comparable conditions possessed any knowledge of the shielding of natural fields, or of the introduction of artificial fields; no subject has reported being aware of the switching on and off of the 10-cps field. Figure 3 shows the result of another experiment; here the temporal sequence is reversed. In addition, later in the record, the 10-cps field was switched on and off periodically to operate as a *Zeitgeber*.

The influence of a constant 10-cps field on the free-running period has been examined in 10 experiments. All these experiments have had the same outcome (Figure 4): τ is shorter with the field in operation than without

Time (hours)

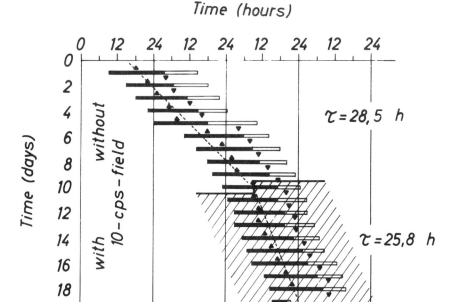

FIGURE 2   Free-running rhythm in a human subject: in the absence of natural and artificial electric and magnetic fields during the first section of data, and under the influence of a weak electric 10-cps field during the second section. Designations as in Figure 1. (From Wever, 1968c)

FIGURE 3   Free-running rhythm in a human subject during the top two sections: under the influence of a 10-cps field during the first section of data, and without any field during the second section. During the third section of data, the 10-cps field was switched on and off periodically ("field Zeitgeber"). Designations as in Figure 1.

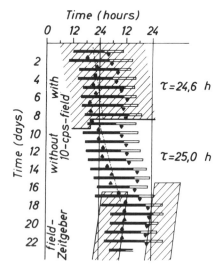

FIGURE 4   Results of 10 experiments on free-running period values obtained without field and with the 10-cps field in operation. Left and right: arithmetical means and (interindividual) standard deviations. The statistical significance of the difference in τ values is indicated.

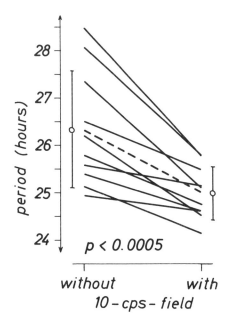

it; statistical analysis shows that the difference is highly significant (p<0.0005). These 10 experiments have been made in one of the two rooms of our underground system. Only this room is shielded from electric and magnetic fields. In the shielded room, the mean period is longer than in the nonshielded room (Wever, 1968a). Therefore, the 10 subjects included in Figure 4 have rather long periods as compared to the mean of all 102 subjects, which includes results from measurements in both rooms. Moreover, Figure 4 shows that the interindividual standard deviation is smaller with the field in operation than without the field (p<0.05).

In all these experiments, the values of some other measurable parameters, in addition to period, change depending on the state of the field. The rhythms of rectal temperature and activity will be considered below. Different parameters were measured in the two rhythms since in the rectal temperature rhythm the continuous course of the function was recorded, whereas only two points, onset and end of activity, were recorded in the activity rhythm.

In Figure 5 results are given for three parameters of the rectal temperature rhythm for 9 of 10 subjects (in one experiment, the temperature record was lacking). The left diagram in the upper row shows a significant positive correlation between the mean value of temperature and the presence or absence of the 10-cps field. The middle diagram shows that the

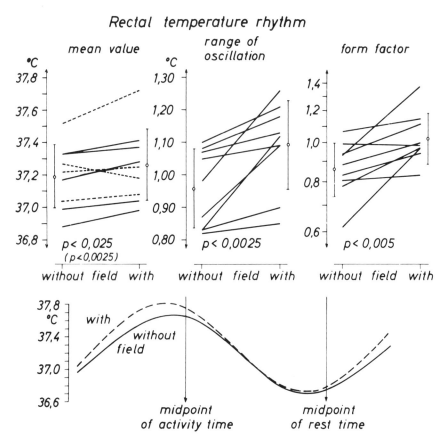

FIGURE 5 Results of 9 experiments with free-running rhythms in human subjects obtained without a field and with a 10-cps field in operation, measuring three parameters of the rectal temperature rhythm. Designations as in Figure 4, with the following additional marks: in the "mean value," results from female subjects are shown as dotted lines, and the number in parenthesis is the significance when only male subjects are considered; the "form factor" is indicated in a logarithmic scale and arithmetical means and standard deviations are calculated from the logarithms of the parameter values. Lower diagram: course of the rectal temperature relative to the activity rhythm, averaged from data obtained without field and from those with the 10-cps field in operation.

range of oscillation is significantly greater with the field in operation than without it. The third parameter, whose changes are shown in the right-hand diagram, is the form factor, defined by the relation between the duration of the descending section and that of the ascending section of the oscillation (Wever, 1968c). This parameter has a significantly larger value with the field than without the field; that is, when the field is switched on, the ascent becomes steeper and the descent becomes flatter.

Variability in the mean value of temperature is greater than that of the two other parameters; consequently the change in this parameter, with and without the field, is less significant than that of the others. The reason may be that in female subjects the menstrual temperature cycle is superimposed on the circadian temperature rhythm, a factor not controlled in these experiments. To test this hypothesis, results are considered separately for male and female subjects. In Figure 5, changes in the mean values of female subjects are shown as dotted lines. Both the line with the largest absolute deviation and that with the reversed slope come from females. Thus, the change in the mean temperature value is more significant when only male subjects are considered, in spite of the diminished number (the change for female subjects only is not significant).

A schematic representation of the mean course of rectal temperature is given in the lower part of Figure 5 for data obtained with and without field. For both conditions, the period is standardized to the same value (in relative units); moreover, the position of the rectal temperature rhythm is given relative to the activity rhythm. The position of the temperature minimum is nearly independent of the state of the field with regard to absolute value as well as with regard to its temporal position relative to the activity period. In contrast, the temperature maximum shifts to a higher value and to an earlier position relative to the activity period when the field is switched on. Although small in amount, both shifts are statistically significant.

Likewise, three parameters of the activity rhythm have been examined (Figure 6). Duration of activity time and rest time is calculated from onset and end of activity. The ratio of these two parameters can be determined independent of period. As the left diagram in Figure 6 (upper row) shows, $a:\rho$ ratio is significantly larger when the field is in operation than without the field. In the next step, the deviations of successive onsets and cessations were considered. Because of the serial dependency of successive periods, these deviations cannot be used for statistical purposes although they are calculated formally as standard deviations. In all cases, the intra-individual deviation of the end of activity is significantly ($p < 0.005$) larger than that of the onset of activity both with and without the field. Two parameters that can be determined independent of period length were derived from the observed deviations. One of these, the second parameter of the activity rhythm, is precision, which is defined as the reciprocal of the variability coefficient; i.e., precision is the quotient of the period value divided by its standard deviation. As shown in the middle diagram, this parameter when calculated with the deviations of onset of activity is generally larger with the field in operation than without it. By conventional

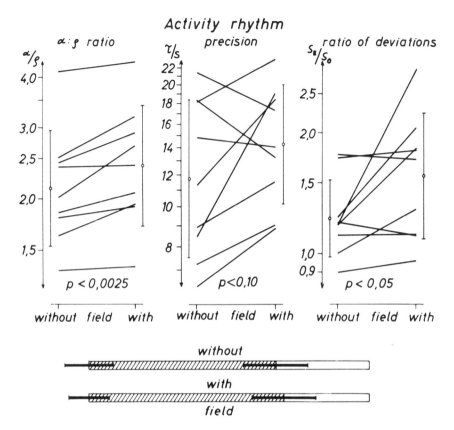

FIGURE 6 Results of 9 experiments with free-running rhythms in human subjects obtained without a field and with a 10-cps field in operation, measuring three parameters of the activity rhythm.
The values of all parameters are indicated on logarithmic scales; arithmetical means and standard
deviations are calculated from the logarithms of the parameter values. Other designations as in Figure 4. Lower diagram: schematic representation of the activity rhythm, with activity time (shaded
bars) and rest time (white bars); black lines indicate intraindividual "standard deviations" of onset
and end of activity, averaged from data obtained without field and from those with the 10-cps field
in operation.

standards, this change is not significant because results of the experiments
with the highest absolute precision value change in the opposite direction
(changes in deviations of onset of activity are significant with $p < 0.05$).
The last parameter (right diagram) is the ratio between the deviation of
end of activity and onset of activity. This parameter also increases significantly when the field is switched on; in other words, with the field on,
deviation of onset of activity becomes relatively smaller and deviation of
end of activity becomes relatively larger.

As a consequence of the differences in the variance of onset of activity and of end of activity, there is a significant negative correlation between the duration of an activity time and that of the following rest time; this negative correlation is larger with the field in operation than without the field. Generally, however, there is no significant correlation between the duration of an activity time and that of the preceding rest time (Wever, 1969b).

A schematic representation of the activity rhythm with and without field, averaged from the results of all experiments, is given in the lower part of Figure 6. As in Figure 5, the period is standardized to the same duration for both cases. This diagram shows that activity time is longer with the field in operation than without it. In addition, averaged intra-individual "standard deviations" of onset of activity and of end of activity reveal that deviations of the onset are smaller than those of the end, and deviations with the field in operation are smaller than those without field.

The results reported above can be examined from two quite different aspects:

It is of general interest that a weak electric field, similar to a natural field of the earth's atmosphere, can influence human beings (Wever, 1970); an effect has been shown on activity and its free-running period, and on rectal temperature. This particular matter will not be discussed further because the mechanism of influence is still unknown; it cannot even be stated with certainty whether the influence of the field is direct or indirect (e.g., via ionization).

The stimulus—independent of its special quality—influences several parameters of circadian rhythms that are correlated to each other in a regular way. Because the mathematical model (Wever, 1965, 1966) also defines correlations between different rhythm parameters, comparison of data from humans with the model's predictions provides a test of applicability of the model. Since change in the τ value dependent on the state of the 10-cps field is significant at a higher level than changes in the other parameters, τ changes will be used for reference in the comparison with the model's predictions.

The model postulates a factor "x," representing external conditions, that controls the oscillation; all oscillation parameters depend on this controlling factor. Dependency of the period on x is illustrated in Figure 7, as computed from the model. The figure shows that the period is always negatively correlated with x; from this it follows that the switching on of the 10-cps field, which induces a period-shortening, corresponds to an increase in the x-value. Figure 7 also indicates that both the period value and the variability of the period under the influence of a standard noise

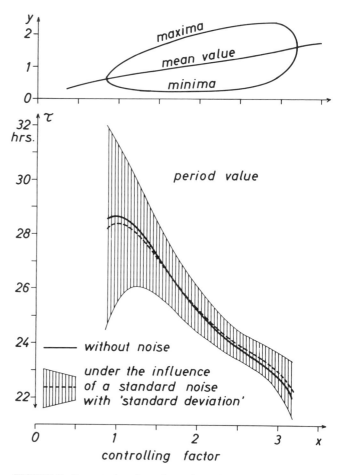

FIGURE 7 Computations from the mathematical model (Wever, 1965, 1966): dependency of the period on the controlling factor x, computed without and with an overlying standard noise. Upper diagram: maximum, minimum, and mean values of the oscillation, depending on the controlling factor x.

decrease with an increasing x-value. Only with very short periods does the variability become larger again. Feedback from the variability to the mean period value can also be recognized in Figure 7. Comparison with the upper diagram shows that the range presented includes the total oscillatory range within which the oscillation is self-sustained (Wever, 1965).

On the basis of the correlation between the state of the 10-cps field in the biological experiment and the x-value in the mathematical model, as

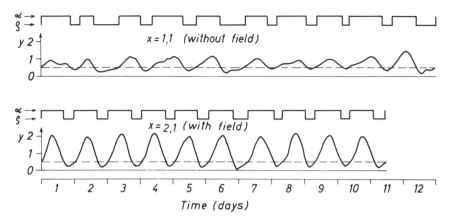

FIGURE 8 Computations from the mathematical model (Wever, 1965, 1966): model oscillations, computed with standard noise and two different x-values, simulating circadian rhythms without and with the 10-cps field in operation. With each x-value is given the original oscillation, with an additional threshold at y = 0.5, and a square wave, derived from the original oscillation by using only the crossovers with the threshold. The original oscillation simulates the rectal temperature rhythm, and the derived square wave simulates the activity rhythm.

derived in Figure 7, the dependency of all other rhythm parameters on the state of the field should be predictable. In Figure 8, two solutions of the model equation are given, the upper with a smaller x-value, corresponding to a circadian rhythm without the field, and the lower with a larger x-value, corresponding to a rhythm under the influence of the 10-cps field. The model oscillations are computed with equal background noise in each case. For both x-values, the original oscillation was drawn as a model for the temperature rhythm. In addition, a model for the activity rhythm was derived by introducing a threshold ("level-threshold hypothesis," Aschoff and Wever, 1962). From each oscillation, 10 periods are drawn. The two oscillations differ from each other with regard to the mean period value; this was the only point put into the computation. Additional parameters of the oscillation are clearly different in the two oscillations, for example, variability, amplitude, and oscillation shape. In contrast to the τ value, these differences are consequences, and not conditions, of the computation.

From the model oscillations, the same parameters can be determined as in the experiments with humans. Table 1 indicates the values of these parameters on the one hand as derived from the model oscillations with x = 1.1 and with x = 2.1 (compare Figure 8), and on the other hand as obtained from experiments with and without field (compare Figures 4, 5, and 6). As Table 1 reveals, the values of all parameters change in the same

TABLE 1   Values of Some Parameters of Circadian Rhythms, Calculated from a
Mathematical Model (Wever, 1965, 1966) or Obtained from Human Experiments

Parameter	Value of the Parameter (without → with 10-cps field)	
	Model	Men
Free-running period	28.45 → 24.95	26.35 → 25.00 hr
Standard deviation of the period	3.82 → 1.04	1.23 → 0.56 hr
Rectal temperature rhythm		
Mean value (level)	0.72 → 1.11	37.19 → 37.26°C
Range of oscillation	0.83 → 1.82	0.96 → 1.09°C
Form factor	0.60 → 0.85	0.86 → 1.03
Activity rhythm		
$a{:}\rho$ ratio	1.79 → 2.27	2.13 → 2.41
Precision	5.89 → 15.20	11.7 → 14.3
Ratio of deviations	1.35 → 1.62	1.22 → 1.56

direction in the model and in the human experiments. The absolute
amounts of all changes are larger in the model than in the biological experiment (model predictions resulting from oscillations with x = 1.6 and
x = 2.1 would show a closer quantitative agreement with the empirical
results, but these oscillations would not show differences as obvious as
those in Figure 8). All experimental results confirm the corresponding
model predictions; with this, the mathematical model is shown to be applicable to free-running human circadian rhythms. Moreover, these results
give the first direct proof of the validity of the level-threshold hypothesis.

All changes in the parameters of human rhythms included in Table 1
were shown to be significant with the exception of the change in precision
(compare Figure 6). In the results with the highest absolute precision
values, the precision is higher without field than with field, in contrast to
all other observations. However, this result may also agree with model
predictions. In contrast to all other parameters mentioned whose values
are negatively correlated with the period value in the total range, precision
is predicted to be not always unidirectionally correlated with the period.
Figure 7 shows that the variability of the period decreases with decreasing
period length within the greatest part of the oscillatory range, but that it
increases again when the period becomes shorter than about 24 hr. Therefore, with short periods that have, in the average, high precision values, the
correlation between period and precision is positive, in contrast to what
occurs with medium and with long periods.

In addition to those parameters summarized in Table 1, some further properties of circadian rhythms can be compared with corresponding model predictions. One of these concerns the coupling between the activity rhythm and the temperature rhythm, which can be measured as the inverse value of the tendency toward internal desynchronization (Aschoff *et al.*, 1967). In the model, the internal coupling between different oscillations depends on the ranges of the oscillations; the larger the ranges, the stronger the coupling. Figure 8 and Table 1 show that the range of oscillation is larger when a 10-cps field is present; therefore, the tendency toward internal desynchronization can be predicted to decrease when the 10-cps field is switched on.

In experiments with humans, real internal desynchronization has never been observed with the field in operation but has been observed many times without the field; this difference is statistically significant (Wever, 1968a). The following figures illustrate the direct effect of the 10-cps field on internal coupling. Figure 9 shows the results of an experiment in which

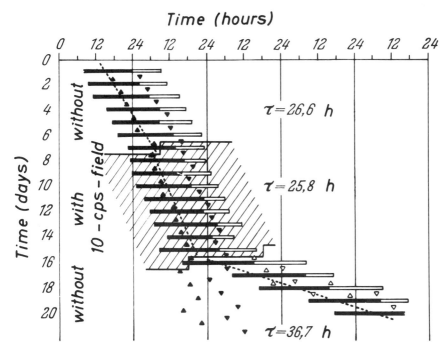

FIGURE 9   Free-running rhythms in a human subject: without field during the first and third sections of data, and under the influence of a 10-cps field during the second section. Internal desynchronization during the third section. White triangles: temporally correct repetitions of corresponding black triangles. Other designations as in Figure 1. (From Wever, 1968a)

conditions were changed twice. Comparison of the first and second sections of data shows once more that the effect of the field is to shorten the period; in the third section (without field), τ lengthens again and internal desynchronization occurs. Figure 10 shows the results of an experiment in which internal desynchronization occurred from the beginning of the experiment, in this case with a remarkable shortening of the activity period. At subjective day 21 (objective day 17) the field was switched on, and on that day internal desynchronization ended. These results confirm the prediction concerning the internal interaction in humans between different functions assigned to different oscillators.

The final properties to be considered concern the influence of a periodically changing field. On the one hand, the field can be switched on and off

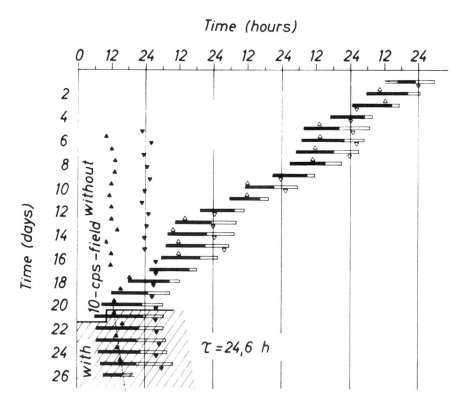

FIGURE 10   Free-running rhythms in a human subject: without field during the first section of data, and under the influence of a 10-cps field during the second section. Internal desynchronization during the first section. Designations as in Figures 1 and 9.

externally, thus operating as a *Zeitgeber* (Figure 3, third section). In 10 experiments with a field-*Zeitgeber*, the human circadian rhythm was entrained for at least some days (Figure 3); the change in phase-angle difference between the rhythm and the field-*Zeitgeber* was consistent with results of experiments with humans under the influence of an artificial light-dark *Zeitgeber* (Aschoff *et al.*, 1969) and with model predictions (Wever, 1969b).

On the other hand, a periodically changing field can be switched on and off "internally," depending on the activity of the subject. This self-controlled change in the external conditions does not alter the autonomous state of the rhythm, but the model predicts that it will have a remarkable influence on the rhythm; that is, to lengthen the period and to reverse the correlation between the free-running period and the strength of the stimulus. With light as the controlling stimulus, these predictions have been confirmed for finches (Aschoff *et al.*, 1968) and for humans (Wever, 1969a). In two experiments with humans, the 10-cps field was switched on only when the subjects were active, automatically and without their knowledge. Here, τ was longer during the portion of the experiment when change in the field was controlled by activity than during the portions without any field. This effect is statistically different from the shortening effect of a constant field (Wever, 1968c). This finding confirms the prediction of the model concerning a self-controlled change of the external conditions.

In summary, investigations of human circadian rhythms under the influence of a sufficient stimulus (10-cps electric field) have confirmed the predictions of a special mathematical model (Wever, 1965, 1966). The model was tested and demonstrated to be applicable to changes of rhythm parameters under both constant and periodically changing conditions, and to the properties of a single rhythmic function as well as to interactions among different functions. Because the differential equation proposed by the model is the simplest possible expression for an oscillator that is both self-sustained and externally driven (subject to both parametric and non-parametric excitation; Wever, 1966), these empirical results mean nothing more than that the human circadian rhythm, when under the influence of a weak electric field, behaves like such a simple oscillator, or like a variety of such oscillators when a variety of biological functions is considered. Additional complications or modifications of the model would not really add to the model's usefulness, in explaining all the behaviors observed. On the other hand, the model discloses properties of the oscillation that are easily overlooked without it. Considered simply, all the results that have been discussed may seem to be independent of each other, but, in fact, a sufficient model demonstrates that they are logically and consistently related.

REFERENCES

Aschoff, J. 1960. Exogenous and endogenous components in circadian rhythms. Cold Spring Harbor Symp. Quant. Biol. 25:11–27.

Aschoff, J. 1964. Die Tagesperiodik licht- und dunkelaktiver Tiere. Rev. Suisse Zool. 71:528–558.

Aschoff, J., U. Gerecke, and R. Wever. 1967. Desynchronization of human circadian rhythms. Jap. J. Physiol. 17:450–457.

Aschoff, J., E. Pöppel, and R. Wever. 1969. Circadiane Periodik des Menschen unter dem Einfluss von Licht-Dunkel-Wechseln unterschiedlicher Periode. Pfluegers Arch. 306:58–70.

Aschoff, J., U. von St. Paul, and R. Wever. 1968. Circadiane Periodik von Finkenvögeln unter dem Einfluss eines selbstgewählten Licht-Dunkel-Wechsels. Z. Vgl. Physiol. 58:305–321.

Aschoff, J., and R. Wever. 1962. Aktivitätsmenge und $a:\rho$-Verhältnis als Messgrössen der Tagesperiodik. Z. Vgl. Physiol. 46:88–101.

Wever, R. 1965. A mathematical model for circadian rhythms, p. 47 to 63. In J. Aschoff [ed.], Circadian clocks. North-Holland Publ. Co., Amsterdam.

Wever, R. 1966. Ein mathematisches Modell für die circadiane Periodik. Z. Angew. Math. Mech. 46:T148–T157.

Wever, R. 1967. Über die Beeinflussung der circadianen Periodik des Menschen durch schwache elektromagnetische Felder. Z. Vgl. Physiol. 56:111–128.

Wever, R. 1968a. Einfluss schwacher elektro-magnetischer Felder auf die circadiane Periodik des Menschen. Naturwissenschaften 55:29–32.

Wever, R. 1968b. Mathematical models of circadian rhythms and their applicability to men, p. 61 to 72. In J. de Ajuriaguerra [ed.], Cycles biologiques et psychiatrie. George & Cie., Genève; Masson & Cie., Paris.

Wever, R. 1968c. Gesetzmässigkeiten der circadianen Periodik des Menschen, geprüft an der Wirkung eines schwachen elektrischen Wechselfeldes. Pfluegers Arch. 302: 97–122.

Wever, R. 1969a. Autonome circadiane Periodik des Menschen unter dem Einfluss verschiedener Beleuchtungs-Bedingungen. Pfluegers Arch. 306:71–91.

Wever, R. 1969b. Untersuchungen zur circadianen Periodik des Menschen, mit besonderer Berücksichtigung des Einflusses schwacher elektrischer Wechselfelder. Bundesminist. Wiss. Forsch., Forschungsber. W 69–31, Weltraumforsch. 212 p.

Wever, R. 1970. The effects of electric fields on circadian rhythmicity in man. Life Sciences and Space Research VIII, p. 177–187, North-Holland Publ. Co. 12th Meet.

DISCUSSION

HASTINGS: Have you attempted to repeat these experiments with any subjects other than humans?

WEVER: We have had a few experiments with the green finch, in which the results were about the same as with humans. The free-running period is shorter with the field in operation than without it.

HOSHIZAKI: Would you please describe the experimental set-up?

WEVER: The subjects lived in an underground bunker, in which there were two apartments. One of the apartments was shielded against external electromagnetic fields and had facilities for introducing any electric or magnetic a-c or d-c field into the room. The electrodes and the coils for introducing fields were set into the wall and were invisible to the subject. The weak field used in the experiments could not be felt by the subject. The voltage gradient in these experiments was about 2 volts per meter.

HAYES: How does the magnitude of the field that you apply to the subjects compare to the magnitude of change one would experience in moving from place to place on earth, or up and down in an airplane?

WEVER: There are 10-cps fields in nature, but they are a thousand times weaker than our artificial field.

GORDON: Is the shortening of τ frequency-dependent?

WEVER: That has not yet been tested.

HALBERG: I believe König, in Munich, has shown that reaction time in humans changes with the application of a 10-cps electric field.

WEVER: He has shown a response to an artificial 10-cps field but he has no significant results as yet. Moreover, his subjects were not shielded against additional natural fields, and he could not absolutely exclude the possibility that his subjects knew about the application of the field. This was different in my experiments.

CROWLEY: Do you have any idea how the electric field is sensed?

WEVER: No, our first problem was just to find a suitable stimulus.

HASTINGS: The electric field did act as a *Zeitgeber* for a few days, you said. At what period length did you attempt to synchronize them?

WEVER: I used a driving period of 23.5 hr (comp. Fig. 3). When the *Zeitgeber* period is nearer the free-running period, it takes too long to determine if entrainment has occurred.

KLAUS HOFFMANN

Splitting of the Circadian Rhythm as a Function of Light Intensity

When locomotor activity of an animal is recorded in constant conditions one of the following results may ensue: Either there is no rhythm discernible (or it soon damps out) and apparent arrythmicity results, or a circadian rhythm can be discerned. In the latter case, the rhythm may have one or several peaks, but usually all components show the same frequency. Predictions based on the assumption that the cycle underlying the circadian rhythm of locomotor activity can be considered one self-sustained oscillator (with one degree of freedom) have been experimentally verified (Aschoff, 1964, 1965, this volume; Hoffmann, 1969).

However, Pittendrigh (1960, 1961, 1967) and Swade (1963) have reported a number of cases in nocturnal rodents in which, after prolonged constant illumination, the rhythm of locomotor activity split into two components. These components showed, at least for a time, distinctly different frequencies. Eventually they resynchronized with each other at a new phase relation. Such findings indicate that several circadian oscillators control the overt rhythm of locomotor activity. Normally they are entrained with each other and show the same frequency, but in some cases they may dissociate, at least temporarily. In this paper, evidence will be presented for the occurrence of two components in the rhythm of activity in a light-active mammal.

MATERIAL AND METHODS

In the experiments reported here, the species studied was the tree-shrew, *Tupaia belangeri*, of the family Tupaiidae. There is some disagreement on nomenclature, and in many studies the names *T. glis* or *T. tana* have been used for this species (compare Martin, 1968; v. Holst, 1969; Hoffmann, 1970). Most taxonomists place the Tupaiidae at the bottom of the Order of primates (e.g., Simpson, 1945; Le Gros Clark, 1959), but their inclusion in the primates has been recently questioned (Martin, 1968). These squirrel-like mammals, which live in tropical Southeast Asia, are strictly diurnal (Martin, 1968; Aschoff *et al.*, 1970) and possess pure cone retinae (Samorajski *et al.*, 1966).

The running-wheel activity of 15 animals was recorded over extended lengths of time by event recorders and by print-out counters. Each animal was isolated in a light- and sound-proofed box. Temperature, humidity, and light intensity were kept constant. Light intensity, however, was abruptly changed at intervals ranging from 20 to 195 days. For further details of method see Wever (1967). The animals were kept in these conditions for at least 3 months, and three of them for more than 2½ years.

RESULTS

When a Tupaia was placed in constant illumination of medium or high intensity its locomotor activity always showed a clearcut circadian rhythm. In all cases except one (see Figure 2), the rhythm was monophasic at the beginning of the experiment (see Figures 1 and 4). Indications of a double-peaked pattern could be found (e.g., Figure 4, left actogram), but always one activity time could clearly be distinguished from one rest time. If, however, the light intensity was decreased below a certain value, the rhythm began to split into two components, which for some time showed different frequencies. Figure 1 gives two examples (compare also Figures 4 and 7). The actograms do not suggest the separation of two peaks of a bimodal activity pattern, but rather the separation of two oscillators that have been in phase, and then shift towards a new phase relationship. Eventually the two components become parallel, with identical frequencies.

Such splitting into two components occurred in all cases if the illumination was reduced sufficiently. Figure 2 shows for all experiments the steps in light intensity leading to splitting. There was some variation in the light intensity at which splitting occurred. Generally this happened when the illumination was reduced to about 1 lux; in some cases, however, splitting

was induced at a higher light intensity, in others only by a further reduction in illumination. Figure 2 indicates that it is not the size of the step-
down in light intensity that initiates splitting, but rather the reduction of
illumination below a certain level. When the light intensity was further
reduced the double pattern persisted in all cases (see Figures 1 and 6),
although the phase relation of the two components, as well as their even-
tual common frequency, could be slightly changed.

In nearly all cases the end points of the two components were the most
distinct and stable points of the activity pattern. They were therefore used
to determine frequency and phase. Figure 3 tabulates the phase angle differences between the two components, as measured from the end of the
first part to the end of the second part, for all cases in which the two parts
of the pattern ran parallel and showed the same frequency. Not all possible phase angles between the two components were observed, the data
cluster around 180° (mean value = 183°). This indicates that below a certain level of light intensity a preferred stable phase relation of about 180°

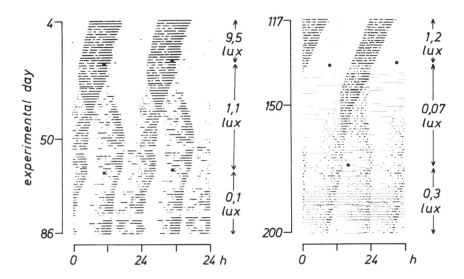

FIGURE 1 Splitting of the circadian rhythm of locomotor activity into two components after
abrupt reduction of the otherwise constant illumination in two Tupaias. Running wheel activity
was registered by an event recorder. Activity is indicated by the vertical pen markings on the line,
which can fuse to give a band of solid activity; successive days are plotted one below the other. To
facilitate visual following of the different components of the rhythm the total record, or a part of
it, has been reproduced on the right, displaced upwards one day. The black dots indicate the times
at which light intensity was changed. At the left the number of days under experimental conditions
is given, at the right the intensity of constant illumination is indicated. (From Hoffmann, 1970)

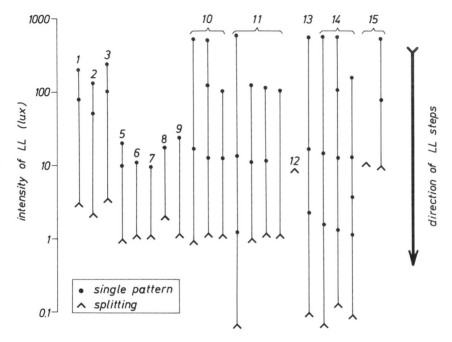

FIGURE 2   Schematic summary of the experimental step-downs in light intensity leading to splitting for all experiments. A black circle indicates that in this intensity the single pattern persisted, a fork means that in this intensity the rhythm began to split as shown in Figure 1. The arrow at the right indicates the direction of the experimental steps. The numbers on top of the graph are animal identification numbers. Animal 4 showed a double pattern immediately at the beginning that persisted to the end of the experiment and is not included.

exists between the two parts. In most cases splitting did not take place immediately after the reduction of light intensity (see Figures 1, 4, and 7). The first indications of splitting were visible only after an average of 11 days (range: 0–30 days).

For six animals an attempt was made to rejoin the two components by raising the light intensity. Two examples of these experiments are given in Figure 4. The animals first showed a circadian rhythm of locomotor activity with a single pattern. A reduction of light intensity to about 1 lux caused the rhythm to split. The first indications of splitting could be discerned only several days after the illumination was reduced. The two components at first showed slightly different frequencies, but finally they ran parallel, with the same frequency. When the light intensity was raised again to the preceding value, the double pattern persisted, although with some modifications. Only after a further increase of illumination were the

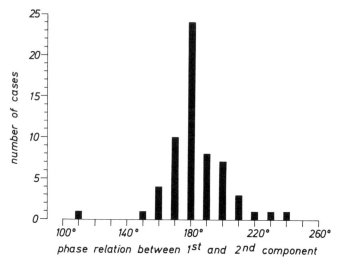

FIGURE 3   Histogram of the phase angle differences between first and
second component of the activity pattern after splitting for all cases in
which the two parts run parallel. The period length has been normalized
to $\tau = 360°$. The endpoints of the two components were used as the
phase reference points.

two components fused to a single pattern again. In these two examples the
two parts fuse inversely to the course of splitting: the first component
joins the second from the rear. This was not a general rule, however; in
the majority of experiments the two parts united in the same sequence in
which they split.

   In four of the six animals the attempt to rejoin the two components of
the activity cycle by raising the light intensity was successful. Figure 5 indicates the steps in light intensity leading to fusion. Junction of the two
parts was achieved only at light intensities of about 100 lux or above in 9
out of 10 attempts, in the remaining case the rhythm remained split at
560 lux. In the other two animals the double pattern persisted even when
the light intensity was raised to between 500 and 600 lux. These results
show that a much higher light intensity is necessary to reform the single
pattern than to maintain it (*cf.* Figure 2). Apparently, a carryover or after-
effect of the previous conditions exists: the pattern shows hysteresis. The
hysteresis is also evident in Figure 6, which indicates the total course of
the experiment for the three animals whose activity was recorded continuously for more than 33 months. During this time they were never exposed
to *Zeitgeber* conditions. The single pattern of activity persisted whenever

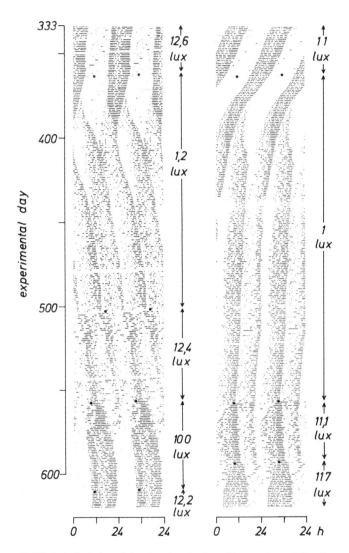

FIGURE 4 Splitting of the circadian rhythm of locomotor activity of
two Tupaias into two components and rejoining into one component, depending on the intensity of previous and present illumination. For further explanation see Figure 1. (From Hoffman, 1970.)

the light intensity was increased. If the illumination was reduced below a certain level, the rhythm split into two components. A further reduction in light intensity did not essentially change this pattern. When the light intensity was again increased fusion took place, but only at intensities well above those at which a single pattern had persisted when this level was approached by a decrease in illumination.

In most experiments only the activity patterns described above were observed. However, in some actograms disintegration into even more circadian components was indicated. An example is given in Figure 7 where,

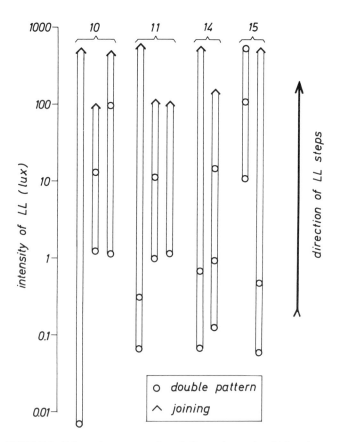

FIGURE 5   Schematic representation of all experiments in which an increase in light intensity resulted in a fusion of the two components into one. The open circles indicate that at this light intensity the pattern persisted; the arrowheads indicate that fusion occurred at the intensities indicated. For further explanations and for comparison see Figure 2.

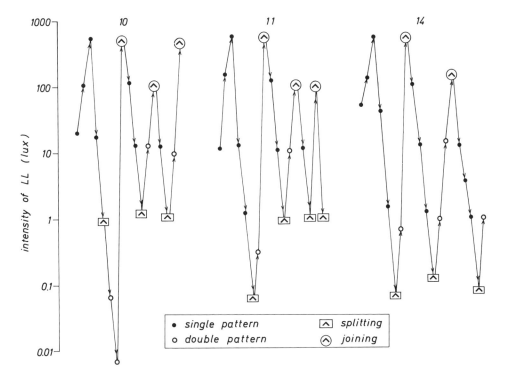

FIGURE 6   Schematic representation of the course of steps in light intensity and their results in three animals that were kept in the experiment for more than 33 months. The direction of the experimental steps is indicated by the small arrows. The effects of these steps on the pattern of the rhythm of locomotor activity are indicated by the symbols given in the key.

after the light intensity was increased to 106 lux and before the two major components fused to a single pattern, there is a suggestion of a third component with a period distinctly longer than that of the two major parts. Indications of more than two circadian components with frequencies different from the two major parts can also be found in the actogram in Figure 8 after the light intensity has been elevated to 98 lux. Later these additional components disappear and only the two major parts persist. Indications of such further disintegration of the circadian rhythm of locomotor activity were found especially in higher light intensities if the double pattern still persisted under these conditions. These instances were rare, however, and the additional components were unstable. The only stable and persistent features in all the experiments were the two major components in the rhythm of locomotor activity, which originated from a single pattern by

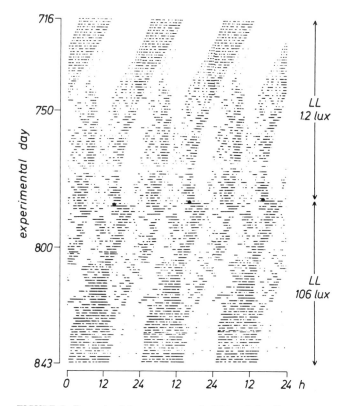

FIGURE 7 Example of the appearance of additional circadian components with different periods in the activity rhythm of a Tupaia. Note that
after the increase of light intensity from 1.2 to 106 lux there is a suggestion of a third component with a longer period passing the other two. For
further explanation see Figure 1.

splitting if the illumination was sufficiently reduced, and could fuse again
into a single pattern, at least in many cases, if the light intensity was raised
adequately.

DISCUSSION

The findings reported here strongly suggest that the rhythm of locomotor
activity in *Tupaias* is controlled by two coupled oscillators (or two groups
of oscillators), which may have two stable phase relations. They are either
in phase, creating a single pattern of activity, or about 180° out of phase,

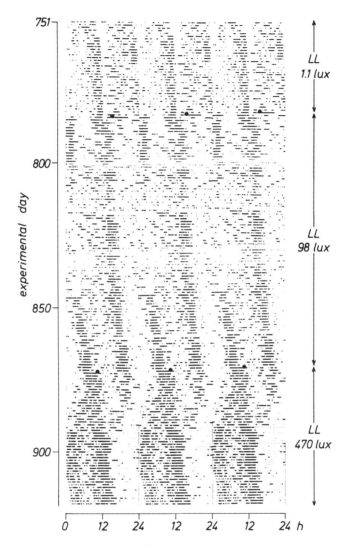

FIGURE 8   Example of the appearance of additional circadian components in the activity rhythm of a Tupaia. Note that after the step-up in
light intensity from 1.1 to 98 lux small bands of activity seem to split off
from one component and run through the pattern with different frequencies. Finally these components are fused with the two major parts.
For further explanation see Figure 1.

causing a double pattern. Lowering the intensity of illumination below a certain value initiates a shift of phase that results in the transition from one stable phase relation to the other. This shift is gradual and the final phase relation is reached only after many transients. The process may be reversed by exposing the animals to high light intensities. The occasional occurrence of still more circadian components with different frequencies under these conditions suggests that the two oscillations themselves consist of several circadian oscillators.

In general, the results support the view that a multioscillator system underlies the overt circadian periodicity, even of a single function and even if it appears as a single self-sustained oscillation, and that these oscillators are themselves circadian. They also indicate that one-oscillator models are not suitable for analyzing the nature of the basic clock, although they have been very useful in describing overall rhythmic behavior, and in reducing some aspects of behavior in the field to simpler circadian laws (e.g., Aschoff, 1964, 1965, 1969, 1970; Wever, 1965, 1967; Hoffmann, 1969).

The facts (1) that both the stable phase relation between the two circadian components and the change in phase relation depend on the previous and the present light intensity and (2) that several days usually elapsed before the first signs of a split could be observed, suggest that endocrine influences that can be modified by light are involved in the internal synchronization of the circadian oscillations. Light effects on endocrine systems are well known (e.g., Wurtman, 1967; Rensing, 1969). The experiments reported here also suggest that the apparent arrhythmicity of locomotor activity found in many animals under certain conditions of constant light are the result of internal desynchronization.

In view of the work of Gaston and Menaker (1968) and Gaston (this volume) on the effect of pinealectomy on the circadian rhythm in birds one might speculate that the pineal is involved in internal synchronization and coupling of the several circadian oscillators underlying the rhythm of locomotor activity in *Tupaia*. Effects of light on pineal function have been described in mammals (e.g., Quay, 1965; Wurtman, 1967; Wurtman, Axelrod, and Kelly, 1968). It must be noted, however, that in the rat no effect of pinealectomy on the circadian rhythm of locomotor activity has been reported (Quay, 1965; Richter, 1967).

ACKNOWLEDGMENTS

The Tupaias used in these studies were provided by I. Eibl-Eibesfeldt, Seewiesen, and D. v. Holst, Munich, to whom I wish to express my gratitude. I also gratefully acknowledge the untiring assistance and cooperation of N. J. Henderson in the evaluation of the data.

REFERENCES

Aschoff, J. 1964. Biologische Periodik als selbsterregte Schwingung. Arbeitsgem. Forsch. Landes Nordrhein-Westfalen, H. 138:51–79.
Aschoff, J. 1965. The phase-angle difference in circadian periodicity, p. 262 to 276. *In* J. Aschoff [ed.], Circadian clocks. North-Holland Publ. Co., Amsterdam.
Aschoff, J. 1967. Circadian rhythms in birds. Proc. XIV Int. Ornithol. Congr., Oxford, p. 81–105.
Aschoff, J. 1969. Phasenlage der Tagesperiodik in Abhängigkeit von Jahreszeit und Breitengrad. Oecologia (Berl.) 3:125–165.
Aschoff, J., E. Gwinner, A. Kureck, and K. Müller. 1970. Diurnal rhythms of chaffinches (*Fringilla coelebs* L.), tupaias (*Tupaia glis* L.) and hamsters (*Mesocricetus auratus* L.) at the Arctic Circle. Oikos Suppl. 13, 91–100.
Gaston, S., and M. Menaker. 1968. Pineal function: The biological clock in the sparrow? Science 160:1125–1127.
Hoffmann, K. 1969. Zum Einfluss der Zeitgeberstärke auf die Phasenlage der synchronisierten circadianen Periodik. Z. Vgl. Physiol. 62:93–110.
Hoffmann, K. 1970. Circadiane Periodik bei Tupajas (*Tupaia glis*) in konstanten Bedingungen. Zool. Anz. Suppl. 33:171–177.
Holst, D. von. 1969. Sozialer Stress bei Tupajas (*Tupaia belangeri*). Z. Vgl. Physiol. 61: 1–58.
Le Gros Clark, W. E. 1959. The antecedents of man. Edinburgh Univ. Press, Edinburgh.
Martin, R. D. 1968. Reproduction and ontogeny in tree-shrews (*Tupaia belangeri*), with reference to their general behaviour and taxonomic relationships. Z. Tierpsychol. 25:409–495, 505–532.
Pittendrigh, C. S. 1960. Circadian rhythms and circadian organization. Cold Spring Harbor Symp. Quant. Biol. 25:159–184.
Pittendrigh, C. S. 1961. On the temporal organization of living systems. Harvey Lect. Ser. 56:93–125.
Pittendrigh, C. S. 1967. Circadian rhythms, space research and manned space flight, Life sciences and space research V. North-Holland Publ. Co., Amsterdam. pp. 122–134.
Quay, W. B. 1965. Photic relations and experimental dissociation of circadian rhythms in pineal composition and running activity in rats. Photochem. Photobiol. 4:425–432.
Rensing, L. 1969. Zur Ontogenese und hormonellen Steuerung circadianer Rhythmen. Nachr. Akad. Wiss. Goettingen, II. Mathem.-Physikal. Klasse. Jahrgang 1969, Nr. 8, pp. 57–70.
Richter, C. P. 1967. Sleep and activity: their relation to the 24-hour clock. Reprinted from "Sleep and altered states of consciousness." Association for Research in Nervous and Mental Disease, Vol. XLV, pp. 8–29. Baltimore. The Williams & Wilkins Company.
Samorajski, F., J. M. Ordy, and J. R. Keefe. 1966. Structural organization of the retina in the tree shrew (*Tupaia glis*). J. Cell Biol. 28:489–504.
Simpson, G. G. 1945. The principles of classification and a classification of mammals. Bull. Am. Mus. Nat. Hist. 85:1–350.
Swade, R. H. 1963. Circadian rhythms in the arctic. Ph.D. Thesis Princeton University. University Microfilms Library Services, Ann Arbor, Michigan 48106, 1967, 64/9145.

Wever, R. 1965. A mathematical model for circadian rhythms, p. 47 to 63. *In* J. Aschoff [ed.], Circadian clocks. North-Holland Publ. Co., Amsterdam.

Wever, R. 1967. Zum Einfluss der Dämmerung auf die circadiane Periodik. Z. Vgl. Physiol. 55:255–277.

Wurtman, R. J. 1967. Effects of light and visual stimuli on endocrine function, p. 20 to 59. *In* L. Martini and W. F. Ganong [eds.], Neuroendocrinology. Vol. II. Academic Press, New York.

Wurtman, R. J., J. Axelrod, and D. E. Kelly. 1968. The pineal. Academic Press, New York.

DISCUSSION

LICKEY: Is there any phase-splitting evident under *L*D conditions?

HOFFMANN: I have never seen any. As far as I know, the only published accounts of clear-cut splitting have been the 10 or 12 cases from Dr. Pittendrigh's lab, and the experiments of Dr. Swade. So while splitting occurs quite regularly in *Tupaia*, it is apparently not too common elsewhere. I have never seen it in birds or lizards, even in constant conditions, and in *L*D I have never seen it at all.

DeCOURSEY: It appears that *Tupaia* rhythms are peculiarly susceptible to splitting. Have you tried systematicaliy to split the rhythm in any other animal?

HOFFMANN: No, we haven't.

PITTENDRIGH: There is one difference between the *Tupaia* data and the results I have for hamsters and two species of *Peromyscus*. In the data Dr. Hoffmann presented, the split showed up after only about 11 days; in the nocturnal mammals, it is as long as two months before the split becomes clear. One reason this phenomenon has not been noted before may be that not many people have put animals in constant light for two months and left them there without doing something else to them. As a

matter of fact, we discovered splitting when I set up an experiment involving 50 hamsters one summer, and went away to fish. The instructions for the experiment got lost, the animals free-ran all summer in LL, and the rhythms split. So if I hadn't gone fishing and left everything alone for two months, we wouldn't have found the phenomenon.

HOFFMANN: I have looked at lizards after two months in constant conditions, and have never seen it there.

MENAKER: Those are certainly beautiful records you've shown us. I am somewhat puzzled by the fact that it is a decrease in light intensity that causes splitting. Somehow I would have expected it to be the other way around. Also, what happens in DD, or can't you keep them in DD?

HOFFMANN: In answer to your first question, Pittendrigh found in nocturnal rodents a splitting in constant light—I don't know of what intensity— and I am not so surprised that in diurnal animals it is decreased light intensity that leads to splitting. If the rhythm is split it remains so when the light intensity is further reduced (compare Figures 1 and 6). We have not tried keeping them in real DD.

PITTENDRIGH: In the cases I know of in nocturnal animals—hamsters and *Peromyscus*—when the components separate, the component that begins the normal activity cycle has a longer free-running period than the component that belongs to the end of the activity cycle. Now I am not quite sure, but it seemed from your figures that the same thing was true for your animals, and I am a little surprised that that should be the case for both nocturnal and diurnal forms.

HOFFMANN: My impression from the actograms is that splitting in *Tupaia* is not the separation of the two peaks of a bimodal pattern, but rather that before the split the two components have nearly the same phase; i.e., the beginnings as well as the ends of the first and second components coincide. After the split only the ends of the two components are distinct, while their onsets are somewhat blurred. During splitting, as long as the two components overlap, the beginnings of the second component are still well marked, and extrapolate to the beginning of activity before splitting. The beginnings of the first component can also be extrapolated to the beginning of activity before the split, though the picture is less clear here because of the blurring of onsets. This holds for nearly all the cases of splitting in *Tupaia* I have seen (see Figures 1, 4, and 7). When the two components fuse again under high light intensity, however, the picture is far more complex and less uniform.

PAVLIDIS: How does the duration of activity in the single rhythm compare to the sum of the durations in the split rhythm? Depending on their relationship, one could support alternative

hypotheses about the nature of splitting.

HOFFMANN: It is difficult to establish that, really, because of the diffuse onsets of activity in the split rhythm. I have made no attempt to calculate it quantitatively, but my general impression is that activity time of each component corresponds roughly to that which was shown before splitting, so that total activity time now seems to be nearly twice as long as before splitting. Intensity of activity per unit time, however, is greatly reduced.

EHRET: We have noted a phenomenon possibly related to rhythm-splitting in free-living populations of *Tetrahymena* cells in continuous cultures. Ordinarily we can use either light (for example, in a "switch-down" of several thousand lux from LL to DD), or food (asymmetric feeding) to synchronize the population, as it moves from the ultradian to the infradian mode of growth. The resulting free-running circadian rhythm of cell division has a τ of about 21 hr under our conditions. Dr. Wille and I were experimenting once with *simultaneous* asymmetric feeding synchronization and light switch-down synchronization and were surprised to find a rhythm with roughly an 11-hr period, which was clearly two circadian populations out of phase with one another. Unfortunately the cultures didn't last long enough for us to decide whether the two sets had significantly different periods. I wonder if anyone working with *Gonyaulax* or *Euglena* has any comment.

HASTINGS: In *Gonyaulax* there occurs a second peak $180°$ out of phase with the first after chloramphenicol is added to the medium. But I don't believe the speculation in that case is that there are two populations out of phase.

ASCHOFF: There is great appeal in considering the rhythm-splitting Dr. Hoffmann has described as evidence for a two-oscillator rhythmic system. But as he pointed out, the splitting phenomenon is quite different from the occurrence of two peaks 180° out of phase.

PITTENDRIGH: That is the point I wanted to make, but I think it can be stated even more strongly. It is clear from several comments that the full force of Dr. Hoffmann's observations has not been appreciated; that is, that there is a difference in free-running frequencies of the two parts. You have not just the separation of a band of activity into two components, but two components running with quite distinct frequencies for a while.

COMMENT

A Split Activity Rhythm under Fluctuating Light Cycles

RICHARD H. SWADE

The accompanying figure shows the record of a deer-mouse having a split activity rhythm with two noteworthy characteristics: relative independence of each part in responding to different levels of illumination, and feedback between the parts.

Before December 26, 1961, the animal had been entrained to a light cycle like that described in the figure legend, except that the maximum intensity of illumination was about 7 lux and the minimum was dark (<< 0.02 lux). The animal showed two peaks of activity in each 24-hr period: one near dusk (the early peak) and the other about 7 hr later (the late peak) in the middle of the low illuminance section of the light cycle. On December 26 the maximum and minimum illumination was increased (see figure legend) and entrainment of the early peak at dusk was lost. The two peaks free-ran in parallel until the late peak encountered dawn (A), when entrainment occurred. However, the early peak continued to free-run for about 6 days (B), whereupon its period

gradually became shorter, until entrainment occurred 4 days later (C).

On February 12 (D) the minimum illumination was again increased. The early peak lost entrainment, while the late peak retained entrainment for 13 days. There was a phase shift of the late peak of about 10 hr—across the photofraction (E). The early peak makes a similar jump across the photofraction from 2 to 6 days later. In the meantime the late peak develops a few periods shorter than 24 hr until the early peak makes its jump across the photofraction. The peaks then show "oscillatory free-runs" (Swade and Pittendrigh, 1967; Swade, 1969) or "relative coordination," and the phenomena described above recur in each cycle of the oscillatory free-run.

In these subsequent cycles the late peak may show some activity both before dawn and after dusk (G). This is probably a further splitting of the activity rhythm, since the periods of the two parts are very different, and there is a variable amount of time between the two parts. The part

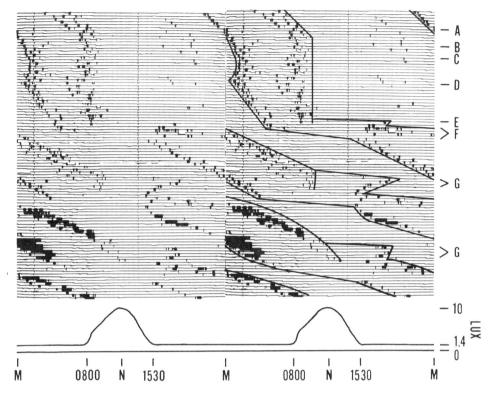

FIGURE 1 Running-wheel activity of *Peromyscus maniculatus*, No. 67, from January 18–April 26, 1962. Light regimen: gradually fluctuating light cycle with a photofraction extending from about 0800 to 1530 and peaking at about 1200 hr (see the representative light cycle plotted at the bottom of the figure). Before D (February 12) the maximum illumination was *ca.* 10 lux at noon and 0.8 lux between 1530 and 0800 hr. After that date the minimum illumination was increased to *ca.* 1.4 lux. M, midnight; N, noon.

near dusk remains shorter than 24 hr until the dawn part disappears or recombines with the dusk part.

Two important points are to be made:

• Within limits each peak of activity can react independently according to the light intensity of the part of the cycle in which it falls.

• There is a feedback between the peaks, such that if the number of hours between them becomes too great or too little (due to the first point above), there

is a phase shift of one peak until a more stable phase angle is attained.

REFERENCES

Swade, Richard H. 1969. Circadian rhythms in fluctuating light cycles: toward a new model of entrainment. J. Theor. Biol. 24:227–39.

Swade, R. H., and C. S. Pittendrigh. 1967. Circadian locomotor rhythms of rodents in the arctic. Am. Nat. 101:431–66.

COMMENT

ARTHUR T. WINFREE

Dr. Hoffmann has surprised us again with the intriguingly complex behavior of a primate's activity rhythm in LL. Drs. Swade and Pittendrigh drew our attention to similar observations on rodents. Let me try to reinforce the argument that these phenomena may indicate a multioscillator basis for circadian rhythmicity, by reporting similar observations on the behavior of 71 nearly identical electronic oscillators, each weakly interacting with all the others.

In a project supported by Dr. Pittendrigh in 1965, I constructed gadgetry by which each oscillator made a tiny black mark on photographic film each time it completed a cycle of about 24 time units (call them "hours"). The marks fall in the familiar format, the events of successive 24-"hour" days being double-plotted horizontally, days stacked one atop the next for several hundred days (Figure 1).

With the constituent oscillators uncoupled from each other, the population showed no rhythmicity, but above a modest level of mutual interaction, mutual synchronization was quickly established: part A of Figure 1 shows most of the black marks occurring simultaneously every 23.6 "hours." But at the arrow, something happens, quite spontaneously and unpredictably, resulting in 20 days of splitting transients in part B. In part C the population has segregated into two stable groups, mutually entrained at a new period, but not synchronous: there are two events per day.

If the intensity of interaction is now reduced, but not below the minimum required for mutual synchronization, then

after 10–100 cycles these two groups suddenly recombine.

Although Figure 1 is too compressed to show it clearly, there are other groups of oscillators, flitting back and forth between the two big groups, with periods ranging ± 1 "hour" about the dominant period. These show up as little tracks of dots slanting across the graph from one day to the next, often vanishing after a week or so.

The period of the dominant rhythm occasionally changes abruptly, and apparently spontaneously, by a fraction of an hour; this may represent capture of one or more of the stray "minority groups."

The ephemeral coherence of "minority groups" with a different period, and the spontaneous changes of period are not yet understood, and although I don't believe it, I cannot now exclude the possibility that they are peculiarities of my model rather than characteristic features of multioscillator systems.

The existence of a threshold intensity of coupling for mutual synchronization, and of a higher threshold for transition to the two-event mode have been analyzed mathematically and found to be characteristic of a wide variety of oscillator populations. Could the influence of LL on circadian clocks heighten their mutual influence or sensitivity?

BIBLIOGRAPHY

Winfree, A. T. 1965. Bachelor's thesis, Dept. of Engineering Physics, Cornell Univ.

Winfree, A. T. 1967. Biological rhythms and the behavior of populations of coupled oscillators. J. Theor. Biol. 16: 15–42.

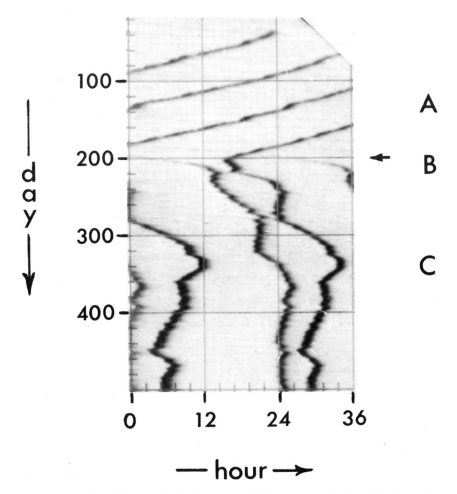

FIGURE 1   500 "days" of double-plotted aggregate rhythm from a population of 71 electronic oscillators with free-running periods 23–25 "hours." Mutual synchronization is maintained by interactions of each with all the others: A, single main coherent group of oscillators, period 23.6 "hours." B, spontaneous splitting transients, lasting 20 "days." C, Two coherent main groups of oscillators, period 24.0 "hours."

GÖRAN WAHLSTRÖM

The Internal Clock of the Canary: Experiments with Self-Selection of Light and Darkness

Results of experiments with self-selection of light and darkness, in which the circadian rhythm of activity in the canary was influenced by an enforced period of darkness (EPD) after arousal, have been presented in detail (Wahlström, 1965). In these experiments there were some indications that light did not work on the "clock" directly, but rather by awakening the bird. This paper presents experiments with an enforced period of light (EPL) prior to awakening, designed to investigate those suggestions. [Some of these results have been briefly presented before (Wahlström, 1966, 1967).]

METHODS

The self-selected circadian rhythm in the canary (*Serinus canarius*) was recorded as described in detail earlier (Wahlström, 1964). Each cage has a separate light source (a common electrical bulb 220V, 75W), which is connected to a perch (the night perch) inside the cage in such a manner that the light is extinguished as long as the bird stays on this perch. The bird can thus choose darkness simply by hopping onto the night perch, and choose light by leaving it. A second perch is used to record the general

152

activity of the bird when the light is on. The birds (usually males) are put singly in these cages and left there at least a month before experimentation is begun. Water and food (commercial canary seed mixture) are available at all times.

The circadian rhythm of activity and rest recorded in this manner usually consists of one activity period and one rest period. The waking-point was used as a reference for calculation of the circadian period. Sometimes short intervals of rest (around 1 hr) were interspersed in the main activity period. These intervals of rest normally occurred more than 2 hr after the first awakening and in such cases do not influence the circadian period (Wahlström, 1964). Thus they have been disregarded and the activity period is considered to be the time from the first awakening until roosting, which marks the start of the main rest period (called overall activity in Wahlström, 1964). Short activity periods during the main rest period have never been observed in the present experiments.

For each experiment a pre-experimental average of activity time, rest time, and circadian period length was calculated from the five periods prior to that in which the treatment started (denoted circadian period 0). The changes in duration of activity, rest, and circadian period induced by the treatment were then followed for at least a week and were calculated as differences from the corresponding pre-experimental average.

In each series of experiments (similar treatments) an average change in activity time, rest time, and circadian period was calculated for several successive periods after the experimental treatment. Nine out of 131 experiments of the series presented in this paper were excluded from the computations, mainly because of technical difficulties.

RESULTS

EXPERIMENTS WITH ENFORCED PERIODS OF LIGHT (EPL) PRIOR TO THE EXPECTED AWAKENING

In this series of experiments the light was turned on from the outside for a predetermined time period (denoted by S in the figures and tables) prior to the expected awakening. The expected waking-point was determined by extrapolation from the waking times of the pre-experimental average. The night perch was not allowed to regulate the light until the expected waking-up time, when the self-selection mechanism was reinstituted. The enforced light lasted either up to the expected waking-point or for a shorter time. In the latter case, there was enforced darkness after the enforced light up to the time of the expected awakening. The duration of the enforced light is denoted by L in the figures and tables. S and L are schematically diagrammed in the inset in the right-hand corner of Figure 1.

The average changes induced by different kinds of EPL-schedules are shown in Figure 1. Circadian period 0 is the one in which the light was turned on; turning on the light ends this period. If the extrapolation defining the starting time of the EPL was correct, the decrease in length of circadian period 0 from the pre-experimental average should correspond to the S value given for the series. This was always the case (Figure 1, panels CP). Circadian period 1 was measured from the time the light was turned on until the point of awakening after the first main rest period. If the rhythm is uninfluenced by the EPL, the increase in circadian period 1 should be equal to the decrease in circadian period 0. This was the case only in series S=4.0/L=0.02 (Figure 1, panel CP). In the other series the waking-point starting circadian period 2 occurred earlier than one would expect by extrapolation from the pre-experimental ones (i.e., the decrease in period 0 was larger than the increase in period 1).

It is also evident from Figure 1 that all the changes induced in the circadian period by the EPL were seen in periods 0 and 1. There was no deviation from the pre-experimental average in circadian periods 2 or 3 (panel CP).

The duration of rest in circadian period 0 was, of course, decreased by an amount equal to the corresponding S value (Figure 1, panels R). In circadian period 1, duration of activity was usually increased, but this increase was less than the decrease in duration of circadian period 0 in all series except S=4.0/L=0.02. Thus, in most series, roosting in circadian period 1 occurred earlier than one would expect by extrapolation from the pre-experimental roostings. In four series (S=0.5/L=0.5; S=2.0/L=2.0; S=4.0/L=4.0; and S=4.0/L=0.5) the duration of the rest period in circadian period 1 tended to be reduced. No definite changes in duration of activity and rest were seen in circadian periods 2 and 3.

The first roostings and the first spontaneous awakenings after the EPL thus occurred earlier as a result of the EPL, but there was no further change in the rhythm. The amount of these resets (advances in phase) can be calculated for each experiment. In Figure 2A, C, and D, the average reset in the waking-point is plotted against the different parameters S and L in the EPL.

Figure 2A shows the results of the experimental series in which S and L were equal; that is, the bird was given light from the start of the EPL up to the expected waking-point. If the EPL lasted 2 hr or less, the reset (phase-advance) was approximately equal to the duration of the EPL; that

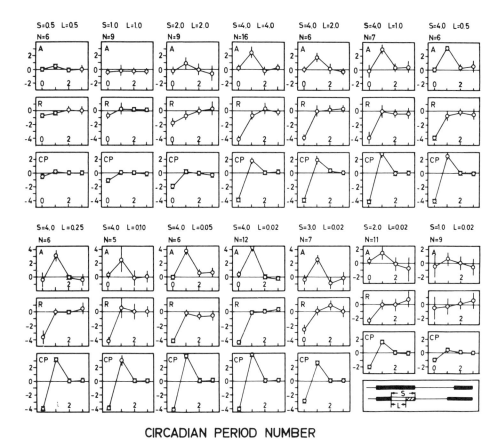

CIRCADIAN PERIOD NUMBER

FIGURE 1   The effects of various schedules of enforced light prior to awakening on the self-selected circadian rhythm. The ordinate is the increase in the measured parameter in hours over a pre-experimental average calculated from five circadian periods. (These averages are given in Table 1.) The different series of experiments are identified by the varied S and L values; S indicates the time prior to the expected waking-point at which the enforced light was turned on, and L is the duration of the light. S and L are schematically diagrammed in the inset in the lower right hand corner; the upper line shows the pre-experimental rhythm and the lower line shows how the rest period (thick line) is interrupted by S and L. Circadian period 0 is the one in which the rest period was interrupted. In each series changes in activity are given in panel A, changes in rest in panel R, and changes in circadian period in panel CP. Values are given ± twice the standard error. N is the number of experiments.

is, there was a circadian period of pre-experimental length from the start
of the light to the next spontaneous waking-point. With an EPL of 4 hr,
the reset was approximately 2 hr.

In another series of experiments the night perch did not turn on the
light when the birds left it. This arrangement produced an enforced period
of darkness (EPD) after the self-selected awakening, which caused a reset
in the opposite direction—that is, a delay of the first awakening after the
EPD. The results of such an experimental series are given in Figure 2B.
They are plotted in a manner similar to those of Figure 2A (data from
Wahlström, 1965). A comparison between Figure 2A and B shows that,
except for the sign, the phase-shift was remarkably symmetrical as a result
of light given prior to the waking-point (Figure 2A) or darkness given after
the waking-point (Figure 2B).

Figure 2C shows the results of the series of experiments in which the
duration of the light (L) was varied, but the time when the light was turned
on (S) was kept constant at four hours before the expected awakening.
There was an approximately linear relation between log duration of light
and induced reset. Figure 2D shows the result of a set of experiments in
which S was varied but L was kept constant at one minute. Varying S was
also found to have a significant, although much smaller, effect on the magnitude of the induced reset. (Note the difference in scale of the ordinate.)

In the series with EPD's there was a negative correlation between the
average duration of pre-experimental rest and the induced delay in the
first waking-point after the EPD. Birds that ordinarily had long rest periods were affected less by the EPD than those that had short rest periods
(Wahlström, 1965). In the present experiments we sought a similar correlation when EPL's were applied; however, no consistent correlation
could be demonstrated between the induced reset in the waking-point and
any of the pre-experimental variables (Table 1). The number of experiments in each series is admittedly few, but no consistent tendency at all
was observed.

The first roosting and the first awakening were both advanced by the
EPL. If the reset in the roosting time were the primary event and the reset
in the waking-point were solely a function of it, one would expect a positive correlation between them and a linear regression coefficient close to
1.0. The linear regression coefficients in all series except S=4.0/L=0.5
deviated significantly ($p < 0.05$) from 1.0 (Table 1). A high positive correlation ($p < 0.001$) was found in two series (S=2.0/L=2.0 and S=4.0/L=0.02).
In the latter series, however, there was no net reset. No positive correlation
at all was found in two of the series (S=4.0/L=0.50 and S=1.0/L=0.02).
No clear pattern emerges; it is unlikely that a direct linear relationship

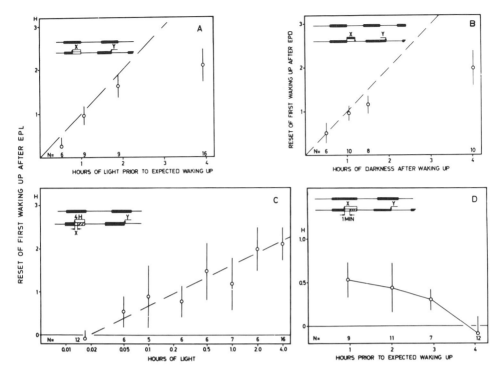

FIGURE 2   The effects of various parameters on the reset of the first waking-point after an enforced period of light (EPL–A, C, and D) or an enforced period of darkness (EPD–B). In A, C,
and D a positive reset indicates an earlier waking-point; in B it indicates a later waking-point. The
upper line of the inset schematically diagrams the pre-experimental rhythm of activity (thin line)
and rest (thick line). In the lower line X denotes the abscissa and Y denotes the ordinate. The bars
indicate ± twice the standard error. N is the number of experiments from which each point was
calculated.

exists between the reset in the roosting and the reset in the waking-point
induced by an EPL. If these changes are related, it must be in a more
complex manner.

The issue of the relation between the two resets is further illustrated
by the results obtained in the three series S=1.0/L=1.0, S=2.0/L=2.0, and
S=4.0/L=4.0. The average resets of roosting time (±S.E.) were 1.23 ± 0.42,
0.94 ± 0.40, and 1.41 ± 0.38, respectively. The corresponding figures for
the waking-point resets were 0.96 ± 0.11, 1.62 ± 0.13, and 2.10 ± 0.19.
Series S=2.0/L=2.0 and S=4.0/L=4.0 showed considerably larger resets in
the waking-point than series S=1.0/L=1.0, but the average resets in the first
roosting time were similar in all three cases. Thus, no definite relations

TABLE 1 Various Correlation (r) and Linear Regression Coefficients (b) with Reset of the First Waking-Up after an EPL as Dependent Variable

Series		N/	Pre-experimental Average of						Correlation and Linear Regression between Waking after an EPL (Y) and Various Independent Variables (X) given Below												
			Activity		Rest		Circadian Period		Pre-experimental Activity			Pre-experimental Rest			Pre-experimental Circadian Period			Reset of First Roosting		
S	L	Birds[a]	Mean	S.E.	Mean	S.E.	Mean	S.E.	r	b	p <	r	b	p <	r	b	p <	r	b	p <
0.5	0.5	6/6	11.81	0.70	12.25	0.68	24.06	0.34	0.23	0.03		0.24	0.03		0.94	0.24	0.01	0.62	0.33	
1.0	1.0	9/9	11.83	0.70	12.13	0.59	23.96	0.24	0.04	0.00		0.04	0.01		0.20	0.09		0.35	0.09	
2.0	2.0	9/7	11.73	0.61	12.26	0.58	23.99	0.26	0.34	0.07		-0.06	-0.01		0.65	0.33		0.97	0.32	0.001
4.0	4.0	16/8	11.97	0.42	11.88	0.41	23.86	0.20	0.09	0.04		0.11	0.05		0.41	0.39		0.58	0.29	0.05
4.0	2.0	6/6	11.98	0.82	11.91	0.73	23.89	0.30	-0.78	-0.23		0.79	0.27		-0.21	-0.17		0.24	0.16	
4.0	1.0	7/5	11.03	0.64	13.12	0.55	24.16	0.33	-0.03	-0.01		0.06	0.03		0.05	0.05		0.44	0.34	
4.0	0.5	6/5	10.65	0.92	13.11	0.84	23.76	0.21	-0.66	-0.24		0.74	0.30		0.09	0.14		-0.10	-0.39	
4.0	0.25	6/4	10.82	0.37	13.23	0.25	24.05	0.26	-0.95	-0.47	0.01	0.75	0.54		-0.60	-0.41		0.88	0.40	
4.0	0.10	5/4	11.44	0.70	12.67	0.72	24.11	0.17	-0.10	-0.05		-0.08	-0.04		-0.75	-1.56		0.53	0.21	
4.0	0.05	6/3	10.68	0.81	13.34	0.76	24.03	0.17	-0.36	-0.08		0.48	0.11		0.43	0.45		0.65	0.39	
4.0	0.02	12/7	11.43	0.45	12.16	0.29	23.59	0.20	0.42	0.10		-0.45	-0.16		0.30	0.16		0.92	0.40	0.001
3.0	0.02	7/6	10.93	0.55	12.46	0.57	23.39	0.22	-0.18	-0.02		0.40	0.04		0.58	0.02		0.20	0.04	
2.0	0.02	11/9	11.52	0.54	11.76	0.47	23.28	0.15	-0.31	-0.08		0.29	0.09		-0.19	-0.19		0.70	0.19	0.05
1.0	0.02	9/9	10.67	0.64	13.00	0.61	23.68	0.28	-0.11	-0.02		0.23	0.04		0.26	0.09		-0.36	-0.07	

[a]Number of experiments/number of birds used.

between the reset in the first waking-point and the reset in the first roosting time after an EPL could be established.

A previous investigation with EPD's applied after the waking-point failed to establish whether light worked directly on the "clock" or indirectly by influencing arousal (Wahlström, 1965). This issue has been further explored in the present experiments.

After an EPL starting 4 hr prior to the expected waking-point, some (but not all) birds return to the night perch within 2 hr. In these birds the next waking-point is less displaced than in those birds that do not return to the night perch. In series S=4.0/L=1.0, one bird had such a voluntary rest period; the reset of the next waking-point was −0.15 hr (i.e., a slight delay), compared to an average phase-advance of 1.40 hr in the other experiments of the same series. The duration of this voluntary rest period was 1.13 hr. In series S=4.0/L=4.0, 8 out of 16 experiments showed a similar pattern—rest periods occurring immediately after reinstitution of the self-selection situation. The reset of the next waking-point in these experiments was 1.62 hr, as compared to 2.58 hr in experiments in which there was no additional rest period (p<0.01; Student's t). The average duration of the voluntary rest periods in these experiments was 2.43 hr. These voluntary rest periods, occurring early in what should have been the activity period, thus reduced the response to the EPL and made the bird wake up later next morning, an effect similar to that of an enforced period of darkness (Wahlström, 1965)

Birds left the night perch after varying time intervals following the onset of the EPL. If this interval was short, it can be inferred that the light caused a more rapid and complete arousal than when the birds left the night perch "reluctantly." Using this criterion, the effect of arousal on the reset induced by the EPL was analyzed (Figure 3). In the left half of the figure the criterion of arousal was that the bird left the night perch during the enforced light (L). In the right half only one bird did not leave the night perch during the light (in series S=4.0/L=0.5), and the criterion of arousal was instead that the bird left the night perch within 10 min of the onset of the EPL. Under both criteria the following waking-point occurred earlier if the birds left the night perch shortly after the light was turned on, presumably because they were more strongly aroused by the light.

The two groups plotted at the extreme left of Figure 3 (series S=1.0/L=0.02) are compared with series S=1.0/L=1.0 in Figure 4. For those birds of series S=1.0/L=0.02 that left the night perch during the single minute of light (open squares), the reset of the first waking-point after the EPL was only slightly less than the reset in those birds of series S=1.0/L=1.0 that left the night perch within the first minute of light. It was of the same magnitude as the reset induced in the remaining two birds of series S=1.0/L=1.0, both of which left the night perch within 10 min of light. However, there was a difference (p<0.01, Student's t) between all birds of series S=1.0/L=1.0 and those birds of series S=1.0/L=0.02 that did not leave the night perch during the single minute of light (filled squares).

Thus, from the point of view of the subsequent sleeping behavior, 1 min of light was nearly equivalent to 1 hr of light if the birds were sufficiently aroused by the shorter pulse to leave the night perch.

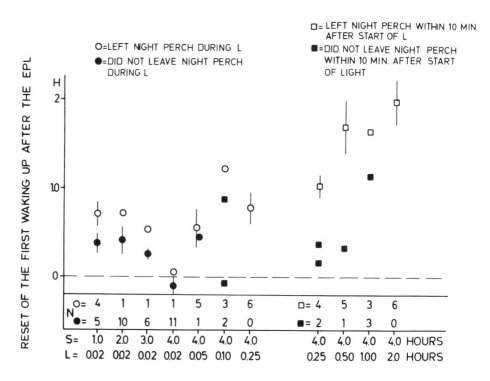

FIGURE 3 The effects of enforced light on the reset of the next waking-point. Open symbols indicate experiments in which light caused an early arousal, and solid symbols indicate experiments in which this was not the case. A positive reset means an earlier awakening the following morning. For groups with more than three experiments, averages are given ± the standard error; for groups with only three experiments, averages are given; with less experiments, individual values are given. N is the number of experiments. The different series are identified by S and L values on the abscissa. The average time the birds stayed on the night perch if they did not leave it within 0.17 hr (filled squares) were in series S = 4.0/L = 0.25, 0.21 hr; in series S = 4.0/L = 0.50, 0.59 hr (left night perch after light was extinguished); and in series S = 4.0/L = 2.0, 0.51 hr.

FIGURE 4 Comparison between the reset of
the first waking-point in birds that got one full
hour of light prior to the expected awakening
(series S = 1.0/L = 1.0; circles) and those that
got only one minute of light at one hour prior
to the expected awakening (series S = 1.0/L =
0.02; squares). The birds that did not leave the
night perch during the 1 min of light in series
S = 1.0/L = 0.02 are indicated by filled squares.

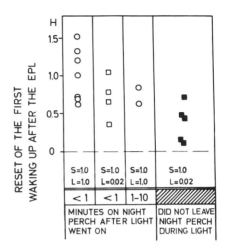

EXPERIMENTS WITH REDUCED LIGHT INTENSITY AFTER AWAKENING

In this series of experiments, light intensity was reduced when the birds
awoke by means of a perforated sheet of metal inserted in the lamphouse.
The holes in the sheet occupied only 40 percent of the area. The sheet was
inserted during the rest period prior to circadian period 0 and taken away
during the rest period of circadian period 6. The results, calculated as deviation from a pre-experimental average of seven circadian periods, are shown
in Figure 5. A reduction of light intensity of this magnitude had no effect
on duration of activity or rest, or on circadian period. The intensity of light
after awakening is thus not critical for the regulation of the self-selected
circadian rhythm, at least in the intensity range employed in the standard
experiment.

DISCUSSION

The "circadian rule" formulated by Aschoff (1960) states that in light-
active animals under constant light conditions, the duration of activity
will increase and the duration of rest and the circadian period will decrease with increasing light intensities. The circadian rule has been incorporated into one theoretical model of circadian rhythms (Aschoff and
Wever, 1962), but the rule is not universally applicable and the model derived from it has been criticized from that standpoint (Hoffmann, 1967).

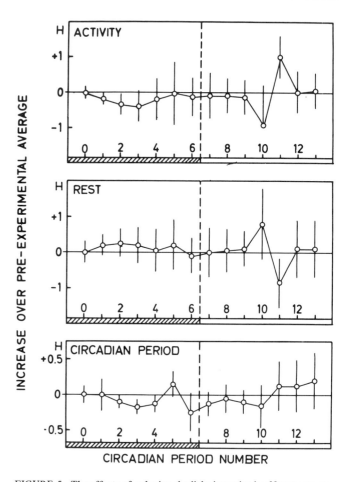

FIGURE 5   The effects of reducing the light intensity by 60 percent on
the self-selected circadian rhythm. A perforated metal sheet was inserted
in the lamphouse during the rest period prior to circadian period 0 and
taken away in the rest period prior to circadian period 7. Presence of the
metal sheet during activity is indicated by the stippled bar. Changes are
plotted relative to an average of seven circadian periods prior to the reduced light. The mean pre-experimental values for these seven birds were
activity time: 12.35 hr; rest time: 11.19 hr; circadian period: 23.53 hr.

In series S=1.0/L=4.0, the changes in duration of activity and rest in the
circadian period after the EPL (period 1) conform to the circadian rule
(Figure 1), if extending the light fraction can be considered equivalent to
increasing light intensity. This was true even if the birds that had a rest
period after the restoration of self-selection were excluded from the calcu-

lations. The effect of the EPL on the length of circadian periods 0 and 1 was also as predicted from Aschoff's generalization—a shortening. The results of series S=2.0/L=2.0 were similar but less clear-cut. In series S=1.0/L=1.0, the duration of the EPL was approximately equal to the reset in phase, and no unequivocal changes in duration of activity and rest were seen (Figure 1), but the period length was significantly shortened (Figure 2A).

The changes induced by an enforced period of darkness (EPD) of more than 3 hr starting when the birds left the night perch (Table II in Wahlström, 1965) can be analyzed similarly. While the duration of rest increased as expected from the circadian rule, the duration of "activity" in the first period after the EPD was also increased (contrary to the rule). However, the duration of the enforced darkness is included in this "activity" time, since "activity" was considered to begin the moment that the birds spontaneously left the night perch. If a correction is made for the duration of the EPD, then activity was decreased by the EPD. The decrease in activity time, and a corresponding increase in rest time, persisted for several cycles after the treatment. Furthermore, the length of circadian period 0 was increased. Similar, but less marked, changes occurred after short EPD's. Thus the EPD can be interpreted as equivalent to reduced light intensity by the criteria of the circadian rule.

If these enforced periods of light and darkness are regarded as short intervals of constant conditions (which they, of course, approach as the duration is increased), the conformity to the circadian rule is not surprising. A direct extrapolation to infinitely long EPL or EPD is, however, not permissible, since the effects of darkness and light on roosting time are not known in detail. Preliminary experiments suggest that the effects are slight, but no systematic study has been made.

In the undisturbed self-selected rhythm of the canary, "lights-on" and arousal are concurrent phenomena. In the present experiments "lights-on" occurred before the time the birds were expected to wake up and the light reset the "clock." But it is still possible that the light has this effect by arousing the bird, and thus only indirectly influences the "clock." This proposition is examined below.

A short (2 hr or less) EPL or EPD induces a change in period length that approximates the duration of the enforced light or darkness (Figure 2A and B): The circadian period was of pre-experimental length from the first light to the next spontaneous awakening. This suggests, but does not prove, that light acts as a waking stimulus that starts a new period. The negative correlation between the pre-experimental duration of rest and the delay induced in the first awakening after the enforced darkness if the EPD was

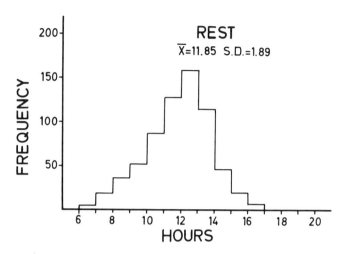

FIGURE 6   Frequency distribution of 676 average rest periods. The averages were obtained from five pre-experimental rest periods in different kinds of experiments.

more than 3 hr long (Wahlström, 1965) supports the suggestion that arousal is the mechanism of the light effect. The sum of the phase-delay and the resting time just prior to the enforced darkness was 14.5 ± 0.23 (S.E.) hr; this sum is not significantly correlated with the delay. Possibly, the bird can only sleep for a certain amount of time; then a new period starts regardless of the EPD. The frequency distribution of the duration of rest periods (Figure 6) further supports this hypothesis. Pre-experimental averages of five circadian periods from a large number of experiments were used. Note the low frequency of rest periods longer than 14 hr.

Also pertinent in this connection are the present experiments in which leaving the night perch readily was used as a crude criterion of arousal (Figures 3 and 4). Those birds that were aroused showed a larger phase-shifting influence of the EPL.

Very short light intervals prior to the EPD produced a graded reduction of the delay of the first waking-point (Wahlström, 1965). At the time of those experiments, the results could not be used as evidence for or against the hypothesis that arousal is an important part of the effects of light on the internal "clock": The birds may have been either asleep or awake during the darkness. The present experiments with EPL (series S=4.0/L=4.0) throw some light on this question. Half the birds in this series used the night perch for more than 2 hr after restoration of the self-selection situation. Thus they voluntarily experienced darkness during a time when there had been enforced darkness in the EPD experiments. The enforced darkness and the voluntarily chosen rest periods shifted the following waking-point in the same direction.

A comparison between series S=4.0/L=2.0, where no voluntary rest period occurred after the end of the EPL, and the experiments with voluntary rest periods in series S=4.0/L=4.0 is also of interest. In the former series there were also 2 hr of darkness, but it was enforced prior to the expected awakening rather than voluntarily chosen after the treatment. The reset in series S=4.0/L=2.0 was 1.98 ± 0.25 (S.E.) hr, and in series S=4.0/L=4.0, when the rest was chosen by the bird, it was 1.62 ± 0.26 hr. The difference is not statistically significant. But if series S=4.0/L=2.0 is compared to the experiments in series S=4.0/L=4.0 (average reset 2.58 ± 0.14 hr) in which the night perch was not used after the end of the treatment, the difference in resetting is significant at $p < 0.05$ (Student's t).

Enforced darkness after enforced light around the normal self-selected waking-point thus influences the "clock" in a manner similar to that of spontaneous rest periods. The enforced darkness probably acts by facilitating sleep. The apparent lack of effect of voluntarily chosen rest periods later in the undisturbed self-selected activity indicates that there could be a critical time around the waking-point when rest periods are effective in preventing resetting of the "clock" by light.

Thus there are several indications that enforced light can influence the self-selected circadian rhythm of activity and rest through an arousal. Rest periods, either those that probably occur during enforced darkness or those that are voluntarily chosen, are also important if they occur close to the potential arousal time. The graded response to light and darkness found in Figure 2C and D could thus depend on two factors that in the present experiments are not independent: arousal by the light and suppression by the following darkness. The situation is similar in experiments with light given prior to an EPD (Wahlström, 1965).

The small effect of a 60 percent reduction in light intensity under self-selection (Figure 5) shows that light intensity during the activity does not critically influence the "clock." The data of Aschoff et al. (1962) on chaffinches (Fringilla coelebs L.) under constant light conditions indicate that a similar reduction of the intensity in that species would decrease activity time, increase rest time, and increase period length by approximately 1.3 hr, 2.0 hr, and 0.7 hr, respectively. These values are much larger than any that could be masked by the scatter in Figure 5.

This difference (in light-active birds) between the effects of intensity under self-selection and under constant light can be explained if light is interpreted as an arousal stimulus. Under self-selection conditions, a reduction in intensity influences the situation only if the bird has already left the night perch. Since arousal seems to participate in the effects of EPL's and EPD's, it is possible that light acts by reinforcing the arousal already going on when the bird voluntarily leaves the night perch. This reinforcing property was hardly affected by the moderate reduction in light intensity used in the present study.

Under constant light the situation is somewhat different. It is unlikely that the light influences the "clock" during the activity time itself, since so little effect was seen under self-selection conditions. For the same reason it is improbable that the phenomena connected with roosting are greatly influenced. Light must then act only during the rest period proper or at the point of termination of rest. The former suggestion cannot be excluded without further experiments, but an effect around the waking-point seems more probable since the effects of different levels of constant light agree qualitatively (a quantitative comparison has not yet been made) with those of EPL's and EPD's.

Since the effects of light under self-selection conditions are apparently related to arousal, the same may be true for constant light conditions. Physically the external light is constant, but this is not necessarily the case from the bird's point of view. An internal cycle could exist such that the arousing property of constant light increases in the later part of the rest period. In the present experiments the arousing effect of 1 min of light (measured by the number of birds leaving the night perch) was larger 1 hr before than it was 2 hr before the expected waking-point. Furthermore, activity studies in the robin (*Erithacus rubecula*) at high latitudes showed "the birds starting earlier and going to rest later (in relation to the sun-time) the shorter the days are" (Palmgren, 1949). This partially compensates for the shorter daylengths in winter, but the duration of rest is nevertheless increased compared to summer. In field studies at high latitudes, similar results have been obtained in various species of light-active birds (Aschoff, 1969). In the winter the birds wake up earlier in relation to sun time, but after a longer sleep. Thus at lower light intensities, the arousing effect of the light could be considered to be stronger. Palmgren (1949) also reported a study of the arousal property of an electric stimulus in one mistle thrush (*Turdus viscivorus*). Less intense shocks were required to arouse the bird close to the spontaneous waking-point than earlier in the night. The reaction to several sensory stimuli, usually auditory, has also been tested during sleep (see review by Oswald, 1962, chapter V). Most of the work has been done on mammals (usually humans) and results are complicated by the recurring states of paradoxical (REM) sleep (Williams, 1967). Nevertheless, reaction times do tend to decrease in the later part of the sleep period. There are several neurophysiological indications—in mammals, at least—that the state of arousal (usually accomplished by stimulation of the reticular formation of the brain) influences the response to sensory stimuli such as light (Lindsley, 1960; Bremer, 1961).

Thus under constant light conditions the bird may become more and more sensitive to the arousing stimulus of light as the rest period goes on. Different light intensities would then be necessary to arouse the bird fully at different times before "spontaneous" awakening, and the arousal thus accomplished could influence the "clock" as it does under conditions of self-selection and EPL's.

The explanatory hypotheses outlined above depend on the assumption of a state of partial arousal, which does not start the "clock" on a new cycle. It is the partial arousal that allows the light under constant conditions to wake the bird and it is the partial arousal that induces the bird to leave the night perch and turn on the light. Only the full arousal caused by light, or perhaps other stimuli (such as sound; Menaker and Eskin, 1966), makes the "clock" start on a new cycle.

ACKNOWLEDGMENTS

This study was supported by the Tri-centennial Fund of the Bank of Sweden. The technical assistance of I. Karlsson, K. Brus, U.-B. Hogberg, and H. Andersson is gratefully acknowledged.

REFERENCES

Aschoff, J. 1960. Exogenous and endogenous components in circadian rhythms. Cold Spring Harbor Symp. Quant. Biol. 25:11–26.

Aschoff, J. 1969. Phasenlage der Tagesperiodik in Abhängigkeit von Jahreszeit und Breitengrad. Oecologia (Berl.) 3:125–165.

Aschoff, J., I. Diehl, U. Gerecke, and R. Wever. 1962. Aktivitätsperiodik von Buchfinken (*Fringilla coelebs* L.) unter konstanten Bedingungen. Z. Vgl. Physiol. 45:605–617.

Aschoff, J., and R. Wever. 1962. Aktivitätsmenge und $a:\rho$-Verhältnisse als Messgrössen der Tagesperiodik. Z. Vgl. Physiol. 46:88–101.

Bremer, F. 1961. Neurophysiological mechanism in cerebral arousal, p. 30 to 50. *In* G. E. W. Wolstenholme and M. O'Connor [ed.], Ciba Foundation symposium on the nature of sleep. Churchill Ltd., London.

Hoffman, K. 1967. Kritik des Erlinger Modells. Nachr. Akad. Wiss. Goettingen, Mat.-Phys. Kl. II. p. 132–133.

Lindsley, D. B. 1960. Attention, consciousness, sleep and wakefulness. Handb. Physiol. 3(Sect. 1):1553–1593.

Menaker, M., and A. Eskin. 1966. Entrainment of circadian rhythms by sound in *Passer domesticus.* Science 154:1579–1581.
Oswald, I. 1962. Sleeping and waking. Physiology and psychology. Elsevier Publ. Co., Amsterdam.
Palmgren, P. 1949. On the diurnal activity and rest in birds. Ibis 91:561–576.
Wahlström, G. 1964. The circadian rhythm in the canary studied by self-selection of light and darkness. Acta Soc. Med. Upsaliensis 69:241–271.
Wahlström, G. 1965. The circadian rhythm of self-selected rest and activity in the canary and the effects of barbiturates, reserpine, monoamine oxidase inhibitors and enforced dark periods. Acta Physiol. Scand. Suppl. 250:1–67.
Wahlström, G. 1966. Resetting effect of enforced light on the self-selected circadian rhythm in the canary. Acta Physiol. Scand. Suppl. 277:216.
Wahlström, G. 1967. The effects of enforced light prior to the waking-up on the self-selected circadian rhythm in the canary. Nachr. Akad. Wiss. Goettingen, Mat.-Phys. Kl. II. p. 123–124.
Williams, H. L. 1967. The problem of defining depth of sleep, p. 277 to 287. *In* S. S. Kety, E. V. Evarts, and H. L. Williams [ed.], Sleep and altered states of consciousness. Res. Publ., Assoc. Res. Nerv. Ment. Dis. No. 45. Williams & Wilkins Co., Baltimore.

DISCUSSION

FARNER: What is the source of your canaries? Are they from domestic stock?

WAHLSTRÖM: Yes, this is domestic stock. We first got them from local dealers; now we get them from Stockholm.

FARNER: Are your results from both sexes?

WAHLSTRÖM: Well, we try to get males—that is, we buy the birds as males—but we get them very young, and sometimes there are mistakes. We occasionally find eggs in the cages, so we always run autopsies on the birds to be sure. But they are predominately males.

FARNER: On what photoperiod regime do you maintain the birds before you put them on self-selection?

WAHLSTRÖM: They are kept initially in LD *12*:12, but they are put in the cage with the self-selector and left there for at least a month—and sometimes as long as a year—before experimentation is begun.

FARNER: Do you investigate the condition of the testes by laparotomy?

WAHLSTRÖM: No; we do look at the testes when we do the autopsies, but we have done no systematic observations on reproductive condition, and don't have any information on the relation between testis weight and spontaneous self-selection.

FARNER: All of these questions were really a search for some basis for the very, very different performance of white-crowned sparrows in self-selection of photoperiod, as I will discuss in my own paper.

R. G. LINDBERG
J. J. GAMBINO
P. HAYDEN

Circadian Periodicity of Resistance to Ionizing Radiation in the Pocket Mouse

Pocket mice (genus *Perognathus*) are facultative homeotherms whose body temperature may fluctuate 3–5°C during active periods and drop to near the ambient temperature during inactivity (torpor). The occurrence of torpor in pocket mice has been well documented (Bartholomew and Cade, 1957; Tucker, 1962; Cade, 1964). Diurnal torpor as an expression of circadian organization in *Perognathus* has been demonstrated through studies of metabolic rates (Chew, Lindberg, and Hayden, 1965). Torpor is a well-defined "marker" for study of circadian periodicity, but it is somewhat difficult to use because all individuals do not undergo torpor, even when kept under identical conditions. Furthermore, duration of torpor, when it occurs, can be extremely variable and is dependent on ambient temperature.

Radiobiological studies have shown *Perognathus longimembris* to be extremely resistant to ionizing radiation (Gambino and Lindberg, 1964; Gambino, Faulkenberry, and Sunde, 1968), and radioresistance is not dependent on heterothermic behavior (Gambino, Lindberg, and Hayden, 1965). The data on which this conclusion is based were derived from irradiations delivered during the normal working day. Reports on other species indicate large differences in response to radiation exposure depending on the time of day at which the exposure occurred (Pizzarello *et al.*, 1964;

Pizzarello, Witcofski, and Lyons, 1963; Straube, 1963; Rugh *et al.*, 1963).

We were curious to know if the high radiation resistance reported for *Perognathus* was related to the timing of exposure. We undertook to test the response of the species to ^{60}Co irradiation delivered at two times of day, the predicted high and low points of the metabolic rate. It is unlikely that metabolic rate *per se* is directly responsible for observed diurnal variations in radiation response, but metabolic activity may serve as a useful temporal marker. If, indeed, diurnal changes in metabolic rate can be shown to be coupled to circadian rhythms in specific radiocritical tissues, a number of informative correlations between timing of radiation exposure and amount of radiation exposure and degree of radiation injury could be made.

In the work reported here, we examined the validity of torpor as an assay of the circadian period of body temperature in pocket mice and as a basis for selecting irradiation times. Next we investigated mitotic activity in the pocket mouse intestinal epithelium as an example of a physiological rhythm that might influence radiation sensitivity. Finally, we exposed pocket mice to ionizing radiation at different times of day. We believe that the results of this study are germane to the broader issue of radiation response in mammals.

VALIDITY OF TORPOR AS A PHASE REFERENCE POINT (ϕR) IN THE STUDY OF CIRCADIAN PERIODICITY

METHODS AND MATERIALS

Animals

Perognathus longimembris less than 1 year old were collected at White Water Canyon and Pearblossom, California, and housed individually in 1-gal jars on a substrate of fine sand for 6 months to 1 year prior to initiation of the experiments. The mice were provided a mixture of rye, millet, oats, wheat, and hulled sunflower seeds *ad libitum*. The animal room was maintained at 20–24°C and under an *LD 12*:12 cycle with the light occurring between 0600–1800 hr PST. The ability of these animals to live without drinking water results in minimal animal wastes and permits very long isolation periods without interruption for tending the cages, a condition particularly desirable for periodicity studies. *P. longimembris* has a maximum longevity in our laboratory of about 4½ years.

Experimental Conditions

Each animal was placed in a $12 \times 12 \times 12$-in. lucite cage fitted with antennae and a 10-in. diameter running wheel. The individual cages were in open-front cubicles in a light-proof, constant temperature room ($\pm 0.5°C$). A bank of fluorescent lights provided approximately 25 ft-c of illumination to each cage in LD regimens. All monitoring and recording equipment, with the exception of preamplifiers, were located outside the experimental room. Enough seeds and fine sawdust bedding were provided to last the duration of the experiment. When it was necessary to enter the experimental room during constant dark conditions (DD), a shielded ruby-red photographic safe light was used. Prior experiments have shown that the mice were not disturbed and could not be entrained by the red light.

Nine female and seven male *P. longimembris* were used to study the range of τ in field-collected animals. They were maintained in DD at an ambient temperature (T_A) of $22°C \pm 0.5$ for 6 weeks following several months in LD 12:12 at $T_A = 20–24°C$.

To investigate temperature compensation of τ, five mice were placed in the constant temperature room in LD 12:12 at $T_A = 21°C \pm 0.5$ following several months' exposure to LD 12:12 at $T_A = 20–24°C$. After seven days the mice were placed in DD with no change in T_A. Twenty-one days later (the experimental room not having been entered), T_A was lowered to $10°C$ ± 0.5 and maintained for an additional 20 days.

Nature of Body Temperature Data

A small calibrated, blocking oscillator-type transmitter that measured changes in body temperature ($\pm 0.2°C$) was surgically implanted in the abdomen (Lindberg, DeBuono, and Anderson, 1965; Chew, Lindberg, and Hayden, 1967). The frequency (counts/second) was sampled for 1 sec every 10 min throughout every experiment.

A computer plot of representative data collected on 2 successive days from *P. longimembris* "free-running" in constant darkness (Figure 1) indicates that each day the animal's body temperature dropped to near ambient temperature for a period of a few hours, and then rose abruptly to a normothermic condition, which was irregularly maintained within $1–3°C$. There is little doubt that for this animal, expression of torpor is a circadian phenomenon. The circadian nature of changes in body temperature in an animal not expressing torpor is less obvious (Figure 2).

In our efforts to define the circadian properties of *P. longimembris*, we have estimated the free-running rhythm (τ) from body temperature mea-

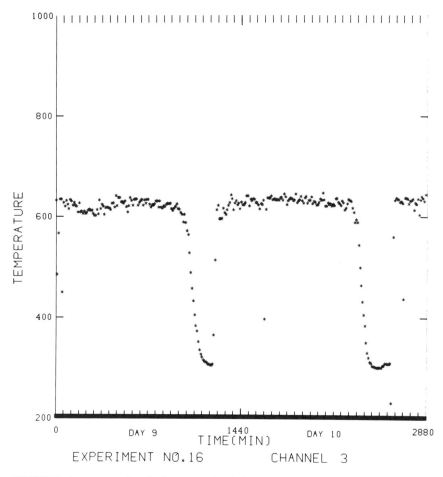

FIGURE 1   A computer plot of a 2-day record of body temperature changes in *P. longimembris* exhibiting torpor while maintained in DD at T_A = 21°C ± 0.5. Temperature expressed as counts/sec; $\sim$17 c/s/°C.

surements in two ways. The first method uses the time of arousal from torpor as the circadian marker (ϕR). This point was chosen because the rate of warming is very rapid and easily defined in either graphic or tabular data, and because data analysis shows time of arousal to be more predictable than time of entry into torpor. The time of arousal is plotted for successive days, a curve fitted to the points, and τ derived from the curve slope (Figure 3).

The second method determines τ by autocorrelation of the total data

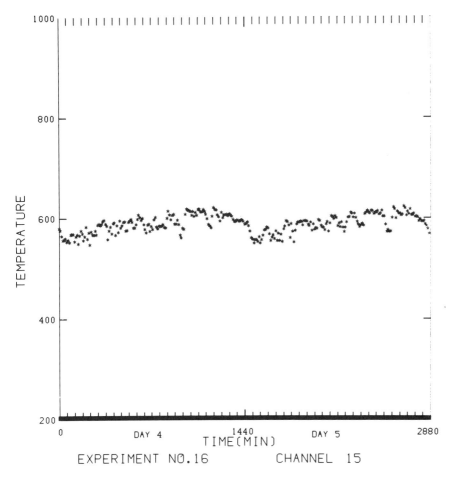

FIGURE 2 A computer plot of a 2-day record of body temperature changes in *P. longimembris* not exhibiting torpor while maintained in DD at T_A = 21°C ± 0.5. Temperature expressed as counts/sec; ∿17 c/s/°C.

collected over 21-day periods. The time of each maximum correlation coefficient is plotted against successive determinations to establish a trend line from which τ can be derived. A 95 percent confidence interval for the slope of the trend line is determined by the Student t-test using the two-tail criterion.

The agreement between the "manual" estimation of τ and the computer determination by autocorrelation was exceptionally good by virtue of the well-defined circadian marker. When torpor was not expressed, measure-

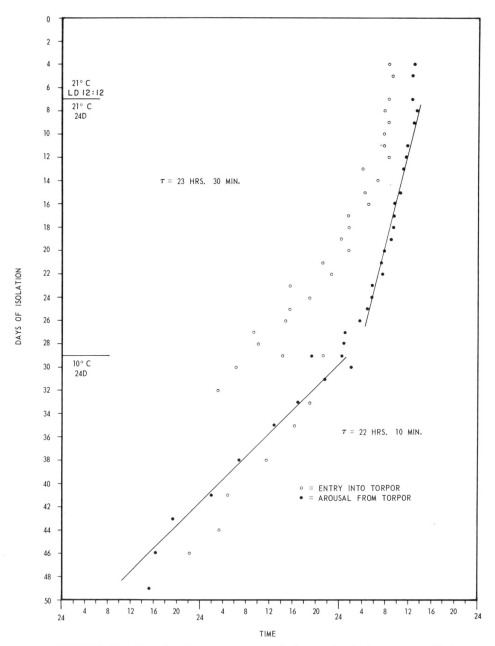

FIGURE 3 The effect of ambient temperature on the free-running rhythm of torpor of *P. longi-membris* in DD. τ is estimated from the time of arousal from torpor. Note that two entries and arousals occurred on day 29; time in torpor increases and arousals do not occur every day at $T_A = 10°C$.

ments of τ by inspection were difficult or impossible, but τ could be determined by autocorrelation. Consequently, when the data permitted, we used autocorrelation as our standard analytical procedure.

RESULTS

Free-running Period in Constant Dark (DD)

Tau was derived by autocorrelation of data collected the second 3 weeks of a continuous 6-week experiment, during which 69 percent of the animals exhibited daily torpor (Table 1). The data show a range of τ in the population of 1.4 hr. This value is well within the range reported for field-collected *Peromyscus* (Rawson, 1959). The standard deviation shown provides a convenient measure of the scatter of data from which the trend line was derived.

Temperature Compensation of τ

Five animals were used to determine whether the heterothermic behavior of *P. longimembris* could be used to demonstrate "temperature compensa-

TABLE 1   Free-running Circadian Period of Body Temperature in Pocket Mice (*Perognathus longimembris*) Maintained in DD at $T_A = 21°C \pm 0.5$ for 6 Weeks

Animal	Torpor Expressed	τ^a (hr)	Dispersion[b] (hr)
Males			
1	0	24.4 ± 0.15	± 0.52
2	+	23.4 ± 0.03	± 0.14
3	+	23.7 ± 0.90	± 0.76
4	+	23.8 ± 0.12	± 0.52
5	+	23.4 ± 0.10	± 0.43
6	0	23.7 ± 0.04	± 0.15
7	+	24.3 ± 0.03	± 0.12
Females			
1	+	23.6 ± 0.03	± 0.12
2	0	23.8 ± 0.09	± 0.40
3	+	23.4 ± 0.04	± 0.18
4	0	23.7 ± 0.12	± 0.51
5	+	23.8 ± 0.12	± 0.51
6	+	23.6 ± 0.06	± 0.23
7	+	24.8 ± 0.04	± 0.16
8	+	23.5 ± 0.02	± 0.07
9	0	23.7 ± 0.04	± 0.15

[a] Confidence interval, $\tau \pm 95\%$.
τ was determined from the final 3 weeks of the experiment by autocorrelation.
[b] Standard deviation of observed values about the trend line.

tion." Tabular data were plotted manually; time of arousal from torpor was used as ϕR to assess the persistence and precision of τ.

One of the five animals failed to demonstrate a stable τ in DD at either T_A. Representative data from one animal is presented in Figure 3; Table 2 is a summary of data from four animals. There was a degradation of the precision of τ at $T_A = 10°C$, and longer periods of time were spent in torpor. However, the arousal of an animal at a time in proper phase with the established τ after 1 or 2 days in continuous torpor was common. The relative independence of τ from metabolic rate is indicated by the ratio of τ at $T_A = 21°C$ to τ at $T_A = 10°C$. In other words, while the animal is torpid its body temperature equals or closely approximates T_A, but τ is relatively unaffected.

All animals were affected by the change in T_A. Two stabilized at a new τ within 1 day. One showed multiple drops (eight) in body temperature on the day following the temperature change, but stabilized at a new τ after 4 days. One animal showed transients for 8 days before stabilizing at essentially the same τ expressed at $T_A = 21°$.

DISCUSSION

The value of torpor as ϕR in determination of τ might be questioned since torpor occurs infrequently during an *LD 12*:12 light cycle at $T_A = 20°C$ when food is abundant, and because the frequency of torpor is greater in winter than in summer. However, diurnal torpor may be induced at any time of the year by lowering the ambient temperature or withholding food, or both. Consequently torpor is generally cited as an adaptive response to environmental extremes. In our laboratory we find that a constant dark environment also increases the frequency of torpor, even when food is plentiful and ambient temperature is moderate (20–25°C).

Using torpor as ϕR we have demonstrated the two primary character-

TABLE 2   Effect of Ambient Temperature on the Free-running Circadian Period of Body Temperature in *Perognathus longimembris* Maintained in Constant Dark

Animal	τ (hr) $T_A = 21°$	τ (hr) $T_A = 10°$	Transients (days)	$\dfrac{\tau 21°}{\tau 10°}$
1	23.5	22.2	1	1.06
2	23.1	22.2	1	1.04
3	23.7	23.5	8	1.01
4	23.1	24.4	4	0.95

Note: τ was determined from time of arousal from torpor.

istics of circadian systems: a precise free-running rhythm (τ) close to 24 hr, and temperature compensation of τ. (Table 1 further demonstrates that by autocorrelation τ is easily determined from animals failing to express torpor.)

Tau has been determined similarly for the following Heteromyids: *Perognathus formosus, P. parvus, P. alticola, P. californicus, P. penicillatus*, and the kangaroo rat *Dipodomys merriami*. It is our conviction that the amplitude of body temperature changes in facultative homeotherms permit more precise determination of τ than is possible in obligate homeotherms. This characteristic, coupled with ease of animal care during isolation, makes pocket mice particularly attractive research subjects.

CIRCADIAN PERIOD OF MITOTIC ACTIVITY IN EPITHELIUM

METHODS AND MATERIALS

Determination of Mitotic Activity

To determine whether a circadian period of mitotic activity could be demonstrated in the radiocritical intestinal epithelium, 40 female *P. longimembris* were picked at random from the colony and divided into eight groups of five mice each. Starting at noon the following day one group was killed every 4 hr by ether. The entire small intestine was removed and fixed in Bouin's solution. Five-micron cross-sections were prepared from a portion of the duodenum 10 mm distal to the pyloric sphincter. The slides were stained with hematoxylin-eosin, and mitotic figures scored on three entire cross-sections for each animal in the group.

Sections for scoring were selected from randomly coded slides at intervals large enough to ensure that the same crypts were not scanned twice.

Determination of Duration of Cell Cycle Stages

To determine the duration of the phases in the cell cycle, 40 female *P. longimembris* were randomly selected from the colony where they had been kept for 2 years. At the beginning of the experiment, each mouse was given an intraperitoneal injection of 10 μCi of tritiated thymidine in 0.1 ml physiological saline. All injections were made within a 30-min period between 0710–0740 hr. The animals were killed at closely spaced intervals from 0.5 to 24 hr postinjection by a blow on the head. A section of the small intestine, extending from the pylorus to the duodenal flexure, was quickly removed and placed in Helly's fixative.

Cross sections (4μ) were cut and prestained by the Feulgen reaction, and autoradiographs prepared by the dipping method (Kodak NTB-2). After

FIGURE 4 Circadian periodicity of mitotic activity in intestinal epithelium, as measured by
the number of metaphase and postmetaphase
mitotic figures per millimeter of gut circum-
ference. Five female *P. longimembris* were sacrificed at 4-hr intervals; mean and standard deviation are shown.

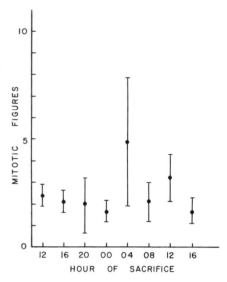

two weeks' exposure the slides were scanned and all labeled mitotic figures
(metaphase–anaphase) scored. The sections were selected to ensure separation by at least 100μ of tissue.

RESULTS

The frequency of mitotic figures determined by hemotoxylin-eosin staining is shown in Figure 4. The mean values obtained at 0400–0500 hr sacrifice time were greater than means obtained at other sacrifice periods
($p \leqslant 0.02$). Circadian periodicity is clearly implied.

Examination of the tritium-labeled mitosis curve of intestinal crypt cells
revealed the presence of labeled figures within 1 hr postinjection (Figure 5).
Between 12 and 16 hr postinjection the percentage of labeled mitosis drops
to and remains near 0. Peak labeling of 80 percent occurs at 6 hr postinjection. The second peak in labeled figures occurs at 20 hr postinjection.
From these data estimates of G_1, S, and G_2 + M phases are 3.5, 7, and
5 hr, respectively. The total cell cycle time, as indicated by the dashed line
on Figure 5, is 15.5 hr.

DISCUSSION

Diurnal fluctuations in the number of cells in mitosis have been reported
in several epithelial tissues (Cameron, 1963; Pilgrim, Erb, and Maurer,

1963; Brown and Berry, 1968). A notable exception has been the epithelium of the small intestine, in which no clear demonstration of daily variation in proliferative activity was found.

When the first reports of a diurnal variation in sensitivity to radiation appeared (Pizzarello *et al.*, 1964), we decided to look at mitotic activity in the gut of the pocket mouse as a step toward elucidating the mechanism of diurnal response to radiation exposure. The pocket mouse appeared to be a likely subject for such a study because it exhibits a pronounced diurnal rhythm in metabolic activity. Our data clearly demonstrated a peak in mitotic activity in jejunal crypt cells of *P. longimembris* at 0400 hr (Gambino and Towner, 1966). Since that time, several groups have reported diurnal fluctuations of mitotic index in a number of radiocritical tissues, including hemopoietic tissues (Ueno, 1968) and intestinal crypt cells of conventional laboratory rodents (Sigdestad, Bauman, and Lesher, 1969).

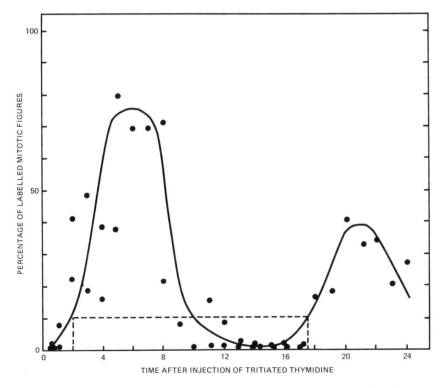

FIGURE 5   Cell cycle time in intestinal epithelium of *P. longimembris*, estimated from the percentage of labeled mitotic figures in jejunal crypts following a single injection of tritiated thymidine. Dotted line is estimated cell cycle time.

It is of considerable interest that the peak of mitoses in C57BL mouse jejunum was found to be 0300 hours (Sigdestad *et al.*, 1969), comparing favorably with the peak reported here for the pocket mouse duodenum. Although a number of hypotheses have been proffered to explain diurnal peaks of activity in the gut, none have been experimentally verified.

The percent mitotic labeling curve of *P. longimembris* intestinal crypt cells is unusual compared with that of conventional laboratory strains of rodents. There is considerable scatter in the ascending limb of the curve. Also, the percentage labeling does not reach 100 percent, but does drop to zero. These peculiarities suggest much greater variation in the duration of various phases in the cell cycle of pocket mouse gut cells than is seen in other animals, and may reflect the characteristic physiology of a facultative homeotherm. For example, the scatter in data points on the ascending limb of the curve suggests variation in the duration of the G_2 phase ranging from 1 to approximately 4 hr. This prolongation of G_2 may indeed be a temperature effect, since the tritiated thymidine was administered just before the metabolic low period of the animals used in the study. This suggests a potent mechanism for synchronizing a portion of the dividing cell population. Torpor may act *in vivo* as cold shock does *in vitro* to synchronize cell division in dividing cell populations.

In spite of the possible complications imposed by a partially synchronized cell population, the 15.5-hr cell cycle time appears to be consistent enough with values reported in other rodents to be reliable. For example, values given for conventional mice and germ-free mice are 11.2 and 13.6 hr, respectively (Lesher, Walburg, and Sacher, 1964). Values found for the G_1 and S phases of all three organisms are extremely close. The difference in total cell cycle time is accounted for in the G_2 + M phases, which are 2 and 2.5 hr in conventional and germ-free mice, respectively, and 5 hr in the pocket mouse.

It is difficult to reconcile a 15.5-hr cell cycle time with a circadian period of mitotic activity in the same tissue. One way the two can be reconciled is to postulate that a portion of the stem cell population is blocked or delayed in G_1 or G_2 and triggered into S or M by appropriate stimuli. In the pocket mouse, lowered body temperature might provide the block, and villus attrition during feeding hours the stimulus. This hypothesis needs verification.

The important relationship demonstrated here, however, is in the timing of the S phase with respect to the peak of mitotic activity. These data place a larger number of mitotically active gut cells in S phase somewhere between 1800 and 2400 hr than at any other time.

RADIOSENSITIVITY AS A FUNCTION OF CIRCADIAN PERIOD

METHODS AND MATERIALS

Animals selected for experimentation had been maintained in the colony for several weeks in *LD 12*:12. The use of torpor as a reliable indicator of the phase of a circadian period permitted selection of experimental subjects by visual inspection. One week prior to irradiation, intensive observations were begun on 300 animals to identify those that were regularly torpid or inactive between 0900–1000 hr. One hundred animals were finally selected and randomly divided into four groups of 25. Two groups were exposed to radiation; two were sham irradiated controls. Sexes were equally represented in all groups.

One hour before exposure, mice were placed in compartmented plastic boxes and transported to the nearby ^{60}Co facility. A dose of 1500 rad ± 5 percent was delivered at 37.8 rad/min calibrated with NBS ion chamber standards and ferrous sulphate dosimeters. The dose was monitored during irradiation by silver activated phosphate glass dosimeters placed in the animal compartments.

One group was exposed at 0900 hr and the second group exposed at 2330 hr of the same day. After irradiation the animals were returned to their jars in the animal room and observed twice daily for mortality over a period of 30 days.

RESULTS

Pocket mice that were irradiated at 0900 hr showed a survival time for 50 percent of the sample (ST_{50}) of 11.6 days. None survived beyond 18 days (Figure 6). Mice administered the same dose at 2330 hr had an ST_{50} of 15.1 days. Thirty days postirradiation, three mice were still alive. None of the sham-irradiated controls died during the same period.

The mean survival times are significantly different at the $p \leqslant 0.02$ confidence level. These results suggest that under the conditions described, pocket mice irradiated during their metabolically active period live significantly longer than those irradiated while their metabolic rate is low.

DISCUSSION

Highest survival occurred in the night-irradiated group of *Perognathus* (i.e., during their metabolic high). These results are contrary to those obtained in Swiss-Webster and C_3H mice, as well as Sprague-Dawley rats, all of

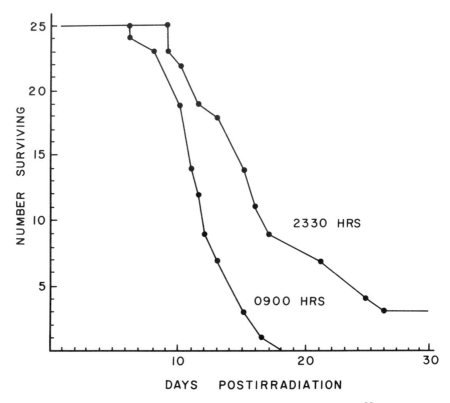

FIGURE 6   Survival in two groups of *P. longimembris* administered whole-body ^{60}Co irradiation at two different times of day.

which were more sensitive to irradiation delivered during their active period (2100–2000 hr) (Pizzarello *et al.*, 1964). An inverse relationship between the level of physiological activity and radiation sensitivity has been proposed (Vacek, 1962). Vacek showed that rats having a lower oxygen consumption during irradiation die sooner than those having a higher oxygen consumption. Since we did not measure oxygen consumption during irradiation, we can only suggest that this relationship between metabolic activity and survival may hold for pocket mice. Other work, however, indicates that minor fluctuations in metabolic rate, as judged by body temperature, may have very little bearing on radiation response unless accompanied by tissue hypoxia (Hornsey, 1957; Fallowfield, 1962).

Although minor rhythmic fluctuations in metabolic rate may have no direct influence on radiation response, metabolic rate may be useful as an index of other cyclic activity within the organism that may be directly related to radiosensitivity at a given phase in a light cycle. *In vitro* and *in vivo* studies on synchronously dividing cell populations have revealed differences in radiosensitivity relative to the phase of the cell cycle (Patt, 1963; Lesher, 1963; Sinclair, 1968; Ueno, 1968). Judging from chromosomal damage and from cell survival measured by colony-forming ability, the S (synthesis) phase is considered to be the most resistant in the mammalian cell cycle.

Our studies of circadian rhythms in cell proliferative activity in gastrointestinal epithelium of pocket mice indicate that the time chosen for night-irradiation coincided with the time when proportionally more cells were in S phase. The higher survival of this group can be simply accounted for if the gastrointestinal epithelium is a radiocritical tissue and if indeed intestinal crypt cells are less sensitive to irradiation during the S phase than at other times in the cell cycle.

ACKNOWLEDGMENT

This study was supported in part by NASA contracts NASw 812 and NAS 2-5037.

REFERENCES

Bartholomew, G. A., and T. J. Cade. 1957. Temperature regulation and aestivation in the little pocket mouse *Perognathus longimembris*. J. Mammal. 38:60–72.

Brown, J. M., and R. J. Berry. 1968. The relationship between diurnal variation of the number of cells in mitosis and of the number of cells synthesizing DNA in the epithelium of the hamster cheek pouch. Cell Tissue Kinet. 1:23–33.

Cade, T. J. 1964. Evolution of torpidity in rodents. Ann. Acad. Sci. Fenn., Ser. AIV. 71:77–112.

Cameron, I. L. 1963. Organismal regulation of mitotic activity in mice. Exp. Cell Res. 32:160–162.

Chew, R. M., R. G. Lindberg, and P. Hayden. 1965. Circadian rhythm of metabolic rate in pocket mice. J. Mammal. 46:477–494.

Chew, R. M., R. G. Lindberg, and P. Hayden. 1967. Temperature regulation in the little pocket mouse, *Perognathus longimembris*. J. Comp. Biochem. Physiol. 21: 487–505.

Fallowfield, T. L. 1962. The influence of hypothermia involving minimal hypoxia on the radiosensitivity of leucocytes in the rat. Int. J. Radiat. Biol. 4:457–464.

Gambino, J. J., B. H. Faulkenberry, and P. Sunde. 1968. Survival studies on rodents exposed to reactor fast neutron radiation. Radiat. Res. 35:668–680.

Gambino, J. J., and R. G. Lindberg. 1964. Response of the pocket mouse to ionizing radiation. Radiat. Res. 22:586–597.

Gambino, J. J., R. G. Lindberg, and P. Hayden. 1965. A search for mechanisms of radiation resistance in pocket mice. Radiat. Res. 26:305–317.

Gambino, J. J., and J. W. Towner. 1966. Papers presented at the Third International Congress of Radiation Research, Cortina D'Ampezzo. Abstr. 354.

Hornsey, S. 1957. The effect of hypothermia on the radiosensitivity of mice to whole-body x-irradiation. Proc. R. Soc., Ser. B. 147:547–549.

Lesher, S. 1963. Radiosensitivity of rapidly dividing cells. Laval Med. 34:53–56.

Lesher, S., H. E. Walburg, Jr., and G. A. Sacher, Jr. 1964. Generation cycle in the duodenal crypt cells of germ-free and conventional mice. Nature (Lond.) 202:884–886.

Lindberg, R. G., G. J. DeBuono, and M. M. Anderson. 1965. Animal temperature sensing for orbital studies on circadian rhythms. J. Spacecr. Rockets 2:986–988.

Patt, H. M. 1963. Quantitative aspects of radiation effect at the tissue and tumor level. Am. J. Roentgenol. Radium Ther. Nucl. Med. 90:928–937.

Pilgrim, C., W. Erb, and W. Maurer. 1963. Diurnal fluctuations in the number of DNA synthesizing nuclei in various mouse tissues. Nature (Lond.) 199:863.

Pizzarello, D. J., D. Isaak, K. E. Chua, and A. L. Rhyne. 1964. Circadian rhythmicity in the sensitivity of two strains of mice to whole-body radiation. Science 145:286–291.

Pizzarello, D. J., R. L. Witcofski, and E. A. Lyons. 1963. Variation in survival time after whole-body radiation at two times of day. Science 139:349.

Rawson, K. S. 1959. Experimental modification of mammalian endogenous activity rhythms, p. 791 to 800. *In* R. Withrom [ed.], Photoperiodism and related phenomena in plants and animals. No. 55. American Association for the Advancement of Science, Washington, D.C.

Rugh, R., V. Castro, S. Balter, E. V. Kennelly, D. S. Marsden, J. Warmund, and M. Wollin. 1963. X-rays: Are there cyclic variations in radiosensitivity? Science 142:53–56.

Sigdestad, C. P., J. Bauman, and S. W. Lesher. 1969. Diurnal fluctuations in the number of cells in mitosis and DNA synthesis in the jejunum of the mouse. Exp. Cell Res. 58:159–162.

Sinclair, W. K. 1968. Cysteamine; differential x-ray protective effect on chinese hamster cells during the cell cycle. Science 159:442–443.

Straube, R. L. 1963. Examination of diurnal variation in lethally irradiated rats. Science 142:1062.

Tucker, V. A. 1962. Diurnal torpidity in the California pocket mouse. Science 136:380–381.

Ueno, Y. 1968. Diurnal rhythmicity in the sensitivity of haemopoietic cells to whole-body irradiation in mice. Int. J. Radiat. Biol. 14:307–312.

Vacek, A. 1962. Whole-body oxygen consumption during irradiation for the survival of rats after exposure to x-rays. Nature (Lond.) 194:781–782.

DISCUSSION

ENRIGHT: You rather glossed over the surprisingly large influence of environmental temperature on the period of the rhythm. A change of environmental temperature of 11°C changed the period by over an hour in some cases.

In addition, in the example you showed (Figure 3), the period was shortened in the colder temperature. While that has been found in plants, as far as I know it has never before been recorded in animals.

ASCHOFF: Oh, yes—Hermann Pohl has published data on that [*Z. Vgl. Physiol.* 58:364 (1968)].

ENRIGHT: That's right. I apologize for my forgetfulness, Dr. Pohl.

LINDBERG: The ratio of period lengths in the two temperature conditions was about 1:06 for the particular example I showed. I think we have a good example of temperature compensation in a mammal. It turns out that the animal is torpid 3 to 4 hr a day under the high ambient temperature—say at 20°C—but if you drop the temperature down to 10°C or lower, the animal is torpid most of the time and only active 3 or 4 hr a day. Total amount of time in torpor gives an approximation of Q_{10} relationships.

ASCHOFF: You mentioned several times the close correlation between heart rate, body temperature, and activity. [Editor's note—These data were introduced during the verbal presentation.] Now in all mammals I know of, large or small, there is a time lag between, at least, body temperature and activity. Wouldn't you expect some phase lag in these animals also?

LINDBERG: Probably not, but it depends, somewhat, on how activity is defined. We monitor activity by counting the number of changes in signal strength of the implanted temperature telemeter that occur in a 10-min sampling interval as a result of animal movement. Consequently, we measure total animal movement rather than a specific behavioral pattern such as wheel-running or perch-hopping. In our experience, the rise in body temperature associated with arousal from torpor in *Perognathus* depends both on increased metabolic rate and animal movement. It is not surprising, therefore, that our data show the two parameters to be closely coupled and in phase.

RENSING: You say that the most mitotic figures—the maximum mitotic activity—are found around 0500 hr. This result is in line with the data on rats and mice. Most tissues have their high at that time even though maximum locomotor activity is earlier in the night, around 2200 hr.

LINDBERG: The work on mitotic activity and the cell renewal system is not really complete. It is largely Dr. Gambino's work; he has recently left our laboratory, but he suggested we mention at least where we are at this point. The pocket mouse, it turns out, has a very slow turnover rate of the cells in the gut compared to other rodents.

HORST O. SCHWASSMANN

Circadian Activity Patterns in Gymnotid Electric Fish

Fish are difficult subjects for circadian rhythm studies, and lack of unequivocal evidence has delayed recognition of the endogenous nature of
fish activity rhythms. Long-term recordings of activity patterns of teleost
fish have now been achieved, however (Lissmann and Schwassmann, 1965;
see review by Schwassmann, 1971).

In studying behavioral adaptations in neotropical freshwater fish, we
found one gymnotid electric fish, *Gymnorhamphichthys hypostomus*, to
be well suited for an investigation of the activity rhythm (Lissmann and
Schwassmann, 1965). Several other gymnotid species have now been
studied to determine whether differences in the frequency of electric
organ discharge are correlated with cyclic behavior changes of circadian
period.

The gymnotid eels of South America and Central America are one group
of freshwater fish that have evolved an electrosensory system as a mechanism for near-orientation. This sense enables them to detect the location
of objects having conductivities different from that of water. The electric
organ, which discharges low-voltage pulses in fairly regular succession, is
derived from muscle tissue, except in the *Sternarchidae* where it consists
of modified nerve fibers. Specialized electroreceptors in the skin detect
local impedance changes due to objects, or to other fish, in the surround-

ing water. The pulse shape and, to a degree, the frequency of electric organ discharge are species-characteristic. One group of these weakly electric gymnotids emits modified sinusoidal pulses at a relatively stable and usually high frequency [Type I of Lissmann (1961)–upper record in Figure 1]. Examples of this group are *Eigenmannia* spp. (250–400 Hz), *Sternopygus* spp. (45–150 Hz), and the *Sternarchidae* (500–2000 Hz). The other group produces very brief biphasic or polyphasic pulses, separated by a long interval, at a relatively low frequency [Type II of Lissmann (1961)–lower record in Figure 1]. Type II species show a relatively stable discharge frequency when resting undisturbed, but produce bursts of higher frequency when swimming around or in response to stimulation. (One member of this group, *Steatogenys*, is exceptional in that it has a relatively stable discharge frequency that varies only within narrow limits and does not exhibit sudden acceleration on stimulation.) The resting levels range from 2 Hz in some species of *Hypopomus* to 60 Hz in *Gymnotus carapo* and some species of *Rhamphichthys*. Increases in frequency permit a faster sampling rate during changing environmental situations.

All gymnotids are active at night and spend the daytime hours hidden in crevices, among plants, or in deep and usually turbid water. At least one species, *Gymnorhamphichthys hypostomus*, buries itself in the sandbed of a creek or stream, from which it emerges after sunset. During field and laboratory studies of this species, it was noted that two distinct levels of electric organ discharge frequency were precisely correlated with the states of activity and rest. A low frequency of 10–15 Hz was present while the fish were "sleeping" in the sand and was replaced by a high discharge level

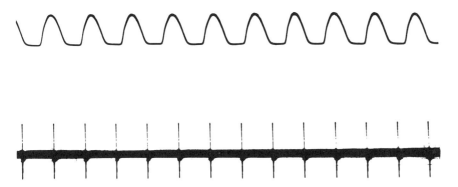

FIGURE 1   Photographed oscilloscope traces of electrical discharges. The upper trace is a record of *Sternopygus* sp. at a frequency of 60 Hz as an example of sinusoidal pulse shape; the lower record is of *Gymnorhamphichthys hypostomus* during activity, discharging at 77 Hz with brief polyphasic pulses.

of 60–100 Hz the moment they emerged and began swimming. Continuous recording of electric discharges provided a suitable method for assaying the activity pattern in this species (Lissmann and Schwassmann, 1965). This same study reported a gradual rise in basic resting frequency during the 2 hr preceding actual emergence from the sand. Field observations revealed that the sudden rise in discharge frequency level is precisely correlated with emergence and swimming activity. Laboratory studies demonstrated entrainment by a light–dark cycle, and free-running periods of less than 24 hr in constant dim light of 0.5 lux.

METHODS OF RECORDING

For continuous recording of discharge frequency, individual fish were confined in separate 25×40-cm plastic containers having a 3-cm layer of beach sand at the bottom and filled with water in which the fish had been kept previously. A large water bath, accommodating eight of these containers, eliminated short-term temperature fluctuations. A thermostatically-controlled heater made it possible to maintain the temperature in a well-insulated darkroom at any desired level between 20 and 28°C with less than 0.5°C fluctuation in the water bath. Carbon rods served as recording electrodes and were placed in the corners of each container. The weak electrical signals produced by the fish were amplified so as to trigger a pulse generator. These standard pulses were converted to direct current by an integrator circuit and fed into the driver amplifier of an eight-channel polygraph. The records obtained represent in a linear fashion the electric organ frequency and its change with time, with zero at the bottom of each record and high frequency at the top. The fish were fed with *Tubifex* worms every 5–10 days. A light trap prevented stray light from entering the room. After completion of an experiment, the records were cut into 24-hr sections and the daily records of each animal were mounted in sequence.

The field studies of *Gymnorhamphichthys* left no doubt that a high discharge level was precisely correlated with the active phase. The same relationship could also be demonstrated for some other species of variable frequency by simultaneously recording electrical output and disturbances of the water made by movement of the fish. A pair of thermistors, one shielded against water movement, was employed in a balanced bridge circuit, a method similar to that of Heusner and Enright (1966). A 3-day sample of such simultaneous records of *Hypopomus* is shown in Figure 2; the voltage changes due to transient imbalances of the bridge in the lower

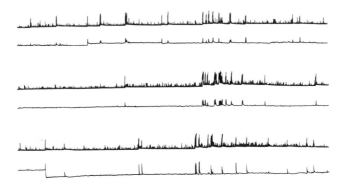

FIGURE 2   Three days of recordings of one *Hypopomus* sp. in LL (70 lux). The upper trace in each of the three double records shows the electric organ discharge frequency. Transient voltage changes in a bridge circuit due to water movement were recorded simultaneously in the lower trace. Bursts of high frequency almost always coincide with locomotor activity.

traces coincide with frequency increases in the upper traces. The record of an animal held in very bright light was selected since the greatly suppressed activity makes the correlation easily detectable.

RESULTS

ACTIVITY PATTERN IN SEVERAL SPECIES OF GYMNOTIDS

Several species were found to exhibit consistent frequency changes that were correlated with a circadian pattern of activity and rest. Two species of *Hypopomus* emit pulses at 2–4 Hz during rest, alternating in a circadian manner with a high level of 10–25 Hz during activity. However, the activity onset in these species is not as sharp as in *Gymnorhamphichthys*, and many high-frequency bursts are superimposed on the upper discharge level. A third species of *Hypopomus* has a resting discharge level of 15–18 Hz, which increases to 20–25 Hz during activity. In *Steatogenys* the basic electric pulse rate during the light phase of the light–dark regimen is 36 ± 2 Hz (Figure 3). During the 2 hr preceding the light–dark transition, the discharge level increases to 44 ± 2 Hz, comparable to the pre-emergence rise in *Gymnorhamphichthys*, and reaches 50 Hz when the fish begin swimming around. Locomotor activity is shown in these records (Figure 3) by trigger failures that cause the pen to drop toward the base line. When the fish is active, it assumes positions such that the signal picked up by the

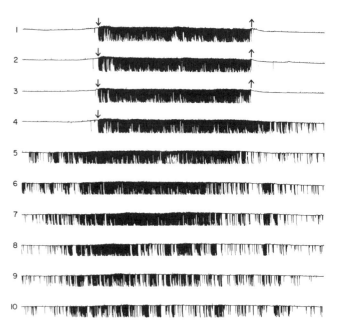

FIGURE 3   Ten days of recorded discharge rate from *Steatogenys* sp. *LD 12*:12 until day 4; constant dim light beginning on day 4. Arrows indicate light transitions; ↓, lights off; ↑, lights on. See text for interpretation of this record.

electrodes in the water is not strong enough to trigger the pulse generator and the pen drops down. Both parameters, the frequency change and the incidence of trigger failures, are discernible for a few days in constant dim light, but the pattern seems to fade gradually. The frequency pattern in *Gymnotus carapo* is similar to that in *Hypopomus*. A high incidence of bursts and a slight increase in basic frequency during activity is present, but here the difference between the two levels is very small. Those species with a steady rate of sinusoidal pulses that were tested—*Eigenmannia*, *Sternopygus*, and two species of *Sternarchidae*—show a pattern of activity and rest when the trigger-level method was used, but this method did not provide a convenient assay of the rhythm under constant light conditions.

DEPENDENCE OF PERIOD AND *a/ρ* RATIO ON LIGHT INTENSITY; MASKING EFFECTS

Several aspects of the activity pattern of *Gymnorhamphichthys* and *Hypopomus* were investigated. The free-running period and the ratio of activity to rest time under two light levels are illustrated in Figure 4A. The dependence of both parameters on light intensity is in accord with Aschoff's generalizations. However, it appears that the reduction of the activity–rest ratio is largely expressed by overt activity, the time the fish spends out of the sand, and that it may not be a true measure of the underlying circadian system. This becomes clear when one takes as the beginning of the active phase not the actual emergence but that particular frequency during the pre-emergence rise at which this fish, under the previous lower light level, began swimming. When these frequencies, shown in the records from days 11–18, are used as onset of activity, the a/ρ ratio is still significantly reduced at the higher intensity, but less so than when overt activity is used as an assay. The negative correlation between period and activity time can still be seen in these data.

We have pointed out previously (Lissmann and Schwassmann, 1965) that light has two different effects on the *Gymnorhamphichthys* activity pattern—besides its effect on the circadian rhythm, it seems to act directly on displayed activity by suppressing emergence from the sand. Further examples of possible "masking" effects are shown in the next figures. Figure 4B is a partial record from another fish of the same species, demonstrating entrainment by a very brief dark pulse, masking of overt activity by the high light level, and a brief free-run in constant dim light. Excerpts of discharge frequency records from two *Hypopomus* in different intensities of constant light are shown in Figure 5. After almost complete suppression of activity in 70 lux, a gradual change to darkness induces strong and sustained activity, which either reverts to the previous phase relation (in A) or seems to persist with the new phase (in B). The pattern appears grossly abnormal and fragmented in both cases. Data from experiments with *Gymnorhamphichthys* and *Hypopomus* concerning the effect of the light level on the period are summarized in Figure 6.

POSSIBLE DEPENDENCE OF PHASE ON "TWILIGHT" DURATION

Experiments with seven *Hypopomus* were performed to determine if phase differences could be induced by varying the duration of "twilight." This study was intended to test a suggestion by Aschoff and Wever (1965; also Wever, 1967) that the slightly changing duration of twilight near the equator could influence photoperiodic induction in tropical regions where annual changes in photofraction are absent or minimal.

In this experiment, the activity patterns displayed under the varied lighting conditions were essentially alike in all seven animals; representative records of one fish are shown in Figure 7. In the beginning, nearly "nor-

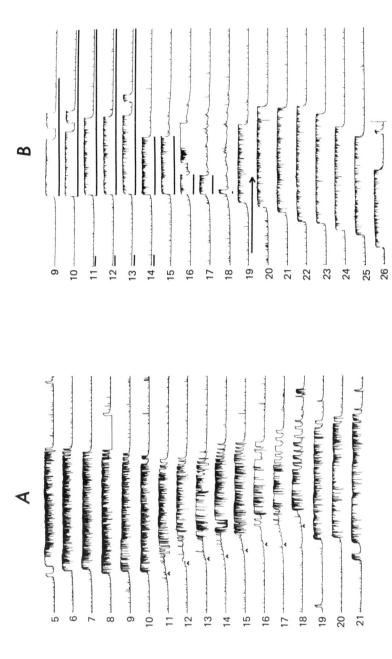

FIGURE 4 Excerpts from frequency recordings of two *Gymnorhamphichthys*. A: free-run in LL (1 lux until day 10 and from day 19; 15 lux from day 11 to 18). Negative masking, particularly of early activity, occurs under the higher light level. Markers under the tracings indicate the frequency at which this fish emerged from the sand under the lower light intensity. B: Pattern in various L/D ratios (L = 180 lux) until day 18; free-run in very dim light from day 19 (arrow). Duration of dark is indicated by black bars. Suppression of overt activity can be seen from days 14–18.

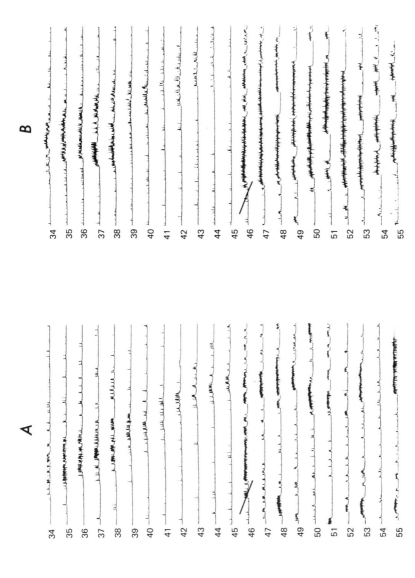

FIGURE 5 Excerpts of activity recordings of two *Hypopomus*; free-run in LL (8 lux to day 38, 70 lux to day 45). A slow reduction of the last light level to 0.1 lux on day 46 causes immediate activity, which reverts to the former phase (record A) or continues with the apparently exogenously triggered new phase (record B).

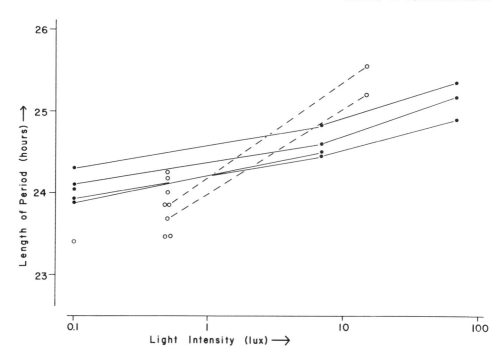

FIGURE 6   Change of free-running period with light intensity. ○, *Gymnorhamphichthys;* ●, *Hypopomus.* Lines connect data obtained from the same individual.

mal" light levels were employed, although the upper level of 200 lux is rather low and occurs only under dense rain forest canopy. In order to detect slight phase-shifts in response to the suspected variable of twilight duration, normally only 5–10 min, this factor was deliberately exaggerated. The midpoints of the light level changes were kept 12 hr apart. Prolonged twilight suppresses early and late activity, probably a masking effect of the high light level of this strong *Zeitgeber* (days 1–4). Raising the light level of the "dark" period reduced the strength of the *Zeitgeber* (day 27). After some initial masking, the peak of activity drifted towards the dark–light transition (days 30–34). Lowering both light levels, without altering the transition time, strengthened the *Zeitgeber* and caused the normal pattern to reappear (days 47–56). The example in Figure 8 indicates that we were observing real phase differences instead of mere masking effects. The weak *Zeitgeber* failed to entrain the rhythm, which drifted through a complete cycle showing the phenomenon of relative coordination.

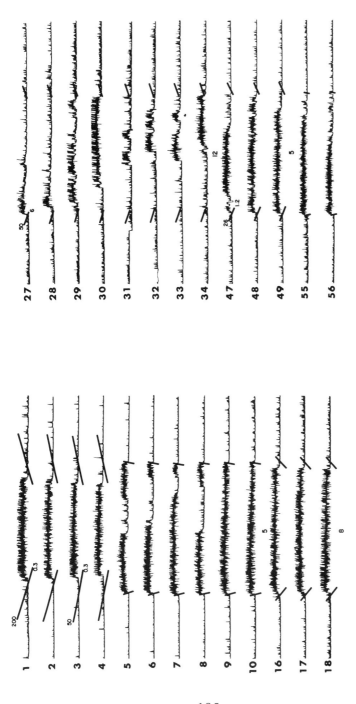

FIGURE 7 Representative samples of the activity pattern of *Hypopomus* in varied duration of simulated twilight. Slanted bars indicate duration and rate of the light transitions; upper and lower light intensities are shown whenever intensity was reset (days 3, 27, 47). Numbers between tracings indicate that so many days of records with the previous pattern were excluded from this figure. The recordings are from one fish, beginning at the upper left and continuing to the right.

195

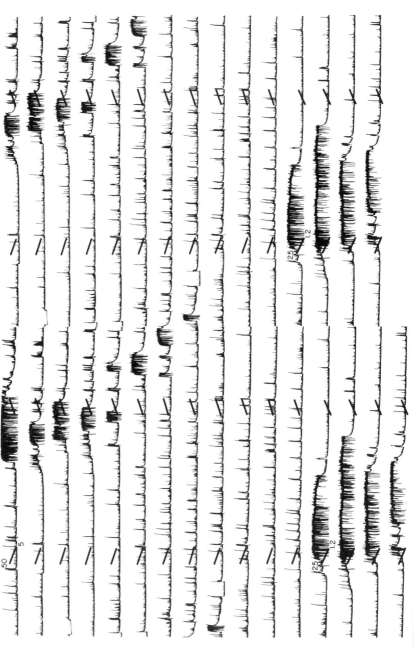

FIGURE 8 Excerpts of the activity pattern of one *Hypopomus* under a weak *Zeitgeber*. The records are continuous and are duplicated on the right to show the relative coordination between weak *Zeitgeber* and the drifting rhythm of overt activity. Light transitions and lux values are indicated.

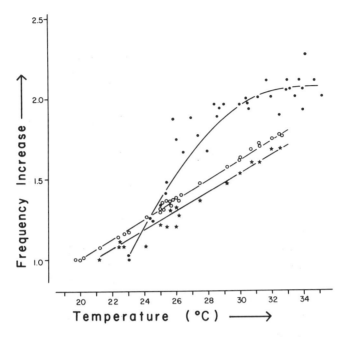

FIGURE 9 Effect of temperature on discharge rate of the electric organ.
○, *Sternarchus leptorhynchus*; •, *Gymnorhamphichthys hypostomus* (low
frequency level in sand); ★, *Gymnotus carapo*. Measured frequencies were
converted to a common factor.

EFFECT OF TEMPERATURE ON DISCHARGE RATE

The discharge frequency of gymnotid fish is directly dependent on temperature; Q_{10} values of approximatley 1.5 have been demonstrated in
several species (Coates, Altamirano, and Grunfest, 1954; Lissmann, 1958;
Erskine, Howe, and Weed, 1966; Enger and Szabo, 1968). As could be expected, the relationship is more precise in those species that exhibit little
variation in frequency than in those species showing variable frequency
(Figure 9). By changing temperature gradually over a period of 5 days and
counting pulses electronically, the following Q_{10} estimates were made:
Sternarchus leptorhynchus, 1.6; *Eigenmannia* sp., 1.5; *Sternopygus* sp.,
1.6; *Gymnotus carapo*, 1.5; *Hypopomus* sp., 1.6; *Gymnorhamphichthys
hypostomus* (low rate in sand), 2.0. In contrast to the marked temperature
dependence of discharge frequency, the free-running period of *Hypopomus* was essentially the same when the water temperature was raised from
22 to 28°C (other conditions constant) in a few preliminary experiments.

DISCUSSION

Continuous recording of electric organ discharges provides a suitable method of assaying circadian activity patterns in those species of gymnotids that emit brief pulses at a variable frequency. Of the species investigated, *Gymnorhamphichthys hypostomus* and two species of *Hypopomus* exhibit strikingly different discharge levels, correlated with a circadian pattern of activity and rest. *Hypopomus* displays additional high frequency bursts that can be shown to coincide with increased swimming activity. Long-term recording of electrical discharges of these fish in constant conditions provided estimates of free-running periods and of the activity/rest time ratio at different light levels. These results were found to be in agreement with Aschoff's generalizations for dark-active animals. However, the data obtained cast doubt on the validity of considering the time and amplitude of overt activity as expressions of the underlying circadian system. The extreme light sensitivity of the activity phase results in strong masking effects even under moderate light intensities. This direct effect of light had been noted in our previous study (Lissmann and Schwassmann, 1965); general aspects of the problem have been treated by Lohmann (1967). Positive masking (sudden and prolonged activity in response to a gradual decrease in light intensity) was also observed in *Hypopomus*. It appears, then, that the activity pattern in these animals not only is under the control of an endogenous circadian rhythm but is also strongly influenced directly by light stimulation. Another feature of the *Gymnorhamphichthys* pattern, the significant rise in low-level discharge frequency, persists at light intensities that prevent emergence from the sand and appears to be a useful parameter more closely reflecting the endogenous rhythm than does overt activity. It would be of interest to know if such metabolic parameters, as amount of activity, influence temporal aspects of the underlying endogenous rhythm (Lohmann, 1967). Another metabolic feature in these electric fish, the discharge frequency of the electric organ, although strongly dependent on temperature, does not appear to affect the free-running period.

The high light sensitivity of the circadian activity pattern in these animals may have obscured phase differences in response to varied durations of artificial twilight. On the other hand, the very fact that they are so highly sensitive to even moderate light strongly suggests that some mechanism of photoperiodic induction could be operating in these tropical fish. In the course of an earlier study at the equator, we noted a similar strong dependence of the nightly emergence times of fruit-eating bats on the varying pattern of decreasing light intensities (Jimbo and Schwassmann, 1967). Our impressions from several years of ecological studies near the equator strongly suggest precise annual periodicities, mainly concerning the migratory and reproductive behavior of many species of fish. The free-running circannual rhythms sometimes noted in populations near the equator seem to be the exception rather than the rule (Schwassmann, 1971).

ACKNOWLEDGMENTS

Several of the experiments were conducted in collaboration with D. G. Flaming and D. Kuffler. I thank Grace G. Kennedy for valuable assistance in the preparation of figures and manuscript. These studies were supported in part by grants from the National Science Foundation (GB-2796) and the United States Public Health Service (NB-7427).

REFERENCES

Aschoff, J., and R. Wever. 1965. Circadian rhythms of finches in light-dark cycles with interposed twilights. Comp. Biochem. Physiol. 16:507–514.
Coates, C. W., M. Altamirano, and H. Grundfest. 1954. Activity of electric organs of knifefishes. Science 120:845–846.
Enger, P. S., and T. Szabo. 1968. Effect of temperature on the discharge rates of the electric organ of some gymnotids. Comp. Biochem. Physiol. 27:625–627.
Erskine, F. T., D. W. Howe, and B. C. Weed. 1966. The discharge period of the weakly electric fish, *Sternarchus albifrons*. Am. Zool. 6:521.
Heusner, A. A., and J. T. Enright. 1966. Long-term activity recording in small aquatic animals. Science 154:532–533.
Jimbo, S., and H. O. Schwassmann. 1967. Feeding behavior and the daily emergence pattern of *Artibeus jamaicensis* (Chiropt. Phyllost.), p. 239 to 253. *In* H. Lent [ed.], Atas do Simpósio sôbre a Biota Amazônica, Vol. 5 (Zoologia): 239–253. Conselho Nacional de Pesquisas, Rio de Janeiro.
Lissmann, H. W. 1958. On the function and evolution of electric organs in fish. J. Exp. Biol. 35:156–191.
Lissmann, H. W. 1961. Ecological studies on gymnotids, p. 215 to 226. *In* C. Chagas and A. P. de Carvalho [eds.], Bioelectrogenesis. Elsevier Publ. Co., Amsterdam.
Lissmann, H. W., and H. O. Schwassmann. 1965. Activity rhythm of an electric fish, *Gymnorhamphichthys hypostomus*. Z. Vgl. Physiol. 51:153–171.
Lohmann, M. 1967. Zur Bedeutung der lokomotorischen Aktivität in circadianen Systemen. Z. Vgl. Physiol. 55:307–332.
Schwassmann, H. O. 1971. Biological rhythms. *In* W. S. Hoar and D. J. Randall [eds.], Fish physiology. Vol. 6. Academic Press, New York. In press.
Wever, R. 1967. Zum Einfluss der Dämmerung auf die circadiane Periodik. Z. Vgl. Physiol. 55:255–277.

II

PHOTOPERIODIC
TIME MEASUREMENTS

ERWIN BÜNNING

The Adaptive Value of
Circadian Leaf Movements

Circadian leaf movements belong to that small group of complicated physiological processes for which until recently no adaptive value was obvious. Indeed, I used to believe they were only an expression of the plant's kindness toward botanists, in allowing them to discover and record circadian rhythms within the plant.

More recently I began to doubt the altruism of such behavior and found it worthwhile to reconsider Darwin's (1880) statement: "That these movements are in some manner of high importance to the plants which exhibit them, few will dispute who have observed how complex they sometimes are." Our recent experiments bear directly on this point; the movements are important for the precision of photoperiodic time measurement.

Photoperiodic reactions result from the fact that the circadian clock controls quantity and quality of responsiveness to light. Photophilic and scotophilic responsiveness alternate in a circadian pattern. Maximum responsiveness to light breaks occurs at the point that I call the subjective midnight point (Bünning, 1969a). In normal day–night cycles, this point generally occurs about 16–18 hr after sunrise. That means, in the case of LD *12*:12, about 4–6 hr after sunset. Light breaks have the strongest effects when presented during this particular circadian phase. The subjective midnight point is also identical with, or close to, the phase of maximum responsiveness to phase shifts.

We must still consider the threshold and saturation intensities for the two separate light reactions of the well known dual light effect in photo-periodism.

In natural day–night cycles the rate of change of light intensities is greatest when the intensity has reached about 10 lux before sunrise, and about 10 lux after sunset (Figure 1). Also, the effects of variable cloudi-ness on light intensity are minimal around dawn and dusk. Thus we under-stand the adaptive value of the fact that in many organisms these twilight regions mark the beginning and end of the daylight period. Threshold values for the plant's perception of light involved in photoperiodism are often as low as 0.1 lux, and saturation is quite often reached with intensi-ties of 10 lux (Bünning, 1969b; Bünning and Moser, 1969a). These thresh-old and saturation values hold with respect to both components of the dual light effect. They have been known for some time from experiments on the influence of light breaks on development. We found them also to hold true with respect to phase shifts of the circadian rhythm in *Phaseolus* and in *Glycine* (Bünning, 1969b). One example is given in Table 1 (see also Figure 2).

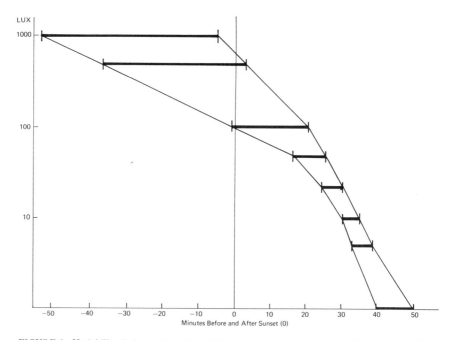

FIGURE 1   Variability in intensities of sunlight reaching a horizontal plane. Measured in Tübingen, March 1–12, 1969.

TABLE 1 *Glycine max.* Circadian Leaf Movements. Length of Periods and Time of Maximum Night-position in *LD 12*:12 with Different Light Intensities. Mean Values and Standard Deviations

Condition	Length of Period (hr)	Maximum Night Position (hr after beginning D)
DD	26.1 ± 0.2	−
LL	26.5 ± 0.4	−
LD, 2000 lux	24.0	4.7 ± 0.4
LD, 10 lux	24.3 ± 0.1	5.8 ± 0.6
LD, 0.6–0.8 lux	24.1 ± 0.3	8.7 ± 1.4
LD, 0.05 lux	26.6 ± 0.8	−

The low threshold values for the two light-reactions could allow moon-light intensities to disturb photoperiodic time measurement. For instance, intensities between 0.5 and 1 lux of white fluorescent light actually reaching the surface of the leaves during the night can inhibit flowering in Biloxi soybeans. The same intensity during the night causes phase shifts of 1–2 hr in *Phaseolus* and in *Glycine*. Table 2 presents some illustrative data.

Moonlight often reaches intensities of 0.5 lux and sometimes nearly 1 lux. Highest values occur under full moon at midnight, the time when plants are most responsive to light (subjective midnight point). However, the night position of the leaves reduces the intensity of moonlight coming from the zenith and reaching the surface of the leaves by 80 or 90 percent, i.e., to values between about 0.05 and 0.1 lux. Thus the threshold values for the light reactions involved in photoperiodism are either not reached at all, or are barely reached (for references, see Bünning and Moser, 1969a).

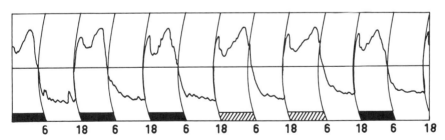

FIGURE 2 *Phaseolus multiflorus.* Circadian leaf movements in *LD 12*:12. During 2 nights (indicated by cross-hatching) plants were in light of 1 lux intensity; note phase shifts during these 2 nights. The first peaks during the dark periods indicate maxima of photonastic downward movements, beginning immediately with the transition from L to D. The second peaks during the dark periods indicate maxima of circadian downward movements. (For experimental details, see Bünning, 1969b).

TABLE 2 *Phaseolus multiflorus.* Circadian Leaf Movements, *LD 12*:12, L:700 lux. Phase Shifts by Weak Light Offered Instead of D. Mean Values and Standard Deviations

Light Intensity during D (lux)	Time of Maximum Night Position (hr after the beginning of D or weak light)
0	7.6 ± 0.2
0.5–1.0	8.2 ± 0.4
1.0–2.0	9.4 ± 0.4

But why are circadian leaf movements sometimes more complicated than a simple upward and downward motion? As an object for my experiments I selected *Parochetus communis*, a tropical plant related to clover. In this species the transition from day to night position is as described by Darwin for *Melilotus officinalis*:

The species in this genus sleep in a remarkable manner. The three leaflets of each leaf twist through an angle of 90°, so that their blades stand vertically at night with one lateral edge presented to the zenith. We shall best understand the other and more complicated movements, if we imagine ourselves always to hold the leaf with the tip of the terminal leaflet pointed to the north. The leaflets in becoming vertical at night could of course twist so that their upper surfaces should face to either side; but the two lateral leaflets always twist so that this surface tends to face the north, but as they move at the same time towards the terminal leaflet, the upper surface of the one faces about N.N.W., and that of the other N.N.E. The terminal leaflet behaves differently for it twists to either side, the upper surface facing sometimes east and sometimes west, but rather more commonly west than east. The terminal leaflet also moves in another and more remarkable manner, for whilst its blade is twisting and becoming vertical, the whole leaflet bends to one side, and invariably to the side towards which the upper surface is directed; so that if this surface faces the west the whole leaflet bends to the west, until it comes into contact with the upper and vertical surface of the western lateral leaflet. Thus the upper surface of the terminal and of one of the two lateral leaflets is well protected.

This behavior fits into another statement of Darwin's: "It is . . . a very common rule that when leaflets come into close contact with one another, they do so by their upper surfaces, which are thus best protected . . . it is obviously for the protection of the upper surfaces that the leaflets . . . rotate in so wonderful a manner. . . ." But what is to be protected in the upper surface of the leaf? We know that the epidermis of the leaf is very important in photoperiodism. Perhaps it is the decisive site of photoperiodic light reception (Schwabe, 1968). Moreover we know that the upper and lower surface of the leaf may show different responsiveness to light. In *Kalanchoe*, light breaks of very low intensities presented to the upper surface of the leaf strongly inhibit flowering, but if presented to the

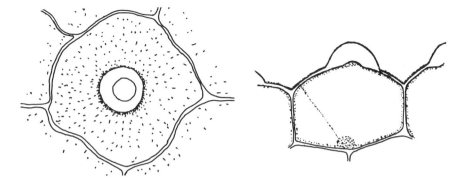

FIGURE 3 *Parochetus communis*. Surface view and cross section of ocelli in the leaf's upper epidermis.

lower surface, their effect is very small (Bünning and Moser, 1966). We have experimental evidence of comparable differences for other species of plants.

Now, let us look at the situation in the leaf of *Parochetus*. In the upper epidermis we find "ocelli" of the type described many years ago by Haberlandt (1905) (Figure 3). Haberlandt, in detailed studies of the anatomical and optical properties of the leaf epidermis, concluded that the upper epidermis functions as a sense organ for light, although experiments to ascertain whether the upper epidermis and the ocelli are important for phototropic movements did not support his hypothesis in this regard. But we should now reconsider these anatomical findings in relation to photoperiodic light reception. In the case of *Parochetus*, we may say that the complicated transition of the leaves to their night position results in a nearly complete shading of the ocelli, although it was not possible to determine whether the upper epidermis is more important for photoperiodism. It should be noted that in *Parochetus*, as well as in *Arachis* and *Trifolium*, the leaves of which behave in a similar way, the night position results in a decrease of zenith moonlight, coming to the upper surface of about 95 percent; i.e., moonlight reaching this surface will be less than 0.05 lux (Bünning, 1969b).

There is still another point. With normal day–night cycles, the transition from the night position to the day position, and from the day position to the night position, occurs often at those times during twilight that are characterized by light intensities between 1 and 100 lux. This is true despite variations in the length of the day, and is brought about by a combination of circadian and photonastic leaf movement. Laboratory records of the movements clearly show this combined effect (Figure 2). In this way, the transitions from one position to the other result in a remarkable increase of the steepness of intensity changes during the photoperiodically decisive parts of twilight noted earlier (Figure 4). Thus the leaf movements contribute to the precise definition of the dawn and dusk signals. Under simulated tropical conditions, the decisive intensity range for these signals, i.e., the range from about 1 to 10 lux, is thus passed within about 5 min (Bünning, 1969b).

To check directly the disturbance of photoperiodic time measurement by moonlight when leaf movements are prevented, we fixed the leaves in their day position with help of wires. In the course of these experiments, which are not yet complete, we found that the movements play an additional role. The prevention of the leaf movements itself influences flower formation, independent of any possible interference from moonlight. Figure 5 shows data from an experiment with *Perilla*. After subjecting this short-day plant to 7 inductive short days, the number of flowers developing during the ensuing weeks was counted. The development of flowers was strongly inhibited when the leaves were fixed in their day position during the inductive days, that is, when the downward movement in the evening was prevented. The strong inhibition of flower formation due to this blockage of leaf movements is shown in Figure 5. We observed the same phenomenon in the short-day plant *Chenopodium amaranticolor*. In this plant, leaves move upward in the evening and return to a nearly horizontal position during the day. There is a remarkable inhibition of flowering when leaf movements are prevented (Table 3). This inhibition is seen not only after fixing the leaves horizontally, i.e., approximately in their day position, but is also observed after fixing them upward vertically, corresponding to the night position. Similar inhibition results from fixation in a downward position.

Influence of gravity on flowering has been shown before (Hoshizaki and Hamner, 1962; Longman, Nasr, and Wareing, 1965), but apparently there is no optimum fixed angle of the leaf in relation to the gravitational field. It appears, instead, that the rhythmic change in leaf positions permits an optimum gravity-dependent diurnal and nocturnal flow of morphogenetically active substances from the leaves toward the buds. These findings require further analysis.

From the data reported in this paper, I feel we are justified in stating that Darwin was correct in his view that leaf movements have an adaptive value. The movements allow a precise perception of the photoperiodically decisive twilight intervals. They also prevent disturbing effects of moonlight. And, finally, the circadian oscillation of leaf positions apparently yields optimum gravitational conditions for flower formation.

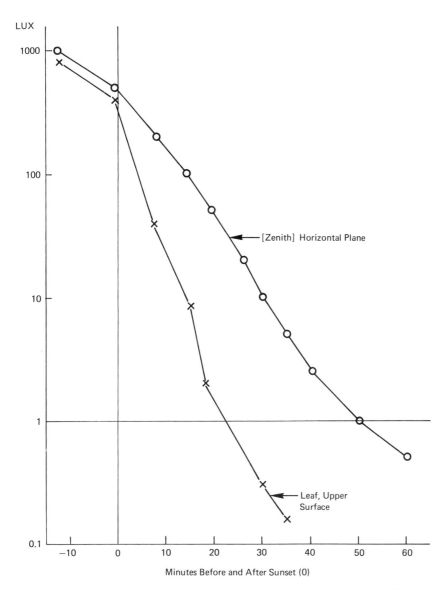

FIGURE 4 *Parochetus*. Intensities of sunlight reaching a horizontal plane, compared with the intensities reaching the upper surface of a leaf allowed to engage in normal movements. (For experimental details see Bünning, 1969b).

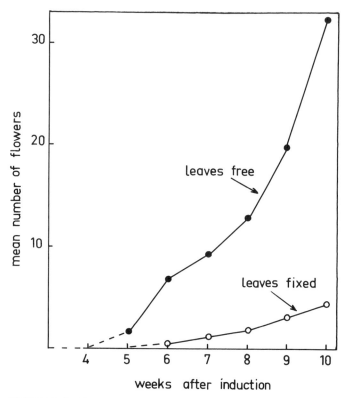

FIGURE 5 *Perilla ocymoides*. Effect on flower formation of fixing the leaves in the horizontal position during the short-day induction. Abscissa: Weeks after 7 inductive short-days (*LD 11*:13). Plants before and after induction in LL. Mean values from 10 plants. (For experimental details see Bünning and Moser, 1969b).

TABLE 3 *Chenopodium amaranticolor*. Relative Number of Flowers after Short-Day Induction. Leaves Free or Fixed[a]

Expt. No.	Weeks after Induction	Leaves Free	Leaves Fixed Horizontally	Leaves Fixed Upward	Leaves Fixed Downward
1	5	100	71	0	0
2	8	100	66	0	0
3	5	100	74	23	36
4	4	100	75	79	61
5	4.5	100	60	74	38

[a]See Bünning and Moser (1969b) for experimental details.

REFERENCES

Bünning, E. 1969a. Common features of photoperiodism in plants and animals. Photochem. Photobiol. 9:219–228.

Bünning, E. 1969b. Die Bedeutung tagesperiodischer Blattbewegungen für die Präzision der Tageslängenmessung. Planta 86:209–217.

Bünning, E., and I. Moser. 1966. Unterschiedliche photoperiodische Empflindlichkeit der beiden Blattseiten von *Kalanchoe blossfeldiana*. Planta 69:296–298.

Bünning, E., and I. Moser. 1969a. Interference of moonlight with photoperiodic measurement of time by plants, and their adaptive reaction. Proc. Natl. Acad. Sci. U.S. 62:1018–1022.

Bünning, E., and I. Moser. 1969b. Einfluss der Blattlage auf die Blütenbildung. Naturwissenschaften 56:519.

Darwin, C., and F. Darwin. 1880. The power of movements in plants. 2nd ed. John Murray, London.

Haberlandt, G. 1905. Die Lichtsinnesorgane der Laubblätter. W. Engelmann, Leipzig.

Hoshizaki, T., and K. C. Hamner. 1962. Effect of rotation on flowering response of *Xanthium pennsylvanicum*. Science 137:535–536.

Longman, K. A., T. A. A. Nasr, and P. F. Wareing. 1965. Gravimorphism in trees. 4. The effect of gravity on flowering. Ann. Bot. 29:459–473.

Schwabe, W. W. 1968. Studies on the role of the leaf epidermis in photoperiodic perception in *Kalanchoe blossfeldiana*. J. Exp. Bot. 19:108–113.

COLIN S. PITTENDRIGH
DOROTHEA H. MINIS

The Photoperiodic Time Measurement in *Pectinophora gossypiella* and Its Relation to the Circadian System in That Species

In previous papers (Pittendrigh and Minis, 1964; Minis, 1965; Pittendrigh, 1966) we have discussed a particular "coincidence" model of the general causal relationship between circadian rhythmicity and the photoperiodic time measurement that Bünning (1936) proposed many years ago. In this model photoperiodic induction occurs when some specific ("inducible") phase-point in the circadian system of the organism coincides in time with light in the environmental light/dark cycle. That phase-point (designated ϕ_i) was thought to be of very limited duration.

Our papers stressed three general points. The first was the dual role of the light cycle: It entrains circadian rhythmicity, thus establishing a determinate phase relationship between the rhythm and the light cycle, and it effects photoperiodic induction. In the strictly formal analysis to which one is still limited in this field these two actions are distinct, and could, therefore, be concretely different: they could, for example, involve two different photoreceptor pigments.

Our second point was that the general strategy underlying any evaluation or test of the model should consist of parallel studies of induction and entrainment seeking parallel peculiarities. We wished to understand the mechanism of entrainment sufficiently well to be able to predict as specifically as possible what light cycles—natural or contrived—would effect induction, and why.

212

The Photoperiodic Time Measurement in *Pectinophora gossypiella*
and Its Relation to the Circadian System in That Species

213

FIGURE 1 *Pectinophora gossypiella*. Upper: Photoperiodic response curve plots the incidence (%) of diapause as a function of photoperiod. Lower: Phase response curve for the insect's circadian oscillations plots phase-shifts ($\Delta\phi$) caused by a standard light pulse (15 min 200 lux) as a function of phase. (See Figure 4 for detailed derivation.)

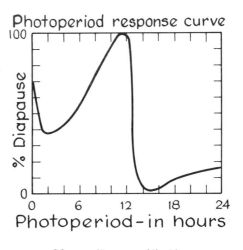

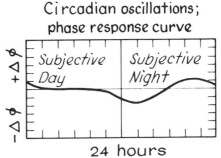

Our third concern was the need to study these two light effects (entrainment and induction) in one and the same species. Principles of entrainment derived from analysis of one species (e.g., *Drosophila*) may be—and indeed have been—useful in attempting to understand induction in another (e.g., *Pectinophora* or *Lemna*), but given the known differences in detail of the phase-response curves characterizing circadian oscillations in different species, really strong predictions testing the relationship between entrainment and induction should properly be pursued in a single species whose phase-response curves for its circadian oscillations are as well-known as its photoperiodic response curve.

We have attempted to do this in the moth *Pectinophora gossypiella*, which is a cotton pest, the pink bollworm (Figure 1). This paper summarizes what we have learned about the circadian system in *P. gossypiella* and how that knowledge bears on the "coincidence" model of photoperiodic induction. It also, incidentally, explicates more fully than Pittendrigh

(1966) did earlier, his understanding of the meaning of Hillman's (1964) data on *Lemna*.

Before summarizing our experimental work it may be useful to emphasize some of the uncertainties and make explicit the fundamental assumptions that bear on almost all experiments in this field.

COMPONENTS OF A CIRCADIAN SYSTEM: RHYTHMS AND THEIR DRIVING OSCILLATIONS

The organism is a circadian system that comprises (1) a diversity of distinct circadian rhythmicities in its several functions; and (2) an underlying circadian oscillation (or oscillations) that functions as a driver or pacemaker for this array of overt rhythmicities. The driving oscillation(s) of the system is coupled to, and hence entrained by, the daily cycle of light and darkness. That cycle (which Aschoff has called the *Zeitgeber*) thus establishes, through the agency of the system's driving oscillation, a determinate and adaptive phase-relationship (ψ) between all the biological rhythms coupled to the driving oscillation and all the rhythmicities of the environment (temperature, moisture, etc.) correlated with the light cycle and, like it, deriving from the earth's rotation.

We use ψ_{RL} to designate the phase relationship (the phase angle difference in hours) between a rhythm (R) and the light cycle (L). ψ_{OL} denotes the phase relationship between the underlying circadian oscillation (O) and the light cycle. For the measurement of ψ_{RL} and ψ_{OL} we need phase reference points (ϕ_r) for the rhythm, the oscillation, and the light cycle. These are most clearly illustrated by reference to the *Drosophila* eclosion rhythm in Figure 2, where ϕ_r (rhythm) is the median of the observed daily distribution of activity, ϕ_r (oscillation) is the phase of the oscillation's phase-response curve where $\Delta\phi$ is maximum, and ϕ_r (light) is the onset of each daily light pulse.

In discussing the total system, the importance of distinguishing between ψ_{RL} and ψ_{OL} is well attested by our use of artificial selection in *Drosophila pseudoobscura* to obtain two strains that differed by several hours in the phase relation (ψ_{RL}) of their eclosion rhythms to the light cycle (Pittendrigh, 1967). We found that selection had acted only on the driven elements in the system immediately timing the emergence rhythm itself; the phase of the driving oscillation (ψ_{OL}) relative to the light cycle remained unchanged in the two stocks with different ψ_{RL}'s (Figure 3).

The moth *Pectinophora* further underscores the importance of distinguishing between ψ_{RL} and ψ_{OL}. This moth displays three easily assayed

The Photoperiodic Time Measurement in *Pectinophora gossypiella*
and Its Relation to the Circadian System in That Species

215

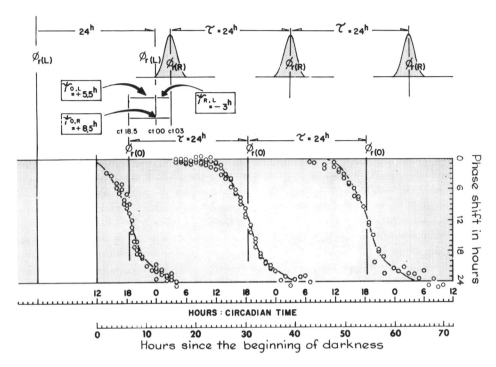

FIGURE 2 *Drosophila pseudoobscura.* The distinction between rhythm and oscillation; between ψ_{RL} and ψ_{OL}. The upper panel shows the periodicity of eclosion activity during the first 72 hr of constant darkness following prior entrainment by a light/dark cycle (12-hr photoperiod); it illustrates a circadian rhythm. The lower panel plots $-\Delta\phi$ values (responses to 15-min white light signals, 1000 lux) as a function of time during the first 72 hr of the free-run in constant darkness: It thus illustrates the time course of the driving oscillation as distinct from the rhythm. The phase reference is indicated for rhythm ($\phi_{r(R)}$), oscillation ($\phi_{r(O)}$) and light cycle ($\phi_{r(L)}$); ψ_{RL} and ψ_{OL} are defined. Following a 12-hr photoperiod the oscillation begins its free-run in darkness at ct 12; the abscissa indicates both the "real" and the "circadian" time of the free-running system.

circadian rhythms: in a population of eggs the hatching of caterpillars is limited to the early hours of each daily photoperiod; pharate adults emerge from their pupal cases later in the day; but the general activity of adults (including, specifically, egg-laying) is strictly nocturnal (Figure 4).

Figure 4 also shows the phase-response curves for each of the three rhythms assayed separately. The phase-response curve plots, as a function of phase, the phase-shift ($\Delta\phi$) caused by an appropriate standard signal, which in these cases was a pulse of 15 min of white fluorescent light ($\sim$200 lux). The phase-response curve, measuring the state of the driving oscillation as a function of time, is, at present, our only assay of its state *vis à vis* that of the rhythm it drives. Figure 4 shows, then, that the oscilla-

FIGURE 3 *Drosophila pseudoobscura.* The effect of selection on ψ_{RL} of the pupal eclosion rhythm. Upper panel: 50 generations of breeding from the earliest and latest emergers in a distribution established two strains ("early" and "late") in which ψ_{RL} differed by 3 hr. Center panel: The $\Delta\psi_{RL}$ is independent of photoperiod. Lower Panel: Selection has not effected ψ_{OL}; the phase of the phase-response curve relative to the last seen light pulse (measuring the phase of the driving oscillation relative to the last seen pulse) is the same in all three strains.

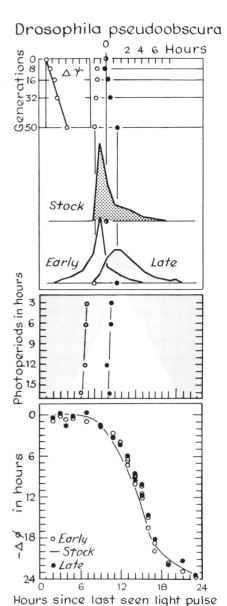

The Photoperiodic Time Measurement in *Pectinophora gossypiella*
and Its Relation to the Circadian System in That Species

217

Pectinophora gossypiella

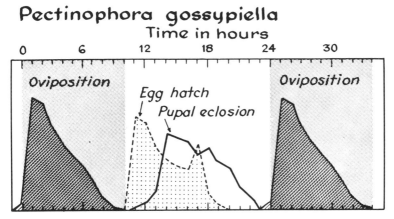

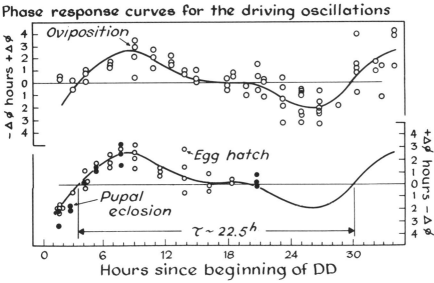

FIGURE 4 *Pectinophora gossypiella.* Upper: The daily distribution of oviposition, egg hatching, and pupal eclosion in a light/dark regime (*LD 14*:10). All three activities persist as circadian rhythms in DD. Lower: Phase-response curves for the three rhythms beginning in each case at the hour following entry into constant dark after entrainment to *LD 14*:10. A single curve has been eye-fitted to the data.

tions controlling all three rhythms are very similar: all have the same general form (they are all type-0 in Winfree's (1970) sense); all have the same amplitude; and all have the same phase relationship (ψ_{OL}) to the light cycle. Again the distinction between ψ_{RL} and ψ_{OL} is meaningful.

However, the similarity of their phase-response curves cannot be taken— of itself—to mean that all three rhythms have the "same" driver. Indeed, as we show elsewhere (Pittendrigh and Minis, 1971), each driving oscillation is different in some respect. The period of the oscillation gating the caterpillar's hatch from the egg is much longer (τ=24 hr) than that of those oscillations in the pharate adult that (a) gate its emergence from the puparium (τ=22.4 ± 0.2) and (b) drive its rhythm of oviposition, which appears to reflect well its general locomotory rhythm (τ=22.6 ± 0.2). And these two driving oscillations in the adult—identical in period—are nevertheless different, as is shown by their differential response to red light. We shall see later that light of 600 nm fails to entrain either rhythm (just as it fails to initiate them), but it nevertheless perceptibly shortens the period of the "free-running" pupal eclosion rhythm and hence the period of its driving oscillation, while it has no effect on the period (τ) of the free-running oviposition rhythm (Pittendrigh and Minis, 1971; Figures 15 and 16).

Selection experiments of the type initially used on *Drosophila* (Pittendrigh, 1967) extend the demonstration of the separability of distinct components in the *Pectinophora* circadian system. Figure 5 summarizes the outcome of nine generations of selection exerted on the pupal eclosion rhythm for early and late emergers. As in *Drosophila*—but even faster—there was a response to this selection on ψ_{RL}: we had early and late strains in which ψ_{RL} differed by 5 hr. Assays then revealed that ψ_{RL} for the egg-hatch rhythm showed a correlated response to the selection pressure exerted on the pupal rhythm: individuals that as adults emerged early from the pupal case had, as caterpillars, hatched early from the egg; late emerging adults had been, as caterpillars, late hatchers. But the adults of the two selected strains showed no difference in ψ_{RL} of their oviposition (adult activity) rhythms. It is as though the hatching of the caterpillar and the emergence of the pharate adult were timed by some identical driven element in the system quite distinct from that which times general activity and oviposition in the adult.

The circadian system in this moth is demonstrably a complex of "distinct" circadian oscillations (sensitive to light) and driven elements coupled to them. Elsewhere (Pittendrigh and Minis, 1971) we discuss the compatibility of this important conclusion for a multicellular circadian system

The Photoperiodic Time Measurement in *Pectinophora gossypiella*
and Its Relation to the Circadian System in That Species

219

FIGURE 5 *Pectinophora gossypiella.* Upper:
The effect of selection for "early" (E) and
"late" (L) emergers in the pupal eclosion
rhythm. ψ_{RL} is changed by the selection.
Middle panel: The daily distribution of egg-
hatching in the two selected strains and the unselected stock (S). Lower panel: Comparable
data for the distribution of oviposition activity.
[See Pittendrigh and Minis (1971) for derivation: the ϕ_r of "stock" has been synchronized
for all three rhythms for ease of comparison of
"early" and "late" in all three rhythms.]

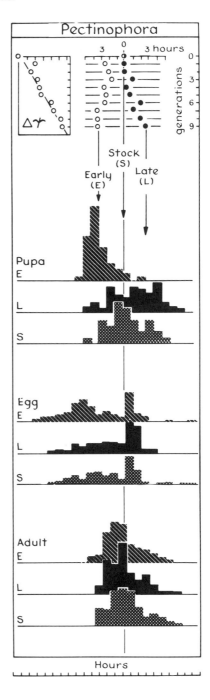

with the view of Sweeney (1969) that in a single-celled system the many separable driven rhythms are controlled by a common (single) driving oscillation.

TESTS OF THE COINCIDENCE OF ϕ_i WITH LIGHT: A FUNDA-MENTAL ASSUMPTION

Figures 6 and 7 illustrate the fundamental generalization about circadian systems that underlies their potential role in photoperiodic phenomena: both ψ_{RL} and ψ_{OL} change in response to change in the duration of light (the photoperiod) in each 24-hr cycle. The dependence of ψ_{RL} on photoperiod applies to all three rhythms studied in *Pectinophora* (Figure 6); in *Drosophila*, where its assay can be more accurately and easily made, the dependence of ψ_{OL} on photoperiod has been fully explored (Figure 7).

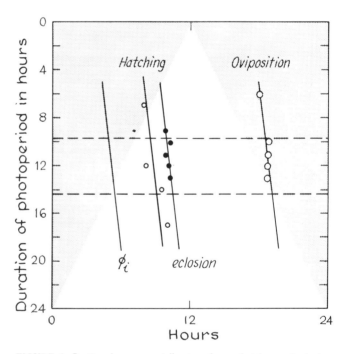

FIGURE 6 *Pectinophora gossypiella.* ψ_{RL} for egg-hatch, pupal eclosion, and oviposition rhythms as a function of photoperiod. The inducible phase (ϕ_i) is postulated to occur 5 hr earlier than the phase-reference point (median of the distribution) of the pupal eclosion rhythm. Dotted horizontal lines mark the range of natural photoperiods at El Paso, Texas, the source of the strain used.

The Photoperiodic Time Measurement in *Pectinophora gossypiella*
and Its Relation to the Circadian System in That Species

221

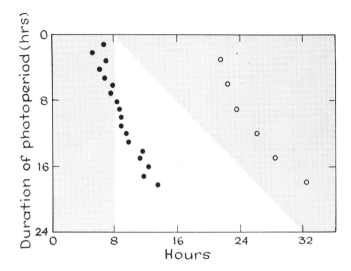

FIGURE 7 *Drosophila pseudoobscura.* ψ_{RL} and ψ_{OL} as a function of
photoperiod. The phase reference point for the rhythm is the median of
the eclosion distribution (solid circles). The phase reference point for the
oscillation is ct 12 (see Figure 2) and is plotted as open circles. The locus
of ct 12 was obtained by measuring the phase of the entire phase response for the oscillation following photoperiods of 3, 6, 9, 12, 15, and
18 hr.

It is obvious from Figures 6 and 7 that some phase-points in the total
circadian system never get out of darkness no matter what the photoperiod
(in the natural range); others similarly never escape the light. Still other
phase-points, exemplified by the time of pupal eclosion in *Pectinophora*
and *Drosophila*, are not illuminated in the entrained steady states effected
by very short photoperiods, but are illuminated in the entrained steady
states realized under longer photoperiods.

The "inducible phase" (ϕ_i) postulated in the "coincidence model" for
photoperiodic induction is conceived as just such a phase-point—it is illuminated in some entrained steady states, but not in others. It must therefore lie somewhere near the beginning or (more likely) the end of the subjective night (see Pittendrigh, 1966): These are the general "phase-regions"
of the oscillating system whose coincidence or noncoincidence with light
is dependent on photoperiod. That, unfortunately, is about as far as one
can go at present in attempting to make the coincidence model as explicit
as possible before designing tests of it.

Since we do not know the particular periodic system of which ϕ_i is part,
we do not know whether it is a phase-point of a driving oscillation (and, if
so, which, because there are clearly several in the system) or whether it is a

point in one of the many driven rhythms in the total system—and again, if so, which? We cannot, in short, directly study coincidence or non-coincidence of ϕ_i and light, and in attempting to study it indirectly we are forced to make a fundamental assumption that is by no means necessarily valid and requires scrutiny.

That assumption is that ϕ_i occurs at some fixed number of hours (k) before or after another phase-point of the system to be chosen for ease and precision of assay. The assumption is valid only if the unidentified oscillatory process of which ϕ_i is part—let us call that the *clock* hereafter—is coupled to and entrained by light cycles in precisely the same way as all the other assayable oscillations and rhythms in the total system. Were that assumption not true, it is virtually certain that ϕ_i would not lie at a fixed point relative to some other arbitrary marker of system-phase.

The only tests of the assumption that we can readily make involve measuring ψ for several components of the same system and asking if the $\Delta\psi$ effected on one component by some experimental procedure is matched by a comparable $\Delta\psi$ in other components. Figure 7 shows that ψ_{RL} for the pupal eclosion rhythm in *Drosophila* is a function of photoperiod; so, too, is ψ_{OL}. While there are significant differences in the shape of the two curves, which we discuss elsewhere (Pittendrigh and Minis, 1971), they are, in fact, slight, and the more easily assayed ψ_{RL} (plus a constant) is a reasonable measure of ψ_{OL}. Similarly, Figure 6, which plots ψ_{RL} as a function of photoperiod for all three rhythms in the *P. gossypiella* system, shows that the three slopes are very similar. [The derivation of the curves is somewhat complex and is developed elsewhere (Pittendrigh and Minis, 1971).]

We have, then, no reason to believe that the clock oscillation of which ϕ_i is part has a different dependence on photoperiod. Therefore, the phase of the easily assayed pupal eclosion rhythm (plus a constant) in any light regime whose T equals 24 hr marks the phase of ϕ_i: to this extent our assumption seems reasonable. In Figure 6 we have indicated ϕ_i lying 5 hr ahead of ϕ_r for the eclosion rhythm. That value was derived from interpretation of experiments reported in the following section; thereafter, its location at that point was a hypothesis to be tested.

Our general assumption of the utility of an arbitrary marker of ϕ_i is, however, on much less secure ground when the period (T) of the experimental light cycle is significantly different from τ. Since this relates to common practice in the experimental study of photoperiodism the point is worth some explicit attention.

Consider one oscillation (L) driving another (A). When the free-running period of the driver exceeds that of the driven member (i.e., when $\tau_L > \tau_A$), the latter oscillation (A) phase-leads the driver (L); conversely,

The Photoperiodic Time Measurement in *Pectinophora gossypiella*
and Its Relation to the Circadian System in That Species

223

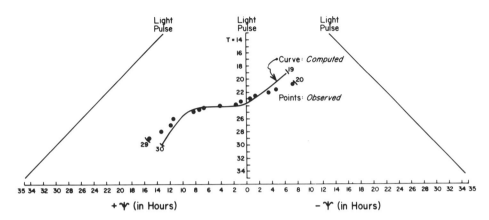

FIGURE 8 *Drosophila pseudoobscura.* ψ_{RL} and ψ_{OL} as a function of T, the period of the light cycle. Entrainment is successful within the range ∿19 to ∿29 hr. Each cycle involves a single 15-min pulse of white light, 1000 lux. The solid curve marks ψ_{OL}; it is a predicted curve (cf. Pittendrigh, 1965) which has been confirmed experimentally (see Pittendrigh, 1967). The points mark the phase (distribution medians) of the rhythm. When $T > \tau$ the rhythm phase leads the oscillation; when $T < \tau$ the rhythm phase lags the oscillation.

when $\tau_L < \tau_A$ the A oscillation phase lags L. This relationship is exemplified by the solid curve plotted in Figure 8; that curve is the predicted phase (relative to the light pulse) of the circadian oscillation driving the pupal eclosion rhythm in *D. pseudoobscura.* It is, in short, a plot of ψ_{OL} as a function of T, the period of the driving cycle. The plotted points are the medians of the observed eclosion rhythms. They constitute a plot of ψ_{RL} as a function of T. The approximation of the points (ψ_{RL}) to the solid curve (ψ_{OL}) is reasonably good, but the discrepancies are significant and meaningful.

Indeed the discrepancy—its sign, but not its magnitude—was predicted, given our knowledge that the observed rhythm in *Drosophila* reflects the behavior of a second rhythmic process driven by a light-sensitive circadian oscillation. [This is the distinction between A and B oscillators in earlier papers from this laboratory, for example, Pittendrigh, Bruce, and Kaus, (1958).] When A (the light-coupled driving oscillation of the system) assumes the long period of an external light cycle (say, 27 hr) and, in turn, drives the B oscillation, it is expected that the driven oscillation (B) will phase-lead the light-sensitive driver (A) just as the latter phase-leads the light cycle that drives it. That is what is observed: the rhythm phase-leads its driving oscillation when $T > \tau$; the rhythm phase-lags its driving oscillation when $T < \tau$. Consequently, in general, when $T \neq \tau$, the more easily observed ψ_{RL} is not (without much further, and unavailable, information) a reliable marker of ψ_{OL} or, for the same reasons, of ϕ_i.

The complexities of a circadian system in a multicellular organism clearly present problems in designing and interpreting experiments directed at specific versions (like the "coincidence" model) of the more general form of Bünning's proposal that circadian rhythmicity underlies the photoperiodic time measurement. We believe, however, that the difficulties are not so great that useful first approximation tests of the coincidence model cannot be made—and we report such tests below.

ENTRAINMENT AND INDUCTION BY ASYMMETRIC SKELETON PHOTOPERIODS

Our initial interest in *P. gossypiella* was prompted by Adkisson's (1964) impressive body of data on the inductive effects of light pulses interrupting the night following various photoperiods in cycles of $T=24$ hr. We noted in our 1964 paper that the general nature of his results, revealing peaks of inductive effectiveness at two widely separated times of the "night," were (far from being an embarrassment to any version of Bünning's general hypothesis) precisely what would be predicted by a combination of the "coincidence" model and the then newer knowledge of the entrainment of circadian systems that we had acquired by study of the *Drosophila* system. We also presented data (Pittendrigh and Minis, 1964; Minis, 1965) supporting the proposition that *Pectinophora* rhythmicity entrains to "asymmetric" skeleton photoperiodic regimes in the same way as does *Drosophila*.

A brief light pulse interrupting a long night, together with the prior longer light pulse (the main photoperiod), constitute what we call a "skeleton photoperiod." The two pulses effectively simulate the action of a single pulse whose duration is equal to the interval between the beginning of the first and the end of the second pulse in the skeleton system (see Pittendrigh and Minis, 1964; Pittendrigh, 1965). When the two pulses are of unequal duration the skeleton is said to be "asymmetric"; when they are equal the skeleton is "symmetric."

Figures 9 and 10a re-examine the issue with new data on both the induction and entrainment effects of asymmetric skeleton photoperiods involving a 10-hr main photoperiod ($T=24$ hr). This time our "night interruption" pulses are of 1-hr duration (*cf.* 15 min in the earlier study), and their action in the entrainment of the system is assayed by observation of the pupal

The Photoperiodic Time Measurement in *Pectinophora gossypiella* and Its Relation to the Circadian System in That Species

225

FIGURE 9 *Pectinophora gossypiella.* The pupal eclosion rhythm in asymmetric skeleton photoperiods. Main photoperiod is 10 hr. Night breaks are of 1-hr duration. The time at which the night break begins (in hours since the beginning of the main photoperiod) is indicated at the right. The median of the daily distribution is marked by a + sign; it is ϕ_r for the rhythm. Top panel: control with no night break.

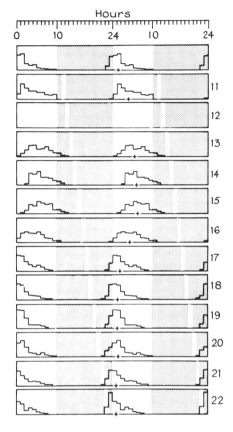

eclosion rhythm. The pupal eclosion rhythm provides a much less variable measure of system phase than the rhythm of egg-laying by adult females that was used earlier.

Figure 10a again confirms Adkisson's (1964) report of the bimodality of the curve that describes the inductive effectiveness of a light pulse as a function of its position during the night: one peak occurs when the light pulse falls a few hours after the end of the 10-hr main photoperiod; the other, much bigger, occurs when the pulse falls a few hours before the onset of the main photoperiod. Figure 10a also shows again, and more clearly this time, that it is not necessary, on that account, to assume that there are two photo-inducible phases in the night, but rather that the data can be interpreted on the basis of a single ϕ_i because of peculiarities in the entrainment of the system to asymmetric skeleton photoperiods. Thus as the night interruption falls later, the steady-state phase of the rhythm shifts

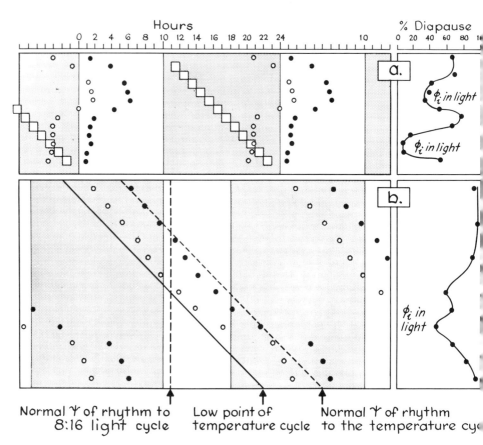

FIGURE 10 *Pectinophora gossypiella.* (a) Entrainment and induction by asymmetric skeleton photoperiods. The phase reference point of the pupal eclosion rhythm (distribution median) is shown by solid circles (see Figure 9 for raw data); the inducible phase, 5 hr earlier, by open circles. Panel at right plots the inductive effect (prevention of diapause) of the same skeleton regimes. (b) Entrainment and induction by environments involving concurrent cycles of light (LD 8:16) and temperature (20–29°C, sinusoidal). The position of the rhythm's phase reference point in LD 8:16 at constant 20° is indicated by the dashed line 1 hr after dawn. Its position in a sinusoidal temperature regime in constant dark is also indicated by a dashed line 7 hr after the lowest point on the temperature curve. Solid circles mark the rhythm's phase reference, and open circles mark ϕ_i, for the 12 environments characterized by 12 different phase relations between the light and the temperature. Panel at right plots the inductive effect (incidence of diapause) of these same environments.

The Photoperiodic Time Measurement in *Pectinophora gossypiella*
and Its Relation to the Circadian System in That Species

227

to the right (ψ_{RL} becomes more negative), just as it does for complete
photoperiods of increasing duration. The first peak of inductive effectiveness occurs when the night interruption ends about 14 hr after the beginning of the 10-hr photoperiod, thus constituting (with the main 10-hr light
pulse) an asymmetric skeleton of a 14-hr photoperiod in which the end of
the night interruption simulates sunset. Skeleton simulation of photoperiods in this manner fails when the pulse falls at 16 and 17 hr, but at
18 hr and thereafter the rhythm assumes a ψ relative to the light regime
such that the night interruption functions as dawn, not as sunset. The
second peak of inductive effectiveness occurs when the pulse is at hour 21,
and the system is entrained to a simulation of a 14-hr photoperiod in just
this way.

In the first peak of inductive effectiveness ϕ_i coincides with light at the
beginning of the main photoperiod; in the second peak of inductive effectiveness ϕ_i coincides with the light of the night interruption itself. As Figure 10a indicates, there is a general correlation, considering the whole
series of night interruptions, between the amount of photoperiodic induction and the proximity to light of ϕ_i (postulated to lie 5 hr ahead of the
phase reference point of the pupal eclosion rhythm).

We should note, however, that there is nothing in our model to explain
the marked difference in amplitude of the first and second peaks of induction.

ENTRAINMENT AND INDUCTION BY CONCURRENT CYCLES OF
LIGHT AND TEMPERATURE

Some encouragement for the coincidence model was also derived from experiments involving concurrent cycles of light and temperature. These were
prompted by the observations of Pittendrigh (1958) on *Drosophila* to the
effect that the system can be entrained by temperature as well as by light
cycles, and that there are marked peculiarities in the system's response to
environments in which the phase relationship between the light and temperature cycles is abnormal. In *Drosophila pseudoobscura* the phase reference point for the pupal eclosion rhythm can be made to occur at any
position within a 12-hr range relative to the light cycle by manipulating
the phase of the temperature cycle relative to that of the light. Bruce
(1960) and Pittendrigh (1960) reported similar behavior in the circadian
rhythmicities of *Euglena* and cockroaches. *Pectinophora* also responds in
the same general way (Figure 10b).

This behavior yields a new testable prediction for the coincidence model of photoperiodic induction: Induction should be effected by noninductive light cycles if the phase of a concurrent temperature cycle is such that ϕ_i is dragged into the beginning of the short photoperiod.

Figure 10b reports the results (in terms of both the phase of the eclosion rhythm and the incidence of diapause) of experiments in which the phase relation of a light cycle to a concurrent temperature cycle is systematically varied. Our expectation, based on the behavior of *Drosophila* and other systems, was upheld: when the low point of the temperature cycle is displaced to the right (later relative to the light cycle) the phase of the eclosion rhythm also shifts to the right. Eventually ϕ_i [lying 5 hr to the left of (earlier than) the eclosion median] is dragged into the light. As the figure indicates, there is again a general correlation between photoperiodic induction and the proximity of ϕ_i to the light: maximum induction occurs when ϕ_i is predicted to be fully coincident with light. There is not, however, any sharp discontinuity in the curve marking the sudden coincidence of a brief, discrete, phase-point with the light; and the general level of induction is also disappointingly low. Nevertheless, the results of Figure 10b do constitute a new kind of general support for the model: Induction is changed without change of photoperiod; it is changed with demonstrable change in the phase relation of the circadian system to the light; and, finally, the phase relation between system and light that effects the greatest induction is one that, presumably, involves coincidence of light and ϕ_i.

There is, to be sure, another way of looking at these data that would relate them to another substantial body of fact: there have been many studies measuring the effect on induction of varying the night temperature, for example, Beck (1962) and Bonnemaison (1966). The general result is that the incidence of diapause increases when the "night temperature" is lowered, either directly relative to fixed daytime temperature, or indirectly by changing the phase relationship between a square temperature wave and the light cycle. Both these procedures will have effects on the phase of the system relative to the light, and, therefore, both can, in principle, yield to explanation along the lines developed in this section. Thus while it is true that the results of Figure 10b can be "explained" without reference to circadian phenomena by saying that the incidence of diapause is maximal when the average night temperature is lowest, the more important point may be that known effects of manipulating night temperatures can be explained in terms of a coincidence model.

Saunders' (1968) strikingly clear results obtained when temperature pulses are applied in the night, as against the daytime, also yield to this general line of explanation as he himself stated.

The Photoperiodic Time Measurement in *Pectinophora gossypiella*
and Its Relation to the Circadian System in That Species

229

ENTRAINMENT AND INDUCTION BY SYMMETRIC SKELETON
PHOTOPERIODS: "BISTABILITY" EXPERIMENTS

As noted above, circadian oscillations will entrain to so-called symmetric skeleton photoperiods consisting of two brief pulses of equal duration in each cycle. These, like the asymmetric skeletons discussed above, simulate complete photoperiods with remarkable fidelity.

Entrainment by symmetric skeleton photoperiods involves a phenomenon unsuspected until computer simulation of entrainment uncovered it; experiment then confirmed its reality. When the skeleton photoperiod is close to $\tau/2$ of the entrained oscillator two alternative, stable, entrained steady-states are possible. Consider the skeleton photoperiod *13*:11; it is also the skeleton of *11*:13 (ignoring the small duration of the pulses themselves). Oscillations that behave like the *Drosophila* system, when exposed to such a regime, can assume either of two phase relationships (ψ_{OL}) to the light cycle: one is the ψ_{OL} characteristic of an 11-hr photoperiod; the other is that characteristic of a 13-hr photoperiod. Which one is actually realized depends on the initial conditions: (1) the phase of the oscillator seen by the first pulse of the entraining cycle; and (2) the interval (11 or 13 hr) between the first two pulses in each cycle. Pittendrigh (1966) has pointed out that the two phase-points illuminated by the two light pulses in each cycle are different in the steady-state characteristic of 11 and that of 13 (see Figure 11a). He also showed that this fact immediately yields an explanation for the remarkably complex but clear and otherwise unexplained results that Hillman (1964) obtained using skeletons of 11- and 13-hr photoperiods to induce flowering in *Lemna*; they are exactly what we would predict from a combination of the coincidence model and the assumption that *Lemna* has a circadian oscillator sufficiently similar to that of *Drosophila* to yield two alternative entrained steady-states under 11- and 13-hr skeletons, depending on initial conditions. Hillman (this volume) elegantly provides the clinching evidence that this is, in fact, the case. This new parallel between entrainment and induction in a system's response to skeleton photoperiods near $\tau/2$ provides, together with the "resonance effects" discussed later, the most powerful evidence that some form of the Bünning hypothesis is valid at least for some organisms.

Following our analysis of the *Lemna* results we extended the technique to *P. gossypiella*. When subjected to the skeletons of 11- and 13-hr photoperiods, the egg-hatch rhythm did indeed display the two alternative entrained steady-states that theory demanded, given the appropriate initial conditions. But we failed, on the other hand, to obtain any significant difference in the incidence of completed development induced by the two

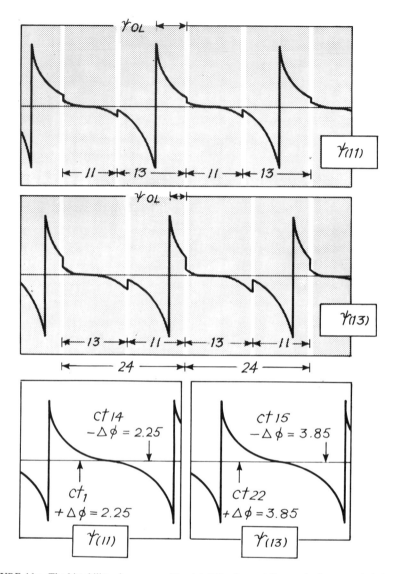

FIGURE 11a The bistability phenomena: The detail for *Drosophila pseudoobscura* and their
bearing on Hillman's (1964) data for *Lemna*. *Drosophila pseudoobscura*. The two, stable, entrained
steady-states effected by the symmetric skeletons (two pulses each of 15-min duration) of 11- and
13-hr photoperiods (T = 24 hr). The ψ_{OL} realized by the 11-hr skeleton [$\psi(11)$] is shown in the
upper panel; $\psi(13)$ is illustrated in the middle panel. The two lower panels show—for $\psi(11)$ and
$\psi(13)$ separately—the two phase-points of the oscillation that coincide with the light pulses; the
phase-points that coincide with light when $\psi(11)$ is realized are not the same as those coincident
with light when $\psi(13)$ is realized. In terms of the general model, ϕ_i is one of those points illuminated when $\psi(11)$ develops.

The Photoperiodic Time Measurement in *Pectinophora gossypiella*
and Its Relation to the Circadian System in That Species

231

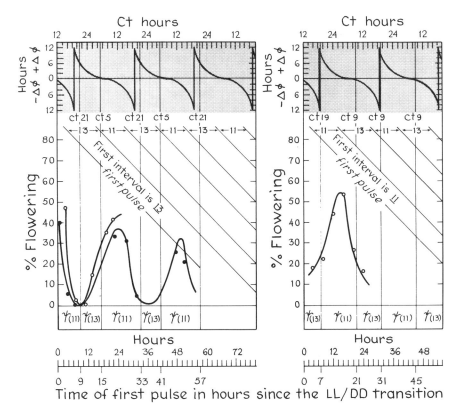

FIGURE 11b Pittendrigh's (1966) interpretation of Hillman's (1964) data on the induction of flowering in *Lemna* by symmetric skeletons of 11- and 13-hr photoperiods. Flowering maxima occur when the initial conditions (the initial relationship between oscillation and light cycle) are such that $\psi(11)$ is predicted and, as Hillman (this volume) now shows, is realized. Whether $\psi(11)$ or $\psi(13)$ develops is determined by (1) the duration of the first interval; (2) the phase-point of the oscillation illuminated by the first pulse. The oscillation begins at ct 12 (the beginning of the subjective night) when the system is transferred from LL to DD. The graph uses the *Drosophila* phase-response curve for illustration, but computer simulations using variations on the amplitude and shape of the curve show the predictions are general.

regimes. At 20°C there was nearly 0 percent development (100 percent diapause) in both regimes (whether ψ_{OL}, as reflected by ψ_{RL}, was that of 13 or 11); at 26°C there was about 20 percent development in both regimes. These frequencies are close to what one gets in constant darkness at the temperatures employed.

It is regrettable, in retrospect, that we gambled on the efficacy of brief 15-min pulses to create the skeletons: they were, in fact, adequate to effect entrainment, but there is now good reason to suspect that our negative re-

sult in the matter of induction could simply have been due to the inade-
quacy of a 15-min pulse to effect induction in any contrived light cycle.
Thus its effect as an inducing agent is often hard to detect when the pulse
is used as a night interruption; and whereas light cycles with a frequency
close to but not equal to τ cause significant differences in induction when
longer light pulses are used, they fail to do so consistently when 15-min
pulses are used. It could well be that by using longer pulses (of several
hours' duration), the "bistability" phenomenon will yet be found in
Pectinophora, as in *Lemna*.

CHANGE IN ψ_{RL} CAUSED BY SELECTION AND ITS EFFECT ON PHOTOPERIODIC INDUCTION

The facts presented are compatible with the coincidence model in general
and, more specifically, with the proposition that ϕ_i is part of a driven oscillation and occurs late in the system's subjective night.

This limited success encouraged us to carry out the experiments summarized by Figure 12 utilizing the two strains of *P. gossypiella*, which, as a
consequence of selection, differ in their ψ_{RL} by 5 hr.

The potential utility of these strains, in our present concern with the
coincidence model for the photoperiodic time measurement, derives from
the possibility that the particular driven oscillation of which ϕ_i is thought
to be part may—as a correlated response to selection on the pupal eclosion
rhythm—have had its phase relation to the light-sensitive driver (and hence
to the light cycle itself) shifted also. If such a shift had, in fact, been
created we would expect marked differences in the photoperiodic responses
of our strains ("early" and "late") to any particular light cycle. Indeed, we
can be explicit—to some extent—about what to expect if such a correlated
shift had occurred. In Figure 12, lower right, we indicate that as in
Drosophila, the oscillation of the *Pectinophora* system begins from a fixed
phase-point at the onset of darkness following light periods of 12 or more
hours. The figure indicates that if ϕ_i occurs 11 hr after that time, it is illuminated when the daily photoperiod is 14 hr but not when it is as short as
12 hr: photoperiodic induction of normal development occurs only in the
longer day. The figure also indicates that we must, however, expect more
induction by short, 12-hr, days in the "late" strain; and, on the other hand,
no difference between the "early" and the unselected line in their response
to that photoperiod.

Experiments made after four and five generations of selection were very
encouraging in support of this view. The response of "late" to photo-

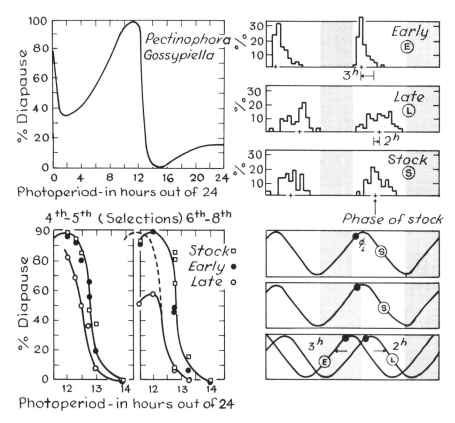

FIGURE 12 *Pectinophora gossypiella.* Photoperiodic induction and ψ_{RL}. Upper left: the photo-periodic response curve. Upper right: the effect of eight generations of selection on ψ_{RL} (cf. Figure 5). Lower right: top and middle panels illustrate the coincidence of ϕ_i with light in *LD 14*:10 and its noncoincidence in *LD 12*:12 (see text); lower panel illustrates the coincidence of ϕ_i and light in *LD 12*:12 in the "late" strain on the assumption that ψ_{RL} of the clock rhythm has also been changed by selection. Lower left: Photoperiodic response curves for early and late as selection progressed. The dotted line portion of the curve for late (right panel) indicates the normal form of the curve—how it would appear had it been merely displaced. The observed value for *LD 11*:13 indicates, however, that the curve is depressed and not displaced.

periods ranging from 12 to 14 hr was different from that of the unselected stock: a consistently smaller fraction of the insects entered diapause. In other words, on all short photoperiods between 12 and 13 hr duration, "late" responded as though the day were significantly longer. Another series of observations made on the strains after six and eight generations of selection gave results that confirmed these striking differences between "stock" and "late" in their response to photoperiods of between 12 and

13 hr: "late" again responded as though each short day were in fact longer. However, in this series of experiments we made observations on the response of both strains to 11-hr days; these observations severely complicate the picture, rendering it intractable to any obvious explanation in terms of the coincidence model. As Figure 12, lower left, shows, we had not—as we thought on the basis of our past experiments—simply displaced the entire photoperiodic response curve to the left by shifting the phase (relative to the light) of that particular driven oscillation of which ϕ_i is part. The photoperiodic response curve in "late" is not displaced; it is depressed; without the observation on the 11-hr day effect the depression of the curve looked like a displacement.

It remains a remarkable fact that the early and late strains, differing markedly in the phase relationship of at least two driven rhythms to the light cycle, also differ markedly in their photoperiodic response. While the meaning of this difference is certainly not made clear by any simple coincidence model for the measurement of night length, it is tempting to attribute it to some consequence of the change in phase between the circadian system and the light cycle. Even that, however, may be hazardous: it is plausible that there were correlated responses to selection in other sections of the insect's genotype having nothing to do with circadian organization at all, and that it is some such disturbance of a physiological system quite unrelated to circadian phenomena that has changed the general level of diapause incidence attainable under any photoperiodic regime. In this context we point out that there is slight evidence (that, of itself, we would otherwise ignore) that the curve for early is perhaps also very slightly depressed.

ENTRAINMENT AND INDUCTION WHEN T IS CLOSE TO τ

The experiments reported in this section begin the positive case, which becomes stronger in subsequent sections, against the coincidence model, in any form, for the explanation of time measurement in this species.

The rationale for these experiments, originally developed in our 1964 paper, derives from our understanding of the mechanism of the entrainment of the circadian system by light cycles. In an entrained steady-state, the system assumes the period (T) of the entraining light cycle; the discrepancy between the period (τ) of the free-running system and T is effected by a discrete phase shift. The sign and magnitude of that shift is given by $T - \tau = \Delta\phi$. Clearly when $T > \tau$ the phase shift effected in each cycle must be a delay; when $T < \tau$ the phase shift in each cycle must be an

The Photoperiodic Time Measurement in *Pectinophora gossypiella*
and Its Relation to the Circadian System in That Species

235

FIGURE 13 Upper panel: Phase-response curve (for 15-min pulses of white light ∿1000 lux) for the free-running oscillation that drives the pupal eclosion rhythm in *Drosophila pseudoobscura*. It illustrates the general facts (1) that sensitivity to light is restricted to or concentrated in the subjective night; and (2) that light falling early in the subjective night causes phase delays, while light in the late subjective night causes phase advances. The two lower panels illustrate the entrained steady states effected by light cycles whose periods (T) are 21 and 27 hr and consist of one 15-min pulse per cycle. When T = 21 (T < τ) the light falls in the late subjective night; when T = 27 (T > τ) it falls in the early subjective night.

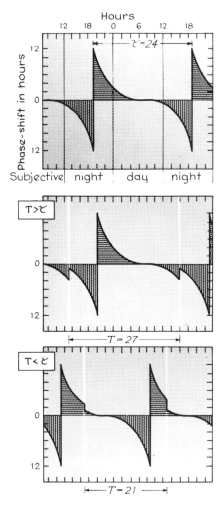

advance. Figure 13 illustrates these relationships schematically and indicates the further point—of importance in this analysis—that when T < τ the light pulse falls in the late subjective night and when T > τ the light pulse falls in the early subjective night. It is a functional necessity for stable entrainment (and an empirical fact for every species studied) that light falling in the early subjective night causes phase delays, but in the late night it causes phase advances. These relationships yield strong predictions, based on the coincidence model, that induction can be effected by a photoperiod of otherwise noninductive duration, simply by a small change in the frequency (or period, T) of the cycle in which it is presented. In

1964 we reported experiments of this type using very short (15-min) photoperiods in light cycles in which T was either shorter or longer than τ; our results in 1964 were inconsistent. Hamner and Enright (1967), adopting the design of our 1964 experiments, found that house finches in T22 and T26 showed significant correlations between their photoperiodic responses and the phase (early or late subjective day as indicated by the activity rhythm) illuminated in the respective cycles.

We have since concluded that in *Pectinophora*, 15-min light pulses are simply inadequate to effect induction at all (see above). The original experiments have therefore been repeated using 8-hr pulses that, in 24-hr

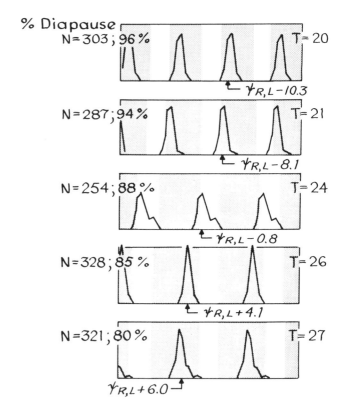

FIGURE 14 *Pectinophora gossypiella*. The dependence of induction on T when T is close to τ (nonresonance effects). The entrained steady-states of the pupal eclosion rhythm effected by light cycles, all involving an 8-hr photoperiod, whose period (T) ranges from 20 to 27 hr. The photoperiodic induction (measured as % diapause) effected by each cycle is indicated at the left.

The Photoperiodic Time Measurement in *Pectinophora gossypiella*
and Its Relation to the Circadian System in That Species

237

cycles, induce very limited amounts of normal development. Using cycles in which T was 20, 21, 24, 26, and 27 hr (all with 8-hr photoperiods), we assayed both their inductive and entrainment effects. Since τ for all three known oscillations is between 22 and 24 hr, we expect the 20- and 21-hr cycles to cause an increased incidence of development—because the light in steady-state entrainment should fall earlier than usual, in the late subjective night, where (presumably) ϕ_i lies; and that the 26- and 27-hr cycles would not differ from the 24-hr cycle—because ϕ_i would simply remain noncoincident with the light which, in these cycles, falls even further to the right on the oscillation's phase-response curve.

The results of these experiments (which were executed with the help of McDonald Horne in our laboratory) are summarized in Figure 14. They are clearly not what the model predicts. There is, to be sure, a clear dependence of the incidence of development (photoperiodic induction) on the frequency of the cycle. But the dependence is the obverse of what the model predicts. On the other hand, the dependence of ψ_{RL} on T is as predicted: In cycles of short T the light does fall in the late subjective night to cause its phase advance, but when it falls there it increases the incidence of diapause instead of decreasing it as theory demands.

We could, of course, note that our earlier discussion of the complexity of circadian systems raised a caveat about interpretations of experiments when T was far from τ. On the other hand, it is, to say the least, not easy to see how that caveat can help us relate the present data (in any way) to a simple model involving the coincidence of light and a discrete ϕ_i. The sign of the effects we found was wrong, and they lacked even a hint of that discontinuity (which the model demands) in the dependence of induction on T when the sign of T$-\tau$ itself changes.

The data in this section, clear in themselves, do not yield to any version of Bünning's hypothesis thus far made explicit.

ACTION SPECTRA

Bruce and Minis (1969) have reported an action spectrum for the initiation by light pulses of the circadian rhythm of egg hatching in *Pectinophora gossypiella*. Their spectrum has the same general features as that reported for the circadian system in *Neurospora* by Sargent and Briggs (1967), and in *Drosophila* by Frank and Zimmerman (1969): there is a broad maximum in the blue and a sharp decline in effectiveness of wavelengths greater than 490 nm; light of 520 nm and above is completely without effect in initiating the circadian oscillation.

Subsequent studies in the Princeton laboratory (Pittendrigh *et al.*, 1970) revealed the fact (not of itself surprising) that photoperiodic induction could be effected by wavelengths (600 nm and above) that are not "seen" by the driving oscillation of the circadian system. As noted in 1964, and remarked on in the present introduction, the functionally separable roles of entrainment and induction might be mediated by concretely different photoreceptors.

There were, however, other aspects of our initial experiments on induction by light of 600 nm that were surprising: using this wavelength from the beginning of egg-life (the moment of oviposition), photoperiods of 12 and 14 hr were discriminated with complete success (100 percent vs. 2 percent diapause). Since earlier studies had shown (1) that the circadian oscillation in the egg does not begin until the midpoint of its development (Minis and Pittendrigh, 1968); and (2) that the oscillation apparently does not "see" light of 600 nm, we were confronted with the surprising prospect that our moth populations, discriminating so successfully between 12 and 14 hr of 600 nm light, were doing so in spite of the fact that their circadian oscillations (if indeed they were running at all) were free-running and unsynchronized; and hence that the phase of the circadian system "seeing" the light has nothing to do with the organism's "perception" of the light's duration. We were confronted, in other words, with the most telling, unequivocally positive evidence that Bünning's proposition in any form is invalid at least in this one species: that the clock measuring photoperiod is neither a circadian oscillator nor any of its slave rhythms.

TIME MEASUREMENT BY POPULATIONS WITH ASYNCHRONOUS OSCILLATIONS

The possibility of that interpretation being true was so important—and to us so unlikely—that we had to consider seriously the following question: Are the initiation and entrainment of circadian oscillations mediated by different pigments yielding different action spectra? Experiments performed by Eichhorn at Princeton (Pittendrigh *et al.*, 1970) to test the efficacy of 600 nm in entrainment settled that matter unequivocally. A 12-hr pulse of 600 nm light is as fully ineffective as an entraining agent (when seen repeatedly in 24-hr cycles) as is the briefer (15-min) single pulse in initiating a rhythm. Figures 15 and 16 illustrate the success of 480 nm and the failure of 600 nm light cycles as entraining agents.

We have, therefore, no escape from the following important conclusion: Populations of *P. gossypiella* in which all known circadian rhythms (and thence their underlying circadian oscillators) are totally asynchronous can

The Photoperiodic Time Measurement in *Pectinophora gossypiella*
and Its Relation to the Circadian System in That Species

239

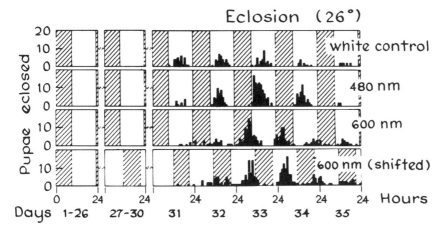

FIGURE 15 *Pectinophora gossypiella*. The failure of 600 nm light to entrain the rhythm of pupal
eclosion. All populations were raised in *LD 14*:10 (white light) for the 26 days. A control group
continues to develop and eventually ecloses in white light (top panel). Another group (second
panel) is shifted on the 27th day into blue light (480 nm) that continues to entrain the rhythm
effectively. The third and fourth groups were transferred on the 27th day into cycles involving red
light (600 nm). The phase of the light cycle is unchanged in the third group but shifted by 12 hr in
the fourth. Both of the groups that experience 600 nm free run, irrespective of the phase of the *LD*
cycle: their synchrony derives from their having been entrained by the white light cycle.

nevertheless measure—with complete efficacy—the difference between 12
and 14 hr of light.

To say that therefore the duration of the light pulse is measured irrespective of its phase relation to any component in the total circadian system is
to go further, and further than is warranted by the data so far presented.
We have already given evidence of a diversity of driving oscillations in the
Pectinophora system and indeed of some differences between them in spectral sensitivity. There remains, then, the important possibility that while
the three known rhythms (and their drivers) were asynchronous in the
populations whose behavior is summarized in Table 1, a separate oscillator,
devoted exclusively to measurement of photoperiod and coupled by a different pigment to the light cycle, was in fact synchronous in the population.

RESONANCE EXPERIMENTS

That possibility raises the interesting prospect that our previous emphasis
on the importance of concurrent analyses of induction and the entrainment of any arbitrary marker of system-phase is not only ill-founded but

Pectinophora gossypiella

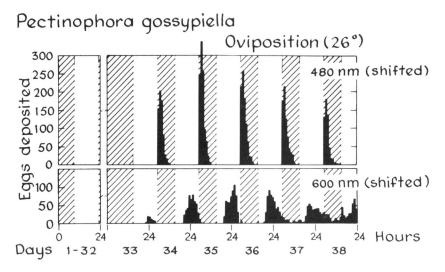

FIGURE 16 *Pectinophora gossypiella.* The failure of 600 nm light to entrain the rhythm of oviposition activity. Moths were raised in *LD 14*:10 (white light) for 32 days and then transferred, as adults, into *14*:10 light/dark cycles involving either 480 nm or 600 nm. In both cases the phase of the monochromatic light cycle was shifted by 6 hr relative to the previous white light cycle. The population in 480 nm is immediately entrained by that light: Oviposition occurs as usual in the dark. The population in 600 nm is unentrained by the new light regime: The residual rhythmicity of the population (synchrony of individuals) derives from the prior white light and is free-running in the red light.

flatly misleading. We are being warned (1) that synchrony of the diverse circadian oscillators in the total system may be due, at least in part, to their common but independent submission to entrainment by white light reaching each of them by direct absorption—not after transduction by a common photoreceptor (eye); and (2) that if their independent coupling to (white) light involves different receptor pigments, some may be synchronized by a particular monochromatic source that leaves the phase of others totally randomized. This prospect of synchrony within the normal system deriving in part from the direct coupling of diverse components to light has become increasingly plausible recently following the several demonstrations (Lees, 1964; Menaker, 1968; Menaker and Keatts, 1968; Adler, 1969) that the eye is bypassed, both in photoperiodic induction and in the entrainment of circadian phenomena.

Given this situation the most promising line of further attack is to use what Dr. Klaus Brinkmann (personal communication) proposes to call "resonance experiments." These are of two kinds. The first, used in plants by Schwabe (1955), Melchers (1958), Bünning (1960), and Bunsow (1960),

The Photoperiodic Time Measurement in *Pectinophora gossypiella*
and Its Relation to the Circadian System in That Species

241

TABLE 1   Results of Exposing Populations of *Pectinophora* to 24-hr Cycles Consisting of 14 hr Monochromatic Light, 10 Dark; or 12 hr Monochromatic Light, 12 Dark; at $25.5 \pm 0.5°C$

Wavelengths	Intensity $(ergs/cm^2/sec)$	N	% Development
14-hr photoperiod ($\pm$ 5 nm)			
White control	40	53	100
480	1.3	52	98
560	1.4	48	98
600	1.4	49	100
640	0.59	46	95
680	0.14	46	83
12-hr photoperiod ($\pm$ 5 nm)			
White control	40	49	2
370	0.13	27	7
380	0.06	25	4
400	0.56	29	3
480	1.3	50	2
540	6.6	57	0
620	0.57	55	2
Constant dark (DD)		142	73
Constant light (LL)	40		15

inserts one protracted dark period (up to 72 hr) into an otherwise 24-hr light/dark cycle and then assays the inductive action of brief interruptions of that protracted night. In the clearest cases [e.g., Melchers (1958) and Bunsow (1960)] there is a marked periodicity of effectiveness as the light is scanned across the long night. Saunders (1970) has recently extended this technique successfully to an insect. The second design—more clearly meriting the designation of "resonance effect"—was introduced into the field by K. C. Hamner (see his discussion of 1964) and his colleagues in the analysis of the photoperiodic control of flowering in Biloxi soybeans and later employed (without modification) by his nephew, W. M. Hamner (1963), in the analysis of the photoperiodic control of gonadal development in the housefinch (*Carpodacus mexicanus*). The design is to use a light pulse in cycles whose period is varied over a very broad range ($n \cdot \tau$, where τ is the period of the circadian system, and n is 2, 3, or 4) by greatly extending the duration of the dark period. The now classical experiments of Nanda and Hamner (1958) and Hamner and Takimoto (1964) on Biloxi soybeans, which is a short-day plant, used 8-hr light pulses. When the plants were exposed to cycles whose period (T) was $n \cdot 24$ hr (and hence close to $n \cdot \tau$), flowering maxima were observed. But when they were exposed to cycles of $n \cdot \tau + \tau/2$, flowering minima were observed. In the first case the recurrent 8-hr light was seen as a "short day" presumably because in each case (whether n was 1, 2, or 3) the light fell during the subjective day; in the second case (when $T = n \cdot \tau + \tau/2$) the short light was seen as a "long day" because it fell in each case (whether n was 1, 2, or 3) in the subjective night.

The facts from both types of resonance experiments are among the strongest evidence in support of some form of Bünning's original hypothesis: The phase relationship between the light pulse and an ongoing circadian periodicity is crucial in the inductive mechanism.

If there is a distinct circadian oscillator devoted to photoperiodic time measurement in *Pectinophora gossypiella*, and if it is coupled to the light cycle by a red-absorbing pigment, its presence should be detectable in resonance experiments using red light—or, for that matter, white light. Adkisson (1964, 1966) has published data from such experiments on

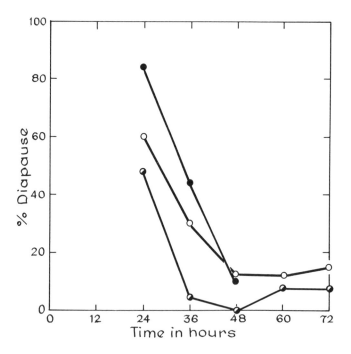

FIGURE 17 *Pectinophora gossypiella*. The available "Resonance Experiments." (See text for explanation.) Open circles: 8-hr light pulses (Adkisson, 1966). Half-filled circles: 12-hr light pulses (Adkisson, 1964). Solid circles: 12-hr light pulses (Pittendrigh and Minis, unpublished data).

The Photoperiodic Time Measurement in *Pectinophora gossypiella*
and Its Relation to the Circadian System in That Species

243

Pectinophora using 8- and 12-hr pulses, and we have obtained data using 12-hr pulses. All are summarized in Figure 17. It is unfortunate that there are not more points on all the curves, but there is no reason to take this limitation about the data sufficiently strongly to avoid the otherwise clear indication there is no resonance effect. Indeed, in our judgment, none exists.

We seem, therefore, to have closed the one avenue of escape from the conclusion that no form of the Bünning hypothesis is a valid explanation for the photoperiodic time measurement in this species. We should, however, point out a final and very different kind of caveat in taking the data of Figure 17 as an unequivocal disposition of the issue, a caveat that does not concern the general power of resonance experiments *per se*. It concerns, rather, their limitations when the experimental object is going through a relatively brief and complex developmental sequence. Two points are relevant: The first is the possibility that the opportunity for photoperiodic induction of complete development may be limited to a relatively small fraction of the developmental sequence, in which case the light may miss it entirely when T is long; the second is that if the susceptible period be limited it may yet require more than one signal to effect induction—and the resonance effect will be still more elusive. In general, resonance experiments are (1) always powerful when a positive resonance effect is found; but (2) powerful when one is missing only if the system remains for a sufficiently long time in its inducible state for the resonance effect to occur.

It is pertinent to note that positive resonance effects have been reported so far from "steady-state" systems of long duration uncomplicated by a sequence of unique developmental stages. They have been reported (1) from mature plants that can be exposed to many cycles of long T; (2) from mature birds of adequate longevity; and (3) from an insect (Saunders, 1970) in which the photoperiodic effect is exerted on the mature adult which again can experience several cycles of long T without the confounding effects of change in developmental state. The negative results reported in Figure 17 lack power to the extent that the inducible stage may well be too limited for resonance to be detected.

The happy circumstance that escape from diapause in *Pectinophora gossypiella* (as in some other insects) is under photoperiodic control does, however, leave open the possibility of performing more meaningful resonance experiments on this insect. The long-enduring steady-state afforded by diapause, lasting for weeks, will permit us to subject the system to a nearly unlimited number of cycles with $T = (n \cdot \tau)$ or $(n \cdot \tau + \tau/2)$. These tests are planned but have not yet been made.

DISCUSSION

There is no escape from one conclusion: The facts bearing on the possible relationship between circadian organization and the photoperiodic time measurement in *P. gossypiella* are complex. A second conclusion is almost as secure: If the circadian system does effect the time measurement, its action in this respect is not adequately represented by our earlier, simple "coincidence" model. There are some facts (asymmetric skeleton photoperiods; conflicting light and temperature cycles) that do yield, more or less, to such a model. But there are others (especially those involving T close to but not equal to τ) that defy it; and still others (the response of the "late" stock) that simply elude it.

We acknowledge here that, in any case, the "coincidence" model has limitations for the *Pectinophora* system that go well beyond any of our experimental data and should have been recognized by us long ago, but were not. These concern the shape of the photoperiodic response curve itself. *P. gossypiella*, like several other insects, has a response curve with an "optimum": the incidence of diapause is maximal when the insect is exposed to photoperiods (T=24) of 11 or 12 hr and there is a decline of diapause incidence on both sides of that duration. To be sure, it is only as photoperiod gets longer that the incidence falls to zero, but the declining incidence when photoperiod shortens from 11 to about 4 is nevertheless substantial (Figure 1), and there is a curious increase again as the photoperiod shortens still further from 4 to 0 hr of light per 24-hr cycle. There is nothing in our present understanding of the mechanism of entrainment of a circadian system that will explain the complexity of these facts.

The greatest complication of all, of course, is the evidence that a perfect time measurement is effected in populations of individuals, all of whose currently demonstrable circadian rhythmicities are asynchronous. Clearly it may well be that the photoperiodic time measurement in *P. gossypiella* is not effected by the circadian system at all; but we recognize the issue remains fully open until strong "resonance" experiments have been redone using 12-hr light signals at 600 nm—in the termination of (as against the induction of) diapause, and until "bistability" experiments have been performed again using pulses of 600 nm light longer than 15 min.

Belief in the prevalence of unity in nature—and the persistent search for it—are among the strongest features of the scientific enterprise. It is only reluctantly that we will conclude that the photoperiodic time measurement everywhere is not the work of a common physiological device—or a limited array of variations on something in common.

The evidence has been indisputable for some time that at least in some

The Photoperiodic Time Measurement in *Pectinophora gossypiella*
and Its Relation to the Circadian System in That Species

245

plants and birds the circadian system does underlie the measurement of night length (even if in a way still to be elucidated). And Saunders' (1970) recent study adds at least one insect to this list of photoperiodic time measurements that are surely mediated by circadian phenomena. The incontestable cases are based on the demonstration of resonance effects and the great complexities of the bistability phenomena in *Lemna* (see Pittendrigh, 1966). It is tempting, therefore, to hope that "strong" resonance experiments on diapausing *Pectinophora* larvae may still show that the circadian system operates in that species, too. However, the fact remains that Lees' (1964, 1965, and this volume) extensive studies of the aphid *Megoura* continue to give no indication that circadian phenomena are involved in the measurement of night length by that insect; indeed they now come as close as one could ask to being positive evidence against the involvement of a circadian system in the photoperiodism of that species.

Evolutionary theory, which is our clearest guide to and explanation of unities in the living world also has another voice: It warns that functional similarity (functional unity) may often be the outcome of convergent evolution, of the fact that natural selection—Darwin's demon—is indifferent to precisely how the functional prerequisite is met.

Daily and annual periodicities are among the most ubiquitous and stringent features of the physical and biological environment of virtually every species on this planet. The demand by selection to meet the challenges and opportunities raised by such periodicities are correspondingly ubiquitous and strong. It should not surprise us, therefore, if the functional advantages of developing innate daily and annual self-sustaining oscillations have been "found" by selection more than once in the history of life, and that the successive solutions to the common problem have involved more than one concrete mechanism. And similarly, the clear functional merit of using night length as an almost noise-free signal of season may well have ensured that selection has found more than one way of effecting that measurement—itself only one way to effect adjustment to seasonal change. In any case, the issue remains intriguingly open in the case of *Pectinophora gossypiella*: the measurement of night length may well yet prove the work of a circadian oscillation, or, like *Megoura*, that of some other nonoscillatory process.

NOTE ADDED IN PROOF

The resonance experiments described in the main body of our paper and regarded as crucial have now been completed. Twenty-seven populations of ∿100 diapausing larvae were subjected to exotic light cycles whose periods ranged from T = 16 to T = 78. Each cycle involved 14 hours of light. The experiment included LD *12*:12 and LD *14*:10 controls. The former (*12*:12) maintained the insects in diapause; the latter (*14*:10) broke diapause. There was no clear evidence of a resonance effect in the action of any of the exotic cycles. But the results also defy explanation by any simple hourglass model. They will be reported more fully elsewhere.

REFERENCES

Adkisson, P. L. 1964. Action of the photoperiod in controlling insect diapause. Am. Nat. 98:357–374.

Adkisson, P. L. 1966. Internal clocks and insect diapause. Science 154:234–241.

Adler, K. 1969. Extraoptic phase shifting of circadian locomotor rhythm in salamanders. Science 164:1290–1292.

Beck, S. D. 1962. Photoperiodic induction of diapause in an insect. Biol. Bull. 122:1–12.

Bonnemaison, L. 1966. Combinaison de photophases et de scotophases avec des températures élevées ou basses sur la production des sexupares de *Dysaphis plantaginea* Pass. (Homoptères, *Aphididae*). C. R. Acad. Sci., Ser. D. 263:177–179.

Bruce, V. G. 1960. Environmental entrainment of circadian rhythms. Cold Spring Harbor Symp. Quant. Biol. 25:29–48.

Bruce, V. G., and D. H. Minis. 1969. Circadian clock action spectrum in a photoperiodic moth. Science 163:583–585.

Bünning, E. 1936. Die endogene Tagesrhythmik als Grundlage der photoperiodischen Reaktion. Ber. Dtsch. Bot. Ges. 54:590–607.

Bünning, E. 1960. Circadian rhythms and the time measurement in photoperiodism. Cold Spring Harbor Symp. Quant. Biol. 25:249–256.

Bunsow, R. C. 1960. The circadian rhythm of photoperiodic responsiveness in *Kalanchoe*. Cold Spring Harbor Symp. Quant. Biol. 25:257–260.

Frank, K. D., and W. F. Zimmerman. 1969. Action spectra for phase shifts of a circadian rhythm in *Drosophila*. Science 163:688–689.

Hamner, K. C., and A. Takimoto. 1964. Circadian rhythms and plant photoperiodism. Am. Nat. 98:295–322.

Hamner, W. M. 1963. Diurnal rhythm and photoperiodism in testicular recrudescence of the house finch. Science 142:1294–1295.

Hamner, W. M., and J. T. Enright. 1967. Relationships between photoperiodism and circadian rhythms of activity in the house finch. J. Exp. Biol. 46:43–61.

Hillman, W. S. 1964. Endogenous circadian rhythms and the response of *Lemna perpusilla* to skeleton photoperiods. Am. Nat. 98:323–328.

Lees, A. D. 1964. The location of the photoperiodic receptors in the aphid *Megoura viciae* Buckton. J. Exp. Biol. 41:119–133.

Lees, A. D. 1965. Is there a circadian component in the *Megoura* photoperiodic clock? p. 351 to 356. *In* J. Aschoff [ed.], Circadian clocks. North-Holland Publ. Co., Amsterdam.

Melchers, G. 1958. Die Beteiligung der endonomen Tagesrhythmik am Zustandekommen der photoperiodischen Reaktion der Kurztagpflanze *Kalanchoe blossfeldiana*. Z. Naturforsch. B. 11:544–548.

Menaker, M. 1968. Extraretinal light perception in the sparrow. I. Entrainment of the biological clock. Proc. Natl. Acad. Sci. U.S. 59:414–421.

The Photoperiodic Time Measurement in *Pectinophora gossypiella*
and Its Relation to the Circadian System in That Species

247

Menaker, M. and H. Keatts. 1968. Extraretinal light perception in the sparrow. II. Photoperiodic stimulation of testis growth. Proc. Natl. Acad. Sci. U.S. 60:146–151.

Minis, D. H. 1965. Parallel peculiarities in the entrainment of a circadian rhythm and photoperiodic induction in the pink boll worm (*Pectinophora gossypiella*), p. 333 to 343. *In* J. Aschoff [ed.], Circadian clocks. North-Holland Publ. Co., Amsterdam.

Minis, D. H., and C. S. Pittendrigh. 1968. Circadian systems. III. Circadian oscillation controlling hatching: Its ontogeny during embryogenesis of a moth. Science 159: 534–536.

Nanda, K. K., and K. C. Hamner. 1958. Studies on the nature of the endogenous rhythm affecting photoperiodic response of Biloxi soybean. Bot. Gaz. 120:14–25.

Peterson, D. M., and W. M. Hamner. 1968. Photoperiodic control of diapause in the codling moth. J. Insect Physiol. 14:519–528.

Pittendrigh, C. S. 1958. Perspectives in the study of biological clocks, p. 239 to 268. *In* Symposium on perspectives in marine biology. Univ. California Press, Berkeley.

Pittendrigh, C. S. 1960. Circadian rhythms and the circadian organization of living systems. Cold Spring Harbor Symp. Quant. Biol. 25:159–184.

Pittendrigh, C. S. 1965. On the mechanism of the entrainment of a circadian rhythm by light cycles, p. 277 to 297. *In* J. Aschoff [ed.], Circadian clocks. North-Holland Publ. Co., Amsterdam.

Pittendrigh, C. S. 1966. The circadian oscillation in *Drosophila pseudoobscura* pupae: A model for the photoperiodic clock. Z. Pflanzenphysiol. 54:275–307.

Pittendrigh, C. S. 1967. Circadian systems. I. The driving oscillation and its assay in *Drosophila pseudoobscura*. Proc. Natl. Acad. Sci. U.S. 58:1762–1767.

Pittendrigh, C. S., V. G. Bruce, and P. Kaus. 1958. On the significance of transients in daily rhythms. Proc. Natl. Acad. Sci. U.S. 44:965–973.

Pittendrigh, C. S., J. H. Eichhorn, D. H. Minis, and V. G. Bruce. 1970. Circadian systems. VI. The photoperiodic time measurement in *Pectinophora gossypiella*. Proc. Natl. Acad. Sci. U.S. 66:758–764.

Pittendrigh, C. S., and D. H. Minis. 1964. The entrainment of circadian oscillations by light and their role as photoperiodic clocks. Am. Nat. 98:261–294.

Pittendrigh, C. S., and D. H. Minis. 1971. Circadian systems. VII. The complexities of a multicellular circadian system. In preparation.

Sargent, M. L., and W. R. Briggs. 1967. The effects of light on a circadian rhythm of conidiation in *Neurospora*. Plant Physiol. 42:1504–1510.

Saunders, D. S. 1968. Photoperiodism and time measurement in the parasitic wasp, *Nasonia vitripennis*. J. Insect Physiol. 14:433–450.

Saunders, D. S. 1970. Circadian clock in insect photoperiodism. Science 168:601.

Schwabe, W. W. 1955. Photoperiodic cycles of length differing from 24 hours in relation to endogenous rhythms. Physiol. Plant. 8:263–278.

Sweeney, B. M. 1969. Transducing mechanisms between circadian clock and overt rhythms in *Gonyaulax*. Can. J. Bot. 47:299–308.

Winfree, A. T. 1970. The effect of light flashes on a circadian rhythm in *Drosophila pseudoobscura*. Ph.D. Thesis. Princeton University.

DISCUSSION

LEES: I am very pleased to hear your evidence indicating that insects may have more than one type of biological clock! I have a question about the wavelength experiments in which you substituted monochromatic light for the whole of the photoperiod in long- and short-day cycles. Am I correct in thinking that monochromatic light at all the wavelengths tested produced the same effect as white light?

PITTENDRIGH: There was some suggestion that the incidence of induction was lower with light of 680 nm. The percentage of diapause was in the nineties for all other wavelengths, and it was 83 percent for 680 nm. But I am not at all sure that means anything, considering the variance in *Pectinophora*.

LEES: What intensities were involved? Do you know anything about the response threshold to monochromatic light of different wavelengths?

PITTENDRIGH: The measured intensities of the monochromatic light (given in Table 1) were all at least an order of magnitude higher than our estimate of the threshold intensity for white light.

BÜNNING: It was never my intention to state my hypothesis for all organisms; I fully agree that there may be some evolutionary convergence in time-measuring mechanisms. But I would like to raise one question about your work, Dr. Pittendrigh. We know that for many organisms—perhaps all organisms—we can, under certain conditions, eliminate photoperiodic control. This holds at least for many insects. In the cabbage butterfly, which I used in my work, for example, it is only necessary to raise temperature by a few degrees to fully eliminate photoperiodic control. That is, at high temperatures you get no diapause whatever, whether the caterpillars are under short-day or long-day conditions. Also red light, which is ineffective in entraining the circadian rhythm of eclosion, eliminates photoperiodic control in this insect. Could it be that the effect you found is simply the elimination of photoperiodic control similar to the temperature effect in the cabbage butterfly?

PITTENDRIGH: That's a fairly easy question to answer, Dr. Bünning, because while there is a temperature effect on the percentage of diapause in *Pectinophora*, it is really very slight. The effect is nowhere near the magnitude of the difference of 2–3 percent diapause for 12 hr of red light *vs.* 80–90 percent diapause for 14 hr of red light. So what we are seeing is clearly not the elimination of photoperiodic control, but the discrimination between a 12-hr and a 14-hr photoperiod.

HASTINGS: Were the action spectrum experiments with the filters done under fluorescent light, and is there any possibility that the angle of perception was such that the light actually reaching the insects was different from the light you were measuring?

PITTENDRIGH: No, the positioning was very carefully controlled. The caterpillars were raised individually in test tubes, and the position of each individual tube on the cabinet floor was noted. There was no geometric spread in variance. The caterpillars were only 30 or 35 cm from the filter; they were really quite close.

HASTINGS: Were any of the other experiments in which red light was ineffective done with red filters, or was a monochromatic light source used?

PITTENDRIGH: All were done with interference filters.

WILLIAM HAMNER: I have two comments to make. The first is about birds. I am finding now that it was a fortunate accident that I started working with house finches. Starlings and bobolinks don't respond like the house finches did. Apparently we are going to find a really broad range of photoperiodic timing devices in all kinds of things besides insects too.

The Photoperiodic Time Measurement in *Pectinophora gossypiella* and Its Relation to the Circadian System in That Species

249

My second comment is a point of information I wanted to direct toward Dr. Lees. In the previous experiments done on action spectra—your own and work you have reviewed—what conditions were used to detect maximal sensitivities to different wavelengths? Were the experiments done with photoperiod differences of 12 hr *vs.* 14 hr, where you would get maximum inhibition or stimulation of diapause, or were they done with more equivocal lighting regimens?

LEES: I am afraid that most action spectrum work with insects and mites—including my own—has been very incomplete. Usually the method has been to extend the photoperiod with monochromatic light, which, of course, is the same thing as curtailing the dark period. However, such an action spectrum refers to only one part of the light/dark cycle. A more satisfactory method is to use brief light interruptions. In my present studies on the aphid *Megoura*, which I will present this afternoon, two points of light sensitivity have been found during the dark period; the main photoperiod is a third point. All three have distinctive action spectra, differing in their wavelength, intensity threshold, and rate characteristics.

WILLIAM HAMNER: I raise the point to deal with one of the criticisms of Dr. Pittendrigh's work, which was that perhaps the insects were perceiving the various wavelengths as darkness. I believe that is a valid criticism for the 14-hr group, but certainly not for the 12-hr group.

PITTENDRIGH: No, it isn't valid for either, because in constant darkness in *Pectinophora* you get 73 percent diapause.

WILLIAM HAMNER: Oh, it is 73 percent. I thought it was 78 percent, which would have been almost within the range of variability.

CUMMING: There is a nice parallel in plants to the question Dr. Pittendrigh raised. The evidence from plants indicates that you need high-intensity light to entrain or phase-shift some circadian oscillations. But there are strong indications that phytochrome is involved in the photoperiodic flowering response, and the phytochrome response reaction is a low-intensity light reaction. Phytochrome is a photoreversible pigment, and you can actually shift the photoperiodic optimum by as much as two hours by altering the amount of the active form of the pigment (P_{fr}) at the end of a light period. But you can't shift the endogenous rhythm by some low-intensity light that does affect the phytochrome pigment. This raises the possibility that there are two systems operating in Dr. Pittendrigh's experimental insects: One would be the circadian rhythm, with a receptor pigment that can be affected by, say, blue or high-intensity light of some sort; the other would be another pigment, responsive to red, involved in photoperiodic time measurement.

PITTENDRIGH: Your comment disturbs me a little because it seems not to address itself to the main point, which is that the distinction between 12 hr and 14 hr is made by insects irrespective of the phase of the oscillation to the red light cycle.

HALABAN: Your insects were not entirely aperiodic, though. They were free-running with a τ of slightly less than 24 hr.

PITTENDRIGH: That is not true. In the

experiments testing for photoperiodic induction with different wavelengths of light, the eggs were laid in constant darkness; they never saw any light. And we know that when the oscillation does develop midway through the egg's development, the phases are at random in the population. That is well-established. The experiments I showed later on were experiments with rhythms that we had previously established with white light; we were then checking whether the various wavelengths of monochromatic light could shift the phase of the previously synchronized populations.

HILLMAN: I have one comment and one question; I'll make the comment first. I think it is a safe statement to make from the literature on plants, that negative arguments on the basis of action spectra have a history of being very difficult and deceptive. You only have to think over the controversy about the phototropic pigment and the controversy over high-energy and low-energy reactions in phytochrome to see why I feel that there is some reason, no matter how subjective, to think that a proof of the sort Dr. Pittendrigh has just given may have some holes in it somewhere, although I can't specify them. Now there is one specific question I would like to ask: Have you

tried to raise these insects from the egg in a rhythm of *LD 12*:12, red light and dark? Would you then get rhythmicity in behavior as well as a photoperiodic response?

PITTENDRIGH: No, that experiment hasn't been done.

HILLMAN: It's a simple one.

PITTENDRIGH: It isn't as simple as you think, you see, because the amount of diapause you get might easily be equivocal.

BÜNNING: I looked once again at our experiments concerning the effect of red light in breaking diapause in the cabbage butterfly (Zeitschr. Naturforsch. 18b:324–327, 1963). The results were very similar to those presented by Dr. Pittendrigh. However, it became clear to me that the red-light effect in the cabbage butterfly is not comparable to the normal photoperiodic effect. Whereas the photoperiodic effect of white or blue light is mediated via the region of the head, red light is effective when applied to other parts of the caterpillar as well. Moreover, the critical daylength of the normal photoperiodic reactions is about 14 hr, but with red light, a daily exposure of 4–8 hr is effective in breaking diapause.

PITTENDRIGH: The bearing of these observations on my own analysis is not clear to me.

WILLIAM S. HILLMAN

Carbon Dioxide Output as an Index of Circadian Timing in Lemna Photoperiodism

The view that photoperiodism is a function of the "biological clock" is becoming widely accepted (Bünning, 1964, 1969; Cumming and Wagner, 1968; Sweeney, 1969). Nevertheless, there are few experimental systems in which relationships between a photoperiodic response and some overt, continuous rhythm—as an indicator of the state of the clock—can be established with ease and precision. The penetrating studies of phytochrome photoreception in higher plant photoperiodism (Borthwick, 1964; Borthwick and Hendricks, 1960) have not been fully matched by investigations of photoperiodic timing, which, with a few notable exceptions, have relied largely on demonstrating oscillations in the photoperiodic response itself. A persistent obstacle has been the frequent unavailability of a reliable overt rhythmic process in plants, precisely measurable under conditions where it might be most useful. As an ironic result, the most definitive evidence relating an overt rhythm to photoperiodic responsiveness was obtained on a *Coleus* in which the role of phytochrome is not understood (Halaban, 1968, 1969). An additional difficulty with higher plants is that long experiments with brief light exposures, as in "skeleton" photoperiodic schedules, are difficult without an energy source other than photosynthesis.

Photoperiodic control of flowering in the duckweed *Lemna perpusilla* 6746 has been extensively studied with respect to light quality, chemical

effects, and other factors (Hillman, 1969; Posner, 1969; Rombach and
Spruit, 1968). In experiments under non-photosynthetic conditions made
possible by axenic culture on media with sucrose, the effects of certain
skeleton photoperiodic schedules suggest the activity of an endogenous
circadian rhythm (Hillman, 1964; Oda, 1969). Until now, however, a continuous measure of the state of the clock was lacking. The experiments reported here show that the rate of CO_2 output serves well for this purpose,
further complementing the already abundant advantages of the *Lemna*
system. Results of these studies strongly confirm the view that an endogenous clock is part of the mechanism of photoperiodism.

MATERIALS AND METHODS

STOCK CULTURES

Lemna perpusilla 6746 was maintained in 125 ml Erlenmeyer flasks on
50 ml of 0.5 X strength Hutner's medium plus 1% sucrose, 600 mg/liter
tryptone, and 100 mg/liter yeast extract. The cultures were kept under
about 200 ft-c (2152 lux) of continuous cool white fluorescent light at
25–27°C. New cultures were started weekly with one 3-frond colony
(Hillman, 1969).

FLOWERING EXPERIMENTS

These proceeded as described earlier (Hillman, 1964, 1967, 1969) except
that cultures were grown on 30 ml of 1.1 X strength Hutner's medium plus
1% sucrose in 25 X 150-mm culture tubes. The red light used was about
one-fifth the "standard" intensity (Hillman, 1967) and gave a total energy
of 8–15 $\mu W \cdot cm^{-2}$ between 550 and 800 nm.

ASSAY OF CO_2 OUTPUT RATE

A schematic summary of the system used is given in Figure 1. Each culture
is started by pouring a single 11- to 16-day-old stock culture, with medium,
into a cotton-stoppered 250-ml narrow mouth reagent bottle containing
150 ml 1.1 X Hutner's medium plus 1% sucrose. Generally, two cultures
are placed in each vessel system. The differences between the CO_2 levels
maintained in vessels with plants and those maintained in precisely similar
vessels lacking only the plants, but with culture bottles and medium, constitute the basic data. Such values are obtained once every 30 min for each

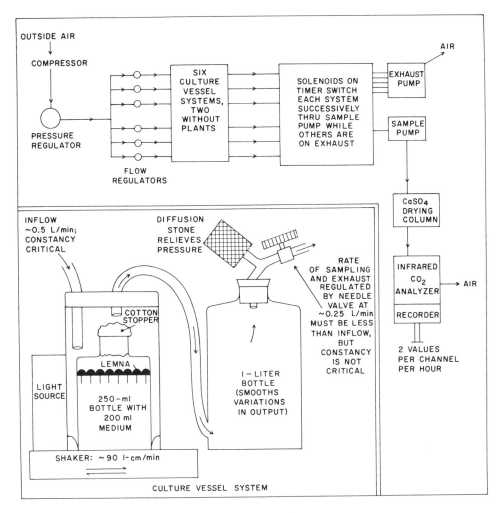

FIGURE 1   Schematic summary of *Lemna* CO_2 monitoring system.

of four sets of plants, with two vessels used as blanks. During the course of the experiments, the ambient (blank) level of CO_2 ranged between about 310 and 370 ppm, depending on weather conditions and time of day. The temperature was 28–30°C in all experiments.

CO_2 output rate was measured under schedules of darkness combined with dim red light. The latter was provided for each vessel by a 4W cool-white fluorescent tube behind a 3-mm thickness of Rohm and Haas 2444 red Plexiglas. As measured with an Agricultural Specialties Company

(Beltsville, Maryland) Model 3052 spectroradiometer, this source at plant level gave about 3.5 μW·cm⁻² total energy between 800 nm and the low wavelength cutoff about 550 nm, with an intensity of .072 μW·cm⁻²·nm⁻¹ in the peak range of 648–660 nm and of .014 μW·cm⁻²·nm⁻¹ at 730 nm. Without the red filter, the intensity was approximately 80 ft-c (861 lux) as measured with a General Electric MR-100 light meter. Because of the complex geometry involved, the actual energies impinging on the plants probably varied several-fold from the values measured.

In spite of the lack of direct contact between the plants and the moving air stream, diffusion through the cotton stopper apparently allows the system to respond rapidly; Figure 2 shows how quickly the effects of shaking or nonshaking are recorded. The shaking frequency used was selected arbitrarily and is probably not optimal, but much more rapid shaking soon damages the plants. For this reason, and since interpretations of the data

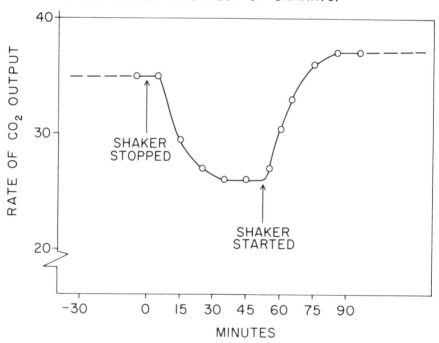

RESPONSE TIME OF LEMNA CO₂−MONITORING SYSTEM AS REFLECTED IN EFFECT OF SHAKING.

FIGURE 2   Response time of *Lemna* CO_2 monitoring system as shown by response to shaking. CO_2 output in arbitrary units.

sought depend only on relative maxima and minima and not on absolute values, little attention was given to the latter. The general range is suggested by the following information on the experiment started July 7 (Figure 9). Blotted fresh weights of five cultures like those used to start the experiment were 0.96, 1.20, 1.25, 1.45, and 1.86 g. At flow rates of about 0.8 liter/min, two of these cultures per vessel gave a CO_2 difference of 40–80 ppm. However, the relationship between total frond number (or fresh weight) and the CO_2 value is not a simple one. During the experiments involving frequent red light, the cultures grow rapidly, probably doubling every 2–4 days, but CO_2 output remains much more level. A reasonable explanation (though without direct evidence) is that, even with gentle shaking, the major part of the activity recorded is that of the fronds in the top layers of the cultures, while those below contribute a decreasing proportion. Whatever its cause, the relatively level CO_2 output under these conditions immensely simplifies both data collection and the location of peaks and troughs; it also, however, strongly emphasizes the purely relative nature of the data and should be borne in mind when considering possible mechanisms.

Many earlier experiments (see Figures 4 and 6) were done with an apparatus differing from that described in several respects, notably in its dependence on random leakage rather than a diffusion stone to relieve the difference between inflow rate and sampling rate, and in the technique for drying the sample stream. The results differed from those with the later equipment chiefly in showing a (spurious) strong underlying uptrend because of excessive pressure buildup, but conclusions with respect to the timing of maxima and minima were unchanged.

NOTATION FOR LIGHT (DARK) TREATMENTS

All numbers refer to hours, and parentheses indicate darkness. Schedules whose sum is 24 hr are understood to be repeated each day. A "skeleton" schedule (Hillman, 1964; Pittendrigh, 1966) is here used to mean 1/4(a)1/4(b), where a + b + 1/2 = 24; 1/4(7 1/2)1/4(16), for example, can be regarded as the "skeleton" of LD *8*:16. All treatments begin from continuous (cool white or dim red) light, and the skeleton photoperiods, in particular, are to be read in the order given. Thus a treatment designated 1/4(13)1/4(10 1/2) means continuous light followed by a 13-hr dark period, then 1/4 hr of light, then 10 1/2 hr of darkness, then 1/4 hr of light, then 13 hr of darkness, and so forth. A treatment designated 1/4(10 1/2)1/4(13) has the same succession of elements *except* that the first dark period experienced is 10 1/2 hr long. Obviously, following con-

FIGURE 3 CO_2 output rate (half-hourly, arbitrary units) of cultures on various LD schedules. Time as number of hours after the start of the second light period. All cultures, from continuous light, were given the succession: dark period 1, light period 1, dark period 2, light period 2, etc. LD $8{:}16$ and LD $1{:}23$ data from experiment started 5/13/69; LD $16{:}8$ data from experiment started 6/17/69.

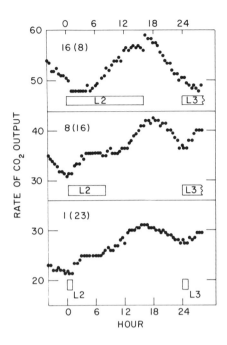

tinuous light, it is immaterial whether a schedule is designated $(7\ 1/2)\,1/4$ $(16)\,1/4$ or $1/4(7\ 1/2)\,1/4(16)$, except for a 15-min difference in the initiation of darkness.

RESULTS

LIGHT/DARK SCHEDULES WITH 24-HR PERIODICITY

If, after an initial 6 to 24 hr equilibration under continuous dim red light, the cultures are started on dark/light (DL) schedules with a 24-hr periodicity, the CO_2 output rate entrains to that periodicity. Portions of two such experiments are shown in Figure 3. These data and those of numerous similar experiments indicate, first, that the dim red light used probably has only the slightest direct photosynthetic effect on the CO_2 output rate, for the effects of turning it on or off depend almost entirely on the cycle time at which it is done. Note, for example, the great perturbation at the end of the 16-hr light period as compared with the small effect at the end of the 8-hr light period. Secondly, in such simple 24-hr cycles the maximum CO_2 output rate depends on the start of the light period for its timing, not the start of the dark period. In Figure 3, the maximum is at approximately (depending primarily on how one handles the perturbation at the close of the 16-hr light period) 15–17, 16–17, and 16–17 hr after the start of 16-, 8-, or 1-hr light period, respectively. Similar conclusions appear to hold for schedules from *LD 1/4*:23 3/4 to *LD 21*:3, with a maximum falling between 14 and 20 hr after the start of the light period. However, some light period lengths lead to greater immediate "light-off" perturbations than others, and these may complicate the picture. In addition, recent experiments suggest that, again with certain light period lengths, true steady-state entrainment may not be reached before at least 4 or 5 cycles have elapsed, so that the data in Figure 3 may not be truly representative.

CIRCADIAN RHYTHMICITY IN DARKNESS

Flowering experiments employing skeleton photoperiods (12, 16) have suggested that transfer from continuous light to darkness initiates a circadian oscillation in *Lemna* cultures. Data such as those in Figure 4 indicate that such an oscillation is also detectable in the rate of CO_2 output, though it does not persist very long. In the experiment shown, the hourly rates of CO_2 output for three sets of cultures are plotted in arbitrary units. (As noted earlier, the spurious general uptrend in these data is an artifact that does not affect the conclusions.) Initially, all three sets were held in dim red light, and the one (channel 3) that remained so throughout the experiment shows a simple linear change of CO_2 output rate with time. The light was turned off on the other two channels at the times indicated by the arrows; subsequently, each rate oscillated for a first maximum, a first minimum, a second maximum, and a second minimum, with timing dependent on the timing of transition to darkness. If the values from the two darkened cultures are summed as functions of the time in darkness, the first maximum is found at hr 9 and the second at hr 30–36 (mean=33), while the first and second minima are at hr 16–18 (mean=17) and hr 40, respectively. The periodicity is clearly circadian. The scatter in the data and the rapid damping prevent more precise evaluation of the period, but this general conclusion has been confirmed in a number of experiments.

The possibility that cultures previously entrained to 24-hr *LD* cycles might show a more persistent oscillation in darkness was tested in several experiments. Typical results are shown in Figure 5, for three sets of cultures given three *DL 12:12* cycles following continuous light. The upper panels show the data for the individual sets in constant darkness, which in the lower panel are summed as functions of time after the third (and last) start of light. Here again, clear first and second maxima and minima, with circadian timing, are evident, but with no greater persistence than in cultures starting from continuous light (Figure 4).

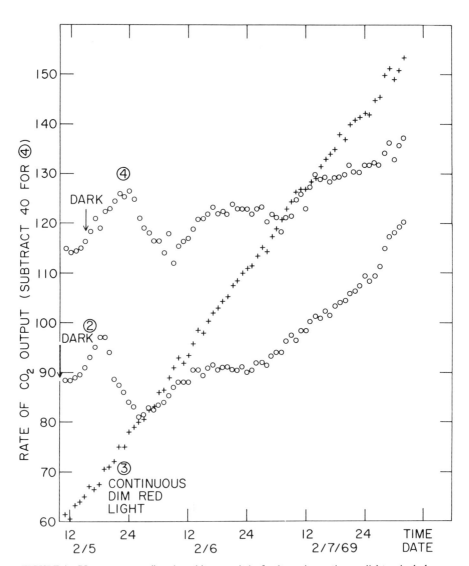

FIGURE 4 CO_2 output rate (hourly, arbitrary units) of cultures in continuous light or in darkness starting at indicated times; see text. Numbers in circles refer to channels; channel 1 was blank.

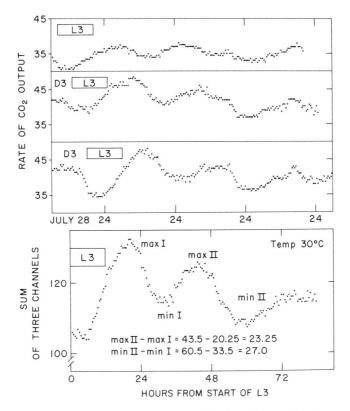

FIGURE 5   Above: CO_2 output rates (half-hourly, arbitrary units) of three sets of cultures. All cultures from continuous light were exposed to the succession: 12 hr dark, 12 hr light, 12 hr dark, 12 hr light, 12 hr dark(D3), 12 hr light(L3). Below: the three records above summed with respect to time from the start of L3.

SKELETON PHOTOPERIODS: COMPARISON OF (13)1/4(10 1/2) WITH (10 1/2)1/4(13)

When plants are transferred from continuous light to six or seven days of a skeleton photoperiodic regime consisting of alternating 10 1/2- and 13-hr dark periods separated by 1/4-hr light pulses, flowering depends on which dark period is given first (Hillman, 1964). The course of CO_2 output was therefore followed under comparable conditions. A typical result for the first three days is shown in Figure 6, in which continuous light ended at 1400 on the first day shown; the 1/4-hr light breaks are indicated by vertical lines. For the first 24-hr period, the patterns are essentially alike. In the second, however, the maximum under 1/4(10 1/2)1/4(13) comes early in

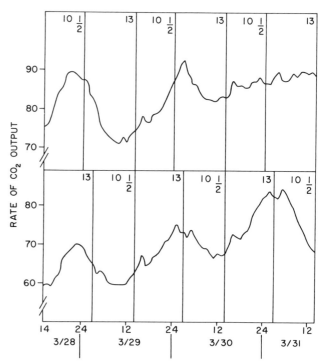

FIGURE 6 CO_2 output rate of two sets of cultures; curves drawn between points representing hourly rates in arbitrary units. Vertical lines
represent ¼-hr light breaks; numbers in panels indicate lengths of dark
period in hour.

the 13-hr dark period, while under 1/4(13)1/4(10 1/2) (lower panel) it is
late in the 13-hr period with some perturbation into the start of the next
10 1/2 hr. By the third 24-hr period the difference is no longer subtle. The
1/4(13)1/4(10 1/2) pattern has established a 24-hr rhythmicity with its
peak surrounding the transition between (13) and (10 1/2), while the other
is almost arrhythmic, with no marked maxima or minima. Thus, two skeleton schedules with markedly different effects on flowering differ also in
their action on CO_2 output, at least for the first few cycles.

While the effects of 1/4(10 1/2)1/4(13) and its inverse differ greatly in
short flowering experiments, the difference is reduced with longer treatments. In a typical experiment, flowering percentage values for 1/4(13)
1/4(10 1/2) were 53, 53, 58, and 66 on days 6, 7, 8, and 9, respectively,
while the corresponding values for 1/4(10 1/2)1/4(13) treatment in the
same experiment were 1, 13, 11, and 39. Hence, it seemed worth examining the CO_2 output in longer experiments also, to see whether the differ-

ence visible in the first few days disappears. Data from a 6-day experiment appear in Figure 7. The results for the first three days resemble those in Figure 6, in spite of the different method and manner of presentation. In particular, on day III, $1/4(10\ 1/2)1/4(13)$ again shows essentially no maxima or minima, while $1/4(13)1/4(10\ 1/2)$ peaks just at or before the 13–10 1/2 transition. By day VI, however, a 24-hr periodicity is evident also in $1/4(10\ 1/2)1/4(13)$. Experiments long enough to determine the stable pattern, if any, finally attained under $1/4(10\ 1/2)1/4(13)$ treatment have not been done, but it is obvious that the arrhythmic condition of days III and IV—the most obvious difference between the two schedules—does not persist indefinitely.

SKELETON PHOTOPERIODS WITH HIGHLY UNEQUAL PORTIONS

Earlier results with *Lemna* flowering (Hillman, 1963), as well as with *Drosophila* (Pittendrigh, 1966) indicate that if the two portions of a skeleton schedule differ sufficiently in length—e.g., $1/4(5\ 1/2)1/4(18)$ instead of $1/4(10\ 1/2)1/4(13)$—then the response does not depend significantly on which dark period is given first. Confirming this, a flowering experiment comparing four treatments, $1/4(10\ 1/2)1/4(13)$, $1/4(13)1/4(10\ 1/2)$, $1/4(5\ 1/2)1/4(18)$, and $1/4(18)1/4(5\ 1/2)$, gave flowering percentage values of 2, 37, 61, and 82 on day VII, and 7, 53, 71, and 78 on day VIII. A 6-day experiment on CO_2 output comparing the same schedules is shown in Figure 8. In the top two panels are the now familiar differences between $1/4(10\ 1/2)1/4(13)$ and its inverse, with the former essentially flat for days III and IV and the latter rapidly entraining the output rate to a 24-hr periodicity. The two lower panels, on the contrary, indicate that both of the schedules with highly unequal portions entrain the output rapidly to a 24-hr periodicity regardless of whether the long or the short dark period is presented first. Within two full cycles in each case, the same phase relationship between schedule and CO_2 output seems to have been established, with the maximum rate roughly 9–11 hr after the start of the long dark period—that is, 14–16 hr after the first light break defining the short dark period.

SKELETON PHOTOPERIODS: INTERRUPTING THE SHORTER DARK PERIOD

Earlier results indicate that if the shorter dark period of a skeleton schedule with relatively equal portions is interrupted by light, then its typical effects no longer appear (Hillman, 1963). To see whether this might be reflected in the CO_2 data, the experiment shown in Figure 9 was per-

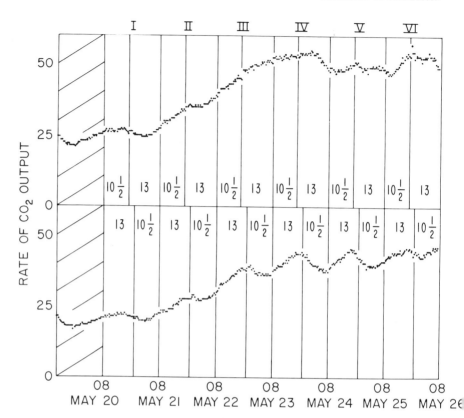

FIGURE 7 CO_2 output rate (half-hourly, arbitrary units) of two sets of cultures. Shaded panels indicate light; vertical lines and numbers as in Figure 6.

formed. In the two upper panels the effects of 1/4(13)1/4(10 1/2) are compared with and without an additional initial 8-hr dark period. In flowering experiments, the initial 8 hr of darkness renders the otherwise favorable schedule highly inhibitory. In Figure 9, it is evident that the initial 8 hr of darkness brings about a situation similar to that caused by 1/4(10 1/2)1/4(13), in which the course of CO_2 output for days III and IV is rendered essentially aperiodic. In the two lower panels, however, the effect of an initial 8-hr dark period is tested on a schedule, 1/4(13) 1/4(5 1/2)1/4(5), which is simply 1/4(13)1/4(10 1/2) with the 10 1/2-hr dark period interrupted. Clearly, this interruption is sufficient to remove whatever ambiguity is present in the original schedule, for the initial 8-hr dark period has essentially no effect on either the rapid establishment of a marked 24-hr periodicity or on the phase relationship attained.

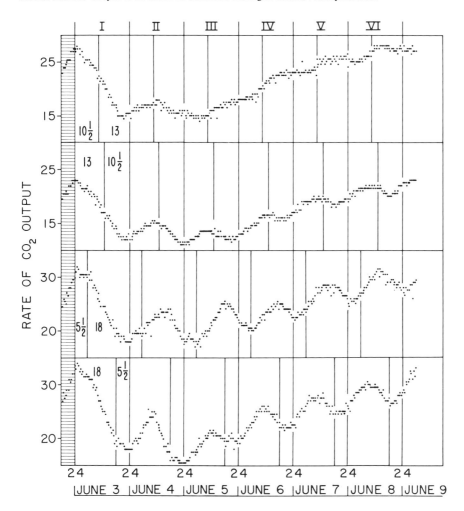

FIGURE 8 CO_2 output rate (half-hourly, arbitrary units) of four sets of cultures. Shaded panels indicate light; vertical lines represent ¼-hr light breaks; numbers in panels indicate lengths of dark period in hour.

DISCUSSION

WHAT IS BEING MEASURED?

Carbon dioxide output rates have been used to follow circadian periodicities in plants mainly by Wilkins (1967) using detached Bryophyllum leaves, though a few other studies exist (Bünning, 1964; Cumming and Wagner, 1968). The work reported here is the first to attempt to correlate such

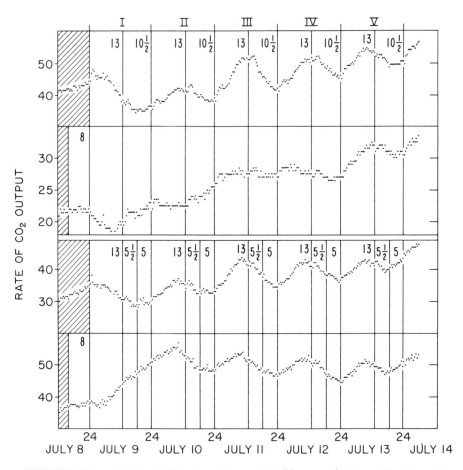

FIGURE 9 CO_2 output rate (half-hourly, arbitrary units) of four sets of cultures. (See Figure 8.)

data with the photoperiodic control of flowering. For this purpose it is probably satisfactory to regard CO_2 output rate as a meter on a "black box"—hopefully, in fact, as "hands on the clock"—but it is worth remembering that the mechanism of the changes measured is unknown. (This is quite aside from the fortuitous quantitative situation, itself incompletely understood, already discussed in Materials and Methods.)

In the data presented, the entrainable variation in CO_2 output is generally between 10 and 20 percent of the total; this may represent any situation from a 0 to 100 percent variation in a reaction responsible for a small part of the total, to a small variation in a single process. The evidence in *Bryophyllum* points towards variation in CO_2 fixation by phosphoenol pyruvate carboxylase as the basis of the periodicity (Wilkins, 1967), but there is no evidence either for or against such a mechanism here. Alternatively, the *Lemna* system might reflect changing respiratory rates (*cf.* Brown *et al.*, 1955), and these in turn might reflect changing rates of frond production or other processes. An account of rhythmic changes in several aspects of respiratory metabolism in *L. gibba* has appeared very recently (Miyata and Yamamato, 1969); although they are not analyzed in relation to the photoperiodic response, these data should prove important in interpreting the present results. The likelihood of rhythms in stomatal activity is reduced by the observation (Dr. H. Ziegler, personal communication, August 1969) that *Lemna* stomates are nonfunctional. In the long run, then, further analysis of the system is obviously desirable; nevertheless, despite all limitations, several significant conclusions can be drawn from the data now at hand.

RHYTHMICITY IN CONTINUOUS DARKNESS: DAMPING AND TECHNICAL PROBLEMS

Experiments such as those summarized in Figures 4 and 5 show that in continuous darkness the CO_2 output rate oscillates through at least two maxima and minima, although it remains linear under continuous dim red light. The phase relationship of a given point on the oscillation to local time is determined either by the time of transition from continuous light to darkness or by the schedule previously used for entrainment. While the cycles do not persist long enough to justify the term "free-running" (Aschoff *et al.*, 1965), their period length is clearly circadian. All these properties of the CO_2 output rate would be predicted if it accurately reflected the kind of endogenous circadian rhythm inferred from the flowering experiments (Hillman, 1964).

Damping in extended darkness unfortunately limits the types of experiments that can be done—excluding, for instance, rigorous phase-response tests—but it is not inconsistent with the original flowering data. Few higher plant systems, in fact, run for more than a small number of cycles under constant darkness, and *L. perpusilla*, in particular, is known to have a nonphotosynthetic light requirement for growth (Hillman, 1964). However, there are at least two grounds for optimism that eventually it will be possible to measure more cycles in darkness than can be done at present. One is the likelihood that the 28–30°C temperature used here is superoptimal (Wilkins, 1965), so that the number of cycles at, say, 21°C might be much greater. The other is that the monitoring system itself can be improved to reduce random noise and external perturbations, both of which limit its usefulness with cycles of small amplitude.

Indeed, perturbation by some external factor(s) is evident in many experiments, such as that shown in the upper panels of Figure 5. While the major fluctuations there are related primarily to the previous LD schedule, there are also discontinuities in all three records at the same local times: 1400 hr on July 29; 0800 and 1800 on July 30; and 1600 on July 31. Modifications such as precise regulation of the voltage of the shaker may serve to eliminate such effects. In addition, the daily outdoor cycle of CO_2 level in the air—with much greater amplitude on some days than on others—reduces precision; a system employing constant CO_2 is being planned. Although the system used here is obviously adequate for work with periodicities either entrained or initiated by light-dark transitions, its openness to perturbation would disqualify it completely if the exogeny or endogeny of the basic timing mechanism (not merely of particular period lengths and phase relationships) were the central question (Brown, 1965).

THE EFFECTS OF SKELETON SCHEDULES ON CO_2 OUTPUT AND ON FLOWERING

The experiments with skeleton schedules are in effect tests of some of Pittendrigh's (1966) formulations of Bünning's theory of photoperiodism. The latter, in its most general form, simply asserts that photoperiodic treatments produce their effect by interacting with an endogenous oscillation or circadian clock that functions as the basic timing mechanism (Bünning, 1964, 1969). More specifically, Pittendrigh has proposed that skeleton schedules having different photoperiodic effects should entrain the endogenous oscillation in different phase relationships. Most specifically of all, a model developed from work with *Drosophila* was the basis of two precise predictions on the actions of skeleton photoperiods. First, schedules with two highly unequal portions should entrain the rhythm in the same phase relationship as full (nonskeleton) photoperiods in which the shorter fraction is the light period. That is, $1/4(6\ 1/2)1/4(17)$ should act like LD 7:17, and this should be true irrespective of which dark period is given first—irrespective, in effect, of what phase of the endogenous oscillation (initiated by the transition to darkness) encounters the first light pulse. The second prediction is that for skeletons with relatively equal portions a "bistability phenomenon" would occur, such that the final steady-state phase relationship would depend entirely on which dark period was seen first—on when in the endogenous oscillation the first light pulse would fall (Pittendrigh, 1966). What is striking is that all predictions, although formulated from a *Drosophila* model, were fully consistent with data on *Lemna* flowering obtained independently. A recent account of work elsewhere (Oda, 1969) confirming the data on flowering still concludes that the original model is fully applicable to *Lemna*. The CO_2 results, however, provide a dimension hitherto lacking.

The least specific prediction–different entrainment by skeletons having different photoperiodic effects–seems fully borne out by the comparison between 1/4(13)1/4(10 1/2) and 1/4(10 1/2)1/4(13). The prediction that skeletons with highly unequal positions would effect the same steady-state phase relationship irrespective of which dark period is given first is also borne out (Figure 8, lower panels). With respect to the skeleton photoperiod 1/4(13)1/4(10 1/2) and its inverse, the model also holds in the sense that the manner of entrainment depends on which dark period is seen first. However, there seems to be no bistability phenomenon of the kind predicted. Rather, the obvious difference between the two schedules is in the rapidity with which one entrains, and the other initially seems to suppress, rhythmicity in the CO_2 output. On the *Drosophila* model, the schedule 1/4(13)1/4(10 1/2) would give a pattern similar to that caused by *LD 11*:13, the corresponding full photoperiod schedule that allows high flowering, while 1/4(10 1/2)1/4(13) should give a pattern similar to that caused by the low flowering *LD 13 1/2*:10 1/2. This effect is certainly not observed within the first few days, so that to this extent the *Drosophila* model seems to fail. However, Figures 7, 8, and 9 suggest that in fact even 1/4(13)1/4(10 1/2) fails to give truly steady state entrainment in six days, while preliminary experiments indicate that the full photoperiodic schedules *LD 11*:13 and *LD 13 1/2*:10 1/2 are among those for which the steady state is also attained quite slowly (see Results). Hence longer experiments might well confirm predictions from the *Drosophila* model.

Two major questions remain unanswered in the foregoing comparison of 1/4(13)1/4(10 1/2) with its inverse. First, explaining why one skeleton photoperiod allows rapid flowering and the other does not will require locating the phase of the rhythm in which light is maximally inhibitory to flowering (*cf*. Halaban, 1968). Attractive as the hypothesis might be, the difference in response cannot be attributed solely to rapid entrainment by one schedule and apparent temporary suppression by the other, though the latter may play some part in the low flowering. It is easy to find schedules–e.g., *LD 16*:8–that entrain rapidly but inhibit flowering completely. The second question concerns the meaning of the apparent temporary suppression itself. There are at least four possible mechanisms: (a) that the CO_2 output rate becomes uncoupled from an underlying clock, which itself persists; (b) that the clock is deranged in some way, perhaps as in the "singular state" discovered in *Drosophila* (Winfree, this volume); (c) that the clocks of individual fronds continue to run but that the population as a whole is desynchronized; or (d) that the apparently flat records result

from a fortuitous succession of resettings that simply do not happen to
show sharp maxima or minima, the only phase markers available. Given
the low resolving power of the method, the last seems perhaps most likely.
Nevertheless, there is a real possibility that the low intensity light breaks
used are insufficient, at certain phases, to reset the clock unambiguously,
and this could bring about condition (b) or (c). In this case, high intensity
skeleton photoperiodic schedules would be expected to act very differently.
Through these and other experiments, it should be possible to distinguish
among the mechanisms suggested.

OTHER CONSIDERATIONS

Overt circadian rhythms in a number of other plant systems persist under
continuous light of various qualities and intensities (Halaban, 1968;
Wilkins, 1967), while CO_2 output in *Lemna* cultures becomes linear under
continuous dim red light. This property, as already noted, was predictable
on the basis of flowering experiments and is perhaps related to the entraining capability of brief red light exposures. However, a point that remains
obscure at present also illustrates the limitation of flowering experiments
alone in analyzing such properties. Before the present system was available,
Lemna flowering work suggested that a single light exposure was effective
in resetting the phase of the rhythm only if it were 4–6 hr long (Hillman,
1969). Present evidence of entrainment even by 15-min light periods seems
inconsistent with such a conclusion, and parallel experiments with the CO_2
system will be needed to resolve the contradiction. It may, of course, be
inappropriate to compare single perturbation experiments with repeated
cycle experiments, particularly when the single perturbation was given at
only one particular phase of the rhythm.

The effects of light other than dim red on the *Lemna* CO_2 rhythm have
yet to be tested; blue light, which permits flowering even when given continuously (Hillman, 1967) will be of particular interest, if complications
due to photosynthesis can be avoided.

Takimoto and Hamner (1965) have interpreted their data on flowering
in *Pharbitis* as evidence for the existence of two circadian components in
the photoperiodic timing mechanism—a "light-on" rhythm initiated by the
darkness-to-light transition, and a "light-off" rhythm. In trying to relate
the present experiments to such a concept, one can note simply that, while
a light-off signal is clearly sufficient to initiate a circadian oscillation
(Figure 4), it seems to be the light-on signal that functions as the primary
synchronizer in entrainment to 24-hr cycles (Figure 3). Further work with

this system may make it possible to relate the Takimoto-Hamner double-rhythm proposal to models of the type proposed by Pittendrigh (1966), which at present seem to fit the data better, although imperfectly.

CONCLUSIONS

The CO_2 output rate of *L. perpusilla* 6746 under combinations of dim red light and darkness appears to reflect the activity of a circadian clock whose properties are those expected from earlier experiments on flowering (Hillman, 1964). The effects of skeleton photoperiodic schedules are consistent with the view that the influence of such schedules on flowering depends on the manner in which they entrain the circadian oscillation, conforming in many, though not in all, respects to predictions from a model based on the *Drosophila* eclosion rhythm (Pittendrigh, 1966). The results thus strongly support the concept that a circadian clock is an important component of photoperiodism. The *Lemna* system provides an experimental subject in which the actions of many factors on the circadian component can be examined independent of other effects.

ACKNOWLEDGMENTS

Research reported in this paper was carried out at Brookhaven National Laboratory under the auspices of the U.S. Atomic Energy Commission. A paper nearly identical with this has appeared in *Plant Physiology* 45:273–279 (1970).

I thank R. Dearing and H. Kelly for technical assistance, C. S. Pittendrigh for repeatedly drawing my attention to the need for studying an overt rhythm in *Lemna*, and R. Halaban and B. M. Sweeney for illuminating discussions of the results. N. Tempel, a member of G. M. Woodwell's ecology group at Brookhaven, was largely responsible for the design and construction of the CO_2 monitoring system. Without his ingenuity and diligence, this work would not have been done.

NOTE ADDED IN PROOF

Continuation of this work now indicates that at least some of the apparent failures of certain schedules to achieve steady-state entrainment rapidly (e.g., p. 256) were due to changes in the medium brought about by growth. This fact may bear significantly on the possibility of finally determining rigorously whether Pittendrigh's (1966) model holds in this system. [See Hillman, W. S.: Nitrate and the course of *Lemna perpusilla* carbon dioxide output under daily photoperiodic cycles. *Plant Physiology* 47:431–434 (1971).]

REFERENCES

Aschoff, J., K. Klotter, and R. Wever. 1965. Circadian vocabulary, p. x to xix. *In* J. Aschoff [ed.], Circadian clocks. North-Holland Publ. Co., Amsterdam.
Borthwick, H. A. 1964. Phytochrome action and its time displays. Am. Nat. 98:347–355.
Borthwick, H. A., and S. B. Hendricks. 1960. Photoperiodism in plants. Science 132: 1223–1228.
Brown, F. A., Jr. 1965. A unified theory for biological rhythms, p. 231 to 261. *In* J. Aschoff [ed.], Circadian clocks. North-Holland Publ. Co., Amsterdam.
Brown, F. A., Jr., R. O. Freeland, and C. L. Ralph. 1955. Persistent rhythms of O_2-consumption in potatoes, carrots and the seaweed, *Fucus*. Plant Physiol. 30:280–292.
Bünning, E. 1964. The physiological clock. Academic Press, New York. 145 p.
Bünning, E. 1969. Common features of photoperiodism in plants and animals. Photochem. Photobiol. 9:219–228.
Cumming, B. G., and E. Wagner. 1968. Rhythmic processes in plants. Annu. Rev. Plant Physiol. 19:381–416.
Halaban, R. 1968. The flowering response of *Coleus* in relation to photoperiod and the circadian rhythm of leaf movement. Plant Physiol. 43:1894–1898.
Halaban, R. 1969. Effects of light quality on the circadian rhythm of a short-day plant. Plant Physiol. 44:973–977.
Hillman, W. S. 1963. Photoperiodism: an effect of darkness during the light period on critical night length. Science 140:1397–1398.
Hillman, W. S. 1964. Endogenous circadian rhythms and the response of *Lemna perpusilla* to skeleton photoperiods. Am. Nat. 98:323–328.
Hillman, W. S. 1967. Blue light, phytochrome, and the flowering of *Lemna perpusilla* 6746. Plant Cell Physiol. 8:467–473.
Hillman, W. S. 1969. *Lemna perpusilla* Torr., Strain 6746, p. 186 to 204. *In* L. T. Evans [ed.], The induction of flowering. Some case histories. Macmillan Co. of Australia, South Melbourne, and Cornell Univ. Press, Ithaca, New York.
Miyata, H., and Y. Yamamoto. 1969. Rhythms in respiratory metabolism of *Lemna gibba* G3 under continuous illumination. Plant Cell Physiol. 10:875–889.
Oda, Y. 1969. The action of skeleton photoperiods on flowering in *Lemna perpusilla*. Plant Cell Physiol. 10:399–409.
Pittendrigh, C. S. 1966. The circadian oscillation in *Drosophila pseudoobscura* pupae: a model for the photoperiodic clock. Z. Pflanzenphysiol. 54:275–307.
Posner, H. B. 1969. Inhibitory effects of carbohydrate on flowering in *Lemna perpusilla*. I. Interaction of sucrose with calcium and phosphate ions. Plant Physiol. 44:562–566.
Rombach, J., and C. J. P. Spruit. 1968. On phytochrome in *Lemna* minor and other Lemnaceae. Acta Bot. Neerl. 17:445–454.

Sweeney, B. M. 1969. Rhythmic phenomena in plants. Academic Press, New York. 148 p.

Takimoto, A., and K. C. Hamner. 1965. Effect of far-red light and its interaction with red light in the photoperiodic response of *Pharbitis nil*. Plant Physiol. 40:859–886.

Wilkins, M. B. 1965. The influence of temperature and temperature changes on biological clocks, p. 146 to 163. *In* J. Aschoff [ed.], Circadian clocks. North-Holland Publ. Co., Amsterdam.

Wilkins, M. B. 1967. An endogenous rhythm in the rate of carbon dioxide output of *Bryophyllum*. V. The dependence of rhythmicity on aerobic metabolism. Planta 72:66–77.

DISCUSSION

K. HAMNER: When your plants are put into DD from continuous dim red light, you find that a circadian rhythm develops. If you postulate that it is induced or initiated by the light-off signal, then your light perturbations (the skeleton photoperiod) could be interacting during the first two or three cycles with the rhythm induced by the previous light-off. It could be this interaction—between the basic light-off rhythm and the light pulses—that is determining your results.

HILLMAN: It is not clear to me from my results that there is definitely one light-off rhythm and one light-on rhythm. All I can say at this point is that I am interacting something with something.

PITTENDRIGH: In *Drosophila* we know that whether it responds to the 11-hr or the 13-hr photoperiod of the skeleton depends not only on the initial interval, but on the phase of the initial pulse relative to the light to dark transition. And your own data on flower induction in *Lemna* show that maximum flowering is a function not only of the first dark interval, but of the phase of the first pulse. I don't know how much work these experiments are; I suspect, however, that like all experiments, they are a lot of work. It would be very nice, though, if you could show that the distinction between a clear rhythm (when the 13-hr skeleton photoperiod comes first) and a flat curve (when the 11-hr one comes first) is not only a function of the 13-hr skeleton coming first, but also a function of the phase of the first pulse. Then you would have a complete analogue for the *Drosophila* system.

HILLMAN: Yes, although it looks to me as if this is not going to be an analogue to the *Drosophila* system in at least one sense. In *Drosophila*, you find two more or less stable states in response to the 11–13 hr skeleton, whereas you don't find that in the *Lemna* CO_2 output rhythm. Here, the state of response to the 11-hr skeleton photoperiod (the flat curve) appears to be confused for a few days, then it goes over into the other state, and you see a rhythm developing like that after the 13-hr skeleton. So it appears that the precise predictions from the *Drosophila* model, such as bistability, may not hold here, although the general prediction—that there will be a different relationship to the two modes of presenting the 11–13 hr skeleton—holds very nicely.

PERRY L. ADKISSON
S. H. ROACH

A Mechanism for Seasonal Discrimination in the Photoperiodic Induction of Pupal Diapause in the Bollworm *Heliothis zea* (Boddie)

Diapause is an adaptive mechanism allowing insects to survive the environmental hazards encountered in adverse seasons—e.g., winter, in temperate climates. Many insects time diapause by monitoring the most precise signal of seasonal change, the daily change in photoperiod (Lees, 1955; Danilevskii, 1965; Beck, 1968). Insects encounter two problems in using this environmental information: (1) They must be able, at a species-critical time during the season, to distinguish between days in which the duration of the daily light and dark periods differs by only a few minutes per day, and (2) they must be able to determine whether daylength is increasing or decreasing.

That at least some insects can measure daylength with great precision has been amply demonstrated in research conducted with *Pectinophora gossypiella* Saunders (Adkisson, 1964, 1966). This insect can distinguish between photoperiods that vary by only a few minutes per day in the relative durations of the light and dark periods. The problem of distinguishing between increasing and decreasing daylengths, however, has not been as thoroughly studied. Corbet (1956) and Norris (1965) showed that certain insects enter diapause in response to changing daylengths, but neither of these authors investigated the problem of seasonal discrimination. Bünning (1964) pointed out that except for the summer and winter soltices, each daylength

A Mechanism for Seasonal Discrimination in the Photoperiodic Induction of
Pupal Diapause in the Bollworm *Heliothis zea* (Boddie)

273

occurs twice within the year. Thus, when an organism uses daylength measurement for seasonal orientation, it must be able to determine whether a particular daylength is within a season of increasing or decreasing daylengths. Bünning also suggested that seasonal discrimination requires considerable photoperiodic information and that the developmental processes controlled by photoperiod must be blocked in more than one place.

Wellso and Adkisson (1966) studied the problem of seasonal orientation in the bollworm in considerable detail. They showed that diapause in this species always occurs in the fall and is induced by a long-day/short-day effect of photoperiod. If diapause is to occur, the adults and eggs of the bollworm must be subjected to longer days than the larvae.

In this report, we will discuss the photoperiodic responses of the bollworm in more detail, showing that

- the adult, egg, and larval stages of the bollworm are sensitive to photoperiod;
- each of these developmental stages has its own critical photoperiodic requirements for triggering the physiological processes involved in diapause;
- diapause occurs in response to the sequential sensitivity of the adult, egg, and larval stages to days decreasing in photoperiod and having certain critical lengths;
- because of the properties named above, diapause is restricted to a relatively narrow range of photoperiods. The combined effect of long and short days on the adult, egg, and larval stages provides the insect with a mechanism for seasonal discrimination.

SEASONAL INCIDENCE OF DIAPAUSE

To determine the seasonal incidence of diapause, Wellso and Adkisson (1966) reared bollworms in an open insectary on artificial diet. The test insects were collected as larvae in the field during September, October, and November. These larvae were placed on a wheat-germ diet (Adkisson *et al.*, 1960) and held until pupation, at which time the incidence of diapause in each test group was determined.

The response curve obtained is shown in Figure 1. The first diapausing pupae appeared during the last of September or the first of October. The maximum incidence of diapause occurred in those individuals that pupated in October. The incidence of diapause declined sharply in early November, reaching a minimum by the middle of the month.

The response curve of the bollworm is bell-shaped. Induction of diapause

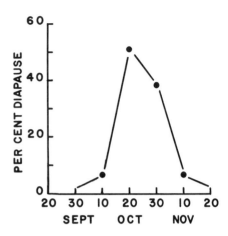

FIGURE 1 Seasonal incidence of diapause in bollworms reared in an insectary at College Station, Texas. Figure drawn from data reported by Wellso and Adkisson (1966).

in this species is restricted to a narrow range of daylengths. The Central Texas daylengths (including civil twilight) experienced by the diapausing bollworms decreased from slightly more than 13 hr at the time of the egg stage of the first diapausing pupae of fall, to slightly more than 11 hr at the time the last diapausing pupae were formed. This change in daylength amounted to a decrease in the light fraction of slightly more than 2 hr per day.

EFFECT OF PHOTOPERIOD ON DIAPAUSE INDUCTION

This study is a continuation of the work of Wellso and Adkisson. The adult, egg, and larval stages of the bollworm were held in photoperiodic regimens of constant, increasing, and decreasing daylengths. Temperature was maintained at 21°C. The larvae were reared on Vanderzant-Adkisson Special Wheat Germ Diet for insects supplied by the Nutritional Biochemicals Corporation, Cleveland, Ohio. Each test was replicated three times.

Results (summarized in Table 1) show that exposure of the bollworm moth, its eggs, and larval progeny to the same photoperiodic regimen did not produce a high incidence of diapause in the subsequent pupae. This was true even when all developmental stages were exposed to photoperiods having very short days. Results obtained with photoperiodic regimens having 13- and 14-hr days might have been predicted from data presented in Figure 1. These daylengths correspond to late summer in Central Texas and are too long to induce diapause. However, the prediction also might be made that if only one developmental stage of the bollworm is responsive

A Mechanism for Seasonal Discrimination in the Photoperiodic Induction of
Pupal Diapause in the Bollworm *Heliothis zea* (Boddie)
275

TABLE 1 Effects of Constant Photoperiod on Induction of Pupal Diapause in the
Bollworm

Photoperiodic Regimen (L:D)			
Adult	Egg	Larva	Percent Diapause
14:10	14:10	14:10	0.5
13:11	13:11	13:11	0.5
12:12	12:12	12:12	16.1
11:13	11:13	11:13	12.6
10:14	10:14	10:14	22.4

to photoperiod, the incidence of diapause produced in 10-, 11-, or 12-hr
days would be relatively great. This was not the case; the photoperiodic
regimen having the shortest day induced diapause in fewer than 25 percent
of the experimental animals.

Results presented in Table 1 clearly show that the induction of diapause
in the bollworm involves more than just exposure to short days. The bollworm must be exposed to daylengths that decrease in sequential order
during the adult, egg, and larval stages. Support for this conclusion is to
be found in Table 2. In these experiments the parents and eggs were exposed to longer days than were the subsequent larval progeny. Exposure of
the parent, egg, and larval offspring to photoperiods longer than 13 hr prevented diapause in the bollworm. This inhibition occurs because the critical
transitional photoperiod for bollworm larvae, i.e., the photoperiod at which
the transition from a long-day to a short-day effect occurs, is LD *13*:11.
Wellso and Adkisson (1966) defined this critical transitional photoperiod
for the bollworm and showed that, regardless of the photoperiodic treatment of parents and eggs, diapause is averted in this species whenever the
larvae are exposed to photoperiods having light fractions of 13 hr or longer.

TABLE 2 Effects of Decreasing Photoperiods on Induction of Pupal Diapause in the
Bollworm

Photoperiodic Regimen (L:D)			
Adult	Egg	Larva	Percent Diapause
14:10	14:10	13:11	0.0
13:11	13:11	12:12	26.1
12:12	12:12	11:13	57.0
11:13	11:13	10:14	20.8

In this study, exposure of parents and eggs to LD *13*:11 followed by exposure of the larval progeny to LD *12*:12 induced diapause in approximately 25 percent of the subsequent pupae. Maximum incidence of diapause occurred when adults and eggs were exposed to LD *12*:12 and the larvae were held in LD *11*:13.

These data (Table 2) amply demonstrate a long-day/short-day effect in the induction of diapause in the bollworm. However, it is important to note that when the adults and eggs were held in an 11-hr day and the larval progeny in a 10-hr day, the incidence of diapause was greatly reduced. We believe this occurred because the daylengths to which the adults and eggs were exposed were too short (below the critical transitional threshold for these developmental stages) to produce maximum induction of diapause. If this is true, there must be at least two physiological processes involved in the photoperiodic control, each with its own critical threshold daylength requirement.

If the data included in Table 2 were graphed, the resulting response curve would have a bell shape similar to that shown in Figure 1 for bollworms reared in an open insectary. Certainly, more is involved in the photoperiodic induction of diapause in the bollworm than a simple shift in the exposure of larvae to days shorter than those experienced by the previous adult and egg stages.

Diapause was largely averted in experiments in which larvae were exposed to longer days than those experienced by the previous egg and adult stages (Table 3). These results suggest that bollworms should not be expected to undergo diapause when daylengths are increasing, as in the spring and early summer.

CRITICAL PHOTOPERIODIC REQUIREMENTS OF ADULT, EGG, AND LARVAL STAGES

Results summarized in Table 4 furnish evidence of the photoperiodic sensitivity of the adult, egg, and larval stages of the bollworm. For example, when the light fraction of the photoperiodic regimen in which the adults were held was reduced from 13 to 12 hr, while both eggs and larvae were held in LD *11*:13, the incidence of diapause more than doubled.

Eggs showed an even greater sensitivity to photoperiod. In experiments in which adults were held in LD *14*:10 and larvae in LD *10*:14, 55.9 percent of the pupae resulting from eggs held in a 14-hr day diapaused. However, when the photoperiod to which the eggs were exposed was reduced from 14 to 10 hr, diapause was almost completely averted. An obvious

A Mechanism for Seasonal Discrimination in the Photoperiodic Induction of
Pupal Diapause in the Bollworm *Heliothis zea* (Boddie)
277

TABLE 3 Effects of Increasing Photoperiod on Induction of Pupal Diapause in the Bollworm

Photoperiodic Regimen (L:D)			
Adult	Egg	Larva	Percent Diapause
10:14	10:14	11:13	26.2
10:14	11:13	11:13	3.5
11:13	11:13	12:12	6.5
11:13	12:12	12:12	0.0
12:12	12:12	13:11	0.0
12:12	13:11	13:11	0.0

conclusion from this series of experiments is that a 10-hr day is shorter than the critical daylength required by the egg to induce pupal diapause. A photoperiodically sensitive process involved in the control of diapause is apparently triggered during the egg stage of the bollworm. If the daylength to which the egg is exposed is longer than 10 hr, diapause may occur, but if it is 10 hr or less, diapause is averted. The 10-hr day appears to be slightly shorter than the critical length required by the eggs (and perhaps also by the adults) for induction of diapause.

The data reported in Table 4 also demonstrate that bollworm larvae are similarly sensitive to photoperiod. When larvae were held in LD *13*:11 following exposure of adults and eggs to LD *14*:10, diapause was averted, but when larvae were held in LD *10*:14, a high incidence of diapause oc-

TABLE 4 Photoperiodic Sensitivity of the Adult, Egg, and Larval Stages of the Bollworm

Photoperiodic Regimen (L:D)			
Adult	Egg	Larva	Percent Diapause
Adults			
13:11	11:13	11:13	22.4
12:12	11:13	11:13	53.4
Eggs			
14:10	14:10	10:14	55.9
14:10	10:14	10:14	1.1
Larvae			
14:10	14:10	13:11	0.0
14:10	14:10	10:14	55.9

FIGURE 2   Percentages of diapausing bollworm
pupae produced by various photoperiodic regimens similar to photoperiods experienced by
the insect at College Station, Texas, from early
September (A) to late November (G). Photoperiodic regimens for adult, egg, and larval stages,
respectively, were as follows: A–LD *14*:10,
LD *14*:10, LD *13*:11; B–LD *13*:11, LD *13*:11,
LD *12*:12; C–LD *12*:12, LD *12*:12, LD *11*:13;
D–LD *12*:12, LD *11*:13, LD *11*:13; E–LD
11:13, LD *11*:13, LD *11*:13; F–LD *11*:13, LD
11:13, LD *10*:14; and G–LD *11*:13, LD *10*:14,
LD *10*:14.

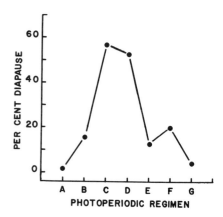

curred. These data reveal a second critical photoperiod sensitivity in the
bollworm—in this instance, in the larvae. When larvae experience days
longer than 13 hr, diapause is averted regardless of the photoperiods experienced by the previous adult and egg stages, because the 13-hr day is
above the upper critical photoperiodic requirements of the larvae for the
induction of diapause (Wellso and Adkisson, 1966). This response provides
the second physiological switch. If daylengths experienced by the larvae
are 13 hr or longer, some physiological process is switched so that the
insect proceeds to the adult stage without pupal diapause, regardless of
the photoperiods experienced by the parents and eggs. If the daylengths
to which the larvae are exposed are shorter than 13 hr, diapause may or
may not occur. Under these conditions, diapause depends on the photoperiods experienced in the previous adult and egg stages.

Thus, in the combined series of experiments reported in Tables 1, 2, 3,
and 4, it is evident that the adult, egg, and larval stages of the bollworm
must be exposed to a decreasing sequence of daylengths if diapause is to
occur. However, even under decreasing daylengths diapause occurs only
when the critical photoperiodic requirements both of the adult and eggs
and of the larvae are met.

Data collected from bollworms reared in selected photoperiods may be
used to retrace the seasonal response curve of Figure 1. Such a response
curve is shown in Figure 2. In photoperiods similar to those experienced
by the bollworm in late August or early September, diapause is averted
because days are longer than the critical photoperiod (*LD 13*:11) required
by the larvae for the induction of diapause. Then, as all the developmental
stages come under inductive photoperiods, the response curve shows a
maximum percentage of diapause in those insects living in 11- and 12-hr
days. As the days continue to decrease in length, the incidence of diapause

A Mechanism for Seasonal Discrimination in the Photoperiodic Induction of
Pupal Diapause in the Bollworm *Heliothis zea* (Boddie)

279

FIGURE 3   A model for explaining the photo-periodic induction of diapause in bollworm pupae. Curve A represents the adult and egg response curve and shows a physiological process that requires days longer than 10 hr. Curve B represents the larval response curve and shows a physiological response that requires days shorter than 13 hr. Diapause occurs only in photoperiods of area C, where the two response curves overlap, that is, when the photoperiod is decreasing from 13 hr of light per day.

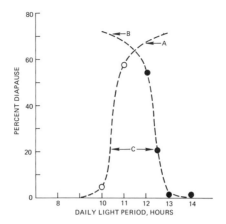

declines, approaching zero when adults and eggs are exposed to 10-hr days; this daylength is below the critical requirements for these two stages. These data, like the seasonal response curve of Figure 1, show that the induction of diapause in the bollworm is restricted to a relatively narrow range of photoperiods.

CONCLUSIONS

Bünning (1969) discussed two types of photoperiodically controlled physiological processes in plants and animals. One type is characterized by a restriction of the process to daylengths that are longer than the critical daylength for the species in question, while in the other type the physiological process is restricted to daylengths shorter than the species-critical value. Plants and animals may have either of the two types or may combine them. In the latter instance, a certain physiological process may occur only when a reaction of the first type (long-day response) is followed by a reaction of the other type (short-day response). This, in effect, provides a physiological response of the long-day/short-day type commonly implicated in the flowering of certain plants (Bünning, 1964). A similar phenomenon has only recently been demonstrated for the induction of diapause in insects (Wellso and Adkisson, 1966).

A model similar to that proposed by Bünning (1969) is presented in Figure 3 to explain the photoperiodic control of diapause in the bollworm. Curve A, plotted from data obtained with photoperiodic treatment of adults and eggs, shows that the developmental processes leading to the induction of diapause can be released in adults and eggs only when the daylengths are greater than 10 hr. Conversely, Curve B, plotted from the larval response curve of Wellso and Adkisson (1966), shows that the larval processes controlling the induction of diapause are blocked by days longer than 13 hr. In fact, diapause may occur only in area C, where the two photoperiodic response curves overlap.

The combining of two daylength measurements by the bollworm provides the insect with a mechanism for seasonal orientation. The long-day/short-day response described herein limits the induction of diapause to the decreasing daylengths of fall. Even then, diapause is restricted to late September, October, and early November, because this is the only time in the year when the lower critical daylength requirement of adults and eggs and the upper critical daylength of the larvae are both met within the appropriate interval.

ACKNOWLEDGMENT

This research was conducted in cooperation with the Entomology Research Division, U.S. Department of Agriculture. Certain of the data reported were used by the junior author in partial fulfillment of the dissertation requirements for the Ph.D. degree at Texas A&M University.

REFERENCES

Adkisson, P. L. 1964. Action of the photoperiod in controlling insect diapause. Am. Nat. 98:357–374.

Adkisson, P. L. 1966. Internal clocks and insect diapause. Science 154:234–241.

Adkisson, P. L., E. S. Vanderzant, D. L. Bull, and W. E. Allison. 1960. A wheat germ medium for rearing the pink bollworm. J. Econ. Entomol. 53:759–762.

Beck, S. D. 1968. Insect photoperiodism. Academic Press, New York. 288 p.

Bünning, E. 1964. The physiological clock. Academic Press, New York. 145 p.

Bünning, E. 1969. Common features of photoperiodism in plants and animals. Photochem. Photobiol. 9:219–228.

Corbet, P. S. 1956. Environmental factors influencing the induction and termination of diapause in the emperor dragonfly, *Anax imperator* Leach (Odonata:Aeshnidae). J. Exp. Biol. 33:1–14.

Danilevskii, A. S. 1965. Photoperiodism and seasonal development of insects. Oliver and Boyd, London. 283 p. [English Transl.]

Lees, A. D. 1955. The physiology of diapause in Arthopods. Cambridge Univ. Press, Cambridge. 151 p.

Norris, M. J. 1965. The influence of constant and changing photoperiods on imaginal diapause in the red locust (*Nomadacris septemfasciata* Serv.). J. Insect Physiol. 11:1105–1119.

Wellso, S. G., and P. L. Adkisson. 1966. A long-day short-day effect in the photoperiodic control of the pupal diapause of the bollworm, *Heliothis zea* (Boddie) (Lepidoptera:Noctuidae). J. Insect Physiol. 12:1455–1465.

BRUCE G. CUMMING

Endogenous Rhythms and Photoperiodism in Plants

Sollberger (1965), in his comprehensive book *Biological Rhythm Research* states that "there are innumerable ways in which rhythms may interact or be controlled. The models are rather exact; whereas the construction of the biological oscillators is largely unknown." He also suggests that "The cybernetic approach has many avenues. . . . Imagine that we let the movements of plant leaves control the intensity or spectral composition of the illumination. What would happen if we put a plant on wheels and let the leaf position direct the carriage (slowly) with respect to the position or movements of a light source?"

In this paper I will first outline some conclusions and suggestions that have been derived from experiments on the control of flowering in *Chenopodium rubrum*: a so-called short-day plant (Cumming, 1959, 1969a). I will then suggest the possible operation, coupling, and interaction of the components of a theoretical mechanical motive-power model analogue that provides a basis for the explanation and prediction of some components of the biological system under consideration.

MATERIALS AND METHODS

Chenopodium rubrum shows extremely sensitive morphogenic responses to differences in the relative duration of light:dark cycles, or to a single period of darkness that interrupts continuous light. The experimental pro-

281

cedure in general use is to germinate seeds and maintain the resulting seedlings either in alternating light:dark cycles, or in constant light that is
interrupted by a single period of darkness. Seedlings that produce a floral
primordium are classified as flowering, the remainder are considered to be
vegetative. Under optimal inductive conditions *C. rubrum* can produce a
microscopically small flower on a seedling that is only 2–3 mm tall. Experiments on flowering can be completed within 2–3 weeks after sowing seeds,
thus economizing on both space and time (Cumming, 1959, 1967a). As
many as 400 6-cm petri dishes may be used in one experiment, each containing about 120 seeds sown on filter paper or agar that incorporates a

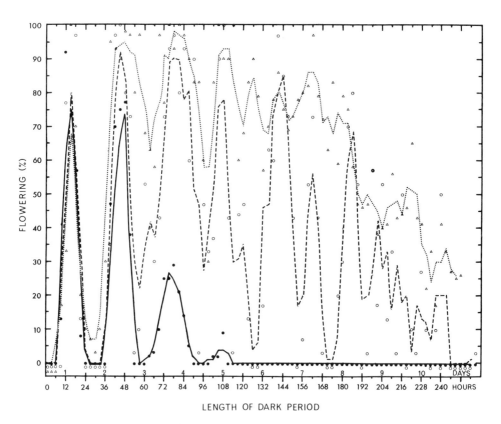

LENGTH OF DARK PERIOD

FIGURE 1   Dependence of rhythmic flowering response in *Chenopodium rubrum* (ecotype
60°47′N) on the duration of a single dark period interrupting continuous cool white fluorescent
light, and glucose or sucrose supplied throughout darkness. Solid circle, solid line, Hoagland's solution; open circle, dashed line, + 0.6 *M* glucose; open triangle, dotted line, + 0.6 *M* sucrose (modified from Cumming, 1967c and reproduced by permission of the National Research Council of
Canada from the *Canadian Journal of Botany* 45:2179 [1967]).

chemically defined medium; chemicals can be added to the medium for prescribed periods, in either light or darkness and then can be washed off after appropriate test periods.

Flowering may be induced by a single dark interruption of constant light as follows: After an initial period of 3 or more days in low intensity white fluorescent light (600 ft-c), during which temperature is varied to stimulate germination, seedlings are transferred to 3000 ft-c fluorescent, or 1000 ft-c incandescent light (Cumming, 1967c). To induce flowering, 4–5 day old vegetative seedlings are placed in darkness and the length of the dark period is varied by returning replicated dishes (containing the seedlings) to light after successively longer periods. Alternatively, a dark period of duration known to induce subsequent flowering can be interrupted by light of a particular quality and intensity, and the effect of this light on flowering compared with the control flowering response (Cumming, Hendricks, and Borthwick, 1965). After removal from the dark, seedlings are left in continuous light for about 1 week and the percentage of seedlings showing a flower primordium is determined using a binocular dissecting microscope. The percentage flowering can be plotted against the time at which light terminated or interrupted the dark period, and a sine curve with circadian periodicity may be obtained. For example, a rhythmic flowering response was evident when the duration of darkness interrupting continuous light was varied between 0 and 5 days (Figure 1), but the amplitude of the last one of the four oscillations was very small (i.e., "damped" out) and there was no flowering after a dark period of more than 5 days' duration. However, oscillations were sustained for a total period of at least 10 days when either sucrose or glucose were supplied during darkness (Figure 1).

ANALYSIS OF PROCESSES LEADING TO RHYTHMIC FLOWERING RESPONSE

If seeds of *C. rubrum* are germinated and the seedlings maintained continuously in daily photoperiodic cycles of 12 hr light:12 hr dark (*LD 12*:12), flower primordia are produced on nearly all seedlings in less than 2 weeks. If, however, the same daily amount of light and darkness is imposed but in twice-daily photoperiodic cycles of *LD 6*:6, there is no flowering at all within the same period. Significantly, however, if seedlings are germinated and maintained in *LD 6*:6 cycles for 5.5 days, but are then transferred to a long dark period (of varied duration), which is then followed by continuous light, the dark period elicits a rhythmic flowering response that, although of smaller amplitude, is very similar to that obtained when darkness interrupts constant light (Cumming, B. G., and King, R. W.; unpublished results). These results are most readily explained by Bünning's (1936) hypothesis of the existence of an endogenous free-running oscillation with a period of about 24 hr, that is, each oscillation of the *C. rubrum* flowering response involves a regular alternation of two half-cycles (skotophile and photophile phases) of approximately 12 hr. The induction and initiation of flowering is apparently effected only when the endogenous skotophile and photophile phases are entrained to or coincide with the external photoperiodic cycle(s). Cycles of 12 hr of darkness and 12 hr of light clearly "satisfied" the endogenous requirement and induced flowering, but the cycles of 6 hr dark and 6 hr light did not. Nor did the latter cycles entrain the circadian rhythm so that there was any rhythmic induction of flowering of higher than circadian periodicity in prolonged darkness. However, because the endogenous circadian rhythm can be set by the transition from 6 hr light to darkness and will be sustained in prolonged darkness, the rhythmic expression of the physiological clock was realized in the prolonged dark period in the same way as if it had been preceded by continuous light.

Changes in photosynthesis and in phytochrome are clearly implicated in the transition from light to darkness. Although there is no photosynthesis during darkness, photosynthate already produced appears to be involved in the processes leading to flowering. Increasing the white light intensity preceding darkness or supplying glucose or sucrose before or during darkness can stimulate flowering (Figure 1, Cumming, 1967c). There is some time dependency in this effect of sugar. If glucose is supplied only during a subjective skotophile phase of a long dark period it is nonstimulatory, but when it is supplied only during a photophile phase it does stimulate flowering (Cumming, 1967c).

The red portion of the spectrum is most effective for maintenance of phytochrome in the physiologically active form (P_{fr}) (Kasperbauer, Borthwick, and Hendricks, 1964); P_{fr} can be converted to the physiologically inactive form, P_r, by far-red (fr) radiation. In effect, P_{fr} is predominant during the exposure of *C. rubrum* seedlings to white light (Cumming, 1963; Cumming *et al.*, 1965). If light of a low R/FR ratio is used to terminate a white light period (and thus reduce the P_{fr} level just before entering a long dark interruption), less flowering results because there is then too little active P_{fr} to function positively during darkness (Cumming, Hendricks, and Borthwick, 1965). The onset of darkness results in gradual reversion of an increasing proportion of the P_{fr} form of phytochrome to P_r (Borthwick and Hendricks, 1960; Butler *et al.*, 1959). It is not clear how rapidly this reversion occurs, but the response of *C. rubrum* ecotype $60°$ $47'$N suggests that, normally, some P_{fr} remains even during a prolonged dark period of several days and that it has a positive physiological role in the induction of flowering during this time (Cumming *et al.*, 1965). In view of this, I had reasoned that it might be possible to induce flowering by replacing a dark interruption of continuous light with a light interruption—if such light were to maintain phytochrome-P_{fr} and photosynthetic energy utilization at a level that duplicates, or substitutes for, the metabolic processes occurring in darkness (on some integrated duration–concentration basis). This has been accomplished experimentally, and it is particularly noteworthy that optimal flower induction has been obtained with light containing enough red irradiation to maintain a physiologically significant level of phytochrome in the active P_{fr} form and to allow some photosynthesis to occur (Cumming, 1969c).

The last environmental transition in the sequences that I have been describing is the change from darkness to light ("dawn" signal). Since the existence of a rhythm in the induction of flowering is revealed by determining the percentage flowering of seedlings that have been in continuous light for up to a week following the dark interruption, it is obviously important to discover what series of processes lead to the final rhythmic flowering response. Certain details are clear. A threshold minimum period of darkness (first skotophile phase from zero to about the 9th hour of darkness: Figure 1) must elapse before the imposition of light can result in the initiation of flowering. During the next 15-hr period of darkness (i.e. the first photophile phase, from about the 9th to the 24th hour of darkness) the institution of continuous light or *LD* 6:6 cycles (that are noninductive when given alone) will result in some flower initiation. An inductive period of darkness (whether of less or more than 24-hr duration) must therefore provide some essential step in the process leading to the initiation of a flower primordium, but this step appears to be either reversible or subject to inhibition, and a flower primordium is produced only if light is instituted at the correct subjective time.

One basis for the rhythmicity that is elicited by prolonged darkness could be that the induction that occurs during the skotophile phase dissipates during the photophile phase—unless light is instituted early enough in that phase; otherwise, a further skotophile phase is required for reinduction, and so on, cyclically. Light given during the skotophile phase could prevent adequate realization of this inductive state or principle by not allowing sufficient time for formation of a stimulatory entity or by stimulating the production of an inhibitor.

An alternative basis of the rhythmicity in flowering could be that some partial inductive process is completed during the first skotophile phase of darkness, but that the whole process leading to flower initiation is only realized if light is imposed in a subsequent photophile phase of darkness. However, if light is instituted during any skotophile phase of darkness this might well be inhibitory to completion of the inductive process.

MECHANICAL MOTIVE-POWER MODEL ANALOGUE OF BIOLOGICAL CIRCADIAN-PHOTOPERIODIC SYSTEM

In thinking over the foregoing it occurred to me that some aspects of the common automobile provide an intriguing analogy (Figure 2). My precedent is the recognized value of previous abstractions—pendulums, springs, wheels of clocks, and so forth—as analogues of biological oscillations (Bünning, 1967; Sollberger, 1965; Sweeney, 1969). The following discussion analyzes some of the partial processes of the biological circadian-photoperiodic system and the proposed components of the mechanical motive-power model analogue.

ENERGY UTILIZATION FOR THE DISPLAY OF PHOTOPERIODICALLY CONTROLLED PHENOMENA (ENGINE)

Endogenous rhythmicity and photoperiodic response can be considered as adaptations of the living organism to cyclic environmental conditions of energy supply, in which changes in the timing of energy input are used as cues for the timing of significant plant responses. The onset of flowering (a development response) involves growth and therefore expenditure of energy, just as leaf movements (motor responses) utilize energy.

In the mechanical model the drive speed of the wheels (display of function: "progressive development") is dependent on the revolutions of the engine that result from the utilization of the energy available from the motive power energy source.

QUANTITATIVE CONTROLS OF PHOTOMORPHOGENESIS MEDIATED BY PHYTOCHROME (ACCELERATOR AND THROTTLE)

Many quantitative effects of phytochrome are evident in plants. Its action has been likened to a pacemaker (Cumming, Hendricks, and Borthwick, 1965) or valve (Gordon, 1964) that exerts some control over metabolic

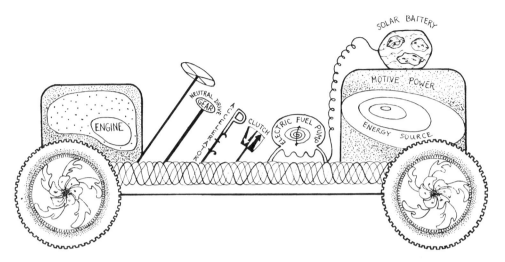

FIGURE 2   Mechanical motive-power model analogue of biological circadian-photoperiodic system.

substrate availability and utilization. Important timing and control features are provided by both the photoreversibility of phytochrome and the dark reversion of the active form of phytochrome, P_{fr} to P_r.

In the mechanical model, the accelerator system, which could include a throttle (or governor), would have a role similar to phytochrome in controlling the rate of utilization of fuel for production of energy to drive a response. The accelerator could be controlled through a photocell having properties of photoreversibility and dark reversion similar to phytochrome.

THE BASIC OSCILLATOR (ELECTRIC FUEL PUMP)

The operation of this biological component remains an open question (Bünning, 1967), since its existence is not proven and because we can only infer its presence from indirect experimental results. Bünning (1967) has discussed reasons for considering that a tension-relaxation model accounts for many of the properties of endogenous circadian rhythmicity. The possibility that circadian rhythmicity originates from changes in energy transduction and is based on alterations in the conformation of flexible lipoprotein ion-exchange membranes, has been discussed elsewhere (Cumming and Wagner, 1968; Wagner and Cumming, 1970).

In the mechanical model, the electric fuel pump would operate on a tension-relaxation principle, employing membranes, springs, or some other system that is capable of time measurement. With the fuel pump in the tension phase, the passage of fuel to the engine via the accelerator system would be maximal, and in the relaxation phase fuel would be restricted. Operation of the fuel pump could be supported by the discharge of a solar battery to which it would be coupled.

INDUCTION AND DISPLAY OF RESPONSES (CLUTCH AND GEAR: "NEUTRAL" AND "DRIVE")

In *C. rubrum*, darkness stimulates flowering, but ordinarily the formation of a floral primordium is dependent on light following darkness at an appropriate time.

   The mechanical model for flowering would include a clutch, which allows a gear to become engaged in drive so that the engine is coupled to the wheels, resulting in progressive development. In uninterrupted light this gear is normally in "neutral" (so far as progressive development is concerned). During darkness the clutch is slipped "in," but only if there is a sufficient duration of darkness (critical dark period). If darkness is sustained beyond this critical period (first skotophile phase), the gear automatically engages in "drive" because of the increase in engine revolutions due to the control exerted by the fuel pump (basic oscillator). Nevertheless, progressive development will not occur unless light is imposed during the photophile phase, because only then is the clutch let "out" so that the drive is directly engaged and the engine is coupled to the wheels. If there is no light, the gear reverts to neutral at the beginning of the next skotophile phase when the engine revolutions decrease because of the fuel pump control. If light is imposed during the skotophile phase, the clutch is let "out" but the gear remains in neutral, as it was in the previous constant light period, so that there is no progressive development.

   The induction of flowering in *C. rubrum* by completely substituting "inductive light" (of intermediate energy and appropriate quality) in place of darkness indicates that some forcing of the biological system is possible. This is consistent with the mechanical analogue, because a common feature in automobiles is that it is possible to force the gear into drive without using the clutch, if the engine revolutions are sufficiently low. This suggests an adaptation of the model to the circadian opening (photophile phase) and closing (skotophile phase) of the leaves of some plants (a motor response), which can occur both in prolonged darkness (Hamner *et al.*, 1962) and in constant light (Hoshizaki and Hamner, 1964). Here, the coupling between energy source and leaf motor response must be very direct even in darkness, and a report that petal movements can be sustained for a longer period in prolonged darkness by supplying sugar is significant (Engelmann and Vielhaben, 1966). In the mechanical analogue for such leaf movements there would be no clutch (automatic gear system), and differences in the motor response would depend upon the engine revolutions, which would be higher during the photophile phase and thus engage drive, and lower in the skotophile phase, when the gear would revert to neutral.

SYNTHESIS, STORAGE, AND AVAILABILITY OF ENERGY TO PERFORM WORK (MOTIVE POWER ENERGY SOURCE)

The timing of energy input and changes associated with its availability and utilization may have profound effects on the phasing of the basic oscillator and other coupled components. In both the biological system and the mechanical model, the sun is the original source of such energy, which, after conversion in photosynthesis and subsequent processes, is available to perform work with different degrees of efficiency, for example, in sugar, starch, or oil (biological), and in petroleum, oil, or coal (mechanical).

REGULATORY CONTROL MECHANISMS (SOLAR BATTERY AND ELECTRICAL SYSTEM)

In the biological system, regulatory control mechanisms may involve hormonal balances, compartmentation, enzyme regulation, feedback relationships, and also changes in electric potentials. Hormonal effects on the flowering of *C. rubrum* are reviewed elsewhere (Cumming, 1969a). Compartment relationships probably play an important role in such regulatory control, because circadian adaptation mechanisms may depend strongly on the compartition of energy transformation within such different organelles as chloroplasts, mitochondria, and membranous structures associated with glycolysis (Barker, Khan, and Solomos, 1966). Mitchell (1967) has compared the proton-translocation phosphorylation of mitochondria, as natural fuel cells, with solar cells, and he has emphasized the importance of spatially oriented chemical reactions. Also of possible relevance as regulatory control mechanisms are the indications of bioelectric fields in plants (e.g., Ebrey, 1967; Scott, 1967; Yamaguti, 1932). Considerable evidence indicates that subtle geophysical factors can influence the regulatory control mechanisms of plants (Brown, 1969). The discovery of highly significant correlations between periodicities in the germination of *Chenopodium botrys* and variations in solar radio flux (indicative of sunspot activity) provides further confirmation that the effects of subtle geophysical factors should be investigated more fully (Cumming, 1967b).

In the mechanical model, the solar battery and electrical system provide the storage system and necessary conduction and communication for the electrically controlled phenomena. The solar battery would be capable of periodic charge and discharge and could be coupled to the fuel pump so that it forces positive operation of the latter. The solar battery and electrical system could also be influenced by subtle geophysical factors. A report by Barber (1962) of correlations over a 40-yr period in the effective storage capacity of lead-acid batteries and the 11-yr sunspot cycle provides evidence of such possibilities.

ACKNOWLEDGMENTS

A National Research Council of Canada grant in aid of research is gratefully acknowledged. I thank my wife, M. Cumming, for drawing the model analogue, E. Kratky for technical assistance, and J. Planck for helpful criticism of the manuscript.

REFERENCES

Barber, D. R. 1962. Apparent solar control of the effective capacity of a 110-v 170 AH lead-acid storage battery in an 11 year cycle. Nature (Lond.) 195:684–687.
Barker, J., M. A. A. Khan, and T. Solomos. 1966. Mechanism of the Pasteur effect. Nature (Lond.) 211:547–548.
Borthwick, H. A., and S. B. Hendricks. 1960. Photoperiodism in plants. Science 132: 1223–1228.
Brown, F. A., Jr. 1969. A hypothesis for extrinsic timing of circadian rhythms. Can. J. Bot. 47:287–298.
Bünning, E. 1936. Die endonome Tagesrhythmik als Grundlage der photoperiodischen Reaktion. Ber. Dtsch. Bot. Ges. 54:590–607.
Bünning, E. 1967. The physiological clock. 2nd ed. Springer-Verlag, New York.
Butler, W. L., K. H. Norris, H. W. Siegelman, and S. B. Hendricks. 1959. Detection, assay and preliminary purification of the pigment controlling photoresponsive development of plants. Proc. Natl. Acad. Sci. U.S. 45:1703–1708.
Cumming, B. G. 1959. Extreme sensitivity of germination and photoperiodic reaction in the genus *Chenopodium* (Tourn.) L. Nature (Lond.) 184:1044–1045.
Cumming, B. G. 1963. Evidence of a requirement for phytochrome-P_{fr} in the floral initiation of *Chenopodium rubrum*. Can. J. Bot. 41:901–926.
Cumming, B. G. 1967a. Early-flowering plants, p. 277 to 299. *In* F. H. Wilt and N. K. Wessells [ed.], Methods in developmental biology. T. Y. Crowell Co., New York.
Cumming, B. G. 1967b. Correlations between periodicities in germination of *Chenopodium botrys* and variations in solar radio flux. Can. J. Bot. 45:1105–1113.

Cumming, B. G. 1967c. Circadian rhythmic flowering responses in *Chenopodium rubrum*: effects of glucose and sucrose. Can. J. Bot. 45:2173–2193.

Cumming, B. G. 1969a. *Chenopodium rubrum* and related species, p. 156 to 185. *In* L. T. Evans [ed.], The induction of flowering. Some case histories. Macmillan of Australia, South Melbourne, and Cornell Univ. Press, Ithaca, New York.

Cumming, B. G. 1969b. Circadian rhythms of flower induction and their significance in photoperiodic response. Can. J. Bot. 47:309–324.

Cumming, B. G. 1969c. Photoperiodism and rhythmic flower induction: complete substitution of inductive darkness by light. Can. J. Bot. 47:1241–1250.

Cumming, B. G., S. B. Hendricks, and H. A. Borthwick. 1965. Rhythmic flowering responses and phytochrome changes in a selection of *Chenopodium rubrum*. Can. J. Bot. 43:825–853.

Cumming, B. G., and E. Wagner. 1968. Rhythmic processes in plants. Annu. Rev. Plant Physiol. 19:381–416.

Ebrey, T. G. 1967. Fast light-evoked potential from leaves. Science 155:1556–1557.

Engelmann, W., and V. Vielhaben. 1966. Wirküng verschiedener Zucker auf die *Kalanchoë*–Blütenblattbewegung. Z. Pflanzenphysiol. 55:54–58.

Gordon, S. A. 1964. Oxidative phosphorylation as a photomorphogenic control. Q. Rev. Biol. 39:19–34.

Hamner, K. C., J. C. Finn, G. S. Sirohi, T. Hoshizaki, and B. H. Carpenter. 1962. The biological clock at the South Pole. Nature (Lond.) 195:476–480.

Hoshizaki, T., and K. C. Hamner. 1964. Circadian leaf movements: persistence in bean plants grown in continuous high intensity light. Science 144:1240–1241.

Kasperbauer, M. J., H. A. Borthwick, and S. B. Hendricks. 1964. Reversion of phytochrome 730 (P_{fr}) to P660 (Pr) assayed by flowering in *Chenopodium rubrum*. Bot. Gaz. 125:75–78.

Mitchell, P. 1967. Proton-translocation phosphorylation in mitochondria, chloroplasts and bacteria: natural fuel cells and solar cells. Fed. Proc. Fed. Am. Soc. Exp. Biol. 26:1370–1379.

Scott, B. I. H. 1967. Electric fields in plants. Annu. Rev. Plant Physiol. 18:409–448.

Sollberger, A. 1965. Biological rhythm research. Elsevier Publ. Co., Amsterdam.

Sweeney, B. M. 1969. Rythmic phenomena in plants. Academic Press, New York.

Wagner, E., and B. G. Cumming. 1970. Betacyanin accumulation, chlorophyll content and flower initiation in *Chenopodium rubrum* L. as related to endogenous rhythmicity and phytochrome action. Can. J. Bot. 48:1–18.

Yamaguti, Y. 1932. Über elektrische Potentialänderungen an periodisch sich bewegenden Primärblättern von *Canavalis ensiformis*, D.C. Bot. Mag. (Tokyo) 46:216–222.

FRANK B. SALISBURY
ALICE DENNEY

Separate Clocks for Leaf Movements and Photoperiodic Flowering in *Xanthium strumarium* L.

Photoperiodism is perhaps the first clear-cut example of biological chronometry to be documented (Garner and Allard, 1920). During the past two decades there has been considerable discussion of the mechanism of time measurement in the flowering of plants sensitive to photoperiod. Does timing occur as in an hourglass, the measure being, perhaps, the time required for the far-red-receptive form of phytochrome to drop to some critical level (Hendricks, 1960; Borthwick and Hendricks, 1960), or is timing under the control of some oscillating timer such as that involved in circadian rhythmicity? Research during the past decade has increasingly supported the latter hypothesis (summary in Salisbury, 1969). Several authors (e.g., Borthwick and Downs, 1964; Takimoto and Hamner, 1965; Hoshizaki, Brest, and Hamner, 1969; Cumming, this volume) have, for example, demonstrated circadian periodicities in sensitivity to red or far-red light perturbations and have shown that LD cycles have optimal inductive effect when the period is about 24 hr.

   If we accept the involvement of an oscillating timer in photoperiodism, the next question might be considered trivial: Is the photoperiodic clock identical with the circadian clock that controls other rhythmic phenomena such as daily leaf movements? Many authors have simply assumed that the two "clocks" were indeed equivalent. Salisbury (1959) compared leaf positions with changing light sensitivity during photoperiodic induction of cocklebur plants treated with cobaltous ion. Timing of flowering was distinctly delayed by cobaltous ion, but leaf movements of treated plants remained the same as in controls. No firm conclusions were drawn, however, since it seemed possible that slight but important delays in the leaf rhythms might lie within the range of experimental error. Bünsow (1960) showed that petal movements in *Kalanchoë* correlated well with changing light sensitivity in photoperiodic induction, and concluded that these movements were a reliable indication of the status of the photoperiodic clock. Halaban (1968) found a fixed relationship between leaf position of *Coleus frederici* and sensitivity to a light interruption during an inductive dark period; maximum sensitivity to the light interruption always occurred 5 hr after the lowest leaf position. Hillman (1970) assumed that in *Lemna*, carbon dioxide metabolism, which he demonstrated to have a circadian periodicity, was an accurate indication of the status of "the" clock in this organism. Recently, Bünning and Moser (1969) wired leaves of *Chenopodium amaranticolor*, *Perilla ocymoides*, and *Nicotiana tabacum* so that they could not display their daily movements, and thereby strongly inhibited the response of the plants to photoperiodic floral induction (see also Bünning, this volume). Although several implications are suggested, one might be that the two clocks are the same. We have repeated the experiment with cocklebur, however, and failed to observe a significant inhibitory effect on flowering when leaves are immobilized.

For the past 10 years we have investigated the photoperiodic clock controlling flowering in *Xanthium strumarium* L. (cocklebur). In recent years we have also studied leaf movements of this plant, on the assumption that leaf positions are a valid indicator of the status of "the" clock. Yet beginning with the cobaltous ion studies mentioned above, few of our experimental results have supported this assumption. We have consequently undertaken a careful re-evaluation of our experimental results and have repeated at least one crucial experiment. The conclusion now seems inescapable that the clock controlling circadian leaf movements in cocklebur operates separately from the clock controlling the photoperiodic flowering response.*

The following paragraphs first review previously published work on the characteristics of the photoperiodic clock in the cocklebur; methods and details have been considerably abbreviated. The second section presents new studies on leaf movements.

---

*We arrived at this conclusion after the Friday Harbor Symposium and have modified our manuscript accordingly.

TIME MEASUREMENT IN FLOWERING OF COCKLEBUR

METHODS

Plants were grown in individual pots and were maintained in the vegetative condition by extending the photophase of each day to 20 hr with light of about 50 ft-c. Before an experiment, plants were trimmed to one leaf; only the half-expanded, maximally sensitive leaf remained on each plant. They were then placed in a darkroom or in controlled-temperature growth chambers and exposed to light or darkness according to the specific requirements of a given experiment. Nine days after experimental treatment, buds were examined under a dissecting microscope and classified according to an arbitrary series of developmental stages of the inflorescence primordium (methods in Salisbury, 1963a; Salisbury and Ross, 1969).

TIME MEASUREMENT AND THE CRITICAL NIGHT

One way to investigate time measurement in cocklebur is to expose several groups of plants to a single dark period, using various lengths for different groups of plants, and to plot floral stage after 9 days as a function of length of the dark period. Figure 1 shows the results of an experiment of this type carried out at seven different dark period temperatures (Salisbury, 1963b). Beginning at about 8.7 to 9.3 hr (the critical dark period in this experiment) at intermediate temperatures, rate of development of the floral bud (floral stage) increased rapidly with increasing night length. Saturation was reached at about 11 hr at these temperatures. It is important to note that from 15 to 30°C the critical dark period was relatively independent of temperature ($Q_{10} = 1.02$), a feature characteristic of circadian rhythms.

TIME MEASUREMENT AND THE LIGHT PERIOD

Appropriate experiments reveal that the light period also may have an important role in photoperiodic time measurement in cocklebur. Plants grown under continuous light will flower in response to a single inductive dark period; if, however, the dark period is less that the critical length (a so-called "phasing dark period," then the importance of time measurement during the intervening light period can be investigated (see Figure 2). In this experiment flowering was completely inhibited when the intervening light period was 5 hr or shorter. Longer light periods increasingly promote flowering.

The intervening-light-period approach allows us to measure a critical light period. This proves to be much shorter than light periods normally encountered in nature, and, thus, with this species, it has been overlooked. Note that photoperiod is acting in this so-called short-day plant exactly as one would expect with a long-day plant: that is, flowering is promoted by increasing durations of light.

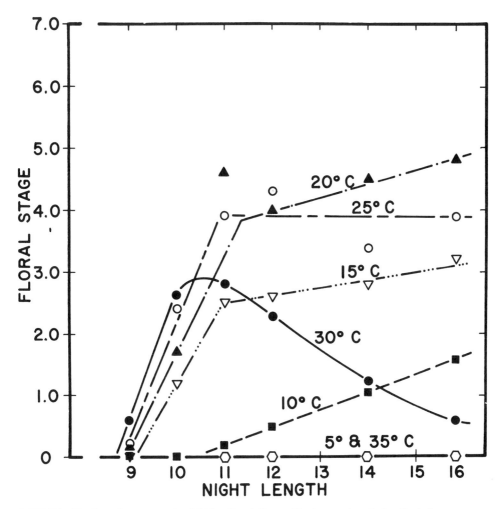

FIGURE 1   The flowering response to night length as influenced by temperature during the dark period; experiment of January 2, 1962. Floral stage is shown as a function of night length under different temperatures. Except for the dark period, the temperatures were 23°C. (From Salisbury, 1963b.)

Modifications of this basic experiment made it possible to study the characteristics of the intervening light period (Salisbury, 1965). Time measurement during the light period (like time measurement during the dark period) was found to be relatively temperature-independent. Furthermore, the intervening light period was effective even when intensities were very low (e.g., those provided by a 25-watt incandescent lamp about 200 cm from the plant), and red light was most effective during the intervening light period, while far-red was least effective. This last observation, of course, implicates the phytochrome system. Moreover, red light is normally inhibitory during the dark period. The plant, then, is rhythmic in its sensitivity to red light: During one part of the cycle, red light inhibits flowering; during the other part, it promotes flowering.

The relationship between length of the intervening light period and

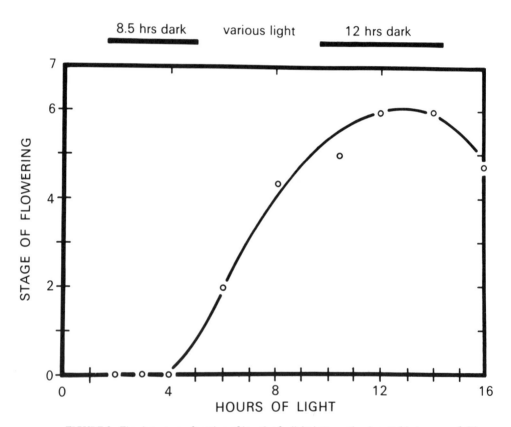

FIGURE 2   Floral stage as a function of length of a light interruption inserted between an 8.5-hr and a 12-hr dark period. Black bars indicate plan of the experiment. (From Salisbury, 1965.)

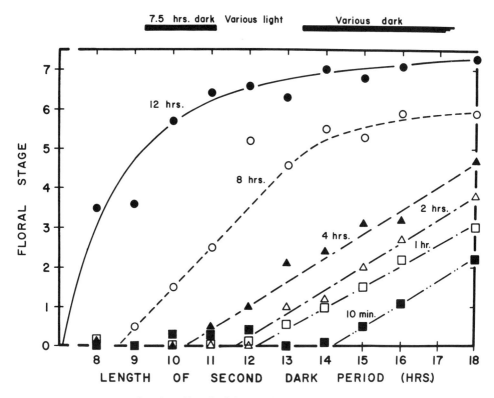

FIGURE 3   Floral stage as a function of length of the second dark period, shown for various intervening light period lengths. Black bars indicate plan of the experiment. (From Salisbury, 1965.)

response to the following test dark period was also studied (Figure 3). As might be expected, longer intervening light periods lead to shorter critical dark periods. This trend continues until, when the intervening light period is 12 hr in length, the critical dark period is considerably less than the usual 8.3 hr—sometimes as short as 6.5 to 7 hr. The significance of this observation will be considered in the light of the following set of experiments.

REPHASING EXPERIMENTS

A second approach to the study of timing in the flowering of cocklebur is probably of greater value than determinations of the critical light or dark periods (Papenfuss and Salisbury, 1967). Short light interruptions (1 min) are given to various sets of plants at different times during an inductive dark period; flowering is examined 9 days later and plotted as a function of the time of the interruption. The control curve in the top half of Figure 4 illustrates such an experiment performed with a 16-hr dark period. The light interruption produces maximum inhibition when it is given 8 hr after the beginning of the dark period. This is true even if the dark period is extended beyond 16 hr. Inhibition is also complete when the interruption occurs after 8 hr in dark periods shorter than 16 hr. But in such cases, light pulses given after intervals of less than 8 hr may also produce total inhibition.

In the experiment illustrated in Figure 4, a light interruption was given to all but the control plants 2 hr after the beginning of the dark period; this interruption was followed by a second interruption given at various times to measure the time of maximum sensitivity. Maximum sensitivity again occurred at 8 hr after the first onset of darkness, although the first light interruption lowers the curve to the right of the point of maximum sensitivity. The bottom half of Figure 4 shows results of determinations of the critical dark period following the light interruption at hour 2. Time measurement, as examined by this approach, does seem to be shifted in response to the light flash, although the shift is only a 1-hr delay rather than the 2-hr delay that might be expected.

Figure 5 shows the results of a modification of the previous experiment (Figure 4, top), in which light interruptions were given at 0 (the control, no interruption), 2, 4, and 6 hr and were followed by another interruption given to various groups of plants at various times. The results of Figure 4 are duplicated with the interruption at hour 2. Time of maximum sensitivity following an interruption at hour 4 remains at 8 hr, but the lowering of the curve to the right (as with the interruption at hour 2) is even more marked. The curve for the interruption at hour 6 is qualitatively different from the other two: The time of maximum sensitivity has now been shifted from 8 to 18 hr. Such a distinct shift, depending on the phase of the light interruption, is what one would expect if the photoperiodic clock were similar or identical to the clock controlling circadian rhythms. A simple hourglass clock might be expected to exhibit only a delay, such as that apparent in the bottom half of Figure 4—although the response could be complicated by the vagaries of phytochrome action.

Figure 6 summarizes several experiments in which plants were given a phasing dark period of constant length (7.5 hr) followed by intervening light periods of various lengths. Light interruptions were then given at different times during the ensuing test dark period to determine times of maximum sensitivity. In terms of real time (counted from the beginning of the phasing dark period), length of the intervening light period has no effect on time of maximum sensitivity during the test dark period until the intervening light period exceeds about 5 hr—that is, maximum sensitiv-

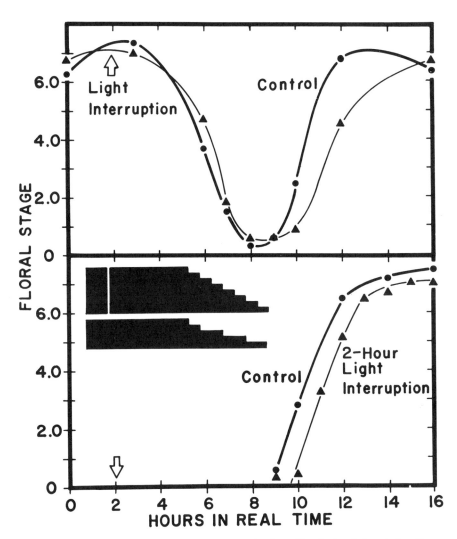

FIGURE 4   Effect of a short light break (1 min), terminating a 2-hr phasing dark period on time of maximum sensitivity (upper) or the length of the critical dark period (lower), as compared with these parameters in controls where no light break at hour 2 was given. Black bars illustrate plans of the experiments, with arrows indicating light interruptions. Light breaks were a mixture of incandescent and fluorescent light at ca. 20,000 lux. Temperature was 22°C. (From Papenfuss and Salisbury, 1967.)

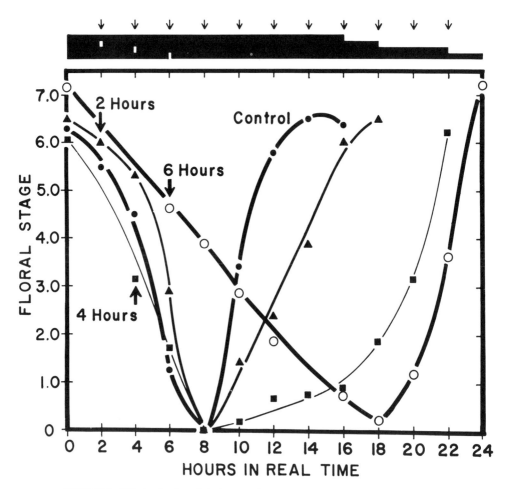

FIGURE 5 Effects of a short light interruption (1 min) given at various times during long test dark periods, following phasing dark periods of 0, 2, 4, or 6 hr. Arrows above the bars at the top indicate times when light interruptions were given and correspond to the data points. Light conditions as for Figure 4; temperatures 25°C light, 21°C dark. (From Papenfuss and Salisbury, 1967.)

ity always occurs at a fixed point (21.5 hr) after the beginning of the phasing dark period as long as the intervening light period is less than about 5 hr. At values greater than 5 hr, the point of maximum sensitivity shifts, assuming a set phase relation, not to the beginning of the phasing dark period, but to the beginning of the test dark period. That is, the point of maximum sensitivity occurs, to a first approximation, 8.5 hr after the beginning of the test dark period (in real time, 8.5 hr plus the length of the intervening light period).

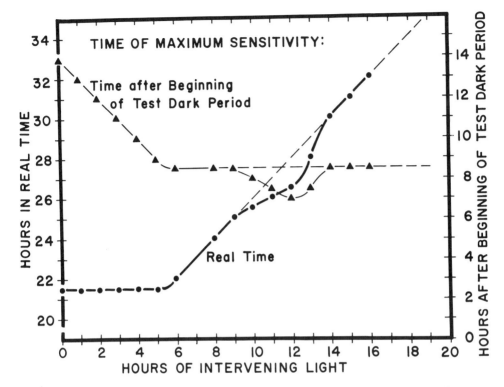

FIGURE 6   Time of maximum sensitivity during the test dark period as a function of the length of the intervening light period. Heavy line corresponds with left ordinate and indicates time of maximum sensitivity in terms of real time. Lighter line shows time of maximum sensitivity in terms of time after beginning of the test dark period (right ordinate). Dashed lines are interpolations and extrapolations. (From Papenfuss and Salisbury, 1967.)

It appears that rephasing is brought about by the beginning of the intervening light period but that the clock tends to go into a suspended state after 5 hr. The interesting dip in the curves, however, indicates that this suspension is apparent only as a first approximation. The dips suggest that the clock tends to oscillate an hour or two into the dark phase, even though plants remain in the light, with this tendency reaching a maximum after 12 hr of intervening light. Following this, the clock returns to the light phase where it remains suspended until the beginning of darkness. This suggestion would account for the short critical dark periods following a 12-hr intervening light period (Figure 3). When plants are placed in the dark after a 12-hr intervening light period, the clock apparently has already moved about 2 hr into its dark phase.

Operation of the photoperiodic clock in cocklebur may be summarized

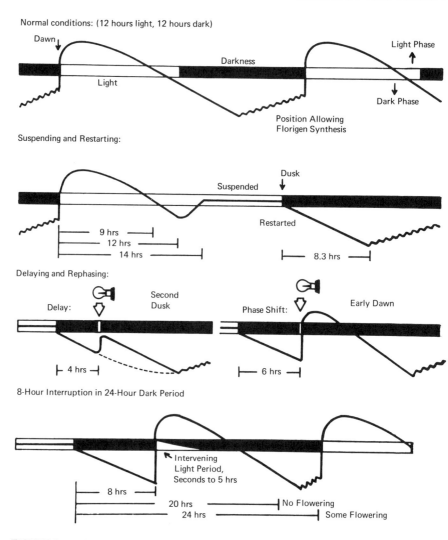

FIGURE 7   A schematic representation of the features of time measurement in flowering of *Xanthium*. (From Papenfuss and Salisbury, 1967.)

as follows (Figure 7): In a normal LD cycle the clock oscillates between a light and a dark phase; phytochrome is the photosensitive pigment that informs the plant as to the nature of the light environment. After the clock has been in the dark phase for approximately 8.3 hr, flowering hormone synthesis is permitted. With long periods of light, the clock apparently can go into a suspended light phase, moving into the dark phase at the beginning of a new dark period. If a light interruption occurs less than 6 hr after the beginning of the dark period, the clock may be delayed somewhat in its operation; if the light interruption comes after 6 hr, the clock begins a new light phase.

THE CIRCADIAN LEAF MOVEMENT CLOCK IN COCKLEBUR

Knowing the above parameters of the photoperiodic clock in cocklebur, it seemed appropriate to search for these same features in the circadian leaf movements of this plant. In 1967 Mr. Ole Christensen, a graduate student in our laboratory, initiated such a study, photographing leaf movements with a time-lapse camera. As the method has now developed, we photograph six plants in front of a grid and then plot the data by examining frames on the film in a microfilm reader. Positions of the leaf tips against the grid are recorded as a function of time. For plants in the dark, a dim, green, luminescent panel behind layers of green and blue Plexiglas is used to silhouette the plants during a time exposure. In control experiments, we have been unable to detect any effect of dim green light on either flowering or leaf movement.

Christensen (1967) was soon able to observe circadian leaf movements in cocklebur under conditions of either continuous light or continuous darkness; his results have been confirmed by Hoshizaki, Brest, and Hamner (1969). Perhaps the most striking observation made on the flowering clock (Papenfuss and Salisbury, 1967) was the marked difference in the effect of a light flash given 4 hr after the beginning of the dark period and that of one given after 6 hr. The leaf movement rhythm was examined for comparable effects, but in no case was it possible to say that the effect of light perturbation at hour 4 was qualitatively different from that of one at hour 6. We could also see no evidence for suspension of the leaf movement clock during long intervening light periods (the rhythm continues under fluorescent light until leaves become senescent). The rhythm is, however, phase-shifted by the onset of darkness following long intervals of light.

Despite mostly negative results, we still thought that the leaf rhythm should be an indication of the flowering clock, and so we retreated into a study of methods, investigating in detail effects of temperature fluctuations (in our earlier experiments, temperature control was imperfect), humidity, and light quality.

The most encouraging result obtained by Christensen was an apparently high level of flowering with light–dark regimes that resulted in more extensive leaf movements. We attempted to duplicate and further document this response by the experiment illustrated in Figure 8. This experiment is the same in principle as that shown in Figure 2, in which flowering is studied

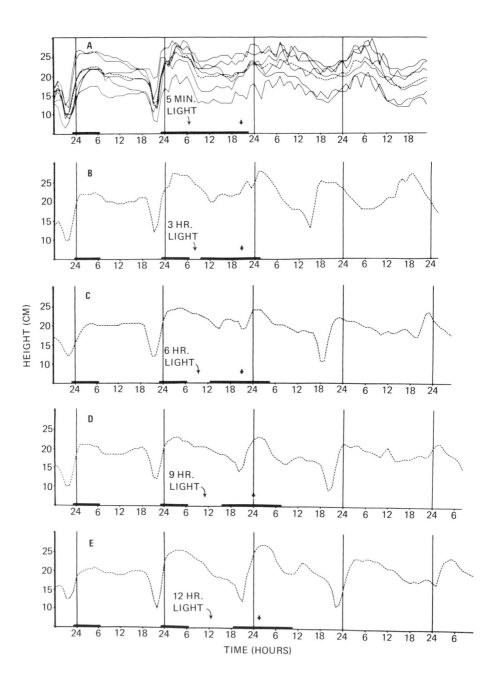

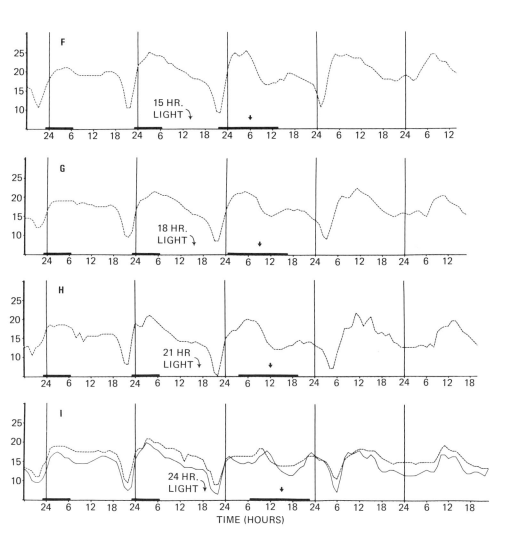

FIGURE 8   Effects of various intervening light periods on leaf movements in *Xanthium*. Phasing dark periods were 7.5 hr long, and test dark periods were 16 hr long. Heavy black arrows indicate the time of maximum inhibition of flowering produced by a light interruption of the test dark period (data from Figure 6). Height figures (ordinates) are those on the grid behind the plants. See text for details.

as a function of different intervening light periods, with the addition of observations on leaf movements. The curves labeled A in Figure 8 are for a 5-min intervening light period. Actual data for all of the six plants are shown. The broken line represents the average leaf position of these six plants; it is clear that this line is representative of the response of all six plants, so curves B–I represent only averages. By the time the experiment shown in I was performed, plants were quite old (13 weeks) compared to those of earlier treatments (7 weeks), and so the experiment was repeated with 10-week-old plants. Results are indicated by the solid line: The leaf movement patterns are essentially identical, even with plants raised under rather different greenhouse conditions before the experiment (early vs. late fall).

In the earlier experiments of Christensen, 5-min intervening light periods had been compared with 12-hr intervening light periods. As Figure 8 clearly shows, the 5-min intervening light periods not only result in poor flowering compared to the 12-hr treatment, but leaf movements also seem to be erratic and strongly damped, while under the 12-hr treatment, the leaf movement pattern is clear and strong. (Flowering is shown in Figure 9.)

A closer examination of this generalization, however, reveals that it is unsubstantiated. Figure 9 summarizes some of the features of Figure 8. Flowering is indicated by the solid circles; results are qualitatively comparable to those of Figure 2. The experiment was repeated with similar plants, giving the open circles shown in Figure 9; this curve duplicated even more closely the results in Figure 2. Leaf movement distances were measured from the lowest point during the inductive test dark period to the highest point. These data are plotted in Figure 9 as solid triangles. The open triangles at 18, 21, and 24 hr indicate leaf movements from the lowest point before the inductive test dark period to the highest point during this period. The pattern of leaf movements during induction clearly does not correlate with the pattern of flowering resulting from this induction.

Halaban (1968) stated that time of maximum sensitivity to a light interruption of an inductive dark period in *Coleus frederici* bore a constant phase relation to the minimum leaf position (always followed it by 5 hr). In Figure 9, the interval of time between the lowest leaf position and the time of maximum sensitivity (as summarized in Figure 6) is plotted (solid squares). Again, it is quite evident that in *Xanthium* no correlation whatsoever exists. The minimum time between low leaf position and time of maximum sensitivity occurs under a light–dark regime that induces maximum flowering—but the maximum time between low leaf position and highest light sensitivity also occurs in a light–dark regime that results in maximum flowering.

It seems, then, that virtually none of the features of the photoperiodic clock as described above correlate with the circadian leaf-movement clock as described here, except for shifting of the leaf movement rhythm in response to dark initiation, which is somewhat similar to delay and rephasing in flowering. The conclusion is unavoidable: The circadian leaf-movement clock is not a valid indication of the flowering clock.

SPECIAL FEATURES OF THE CIRCADIAN CLOCK

Study of the circadian clock in cocklebur has illuminated certain features that may be quite unrelated to the photoperiodic clock but are of interest

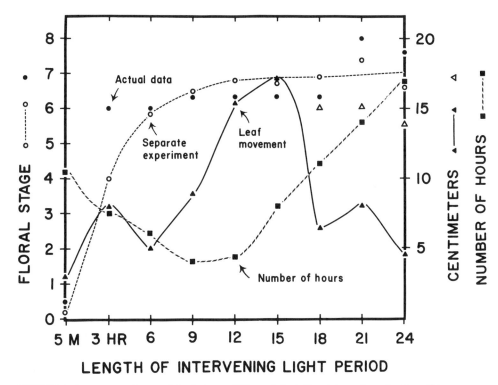

FIGURE 9 A summary of some of the features of Figure 8, including data on flowering from the experiment of Figure 8 plus results of modified replication. Solid circles represent flowering data from the experiment of Figure 8; open circles and the broken line show data from a comparable experiment in which all plants were treated at once instead of during several weeks (compare Figure 2). Solid curve labelled "leaf movement" shows distance moved from lowest to highest points preceding the test dark period (triangles). Curve connecting solid squares shows the number of hours between the lowest leaf position and the following time of maximum sensitivity to a light interruption (in flowering). See text for further explanation.

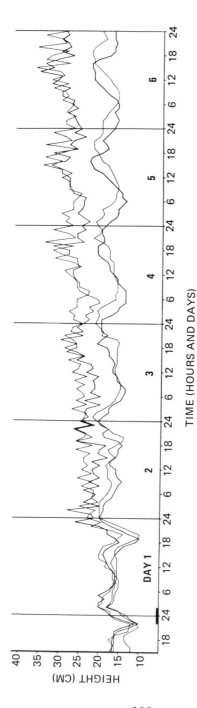

TIME (HOURS AND DAYS)

ENDOGENOUS RHYTHMS UNDER 2 LIGHT QUALITIES

FIGURE 10 Leaf movements of two-branched plants under continuous light. Branches represented by the upper two curves were illuminated with incandescent light; lower curves with fluorescent light. The two dashed lines represent the two branches of one plant; the solid lines represent another plant.

308

in themselves. It is shown by Lörcher (1958) that leaf movements in bean plants continued much longer under fluorescent light than under a mixture of fluorescent and incandescent. Apparently the far-red wavelengths caused a damping of the circadian cycles. We have repeated these experiments with cocklebur, and we find that leaf rhythms under incandescent light become extremely erratic and damp out (Figure 10). Two-branched plants were used; the branch illuminated only with fluorescent light exhibited a clear rhythm for 6 days after the dark period, but the rhythm in the branch illuminated only with incandescent light soon became erratic. One can still discern, however, a component that remains synchronized with the other branch. In experiments in which entire plants are illuminated with incandescent light, this component is seldom seen. These observations implicate phytochrome in the circadian clock of this plant; since this pigment is already strongly implicated in the photoperiodism clock, this phenomenon warrants further study.

Hoshizaki, Brest, and Hamner (1969) have discussed a "light-on" and a "light-off" rhythm in cocklebur. The data of Figure 8 indicate that the leaf movement rhythm is strongly influenced by the test dark period, but it is impossible to tell whether the rhythm was being rephased by the beginning or the end of the dark period since it was held at a constant length. The dip in the leaf position, the most striking feature of the curves in Figure 8, comes just before the beginning of the 7.5-hr phasing dark period (in a 24-hr cycle). With the 5-min intervening light period, the rhythm is greatly disturbed, and the sharp dip no longer appears. The slight dip that does occur starts immediately following the intervening light period—that is, it is advanced about 13 hr. With the 3-hr intervening light period, the dip is very clear on the second day and advanced about 6 hr, following the beginning of the test dark period by about 30 hr. This interval of time between the beginning of the test dark period and the following dip becomes shorter with longer intervening light periods, but its position in real time is, nevertheless, increasingly delayed as the test dark period comes later in real time. This might be a manifestation of the light-on or the light-off rhythm. In any case, there is no such shift in the flowering response to the test dark period.

REFERENCES

Borthwick, H. A., and R. J. Downs. 1964. Roles of active phytochrome in control of flowering of *Xanthium pensylvanicum*. Bot. Gaz. 125:227–231.
Borthwick, H. A., and S. B. Hendricks. 1960. Photoperiodism in plants. Science 132: 1223–1228.

Bünning, E., and I. Moser. 1969. Einfluss der Blattlage auf die Blütenbildung. Naturwissenschaften 56:519.

Bünsow, R. C. 1960. The circadian rhythm of photoperiodic responsiveness in
Kalanchoë. Cold Spring Harbor Symp. Quant. Biol. 25:257–260.

Christensen, O. V. 1967. Leaf-movement rhythms and flowering in *Xanthium*.
M.S. Thesis. Colorado State University, Fort Collins, Colorado.

Garner, W. W., and H. A. Allard. 1920. Effect of the relative length of day and night
and other factors of the environment on growth and reproduction in plants. J. Agric.
Res. 18:553–607.

Halaban, R. 1968. The flowering response of *Coleus* in relation to photoperiod and the
circadian rhythm of leaf movement. Plant Physiol. 43:1894–1898.

Hendricks, S. B. 1960. Rates of change of phytochrome as an essential factor determining photoperiodism in plants. Cold Spring Harbor Symp. Quant. Biol. 25:245–248.

Hillman, W. S. 1970. Carbon dioxide output as an index of circadian timing in *Lemna*
photoperiodism. Plant Physiol. 45:273–279.

Hoshizaki, T., D. E. Brest, and K. C. Hamner. 1969. *Xanthium* leaf movements in light
and dark. Plant Physiol. 44:151–152.

Lörcher, L. 1958. Die Wirkung verschiedener Lichtqualitäten auf die endogene Tagesrhythmik von *Phaseolus*. Z. Bot. 46:209–242.

Papenfuss, H. D., and F. B. Salisbury. 1967. Aspects of clock resetting in flowering of
Xanthium. Plant Physiol. 42:1562–1568.

Salisbury, F. B. 1959. Growth regulators and flowering. II. The cobaltous ion. Plant
Physiol. 34:598–604.

Salisbury, F. B. 1963a. The flowering process. Macmillan Co., New York.

Salisbury, F. B. 1963b. Biological timing and hormone synthesis in flowering of
Xanthium. Planta 59:518–534.

Salisbury, F. B. 1965. Time measurement and the light period in flowering. Planta
66:1–26.

Salisbury, F. B. 1969. *Xanthium strumarium* L. Chapter 2. *In* L. T. Evans [ed.], The
induction of flowering. Some case histories. Macmillan Co. of Australia, South
Melbourne, and Cornell Univ. Press, Ithaca, New York.

Salisbury, F. B., and C. Ross. 1969. Plant physiology. Wadsworth Publ. Co., Belmont,
California.

Takimoto, A., and K. C. Hamner. 1965. Studies on red light interruption in relation to
timing mechanisms involved in the photoperiodic response of *Pharbitis nil*. Plant
Physiol. 40:852–854.

DISCUSSION

K. HAMNER: I have just one comment, with respect to the effect of light perturbation after 7.5 hr of darkness. This effect corresponds exactly to our results with *Pharbitis*, although we interpreted it differently. That is, we found that if this light pertubation is longer than 5–6 hr, it erases the underlying rhythm induced by the previous dark period.

LEES: I don't know much about this, but can you suspend an oscillation in the way you describe in your model (Figure 7)?

SALISBURY: I don't know. I put the suspended state in my model because *Xanthium* is an unusual plant. It does respond to a single dark period, and I thought this perhaps would explain that unusual response.

PITTENDRIGH: In *Drosophila*, the inference is that if the light has been on, in that case, 12 hr or longer, when you turn the light off the oscillation always starts from a fixed phase point. It is as if the whole system is frozen, and it goes into motion again only when you turn the light off.

SALISBURY: That's what I said, but in different words.

WILKINS: I am interested in the effectiveness of different wavelengths of light and in the fact that you were able to get data similar to those of Lörcher in Bünning's lab. We have recently determined an action spectrum for the phase-shifting of the rhythm of activity of phosphoenopyruvic carboxylase in *Bryophyllum*. We found that the rhythm is totally unaffected by all wavelengths of light other than those between 600 and 700 mμ. There is no ultra-violet effect; there is no reversing effect by far-red. So it looks as though there are different pigments involved in different higher plants. In *Bryophyllum* we cannot yet say that phytochrome is the acceptor molecule for the light stimulus.

BÜNNING: Dr. Wilkins, I think we shouldn't forget that rhythms in different tissues may be mediated differently. I mentioned in my paper that photoperiodic photoreception is apparently restricted to the epidermis, but your rhythm is a mesophyll rhythm. Isn't that right?

III

PHOTORECEPTION IN RHYTHMS AND PHOTOPERIODISM

MICHAEL MENAKER

Synchronization with the Photic Environment via Extraretinal Receptors in the Avian Brain

The role of photoperiod in the regulation of reproduction has been extensively investigated during the past 40 or 50 years because of its obvious biological significance and the relative ease of manipulating it in the laboratory. Much of this research has been directed toward a description and analysis of the complex sequence of internal events that are triggered by the photoperiodic stimulus. Considerably less effort has been made to clarify the ways in which light is perceived and the length of the day measured. This is particularly true of the photoperiodic responses of vertebrates.

Work with a few species of mammals indicates that retinal photoreception plays a significant role (Bissonnette, 1935, 1938; Clark, McKeown, and Zuckerman, 1939; Hoffmann and Reiter, 1965; Reiter, 1968) but in none of the other vertebrate classes has the receptor responsible for the perception of photoperiodic information been specifically identified. Daylength measurement (photoperiodic timing) has not been adequately explained even in plants or insects—the groups with which most of the research has been done (Danilevski, 1965; Hillman, 1964; Pittendrigh, 1966; Pittendrigh and Minnis, 1964, this volume). Among the vertebrates, work on this problem has been confined to the birds and only the barest of beginnings has been made (Hamner, 1963; Hamner and Enright, 1967; Menaker, 1965; Menaker and Eskin, 1967). What has been done indicates that there is a relationship between circadian rhythms and the photoperiodic control of reproduction. The existence of this relationship, as yet not fully defined, lends interest to the discovery that, at least in one species, extraretinal photoreceptors in the brain are involved in the perception of environmental light cycles that regulate both of these phenomena. Much of the remainder of this paper is devoted to a review of the work in my laboratory during the past three years that supports the above contention. Although this work is far from complete, the information already obtained does provide an indication of the surprising complexity of the apparatus involved in the perception of light cycles.

ENTRAINMENT (SYNCHRONIZATION) OF THE BIOLOGICAL CLOCK

In the house sparrow, *Passer domesticus*, the circadian rhythm of locomotor activity "free runs" in constant darkness with a period of approximately 25 hr. This rhythm can be entrained (or synchronized) to an artificially imposed light cycle as long as the period of the entraining regimen does not fall outside the range of 17–28 hr (Eskin, this volume). These features of the rhythmic behavior of sparrows maintained in activity cages with controlled lighting are illustrated in Figure 1. Both persistent rhythmicity in constant darkness ("free run") and entrainment to light cycles are characteristic of all circadian rhythms which have been investigated, at all phylogenetic levels from protists to chordates, and at several levels of organization within multicellular organisms (Long Island Biological Association, 1960).

When an organism is entrained by a light cycle the period of which differs significantly from that of its free run in constant darkness, the change in period that is produced is readily detectable and provides an excellent assay of photoreception (*cf.* upper and lower sections of Figure 1). In fact, although Figure 1 is indistinguishable from the record of a normal bird, it was produced by a bird that had been blinded by bilateral enucleation prior to the start of the experiment. Numerous experiments of this kind, coupled with controls that exclude all periodically fluctuating variables associated with the light cycle with the exception of visible light itself, demonstrate that sparrows possess an extraretinal photoreceptor coupled to the clock controlling the circadian rhythm of locomotor activity (Menaker, 1968a). For convenience we have called this receptor the ERR_e (extraretinal receptor for entrainment) although we are quite aware that more than one site may be involved.

The ERR_e is surprisingly sensitive. Cycles in which the light portion con-

FIGURE 1 Continuous record (93 days) of the perching activity of a blinded sparrow. Each horizontal line is the record of a single day's activity. Each day's record is pasted below the record of the preceding day. On the first day of the record the bird was placed in the light regimen diagrammed at the top of the figure and it achieved steady-state entrainment after 4 or 5 days. At 1 (right hand margin) the light cycle was shifted to the right (a "delay" of 8 hr). Five or six days later steady state entrainment was achieved at the same phase relative to the new light cycle (onset of activity roughly coincident with "light on"), which is, however, a new phase relationship with real time. At 2 the light cycle was discontinued and the bird allowed to free run in constant darkness.

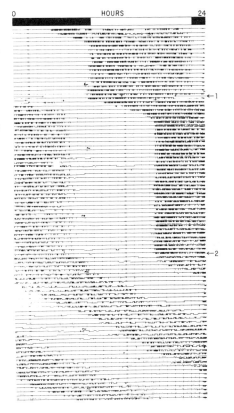

sists of green light of about 0.1 lux (approximate intensity of full moonlight) entrain 50 percent of the blinded birds exposed to them (Figure 2). This level of illumination is therefore a crude approximation of the threshold for the response in blind birds. However, *all* sighted birds entrain to such cycles. In addition, sighted birds become arrhythmic (continuously active) in bright constant light (between 50 and 500 lux) whereas blinded birds do not (Figure 3). These facts indicate that the eyes, while not necessary for entrainment, do convey information about environmental light conditions to the clock when they are present (Menaker, 1968a). In the lizard *Sceloporus olivaceus*, which also has an ERR_e, a similar situation exists (Underwood and Menaker, 1970b) (Figure 4).

In *P. domesticus* the ERR_e is located in the brain. We have demonstrated this in experiments in which we have manipulated the entrainment response of blinded birds by changing the amount of light reaching the brain. We have shown by direct physical measurement that plucking feathers from

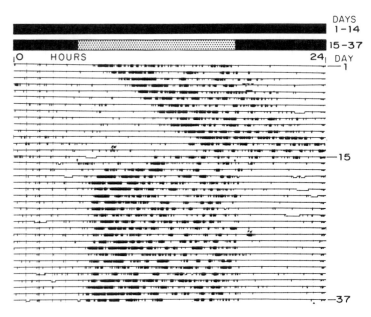

FIGURE 2   Entrainment of a blinded sparrow to a light cycle in which
the light (indicated by stippling in diagram) was approximately 0.1 lux.
The bird was in constant darkness from day 1 through day 14 and was
exposed to the light cycle from day 15 until the end of the record. Other
details are as in Figure 1. (From Menaker, 1968a.)

the top of the head increases the amount of light reaching the brain by a
factor of 10^{2-3}, whereas India ink, deposited by injection under the skin
of the head decreases the intensity reaching the brain by a factor of 10
(Menaker *et al.*, 1970). If a blinded sparrow is placed in a light cycle of
subthreshold intensity (to which it does not entrain) entrainment can subsequently be produced by plucking feathers from the top of its head. If
the bird then receives an injection of India ink under the head skin it will
once again free run in the presence of the light cycle (Menaker, 1968b)
(Figure 5).

Removal of the pineal organ does not abolish the entrainment response
of blind birds (Gaston and Menaker, unpublished experiments) (Figure 6).
This result makes it clear that the pineal is not the exclusive site of extraretinal light perception, but does not exclude the possibility that the pineal
may be one component of a multiple $E R R_e$. Further, it does not bear on
the separate and still unresolved question of whether the avian pineal is
photoreceptive.

Pinealectomy does have a profound effect on the free-running circadian rhythms of locomotor activity (Gaston, this volume; Gaston and Meanaker, 1968) (Figure 7) and body temperature (Binkley, 1971) (Figure 8). In constant darkness, pinealectomized sparrows are continuously active (arrhythmic) and the normally present rhythm of body temperature also disappears. The pineal is thus involved in a centrally important aspect of circadian rhythmicity-persistence in constant conditions. Its role in the sparrow clock system does not appear to be primarily photoreceptive.

Wetterberg, Geller, and Yuwiller (1970) have suggested that the harderian gland may be an E R R in neonatal rats. In *P. domesticus* the harderian gland is normally removed during enucleation (Duke-Elder, 1958). There is, however, the possibility that some harderian tissue may remain in the

FIGURE 3   Maintenance of rhythmicity by a blind bird held in bright constant light. To aid in visual inspection the record has been photographically duplicated and arranged on a 48-hr time base. *LD 12*:12 (500:0) means a light regimen with 12 hr of light of 500 lux intensity and 12 hr of darkness per cycle; DD, constant darkness; LL, constant light of the designated intensity. Note the large change in free-running period and in the proportion of activity time to rest time when the bird goes from DD to LL of 50 lux ("Aschoff's rule"). Upon raising the light intensity to 500 lux, small changes occur in both of these parameters, but the bird does not become arrhythmic as would a sighted bird. Further increase to 2000 lux produces little, if any, change. Other details as in Figure 1. (From Menaker, 1968a.)

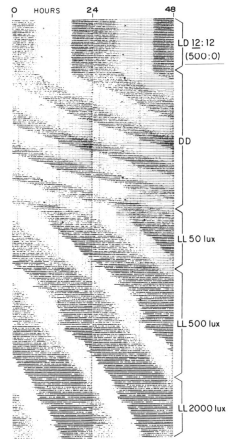

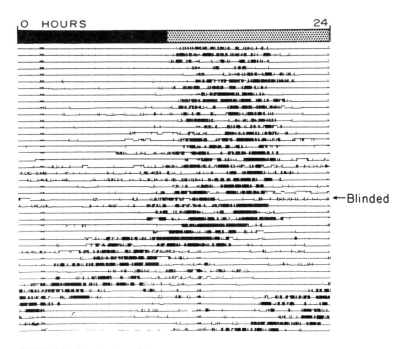

FIGURE 4   Contribution of the lateral eyes to perception of an entraining light cycle in the lizard *Sceloporus olivaceus*. The normal (unoperated)
lizard was entrained to a light cycle in which the light (indicated by stippling in the diagram) was approximately 0.1 lux. After removal of the
lateral eyes, the lizard free ran through the light cycle with a circadian period of about 23 hr. Other details as in Figure 1. (Figure from Underwood
and Menaker, 1970b.)

orbit unless specific steps are taken to remove it. In several experiments
we have surgically removed all tissue that remained in the orbit following
enucleation. Birds so treated entrained normally to light cycles (Silver and
Menaker, unpublished experiments).

PHOTOPERIODIC CONTROL OF TESTICULAR RECRUDESCENCE

In the field, the testis of *P. domesticus* undergoes a dramatic annual cycle
of spermatogenic activity that is reflected by changes in weight. Figure 9
shows the timing and amplitude of this cycle in the Austin, Texas, population (Underwood, 1968). Recrudescence in postrefractory birds can be
stimulated by artificial long photoperiods in the laboratory (from October

FIGURE 5 Demonstration that the ERR_e of the house sparrow is in the brain. For the entire duration of the experiment the blinded bird was exposed to a light cycle (Top of the figure: black bars indicate darkness, stippled bar indicates green light of approximately 0.02 lux). Arrows at the right indicate the days on which various experimental treatments were performed: At 1, feathers were plucked from the bird's back; at 2, feathers were plucked from the head (the bird now becomes entrained to the light cycle until 7 or 8 days before 3); at 3, feathers, which had by now regenerated, were again plucked from the head; at 4, India ink was injected under the skin of the head; at 5, some of the head skin was removed and the ink deposit was scraped from the skull. Other details as in Figure 1. (From Menaker, 1968b.)

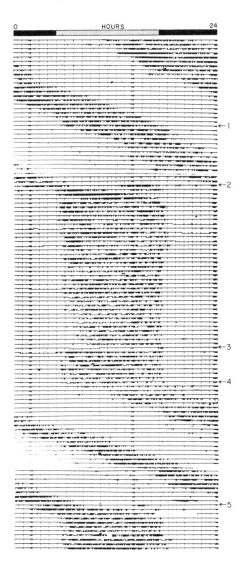

through January). The light intensity required to produce a measurable response is 7–10 lux (Bartholomew, 1949), more than 100 times that required to produce entrainment in sighted birds.

The testes of sparrows blinded by bilateral enucleation also recrudesce in response to artificial long days (Menaker and Keatts, 1968) (Figure 10). Further, there are no significant differences in either the rate or the extent

FIGURE 6 Entrainment of a blind, pineal-ectomized bird. The light cycle diagrammed at the top was begun at 1 and discontinued at 2. Before 1 and after 2 the bird was in constant darkness (for further discussion of the arrhythmicity of pinealectomized birds in constant darkness see Figure 7). Other details as in Figure 1.

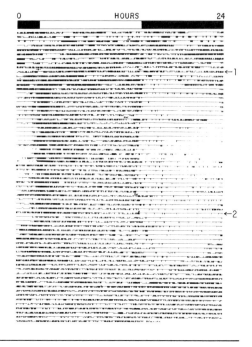

FIGURE 7 A: The effect of pinealectomy on perching activity in *P. domesticus*. The bird is in DD and constant temperature (about 23°C) throughout the record. On day 19 the pineal organ was removed as indicated by the large arrow. Within 2 days after the operation no circadian rhythm is discernible. B: The effect of a sham pinealectomy on the rhythmic activity of a bird in DD. The operation was performed on day 29 (large arrow). The free-running period before the operation was about 23 hr 38 min; subsequently it was about 23 hr 40 min. Also note the phase shift of approximately 2½ hr resulting from the surgical procedure. C and D show entrainment patterns of two pinealectomized sparrows. The beginning and end of the daily light period are marked with arrows. The dense black bars during the light fraction indicate intense perching activity. In C, days 1 to 15 demonstrate arrhythmic activity in DD. On days 16 to 29, the bird received 8 hr of light followed by 16 hr of darkness per 24-hr period, and from day 39 to 59, the bird was once again in DD. A positive phase angle of about 1 hr is evident after the third day of the light cycle. After the light cycle was discontinued, about 8 days were required for this bird to re-establish an arrhythmic pattern. In D, days 1 to 4 show arrhythmic activity in DD, on days 5 to 26 the bird was on LD 8:16; and on days 27 to 59 the bird was in DD. Notice the "decay" of rhythmicity on days 27 to 33, illustrating the transition to arrhythmicity more clearly than does C. The pattern of this decay, with activity onsets occurring earlier and activity terminating later each day is characteristic of pineal-ectomized birds released in DD from LD entrainment. None of the birds used were blinded. Other details as in Figure 1. (From Gaston and Menaker, 1968.)

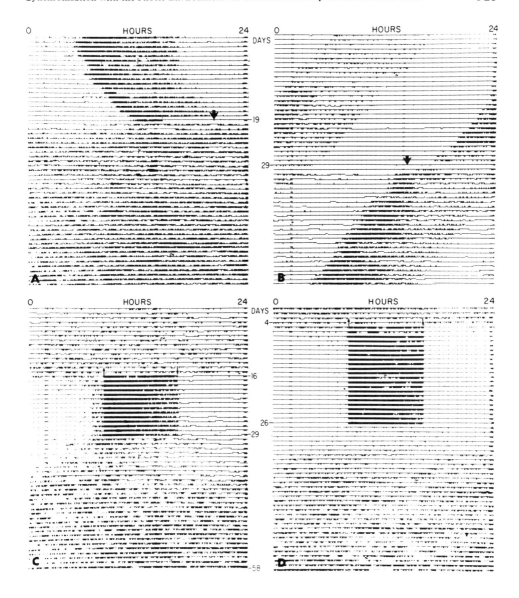

FIGURE 8 Effect of pinealectomy on the circadian rhythm of body temperature in the
house sparrow. The bird was in constant darkness throughout the experiment (except during
surgery). As in the activity records, each horizontal line is a 24-hr record and each day's
record is placed below that of the previous day.
Body temperature was continuously recorded
by telemetry. On the day indicated, the pineal
organ was removed (gap in record). The amplitude of the body temperature rhythm before
pinealectomy was about 4°C. After pinealectomy the body temperature remained continuously high. (The gap 5 days after surgery
and the apparent depression of the temperature
for over 24 hr beginning 7 days after surgery are
both due to apparatus failure.) (Reprinted with
permission from Binkley, 1971.)

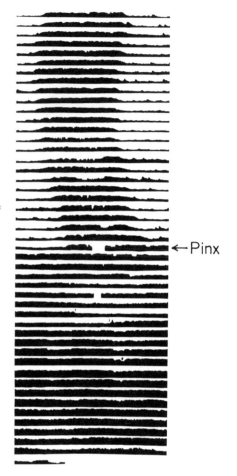

←Pinx

of growth when blind birds are compared with sighted ones (Underwood
and Menaker, 1970a) (Figure 11). These data demonstrate that in *P. domesticus*, as in the duck (Benoit, 1959, 1962, 1964; Benoit and Assenmacher,
1953; Benoit, Assenmacher, and Walter, 1953; Benoit and Ott, 1944),
extraretinal photoreception is involved in photoperiodic induction of testicular recrudescence. For convenience we have chosen to call the receptor
involved the ERR_p (extraretinal receptor for photoperiodism). This designation does not imply identity (or lack of it) of the ERR_p with the ERR_e
or localization of the ERR_p at a single site.

The data in Figure 11 suggest that the eyes may not participate in
photoperiodic photoreception and a further experiment proves that this

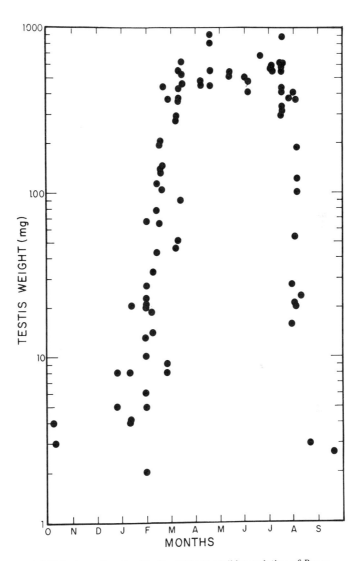

FIGURE 9   Annual cycle of testis weight in a wild population of *Passer domesticus* in Austin, Texas. Each point represents the combined weight of both testes from a single freshly captured sparrow. Note that testis weight is plotted on a logarithmic scale. (Reprinted with permission from Underwood, 1968.)

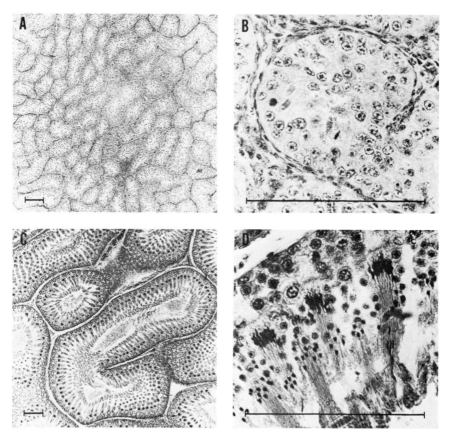

FIGURE 10   Histological sections from the testes of blinded birds exposed to *LD 6*:18 (A and B) and *LD 16*:8 (C and D). B and D have been photographed at approximately 10 × the magnification used in A and C. Bars indicate 10μ in all cases. Note the tremendously hypertrophied tubules in C as compared with A, as well as the mature sperm present in D. (From Menaker and Keatts, 1968.)

is in fact the case (Menaker *et al.*, 1970). Postrefractory (photosensitive) sparrows were divided into two groups. Feathers were plucked from the heads of one group while the other group received India ink injections under the skin of the head. None of the birds were blinded. Both groups were held for 39 days on a long photoperiod (16 hr of light/day) at a light intensity of approximately 10 lux (the threshold for the response in un-treated birds). The brains of the plucked birds were thus exposed to light intensities above the threshold value; the brains of the injected birds were exposed to light intensities below the threshold value; the eyes of both groups were exposed to light intensities at the threshold value. If the eyes

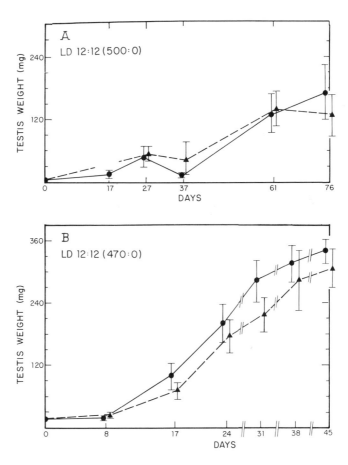

FIGURE 11 Testicular recrudescence of blind and normal sparrows in
response to a cycle consisting of 12 hr of darkness and 12 hr of light
($\simeq$ 500 lux) at two times of year. The birds were first exposed to the experimental photoperiod on day 0 and samples were subsequently collected at the times shown on the abscissa. The experiment plotted in A
began November 15, that plotted in B began February 7. Points are means
of the combined weights of the two testes of all birds in each sample.
Dashed lines connect the means of the groups of blind birds; solid lines
the means of groups of normal birds. The points plotted at day 0 are the
mean testis weights of control groups killed immediately before beginning
each experiment. (From Underwood and Menaker, 1970a.)

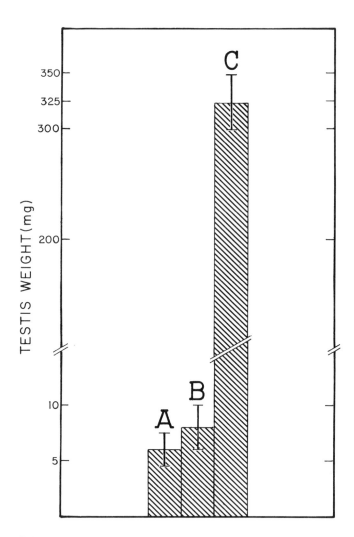

FIGURE 12 Effects of manipulating the intensity of light reaching the
ERR_p on testis weight of sparrows with intact eyes. A, control group sacrificed at the start of the experiment; B, India ink injected group; C, group
with head feathers plucked. See text for additional detail. (Figure modified from Menaker *et al.*, 1970.)

are involved in photoperiodic photoreception both groups should show some testis growth. If the eyes are not involved, only the testes of the plucked group should respond to the photoperiodic stimulus. The results, shown in Figure 12, indicate that the eyes do not participate in the perception of those aspects of environmental light cycles which control testicular recrudescence. These data also demonstrate that the ERR_p, like the ERR_e, is located in the brain. Similar results were obtained when the experiment was repeated with pinealectomized birds, indicating that the pineal is not the sole ERR_p (and, incidentally, is not hormonally involved in recrudescence).

DISCUSSION

Photoreception by structures that are not organized, in any obvious way, to perform this function occurs widely among invertebrates (Bullock and Horridge, 1965; Millot, 1958). Brain photoreceptors have been shown to mediate entrainment of circadian rhythms and the photoperiodic control of several processes in insects (Lees, 1964; Truman, this volume; Williams, 1963; Williams and Adkisson, 1964; Williams, Adkisson, and Walcott, 1965). Extraretinal photoreception has been convincingly demonstrated in some members of all classes of vertebrates in a variety of situations in which perception of the cyclicity of the environmental light regimen is of importance (Adler, 1969, this volume; de la Motte, 1963; Dodt, 1963; Fenwick, 1970; Hoar, 1955; Lisk and Kannwischer, 1964; Scharrer, 1928; Underwood and Menaker, 1970b; Young, 1935; Zweig, Snyder, and Axelrod, 1966). The evidence for the existence of extraretinal light perception is least convincing in the mammals. Reports from two laboratories indicate the existence of extraretinal photoreceptors that control some aspects of pineal biochemistry in neonatal rats (Wetterberg, Geller, and Yuwiller, 1970; Zweig, Snyder, and Axelrod, 1966) but which apparently do not persist into adulthood. A single paper constitutes the entire literature on extraretinal photoreception in adult mammals and the need for further work is clear (Lisk and Kannwischer, 1964). It is worth noting in this connection that most laboratory mammals are nocturnal, whereas extraretinal photoreception has been demonstrated most often in diurnal members of the other vertebrate classes.

It seems worthwhile to take stock of the many intriguing questions, as yet unanswered, that are raised by the demonstration of extraretinal photoreception in a wide variety of vertebrate species: No vertebrate extraretinal photoreceptor has been specifically localized, and, as yet, nothing is known of the neural or hormonal inputs to or outputs from any E R R. We do not know how these brain structures interact with the eyes, the pineal gland, or the hypothalamus, although the available evidence suggests that they do interact with all three. It may be that the whole vertebrate brain is photoreceptive, that only a single structure is present, or that there are multiple discrete E R R's. Our work on sparrows suggests that they possess at least two brain photoreceptors with very different properties; a sensitive E R R$_e$ that, together with the eyes, mediates entrainment of the circadian system and a much less sensitive E R R$_p$ that, operating independently of the eyes, mediates testis growth. Support is lent to this speculation by the work of Homma and Sakakibara (this volume) who have shown that in the brain of the Japanese quail there are two distinct regions that act as E R R$_p$. No action spectra are available for vertebrate extraretinal photoreceptors, and there is no evidence of any kind concerning the photopigments that are involved. We lack even a basis for speculation concerning the adaptive significance of extraretinal photoreception in organisms with elaborately specialized eyes.

Perhaps the most important contribution that the study of extraretinal photoreception can make is to focus attention on a level of control "above" the hypothalamus. Just as information concerning color, form, and motion is processed to a remarkable extent in the retina itself (Granit, 1955; Lettvin et al., 1959), it may well be that a great deal of information-processing is being carried out in brain photoreceptors and that when this is understood, their adaptive significance will become clear. It seems likely that the control mechanisms that operate above the hypothalamic level will turn out to be every bit as complex as those below it.

ACKNOWLEDGMENT

This investigation was supported in part by U.S. Public Health Service Research Grant HD 03803 from the National Institute of Child Health and Human Development, and by a Career Development Award 1 K04 HD 90327-01 from the Department of Health, Education, and Welfare.

REFERENCES

Adler, K. 1969. Extraoptic phase shifting of circadian locomotor rhythm in salamanders. Science 164:1290–1292.
Bartholomew, G. A. 1949. The effect of light intensity and day length on reproduction in the English sparrow. Bull. Mus. Comp. Zool. 101:433–476.

Benoit, J. 1959. The control by visible radiations of the gonadotropic activity of the duck hypophysis. Recent Prog. Horm. Res. 14:143–164.

Benoit, J. 1962. Opto-sexual reflex in the duck; physiological and histological aspects. Yale J. Biol. Med. 34:97–116.

Benoit, J. 1964. The role of the eye and the hypothalamus in the photostimulation of gonads in the duck. Ann. N.Y. Acad. Sci. 117:204–215.

Benoit, J., and I. Assenmacher. 1953. Rôle des photorécepteurs superficiel et profond dans la gonadostimulation par la lumière chez les oiseaux. J. Physiol. (Paris) 45:34–37.

Benoit, J., I. Assenmacher, and F. X. Walter. 1953. Dissociation expérimentale des rôles des récepteurs superficiel et profond dans la gonadostimulation hypophysaire par la lumière chez le canard. C. R. Soc. Biol. 131:89.

Benoit, J., and L. Ott. 1944. External and internal factors in sexual activity. Effect of irradiation with different wave lengths on the mechanisms of photostimulation of the hypophysis and on testicular growth in the immature duck. Yale J. Biol. Med. 17:22–46.

Binkley, S. 1971. The pineal organ and circadian organization in the house sparrow. Ph.D. Thesis. University of Texas at Austin.

Bissonnette, T. H. 1935. Relations of hair cycles in ferrets to changes in the anterior hypophysis and to light cycles. Anat. Rec. 63:159–167.

Bissonnette, T. H. 1938. Influence of light on the hypophysis. Endocrinology 22:92–103.

Bullock, T. H., and G. A. Horridge. 1965. Structure and function in the nervous systems of invertebrates. W. H. Freeman and Co., San Francisco.

Clark, W. E. LeGros, T. McKeown, and S. Zuckerman. 1939. Visual pathways concerned in gonadal stimulation in ferrets. Proc. R. Soc., Ser. B. 126:449–260.

Danilevskii, A. S. 1965. Photoperiodism and seasonal development of insects. Oliver and Boyd, Edinburgh.

de la Motte, I. 1963. Studies on the comparative physiology of photoreception in blind fish. Naturwissenschaften 50:363.

Dodt, E. 1963. Photosensitivity of the pineal organ in the teleost *Salmo irideus* (Gibbons). Experientia 19:642–643.

Duke-Elder, S. 1958. System of opthalmology. Vol. I. The eye in evolution. C. V. Mosby Co., St. Louis.

Fenwick, J. C. 1970. Effects of pinealectomy and bilateral enucleation on the photo-tactic response and on the conditioned response to light of the goldfish *Carassius auratus* L. Can. J. Zool. 48:175–182.

Gaston, S., and M. Menaker. 1968. Pineal function: the biological clock in the sparrow? Science 160:1125–1127.

Granit, R. 1955. Receptors and sensory perception. Yale Univ. Press, New Haven, Connecticut.

Hamner, W. M. 1963. Diurnal rhythm and photoperiodism in testicular recrudescence of the house finch. Science 142:1294–1295.

Hamner, W. M., and J. T. Enright. 1967. Relationships between photoperiodism and circadian rhythms of activity in the house finch. J. Exp. Biol. 46:43–61.

Hillman, W. S. 1964. The physiology of flowering. Holt, Rinehart and Winston, New York.

Hoar, W. S. 1955. Phototactic and pigmentary responses of sockeye salmon smolts following injury to the pineal organ. J. Fish. Res. Board Can. 12:178–185.

Hoffmann, R. A., and R. J. Reiter. 1965. Pineal gland: influence on gonads of male hamsters. Science 148:1609–1611.

Lees, A. D. 1964. The location of the photoperiodic receptors in the aphid *Megoura viciae* Buckton. J. Exp. Biol. 41:119–133.

Lettvin, J. Y., H. R. Maturana, W. S. McCulloch, and W. H. Pitts. 1959. What the frog's eye tells the frog's brain. Proc. IRE 47:1940–1951.

Lisk, R. D., and L. R. Kannwischer. 1964. Light: evidence for its direct effect on hypothalamic neurons. Science 146:272–273.

Long Island Biological Association. 1960. Biological clocks. Cold Spring Harbor Symp. Quant. Biol. 25:575.

Menaker, M. 1965. Circadian rhythms and photoperiodism in *Passer domesticus*, p. 385 to 395. *In* J. Aschoff [ed.], Circadian clocks. North-Holland Publ. Co., Amsterdam.

Menaker, M. 1968a. Extraretinal light perception in the sparrow. I. Entrainment of the biological clock. Proc. Natl. Acad. Sci. U.S. 59:414–421.

Menaker, M. 1968b. Light perception by extra-retinal receptors in the brain of the sparrow. Proc. 76th Annu. Conv. Am. Psychol. Assoc., p. 299–300.

Menaker, M., and A. Eskin. 1967. Circadian clock in photoperiodic time measurement: a test of the Bünning hypothesis. Science 157:1182–1185.

Menaker, M., and H. Keatts. 1968. Extraretinal light perception in the sparrow. II. Photoperiodic stimulation of testis growth. Proc. Natl. Acad. Sci. U.S. 60:146–151.

Menaker, M., R. Roberts, J. Elliott, and H. Underwood. 1970. Extraretinal light perception in the sparrow. III. The eyes do not participate in photoperiodic photoreception. Proc. Natl. Acad. Sci. U.S. 67:320–325.

Millot, N. 1958. The dermal light sense, p. 1 to 36. *In* J. D. Carthy and G. E. Newell [ed.], Invertebrate receptors. Academic Press, New York.

Pittendrigh, C. 1966. The circadian oscillation in *Drosophila pseudoobscura* pupae: A model for the photoperiodic clock. Z. Pflanzenphysiol. 54:275–307.

Pittendrigh, C. S., and D. H. Minis. 1964. The entrainment of circadian oscillations by light and their role as photoperiodic clocks. Am. Nat. 98:261–294.

Reiter, R. J. 1968. Morphological studies on the reproductive organs of blinded male hamsters and the effects of pinealectomy or superior cervical ganglionectomy. Ant. Rec. 160:13–24.

Scharrer, E. 1928. Die Lichtempflindlichkeit blinder Elritzen (Untersuchungen über das Zwischenhirn der Fische I). Z. Vgl. Physiol. 7:1–38.

Underwood, H. 1968. Extraretinal light perception: photoperiodic stimulation of testis growth in the house sparrow (*Passer domesticus*). M.A. Thesis. University of Texas at Austin.

Underwood, H., and M. Menaker. 1970a. Photoperiodically significant photoreception in sparrows: Is the retina involved? Science 167:298–301.

Underwood, H., and M. Menaker. 1970b. Extraretinal light perception: entrainment of the biological clock controlling lizard locomotor activity. Science 170:190–193.

Wetterberg, L., E. Geller, and A. Yuwiller. 1970. Harderian gland: an extraretinal photoreceptor influencing the pineal gland in neonatal rats? Science 167:884–885.

Williams, C. M. 1963. Control of pupal diapause by the direct action of light on the insect brain. Science 140:386. (Abstr.)

Williams, C. M., and P. L. Adkisson. 1964. Photoperiodic control of pupal diapause in the silkworm, *Antheraea pernyi.* Science 144:569.

Williams, C. M., P. L. Adkisson, and C. Walcott. 1965. Physiology of insect diapause XV. The transmission of photoperiod signals to the brain of the oak silkworm, *Antheraea pernyi.* Biol. Bull. 128:497–507.

Young, J. Z. 1935. The photoreceptors of lampreys. II. The functions of the pineal complex. J. Exp. Biol. 12:254–270.

Zweig, M., S. H. Snyder, and J. Axelrod. 1966. Evidence for a nonretinal pathway of light to the pineal gland of newborn rats. Proc. Natl. Acad. Sci. U.S. 56:515–520.

KAZUTAKA HOMMA
YOSHIKAZU SAKAKIBARA

Encephalic Photoreceptors and Their Significance in Photoperiodic Control of Sexual Activity in Japanese Quail

The use of radioluminous paint (RLP) as a new means of photic stimulation in biological research was first reported by Kato, Kato, and Oishi (1967). They found that application of the RLP to the skulls of adult male Japanese quail prevented testicular atrophy when the environmental photoperiod was shifted from LL to LD 8:16. They also reported that this response was abolished after pinealectomy and suggested that the pineal might be a deep photoreceptor participating in the photogonadal reflex in this species (Oishi and Kato, 1968).

The purpose of our experiments with Japanese quail reported here was to localize and characterize such deep photoreceptors and to evaluate their importance relative to the retina in the control of gonadal development. We have also investigated whether the pineal is involved, as a photoreceptor, in the photogonadal reflex of developing Japanese quail.

MATERIALS AND METHODS

Atomloihi #3000 (an RLP) was used as a photic stimulator. The emission spectra from the orange and the blue Atomloihi paints are shown in Figure 1. The brightness of this RLP when it is shipped from the factory

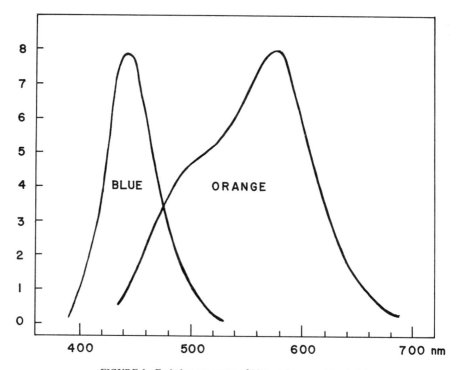

FIGURE 1   Emission spectrum of blue and orange Atomloihi.

in powder form is 0.3 candle/m², which decreases to approximately 0.1 candle/m² or slightly less when the powder is solidified by mixing with equal amounts of binder (R-2).

For local illumination within the brain by R LP, small oval plates, made of solidified R LP, were prepared. The size of the plate was either 1.5×1.0 mm or 3.0×2.0 mm (long axis × short axis of the oval surface) and 0.8 mm in thickness. Each plate was mounted on twisted fine tungsten wire to help in handling during manual implantation. This type of plate was designed for photic stimulation from the surface of specific locations as well as for implantation into narrow cavities. For implantation into brain tissue, an R LP rod was prepared. The rod consisted of a small ball of solidified R LP, less than 0.8 mm in diameter, attached to a glass capillary. This formed a detachable connection with the implanting needle mounted on the stereotaxic instrument used to implant the R LP in a predetermined location. The position of the implanted R LP was determined by x ray of the live birds and was confirmed histologically after autopsy. Gonadal tissues were fixed in Bouin's solution and examined microscopically after H-E staining.

The Japanese quail used in this study were originally bred at Nagoya University. The line has recently been transferred to the University of Tokyo. The lighting regime and the schedule for surgery are given in Table 1. Chicks of both sexes were maintained under LL for the first two weeks after hatching and then transferred to LD 8:16. Male quail were operated on at 4 weeks of age and autopsied at 7 weeks of age. Females were operated on at 4 or 5 weeks of age and autopsied after observation periods of varying lengths; egg production and laying pattern were recorded throughout the observation period.

During the LD 8:16 photoperiod, the light was on from 1100 to 1900. The birds were tended and all surgery was done during the light portion of the regime. The light intensity of the room at floor level was 31 to 56 lux (incandescent bulbs).

RESULTS AND DISCUSSION

1. *Effects of orange* RLP *on testicular development in normal and pinealectomized Japanese quail* Orange RLP plates were implanted inside the skull close to the pineal to test its photosensitivity. In contrast to the observations of Oishi and Kato (1968), this treatment caused no detectable enhancement of gonadal development in developing males.

When the RLP was implanted in the deeper locations of the brain along

TABLE 1 Schedule of Experiment

Sex and Age	Photoperiod	Care and Treatment of Birds
Male		
Hatch	LL	Raised in chick brooder at 35°C
2 wks	LD 8:16	Group-caged at room temperature
4 wks	LD 8:16	Individually caged. Implantation of RLP with or without pinealectomy
7 wks	LD 8:16	Autopsy
Female		
Hatch	LL	Raised in chick brooder at 35°C
2 wks	LD 8:16	Group-caged at room temperature
4–5 wks	LD 8:16	Implantation of RLP
10–20 wks	LD 8:16	Recording of egg production and laying time Autopsy

the *fissura longitudinalis cerebri* (F L C), adjacent to the hypothalamic area, the birds showed rapid and uniform testicular development as if they had been exposed to stimulatory environmental photoperiods (Table 2; weights given are the combined weights of the left and right testes); the nonimplanted controls showed no significant testicular development. Implantation of R L P into F L C of pinealectomized males caused a response similar to that observed in the implanted normal birds (Table 2).

The results of these experiments led us to conclude that the pineal does not play an important role as a photoreceptor or as an endocrine organ regulating gonadal development. This confirms the results of our previous observations, which indicated that pinealectomy had no effect on testicular development when quail were raised under four different photoperiods (Homma, McFarland, and Wilson, 1967).

In one experiment, the R L P was unilaterally implanted into the eye (see Table 2). Testicular growth in this group was slightly more rapid than in the pinealectomized and in the sham-pinealectomized groups, but histological examination showed no significant stimulation by R L P. We also tried bilateral intraocular implantation using larger orange R L P plates. In all these experiments we failed to induce any significant gonadal development. Thus, with regard to gonadal development, the eye appears less sensitive to weak light stimuli than the deep encephalic photoreceptors. In photoperiodic experiments with chickens (Morris, 1967), background light

TABLE 2   Effects of Implantation of Orange R L P Plate and Pinealectomy on Testicular Development

Group	Number of Birds	Body Weight (g) Mean ± S.E.		Testicular Weight (mg) Mean ± S.E.
		4 Weeks	7 Weeks	
Initial Control	5	54.24 ± 2.13	–	5.22 ± 0.72
Sham Pinealectomy	5	58.25 ± 2.10	85.64 ± 3.42	46.88 ± 12.89
Sham Pinealectomy and RLP implantation in F L C[a]	6	61.40 ± 2.60	91.67 ± 4.14	1,384.88 ± 207.38
Pinealectomy	6	58.42 ± 2.41	87.00 ± 2.32	37.83 ± 10.94
Pinealectomy and R L P implantation in F L C	6	57.23 ± 2.88	87.03 ± 2.30	965.02 ± 137.66
Intraocular R L P implantation	5	54.54 ± 4.70	80.96 ± 2.78	50.00 ± 11.29

[a]*Fissura longitudinalis cerebri.*

TABLE 3   Testicular Development after Implantation with Orange R L P or Control
Plates in F L C

Group	Number of Birds	Body Weight (g) Mean ± S.E.		Testicular Weight (mg) Mean ± S.E.
		4 Weeks	7 Weeks	
Intact control	5	63.0 ± 1.2	86.3 ± 1.8	30.8 ± 9.7
Pinealectomy and				
R L P Implantation	4	53.2 ± 4.2	83.0 ± 5.6	893.9 ± 110.3
Implantation of plates made of binders and				
chemicals (no light)	9	55.2 ± 2.4	85.6 ± 3.1	25.8 ± 10.5
Implantation of plates emitting β-radiation				
(no light)	8	61.3 ± 10.4	90.3 ± 2.8	26.3 ± 5.4

of 0.4 lux or less does not differ from complete darkness. A similar threshold may well exist for the quail. If so, light from the intraocular R L P, which cannot be brighter than 0.1 candle/m^2, is too weak to effect a photoperiodic response.

2. *Examination of the possibility that the gonadal development induced by* R L P *is not a result of photic stimulation*   It is clear that the implantation of R L P into the F L C can cause significant gonadal development in the Japanese quail. However, whether such a response has been induced by purely photic stimuli or by other factors (e.g., mechanical damage or stimulation of the brain associated with the implantation procedure, β-emission, or leaks of chemicals from the paint) requires further investigation.

To test these possibilities, the previous experiments were repeated by implanting R L P into the F L C. In another group of birds, a test plate composed of binder and metal-sulfide compound (the pigment component of the R L P) but with no radiation source and therefore no light emission, was implanted in the same location. In a third group we implanted plates emitting β rays, but no visible light, also in the F L C. In the latter two groups, the results were negative as judged by testicular development (Table 3).

In addition, we never observed any positive gonadal development as a result of implanting control plates in several locations of the brain other than the F L C. Thus, it is clear that the effective stimulus for induction of gonadal development was purely photic in nature.

TABLE 4 Effects of Wavelength and Locations of Implantation of Blue and Orange R L P

Experiment Grouping[a]	Treatment		Number of Birds	Testicular Weight (mg) Mean ± S.E.
	Implanted Location	Color of R L P		
A	On sphenoid bone	Orange	5	1,575.2 ± 171.7
	On sphenoid bone	Orange	1	452.1
	Nonimplanted control		6	231.8 ± 101.1
B	On sphenoid bone	Blue	5	79.6 ± 45.2
	On sphenoid bone	Blue	1	1,100.8
	Nonimplanted control		3	57.4 ± 23.5
C	On sphenoid bone	Blue	4	17.4 ± 2.4
	On sphenoid bone	Blue	1	205.3
	Nonimplanted control		4	91.9 ± 28.3
D	F L C[b]	Blue	7	48.6 ± 12.6
	F L C	Blue	5	23.6 ± 6.2
	F L C	Blue	1	1,756.3
	Intraocular	Blue	5	29.1 ± 11.0
	Nonimplanted control		5	30.8 ± 9.7

[a] All quail of one group were of the same hatch.
[b] *Fissura longitudinalis cerebri.*

3. *Photic stimulation through the orbital cavity and effects of wavelength* In the next series of experiments, we asked whether R LP can stimulate the encephalic photoreceptor via the orbital cavity (Benoit's classical route). Orange R LP was implanted near the opening of the optic foramen into the space between the sphenoid bone and muscular layers behind the eyeball, keeping the eye intact. This treatment resulted in positive responses comparable to that of implantation in the F LC, except for one case, in which the plate was apparently dislocated after the implantation and was found anchored to the upper margin of the orbital cavity (experiment A, Table 4).

When blue R LP was implanted behind the eye, one case of marked testicular development and one case of moderate development were obtained out of 11 birds (experiments B and C). The results of implantation of blue R LP into the F LC were similar to those obtained by placing it behind the eye. As shown in experiment D, there was only one positive case out of 13 birds. The results of experiments B through D indicate that the encephalic photoreceptor can respond to blue as well as to orange light, although the response to blue is erratic. As with orange R LP, intraocular implantation of blue R LP was ineffective in all cases tested (experiment D).

The fact that blue RLP is less reliable than orange in producing testis growth may result either from the greater penetrance of the longer wavelengths or from the photochemical properties of the receptor. As yet we have no way of distinguishing between these alternatives.

4. *Location of encephalic photoreceptors* On the basis of all our implant data taken together, it appears that RLP implanted in either the FLC or against the sphenoid bone could be stimulating a single photoreceptor in the hypothalamic area. This conclusion agrees well with the original observations of Benoit (1959, 1962), although the method of local illumination was different.

Direct implantation of RLP into the olfactory lobe of the quail, however, does not cause uniform testicular development (Table 5), although in the duck this area is photosensitive (Benoit, 1959). The size of this lobe in quail is so small that the negative responses may well be due to surgical damage during the operation. Considering this, the approximately 50 percent positive response permits the conclusion that the olfactory lobe is one of the encephalic photoreceptors participating in the photogonadal reflex in the quail.

In our recent experiments, marked and uniform gonadal development occurred after unilateral implantation of a small orange RLP rod into the central area of the optic lobe, 2–4 mm lateral to the midline of the brain. This is a new location, at which the presence of encephalic photoreceptors has not been reported. Complete understanding of these results awaits further investigation; it is possible that light impinging on the hypothalamus

TABLE 5 Effects of Implantation[a] on Testicular Development of the Growing Japanese Quail

Bird Number	Final Body Weight (g)	Testicular Weight (mg)
0479	105.0	1,614.4
0879	93.0	52.9
0880	78.5	11.4
0881	90.5	844.4
0882	96.0	2,051.4
0884	95.5	939.9
0889	91.5	64.6
0915	85.2	56.5
0917	112.9	911.8
0919	98.4	15.8
0920	82.5	677.9

[a]Orange RLP was implanted into the olfactory lobes.

from the implanted site brought about the positive response. In pinealectomy, or implantation of R LP into the F LC, loss of birds from the surgical
procedure was negligible, but implantation of the rod into the central portion of the optic lobe often resulted in death when the implantation was
done from either the lateral or the temporal side. It should be noted that
photic stimulation of the optic tectum by orange or blue R LP never resulted in a positive response.

5. *Encephalic photoreception and female sexual activity*   Recently we
have devoted special attention to the role of encephalic photoreceptors in
female sexual activity. Little work has been done with female birds due to
technical difficulties and the complexities of the gonadal system.

Female quail were raised on the schedule shown in Table 1. F LC implantation of orange R LP plate advanced the initiation of egg-laying by 2
weeks compared to the nonimplanted controls (Figure 2).

At the end of 13 weeks of observation, R LP was removed from the implantation site in those five birds that initiated egg-laying earlier than the
others. Unexpectedly, the birds continued egg-laying as before. Moreover,

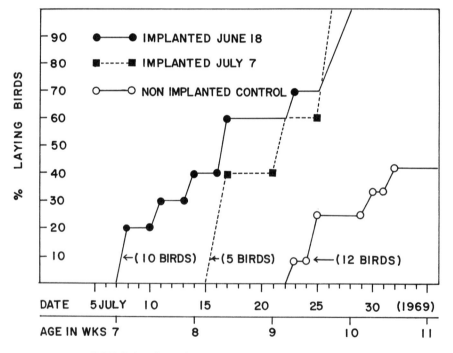

FIGURE 2   Effects of orange RLP implantation on egg-laying.

the removal of the R LP did not alter the laying pattern; i.e., the birds continued to lay at approximately 24-hr intervals. This seems to indicate that the implanted R LP triggers the process of sexual maturation but has little effect on the maintenance and timing of egg-laying. In similar experiments with males, removal of the R LP usually results in a decrease in sexual activity and atrophy of the testes. The detailed mechanism of the central regulation of gonadal maturation is an attractive subject for future study in photobiology.

REFERENCES

Benoit, J. 1962. Hypothalamo-hypophyseal control of the sexual activity in birds. Gen. Comp. Endocrinol. Suppl. 1:254–274.

Benoit, J., and I. Assenmacher. 1959. The control by visible radiations of the gonadotropic activity of the duck hypophysis. Recent Prog. Horm. Res. 15:143–164.

Homma, K., L. Z. McFarland, and W. O. Wilson. 1967. Response of the reproductive organs of the Japanese quail to pinealectomy and melatonin injections. Poult. Sci. 46:314–319.

Kato, M., Y. Kato, and T. Oishi. 1967. Radioluminous paints as activator of photoreceptor systems: studies with swallow-tail butterfly and quail. Proc. Jap. Acad. Sci. 43:220–223.

Morris, T. R. 1967. The effect of light intensity on growing and laying pullets. World's Poult. Sci. J. 24:246–252.

Oishi, T., and M. Kato. 1968. Pineal organ as a possible photoreceptor in photoperiodic testicular response in Japanese quail. Mem. Fac. Sci. Kyoto Univ., Ser. B, Vol. II, 12–18.

KRAIG ADLER

Pineal End Organ: Role in Extraoptic Entrainment of Circadian Locomotor Rhythm in Frogs

The response of blinded animals to light has been attributed to a "dermal light sense," an imprecise term used to designate photoreceptors of varied and often unknown anatomy, not even always located in the dermis. A variety of functions has been claimed for such extraoptic photoreceptors (Steven, 1963), but in vertebrates they seem to be involved in at least four different responses. Pigmentary adaptation in frogs can be controlled extraoptically via photoreceptors in the pineal body (Bogenschütz, 1965; Oshima and Gorbman, 1969). In house sparrows, perception of light controlling both entrainment of the locomotor rhythm and testicular recrudescence can occur via some extraoptic site in the head (Menaker, 1968; Menaker and Keatts, 1968); apparently the eyes are not involved in the testicular response at all (Menaker, this volume). The locomotor rhythms of salamanders and frogs are phase-shifted and the animals can compass-orient in the wild via light perceived extraoptically (Adler, 1969; see also the review by Adler, 1970). Only in the first of these instances has the photoreceptor (or photoreceptors) previously been located.

In vertebrates, the existence of various brain-associated structures that might be light sensitive has been known to anatomists for some time (see Studnička, 1905, for an early review), but their specific functions have remained in doubt. This is particularly true for the intracranial pineal body (or epiphysis cerebri) and, in frogs and toads, the extracranial pineal end organ (also called stirnorgan or frontal organ). A photoreceptive function for these structures has been hypothesized on the basis of electron micrographs (Eakin, 1961; Kelly, 1965) and electrophysiological records (Dodt and Heerd, 1962; Dodt and Jacobson, 1963). Experiments described in this paper suggest that the pineal end organ of frogs is indeed photoreceptive and that light perceived by it entrains the circadian locomotor rhythm in blind frogs.

MATERIALS AND METHODS

Adult male green frogs (*Rana clamitans*) were collected in Sylvatica Pond at Mountain Lake Biological Station, Giles County, Virginia; all experiments reported here were performed at the station between June and August 1969. Immediately after capture, animals were operated on and placed in clear plastic petri dishes (15 cm diameter, 8 cm high) in the recording apparatus (see Adler, 1969, note 4, for details). The animals were blinded by pushing the eyes gently from beneath and removing them with scissors; no anaesthetics or cold treatment were used. Most frogs bled profusely but the socket healed in a few days. Only one frog died during the course of the experiments; most appeared in good health two months after surgery. The pineal end organ was removed by cutting out the piece of skin—a 2×2-mm patch on the top of the head—that contained this organ. To cover the intracranial pineal body, a piece of aluminum foil was placed under the skin on the top of the head by inserting the foil through an incision made transversely between the tympanae; the incision generally healed in a few days. All animals were autopsied after experimentation to determine if the operations had been successful. Animals were force-fed raw calves' liver once a week.

Temperature (17.5° C) and relative humidity (90 percent) were kept constant; some very small, gradual fluctuations (±1° C; ±2 percent relative humidity) occurred but none correlated with the imposed changes in illumination. The light source was a 30-watt daylight fluorescent bulb (GE F30T8D) with a major transmission peak at 600 nm and a minor peak at 450 nm. Light was measured with an Isco spectroradiometer, model SR; intensity was determined in μw/cm^2 from 400 to 700 nm, summed, and converted to lux. Light intensity was measured within a petri dish containing an animal 130 cm from the light source. All times are Eastern Daylight Savings.

344

KRAIG ADLER

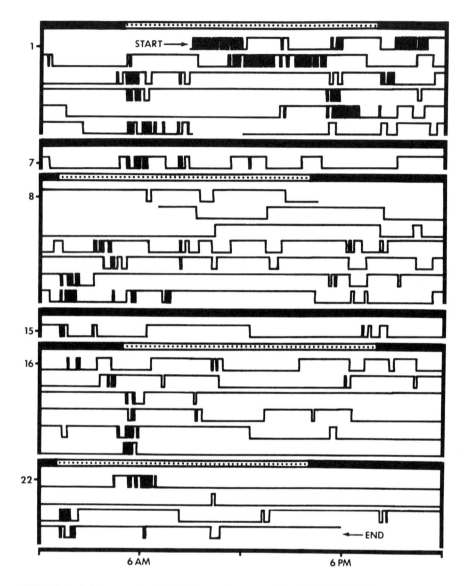

FIGURE 1 Activity record of a blinded green frog exposed to LD 15:9 (290:17 lux). In this and other figures, darkness (or lowest light intensity) is represented by a black bar directly over the time interval concerned; light is indicated by a hollow, dotted bar. These symbols show how the phase of the light cycle, with respect to real time, was changed in the course of the experiment. Breaks in the record indicate equipment failure. Time in days is shown along the ordinate; time in hours on the abscissa. The chart was cut at 24-hr intervals and each day's record was placed beneath the previous day's record; figures are tracings of the original charts. Note that the frog's activity is closely synchronized with "lights-on."

RESULTS

Extraoptic Phase Shifting

Frogs can perceive light extraoptically and adjust their locomotor activity in response. Figure 1 is the locomotor activity record of a bilaterally blinded green frog exposed to a light cycle of 15 hr of light (290 lux) and 9 hr darkness (17 lux); or *LD 15:9, 290*:17. For the first six days, the "lights-on" signal occurred at 5:00 a.m., about the time of onset of activity. There is also a second interval of activity about 5:00 p.m. on some days. On day 7, the animal was given constant low-intensity light (17 lux); the timing of the activity was unaffected. On day 8 the light cycle was advanced by 4 hr so that the light came on at 1:00 a.m. The frog's activity advances in about four transient cycles until it regains its former synchrony with "lights-on." After a day of constant dim illumination, the light cycle is delayed by 4 hr, followed by a final 4-hr advance. In each instance the blind animal's activity becomes synchronized with "lights-on" after a few days of transients.

Identical or similar experiments were performed with 17 blinded green frogs. In each instance the animal's activity was closely synchronized with "lights-on." The timing of the activity rhythm always correlated with the imposed experimental light cycle and not with any temperature, humidity, or noise cycles in the laboratory or in the external environment.

Entrainment of the Rhythm

Entrainment of the locomotor rhythm can be accomplished via an extraoptic photoreceptor. The bilaterally blinded frog whose record is shown in Figure 2a was first allowed to free-run in constant darkness (DD); pre-entrainment period lengths for 10 such frogs in DD varied from 22.5 to 23.8 hr, with a mean of 23.0 hr. On day 4, a light cycle was imposed (*LD 14*:10, *290*:0 lux) and after four or five transient cycles the onset of activity became synchronized with "lights-on." On days 11–15 the frog was allowed to free-run in DD; postentrainment periods for the same 10 frogs varied from 22.5 to 23.7 hr (mean, 23.2). Such a response demonstrates entrainment since (a) the free-running period differs from that of the imposed *L*D cycle, (b) the phase of the postentrainment free-run differs from an extrapolation of the phase of the pre-entrainment free-run, and (c) the entrained locomotor rhythm assumes the period of the imposed *L*D cycle.

Records of clearly defined free-running rhythms were obtained from normal-sighted frogs only with difficulty. In Figure 2b the imposed DD

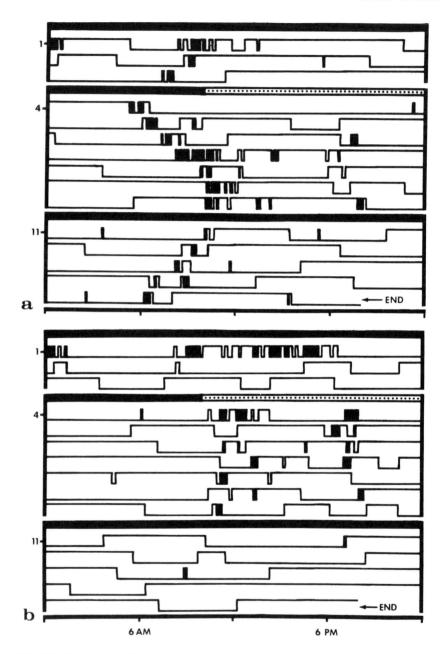

FIGURE 2 Entrainment of locomotor rhythms in blinded and sighted green frogs. A blinded frog
(a) and a normal-sighted frog (b) show entrainment to an imposed photoperiod of LD *14*:10
(*290*:0) of a rhythm previously free-running in DD; see text. Other details as in Figure 1.

and LD regimes are the same as in Figure 2a. There is some indication of a pre-entrainment free-run but little if any evidence of a postentrainment free-run. Similar results were obtained from nine other normal frogs. Estimates of period lengths cannot be made with any confidence, but they do not appear to differ significantly from those of eyeless frogs. Although their free-running rhythms are poorly expressed, visually normal frogs quickly synchronize their activity with the LD cycle.

Location of Extraoptic Photoreceptor

Two experiments were performed to localize the extraoptic site of light perception. In the first (Figure 3a), the frog was bilaterally blinded and the extracranial pineal end organ removed. In addition, a double layer of heavy aluminum foil was inserted beneath the skin on top of the head to shield the intracranial pineal body; the foil extended from just behind the external nares to a line drawn between the tympanae and laterally to the medial edge of the eye sockets. The animal was then exposed first to constant darkness (DD), and then to a light cycle (LD 14:10, 290:0 lux) in a sequence similar to that shown in Figure 2. This eyeless frog, lacking a pineal end organ and with the brain covered, continued to free-run throughout the experiment—its locomotor rhythm apparently was not entrained by the light cycle present on days 4–8 (Figure 3a). Free-running periods of eight such frogs, measured over days 1–12, varied from 22.9 to 23.8 hr, with a mean of 23.5 hr.

In the second experiment, the frog was eyeless and lacked a pineal end organ, but no aluminum cover was inserted under the skin on top of the head, and the pineal body was therefore normally exposed (Figure 3b). This animal also continued to free-run, disregarding the light cycle. During the afternoon, there was a second free-running activity interval for this particular individual (see also Figure 1). In identical experiments, eight such frogs had free-running periods ranging from 22.8 to 24.0 hr, with a mean of 23.1 hr, measured over days 1–12. Five others exposed to a light cycle (LD 14:10, 290:0 lux) for 10 days continued to "free-run." Taken together, these data suggest that blinded frogs lacking pineal end organs are truly free-running under an imposed light cycle and are not merely in transience approaching a different entrained steady-state phase angle to the light cycle; longer runs in LD, however, would more clearly establish this point.

The experiment shown in Figure 2a was actually a double sham control for those shown in Figure 3a–b. In addition to the blinding operation, a small (2 mm-square) piece of skin was removed from the shoulder of the

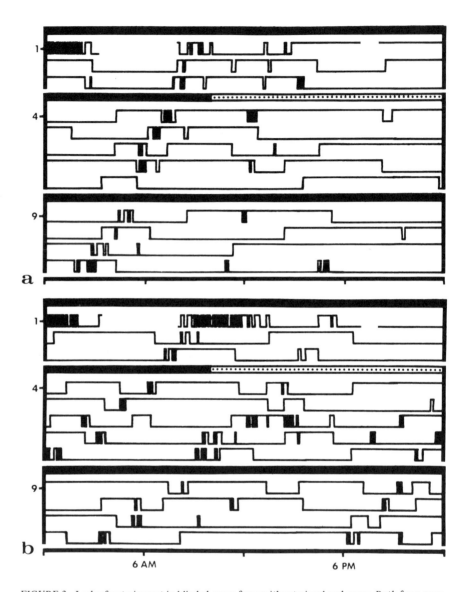

FIGURE 3   Lack of entrainment in blinded green frogs without pineal end organ. Both frogs were blinded and had the extracranial pineal end organ removed; aluminum foil was inserted under the skin on top of the head in (a) but not in (b). Note that the free-running rhythms were unaffected by the imposed light cycle (*LD 14*:10, *290*:0) and both animals behaved as if continuously in DD throughout the record. Other details as in Figure 1.

frog (as a control against which to assess the effects of removing a similar-sized piece of skin containing the pineal end organ from the top of the head). A piece of aluminum foil was inserted under the skin on the back (as a control for the effects of inserting foil under the skin of the head). Each control has been performed separately and in these animals, as well as in those which are merely eyeless, entrainment to the imposed LD cycle occurs. Frogs with eyes intact but lacking pineal end organs have not yet been tested.

In some instances experiments have been performed simultaneously in the same enclosure on individuals possessing pineal end organs and on individuals lacking them; in such cases, the former respond to the light cycles by adjusting their locomotor rhythms and the latter do not. This is strong evidence that the animals are not responding to temperature changes associated with the lighting regime.

DISCUSSION

These data suggest that entrainment of the locomotor rhythm of green frogs can be mediated extraoptically and that the pineal end organ is involved in this response. However, the experiments reported here do not distinguish between two alternative hypotheses: (a) that the pineal end organ is the extraoptic photoreceptor involved in entrainment, and (b) that while the presence of the pineal end organ is necessary for entrainment, photoreception occurs elsewhere. In view of the ultrastructural and neurophysiological data, the first possibility seems much more reasonable. The data also suggest that the intracranial pineal body (which rests on the diencephalon) is not involved as a photoreceptor in entrainment of the locomotor rhythm. The pineal body may, of course, be involved at light intensities above 290 lux or at different wavelengths; these possibilities are currently under investigation.

ACKNOWLEDGMENTS

I thank M. Menaker and R. E. Gordon for criticizing the manuscript, J. J. McGrath and J. A. M. Brown for technical advice, and G. Waggoner and J. Johnston for help in collecting frogs. The research was supported through a postdoctoral research fellowship at Mountain Lake Biological Station (NSF GB-7416) and a biomedical support grant (NIH FR 07033-04).

REFERENCES

Adler, K. 1969. Extraoptic phase shifting of circadian locomotor rhythm in salamanders. Science 164:1290–1292.

Adler, K. 1970. The role of extraoptic photoreceptors in amphibian rhythms and orientation: a review. J. Herpetol. 4:99–112.

Bogenschütz, H. 1965. Untersuchungen über den licht bedingten Farbwechsel der Kaulquappen. Z. Vgl. Physiol. 50:598–614.

Dodt, E., and E. Heerd. 1962. Mode of action of pineal nerve fibers in frogs. J. Neurophysiol. 25:405–429.

Dodt, E., and M. Jacobson. 1963. Photosensitivity of a localized region of the frog diencephalon. J. Neurophysiol. 26:752–758.

Eakin, R. M. 1961. Photoreceptors in the amphibian frontal organ. Proc. Natl. Acad. Sci. U.S. 47:1084–1088.

Kelly, D. E. 1965. Ultrastructure and development of amphibian pineal organs. Prog. Brain Res. 10:270–287.

Menaker, M. 1968. Extraretinal light perception in the sparrow. I. Entrainment of the biological clock. Proc. Natl. Acad. Sci. U.S. 59:414–421.

Menaker, M., and H. Keatts. 1968. Extraretinal light perception in the sparrow. II. Photoperiodic stimulation of testis growth. Proc. Natl. Acad. Sci. U.S. 60:146–151.

Oshima, K., and A. Gorbman. 1969. Pars intermedia: unitary electrical activity regulated by light. Science 163:195–197.

Steven, D. M. 1963. The dermal light sense. Biol. Rev. Camb. Philos. Soc. 38:204–240.

Studnička, F. K. 1905. Die Parietalorgane, p. 124 to 209. In A. Oppel [ed.], Lehrbuch der vergleichenden mikroskopischen Anatomie der Wirbeltiere. Vol. 5. G. Fischer, Jena (whole vol., pp. 1–254).

JON W. JACKLET

A Circadian Rhythm in
Optic Nerve Impulses from
an Isolated Eye in Darkness

Spontaneous impulse activity in vertebrate and invertebrate retinal neurons has often been observed, yet a circadian rhythm in this activity had not been described until 1969 (Jacklet, 1969b). Rhythmic migration of the pigments of arthropod compound eyes (Welsh, 1941; Fingerman and Lowe, 1957), daily fluctuation in the electroretinogram of some beetles (Jahn and Crescitelli, 1940), and circadian rhythms of the electroretinogram and the activity of "sustaining" fibers of the optic nerve of the intact crayfish (Aréchiga and Wiersma, 1969) have also been demonstrated.

The following describes the circadian rhythm of optic nerve potentials recorded from the isolated eye of *Aplysia* (Jacklet, 1969b) and the results of manipulations of this rhythm *in vitro* and *in vivo*.

METHODS

Three species of the sea hare *Aplysia* were used: *A. californica* and two species from Florida, *A. willcoxi* and *A. dactylomela*. The animals were kept in *LD 12*:12 (165:0 fluorescent; 300:0 incandescent) or LL (195) in either natural seawater at 14° C or Instant Ocean at 15° or 22° C. The eye–optic-nerve complex was prepared for electrophysiological recording by freeing the eye, by dissection, from the peripheral integument of the head and severing the central connections of the optic nerve at the cerebral ganglion. The dissections were usually carried out (15 min/eye) during the last few hours of the light phase of the *LD* cycle, and exposure of the eye to intense illumination was minimized. The isolated eye–optic-nerve preparation was placed in a thermostatically regulated recording chamber of 100-ml capacity. Filtered (0.22 μ Millipore) seawater (pH 8.2), Instant Ocean (pH 8.0), or a nutrient culture medium (pH 7.8–8.0) (Strumwasser and Bahr, 1966) was used as the organ culture solution. The electrical activity of the optic nerve was led off by a suction electrode, amplified by a Tektronix 122, displayed on a C R O, recorded continuously on a Grass polygraph, and counted automatically on a Hewlett-Packard counter-printer. The recording chamber was sealed inside a black box containing a lamp permitting either complete darkness or controlled lighting.

CIRCADIAN OSCILLATIONS IN IMPULSE FREQUENCY

The compound action potentials (Jacklet, 1969a) that occur spontaneously in the optic nerve in complete darkness and those evoked by steady illumination appear to be identical in wave form and amplitude and thus represent the same population of neural elements. Figure 1 shows the activity in darkness and during illumination. The steady-state frequency dur-

A.d.   20°C

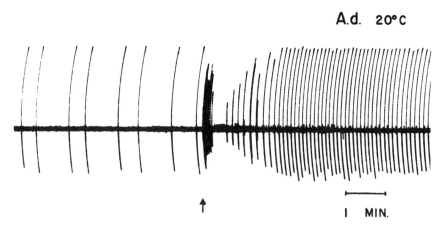

↑   I MIN.

FIGURE 1 Spontaneous and evoked compound action potentials from the optic nerve of an isolated eye–optic-nerve preparation. The spontaneous potentials occurred in complete darkness. Illumination with 600 lux of white light beginning at the arrow evoked a high frequency burst of activity, followed by a pause and then firing at a steady frequency. A.d., *Aplysia dactylomela*.

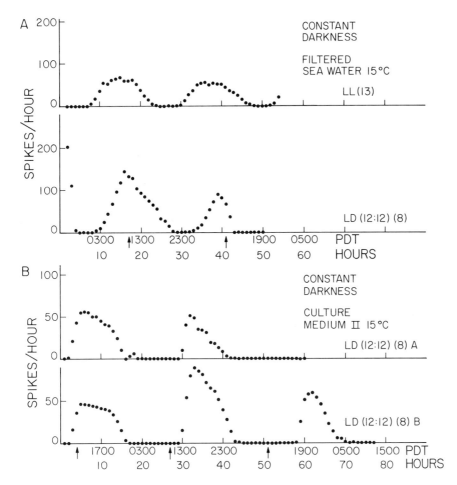

FIGURE 2 The frequency of impulses over several days in total darkness. (A) Both eyes were tested in the same chamber in filtered seawater. The upper graph is that of an LL (13 days) eye; the lower is that of an LD 12:12 (8 days) eye. The black arrows on the time axis indicate the projected dawn for the LD eyes in (A) and (B). (B) A pair of LD 12:12 (8 days) eyes from the same animal in culture medium. (Reproduced with permission from Jacklet, 1969b.)

ing illumination is proportional to the log of the light intensity (Jacklet, 1969a) so that the optic nerve appears to be effectively coding for intensity during illumination.

The spontaneous optic nerve activity in the isolated preparation is entrained by the photoperiod that the intact animal experienced. In darkness the frequency of spontaneous optic nerve activity fluctuates in a circadian fashion; the peak on the first day of isolation (Figure 2) occurs around the

projected dawn of the entraining photoperiod. It remains active throughout the projected light period of the first day, is quiet during the projected
dark period, and becomes active again the next day. On the second and
succeeding days the rhythm drifts out of phase with the projected photoperiod into a "free-run."

The circadian nature of the rhythmicity in impulse frequency is demonstrated by its non-24-hr free-running period, τ. The rhythm in the isolated
eye obeys the circadian rule (Aschoff, 1960) for a diurnal animal: Constant dim light conditions decrease τ and constant dark conditions increase τ.

Period length is influenced by factors other than the intensity of light.
Figure 2 shows that the period of optic nerve activity recorded in seawater is shorter than that in culture medium. The results for two species
of *Aplysia* are summarized in Table 1. The average period for both species
in culture medium is 27 hr regardless of the temperature differences and
LD history. Preparations in seawater have a period of about 22 hr. Prelim-

TABLE 1   Tabulation of Period Length (τ) (time interval between 50% of maximum
amplitude of successive increasing spike frequency curves.) First period is between the
first and second complete oscillation *in vitro*; 2nd period between second and third
oscillations. Med. II is culture medium II (Strumwasser and Bahr, 1966). S.W. is filtered seawater. All LD are *12*:12.

Species	Prior Condition	(Days)	Solution	Temperature	In vitro Period (hr) at 50% Max. 1st	2nd	$\overline{X}$
A. californica	LL	(3)	Med. II	15°C	26.3		
		(12)			28.5		27.4
	LL	(13)	S.W.		22.0		
		(4)			23.0		22.5
	LD	(12)	Med. II		27.3		
		(8)			27.0		
					27.3	28.3	27.5
	LD	(6)	S.W.		20.3		
					23.5		21.9
A. dactylomela		(17)	Med. II	20°C	27.0	27.0	
					25.0	28.5	
		(5)			29.5		
		(4)			25.3		27.1

inary experiments on preparations in *Aplysia* blood indicate that there, too, the period is about 27 hr. The pH of all solutions is in the range pH 7.8–8.2 and is not responsible for the observed differences in period.

SENSITIVITY

The circadian rhythm persists in constant dim light. Presumably, the activity in dim light is the sum of spontaneous activity and activity evoked by the dim illumination; alternatively, the activity could be due to circadian fluctuation in the light sensitivity of the eye. The following experiment provides some insight into the mechanism of rhythmicity in dim light: If a 15-min pulse of light is given every 2 hr in otherwise constant dark conditions (Figure 3), the rhythm in spontaneous activity is still quite evident. In addition, the sensitivity to the light pulse (frequency of impulses evoked by illumination) changes over a 24-hr period. These sensitivity fluctuations may persist with a circadian period; this possibility has

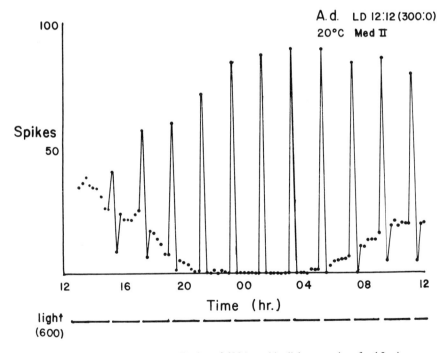

FIGURE 3   Sensitivity of eye *in vitro*. Flashes of 600 lux white light were given for 15 min every 2 hr in otherwise constant darkness. Counts were taken at 15-min intervals. Time is EST. Interruption of light bar indicates the flash duration.

not been adequately tested. It is apparent from Figure 3, however, that the light sensitivity of the eye is inversely related to the spontaneous impulse frequency. The rhythm in dim light cannot be due to changing sensitivity to light, but must be due to the spontaneous potentials. The interaction of the spontaneous and evoked potentials in dim light would tend to reduce the amplitude of the circadian rhythm but raise the overall level of activity.

PHASE SHIFTS *IN VIVO*

Exposure of the intact animal to one period of a phase advanced or delayed light–dark cycle will shift the phase of the eye–optic-nerve rhythm. Depending on the intensity and duration of the new photoperiod, the rhythm may or may not undergo transients in reaching the new phase. Eyes removed from animals subjected to a 4-hr advance in *LD* 12:12 conditions show essentially no transients, but if the light duration is reduced to 2 hr, transients do occur. (At this time, there are not data sufficient to plot a phase response curve.)

The two eyes from the same animal tend to oscillate in phase (see Figure 2) if they are isolated in culture under the same conditions. In order to test the coupling between the eyes, experiments were performed in which the eyes (*in vivo*) were subjected to photoperiods of different phases. The results of such an experiment are shown in Figure 4. The animal was subjected to 4 days of *LD* 12:12. On the fifth day one eye was capped with a plastic cover which allowed little or no light to reach it. Then the photoperiod was advanced 4 hr, so that the uncapped eye would see one period at the new phase. At the end of the sixth day the eyes were isolated; the recordings of the seventh day under constant dark conditions are shown in Figure 4. The capped eye remained in phase with the old photoperiod and the uncapped (advanced) eye advanced to the new phase. The oscillations are less clear-cut under these than under standard conditions and may be due to transients or to loose coupling between the eyes. For the most part, however, the oscillations are independent of each other. This experiment was performed successfully on three animals but none of the recordings gave a clear picture of the oscillations on the second day in isolation. Experiments are in progress to clarify the coupling relationships between the eyes.

PHASE SHIFTS *IN VITRO*

The eye–optic-nerve oscillations will "free-run" in a predictable manner in organ culture. Brief pulses of light of 15 min or so appear to have very

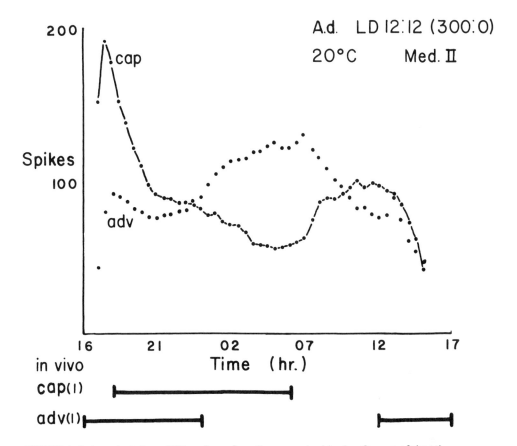

FIGURE 4   Independent phase shifting of eyes from the same animal *in vivo*. One eye of the animal was capped (cap) to shield it from light, then the light schedule was advanced 4 hr to expose the other eye to a phase shift. After 1 day of the shifted light schedule, the eyes were isolated and the optic nerve activity recorded in constant darkness in the same chamber. Clearly only one eye is shifted to the new phase. Time is EST. Black bars indicate the dark portion of *LD* cycle prior to isolation of the eyes.

little effect on the phase of the oscillations (see Figure 3) but a 4-hr light pulse is sufficient to produce a partial phase-shift *in vitro* (Figure 5). Presumably, repeated exposure to a properly phased light would bring about a complete shift. These results show that the phase-shifting apparatus and the oscillator are independent of the rest of the animal.

   Figure 5 shows the results for four eyes: two from *A. dactylomela* and two from *A. willcoxi*. The two species differ considerably in the slope of the curve of increasing frequency of impulses. This difference in rhythm dynamics (*A. willcoxi* always has a steeper slope) is substantiated by other

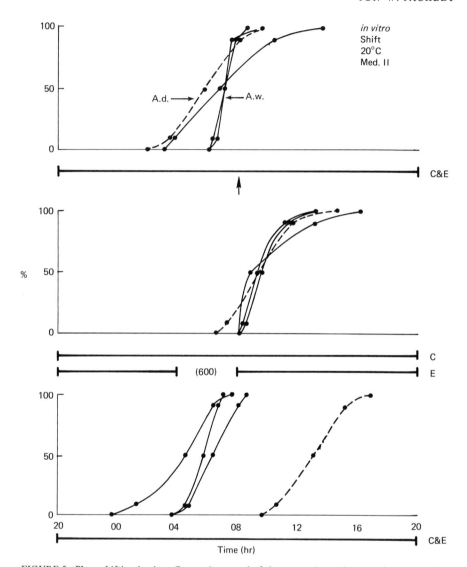

FIGURE 5   Phase shifting *in vitro*. Composite record of three experimental eyes and one control eye over 3 days *in vitro*. The top graph is the activity recorded on the first day in constant darkness. The arrow indicates projected dawn. The second graph shows the activity recorded on the second day in darkness, but does not show the activity evoked in the experimental eyes by the 4-hr pulse of light (600 lux) from 0400–0800. The third graph shows the activity recorded on the third day for the three experimental eyes (solid line) and the control (dotted line). The light pulse has shifted the phase of the three experimentals. Time is EST; bar under graph is darkness; C is control; E is experimental. The spike activity is normalized and only the increasing frequency of impulses from 0 to 100% of maximum is plotted. A.d. is *Aplysia dactylomela* (two eyes); A.w. is *Aplysia willcoxi* (one experimental and one control eye).

experiments. The difference appears to be correlated with the general level of locomotor activity of the species: *A. willcoxi* is an active swimmer and *A. dactylomela* is nonswimming and sluggish.

DISCUSSION

The optic nerve activity in constant conditions is clearly circadian and obeys the circadian rule for diurnal animals. In addition, the period length depends on the *in vitro* culturing solution. In seawater it is about 22 hr but in culture medium it is 27 hr. This difference appears to be due to substances that are present in *Aplysia* blood but absent from seawater. Whether these differences are due to some influences on protein formation or structure, as has been suggested to explain the effects of phase-shifting chemicals (Hastings and Keynan, 1965), must await further investigations.

The rhythm can be completely phase-shifted in one trial if the phase of the *LD 12*:12 *Zeitgeber* is advanced or delayed 4 hr. Transients (see Bünning, 1967) appear in reaching the new phase if the duration of the light pulse is shortened. The rhythm in one eye can be phase-shifted *in vivo* independently of the other eye and *in vitro* independently of the rest of the animal. Thus, in the animal, the eye oscillators are, at most, only slightly influenced by each other or by other oscillators in the animal. This fact is readily understandable in physiological terms, since in preparations in which the optic nerves are intact to the cerebral ganglion no activity from the contralateral optic nerve is detectable in the ipsilateral optic nerve (Jacklet, 1969a). Of course, the eyes could influence each other humorally. The central nervous system could also influence the eyes by a humoral mechanism, since many of the central neurons of *Aplysia* are believed to be neurosecretory (Coggeshall *et al.*, 1966). In addition, a seasonal rhythm in induction of egg-laying by a neural extract has been demonstrated (Strumwasser, Jacklet, and Alvarez, 1969). The activity of the optic nerve might, in turn, influence the central nervous system and specific rhythmically active neurons (Strumwasser, 1965, 1967; Lickey, 1969), either synaptically or by neurohumors, or both, since there are secretory cells in the retina (Jacklet, 1969a).

The data from eyes taken at various times of the year suggest that the onset and peak of optic nerve circadian activity always has the same phase relationship to the D/L transition of the *LD* cycle and does not vary in a yearly cycle like the parabolic burster neuron rhythm (Lickey, 1969). Thus, the animal could use the optic nerve rhythm as a reliable index of the D/L transition regardless of the season.

Some recent experiments have shown that entrainment of a specific neuron (the parabolic burster; Strumwasser, 1965) by light cycles does not require that the eyes be present (Lickey, this volume). On the other hand, the eyes do appear to be necessary for entrainment of the locomotor rhythm (Strumwasser, personal communication) previously shown to be circadian (Strumwasser, 1967; Kupfermann, 1967).

The influence of the eye on the activity of the animal and its central neurons may be related to the intensity of the environmental light. Possibly there are alternate pathways for light influences on central neurons, depending on light intensity; for example, some pigmented central neurons are directly sensitive to light (Arvanitaki and Chalazonitis, 1960). The eye rhythm, being especially sensitive to light, could be phase-controlled at low intensities and, in turn, set the phases of various other rhythms.

ACKNOWLEDGMENTS

The initial observations on the eye rhythm were made at California Institute of Technology and supported by NIH grant NB 07071-02 and NASA grant 05-002-031 to Felix Strumwasser. Later studies were done at SUNYA and supported by NIND & S grant 1 ROINB 08443-01 and Research Foundation of SUNYA Award to JWJ. I gratefully acknowledge the technical support of R. Alvarez, J. Gilliam, and J. Conners.

REFERENCES

Aréchiga, H., and C. A. G. Wiersma. 1969. Circadian Rhythm of responsiveness in crayfish visual units. J. Neurobiol. I (1):71–85.

Arvanitaki, A., and N. Chalazonitis. 1960. Photopotentials d'excitation et d'inhibition de differents somata identifiable (Aplysia) activations monochromatiques. Bull. Monaco Inst. Oceanogr. 57(1164):3–83.

Aschoff, J. 1960. Exogenous and endogenous components in circadian rhythms. Cold Spring Harbor Symp. Quant. Biol. 25:11–28.

Bünning, E. 1967. Setting the clock by light-dark cycles, p. 62–76. In E. Bünning, The physiological clock. 2nd ed. Springer-Verlag, New York.

Coggeshall, R. E., E. Kandel, I. Kupfermann, and R. Waziri. 1966. A morphological and functional study on a cluster of identifiable neurosecretory cells in the abdominal ganglion of Aplysia californica. J. Cell Biol. 31:363–368.

Fingerman, M., and M. Lowe. 1957. Twenty-four hour rhythm of distal retinal pigment migration in the dwarf crawfish. J. Cell. Comp. Physiol. 50:371–379.

Hastings, J. W., and A. Keynan. 1965. Molecular aspects of circadian systems, p. 167 to 182. In J. Aschoff [ed.], Circadian clocks. North-Holland Publ. Co., Amsterdam.

Jacklet, J. W. 1969a. Electrophysiological organization of the eye of Aplysia. J. Gen. Physiol. 53(1):21–42.

Jacklet, J. W. 1969b. Circadian rhythm of optic nerve impulses recorded in darkness from isolated eye of *Aplysia*. Science 164:562–563.

Jahn, T. L., and F. Crescitelli. 1940. Diurnal changes in the electrical response of the compound eye. Biol. Bull. 78:42–52.

Kupfermann, I. 1967. A circadian locomotor rhythm in *Aplysia californica*. Physiol. Behav. 3:179–181.

Lickey, M. E. 1969. Seasonal modulation and non-24-hour entrainment of a circadian rhythm in a single neuron. J. Comp. Physiol. Psychol. 68:9–17.

Strumwasser, F. 1965. The demonstration and manipulation of a circadian rhythm in a single neuron, p. 442 to 462. *In* J. Aschoff [ed.], Circadian clocks. North-Holland Publ. Co., Amsterdam.

Strumwasser, F. 1967. Neurophysiological aspects of rhythms, p. 516 to 528. *In* G. C. Quardon, T. Melnechuk, and F. O. Schmitt [eds.], The neurosciences: A study program. Rockefeller Univ. Press, New York.

Strumwasser, F., and Bahr, R. 1966. Prolonged *in vitro* culture and autoradiographic studies of neurons in *Aplysia*. Fed. Proc. Fed. Am. Soc. Exp. Biol. 25:512.

Strumwasser, F., J. Jacklet, and R. Alvarez. 1969. A seasonal rhythm in the neural extract induction of behavioral egg-laying in *Aplysia*. Comp. Biochem. Physiol. 29: 197–206.

Welsh, J. 1941. The sinus gland and 24-hour cycles of retinal pigment migration in the crayfish. J. Exp. Zool. 86:35–49.

DISCUSSION

MENAKER: Why do you suppose the eyes have rhythms at all?

JACKLET: The *Aplysia* eye is very unusual—it isn't the ordinary kind of eye you find in a frog or a mammal. It is fairly simple; it has retinal cells, support cells, and—what I think is significant—secretory cells at the base of the retina. It may be that this eye is not so much an eye as it is an endocrine organ, setting the phase of other rhythms by some secretory activity.

RENSING: There is a marked difference in period length in the salt water and the medium. Have you tried to change the potassium or sodium concentration in the medium to see if that affects the period?

JACKLET: No, I haven't looked at that further. It clearly needs to be done, though.

STRUMWASSER: I'd like to ask Dr.

Jacklet if he would comment on why the preparation runs down. Is it the eye itself running down, or is it a problem with the suction electrode technique? Ganglia maintained in the organ culture medium are electrically active and have normal resting potentials, spikes, and so on for up to six weeks [Strumwasser, F., and R. Bahr. 1966. Prolonged *in vitro* culture and autoradiographic studies of neurons in *Aplysia*. Fed. Proc. 25:512].

JACKLET: I ran some of these experiments in a tunnel electrode, where there is no pressure on the optic nerve at all. Those also ran down in about three days. The preparation is still light-sensitive, but the rhythm is no longer pronounced after that time.

HOSHIZAKI: The shape of the rhythm, as I recall, was different in continuous light and under a light/dark cycle. In

LD, the curve had a rather sharp peak, but in LL it was flat. Would you like to comment on this? Have you any idea why this is so?

JACKLET: In general, the amplitude of the rhythm in LL is smaller than in LD. As to why it is so, I have no good answer. I could speculate that the reduction in amplitude and steepness of the rhythm is due to the lack of a synchronizing effect on the members of the oscillator population in the absence of a *Zeitgeber*.

WILLOWS: Did you, in conjunction with your earlier work with Strumwasser, ever find a connection between this rhythm and the known rhythm in the parietovisceral ganglion?

JACKLET: Quite early in that study we looked for synaptic influences—that is, illuminating the eye and monitoring the parabolic burster. And those experiments were all negative. Although this possibility has not been exhaustively explored, it looks as though there is no direct synaptic input to the burster from the eye. There is a lot of distance between the eye and the central neuron in this case.

WILLOWS: Could there be a direct effect of light on the burster, not necessarily involving any of the fibers you are recording from?

JACKLET: Yes.

ASCHOFF: You mentioned that there are several secretory cells and thousands of receptor cells in this preparation. Now if the disappearance of the rhythm is due to desynchronization among these cells, it would be interesting to give a light signal or some other stimulus to resynchronize them again after a few cycles. Did you try this?

JACKLET: Yes, eyes that have ceased to show spontaneous potentials or that have a reduced amplitude of the rhythm after two or three days have been subjected to light. This does not bring the rhythm back, although the eye may still respond to light.

RENSING: Do you find any correlation between the spike frequency and the circadian frequency? That is, was the discharge frequency perhaps lower with the 27-hr period than the 22-hr period?

JACKLET: No, there was no correlation readily apparent.

ROBERTS: Could you say something about where the neurosecretory cells are located? Did you try to remove them?

JACKLET: I don't think that is possible. It's a pretty small eye; I doubt you could remove the secretory cells without completely wrecking the eye. The receptors are in a layer next to the optic nerve, and the secretory cells are all around them.

RENSING: Do you have any idea what is being secreted by the neurosecretory cells? Could it be dopamine or catecholamine secretion, for instance?

JACKLET: I don't have any evidence on that yet, but that is one of the directions I want to go.

S. A. GORDON AND G. A. BROWN

Observations on Spectral Sensitivities for the Phasing of Circadian Temperature Rhythms in *Perognathus penicillatus*

Several circadian phenomena in plants are phased by red light (Sweeney, 1969). In view of the sensitivity of mammalian cells and organelles to this spectral region (Gordon, 1964), we though circadian mammalian rhythms might respond similarly. The present experiments were undertaken to test this hypothesis. Temperature rhythm in the pocket mouse was chosen as a test system on the basis of the extensive studies of Lindberg and co-workers (1968): The maintenance requirements of the pocket mouse are minimal, its body temperature can be monitored readily by telemetry, and the circadian variation of body temperature and metabolism in this animal, particularly under stress, is sharply defined.

This report describes preliminary stages of work in progress. However, a spectral region relatively effective in phasing the temperature rhythm in pocket mice can be tentatively identified.

MATERIALS AND METHODS

Adult male and female *Perognathus penicillatus* Pricei, weight approximately 20 g, were collected from the field in April (Pet Corral, Tucson, Arizona). They were maintained on shelled sunflower seed supplemented every 1 or 2 weeks with a fragment of raw lettuce or carrot. No liquid was supplied.

To monitor rhythms in gross motor activity as influenced by exposure to light, mice were placed in activity cages of the interrupted beam type. The light source was a tungsten filament lamp behind an interference filter with transmittance peak at 1.1μ, half-width 19 nm. Beam interruptions were recorded as pulses on a strip chart, which were then summed manually on a continuous sequence of half-hour intervals.

For the telemetry of body temperature, a temperature-sensitive transmitter (Northrop Space Laboratories) was implanted in the peritoneal cavity. The incision was sutured with wire clips. About 1 week was allowed for recovery from surgery before experimenting began. The transmitters, pulse frequency thermistor modulated (Lindberg, De Buono, and Anderson, 1965), were about 1.5 cm long, 1 cm wide, and 0.75 cm thick, encapsulated in wax, and weighed 1.5–2 g. The relation between signal frequency and temperature was sufficiently linear that only signal frequency was recorded.

Animals were caged individually in cylindrical polystyrene containers, 13 cm in diameter and 16 cm high, with parallel inserts of flat plastic so that the animal was confined on two sides between walls 10.3 cm apart. Several inches of washed quartz sand covered the bottom of the cage. Three loop antennae were wound about the cage to ensure signal strengths adequate for pulse-triggering regardless of the position of the animal. The limited range of the transmitter kept cross-talk to an insignificant level when the cages were arranged along the focal curve of the spectrograph. Figure 1 shows a block diagram of the instrumentation.

Each animal was scanned for 0.2 min at 12-min intervals. The number of pulses per scan time was printed on paper tape. These data were entered on punch cards manually and smoothed by the method of moving averages, using successive sets of 7 data points fitted by cubic polynomial (Kendall and Stuart, 1966). This fit was chosen empirically to reduce the noise level of the raw data without masking trend changes. The smoothed data were plotted by computer as continuous graphs of frequency (pulses per 0.2 min) versus scan number.

The Argonne Biological Spectrograph (Monk and Ehret, 1956) was used to test spectral response. Cages were placed at 25-nm spacings on the focal curve; each cage was isolated by a partition of black masonite to reduce the level of spectral contamination. The horizontal beam from the grating was reflected downward by 10-cm wide front-surface mirrors into the top of the cage. Dull black nylon bolting cloth was held in place over the top of the cage by heavy rubber bands. By varying the number of

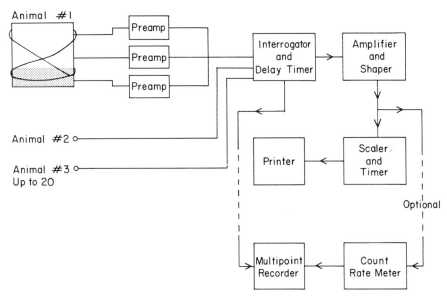

FIGURE 1   Block diagram of the instrumentation used for telemetry of temperature signals from
implanted transmitters (see Jaroslow and Eisler, 1968).

cloths and their arrangement, irradiance was adjusted so that incident energies at the various irradiation loci were the same within 10 percent. Calibrated thermopiles and a high impedance electrometer were used to determine irradiance levels. The mean air temperature at the focal curve was
$19 \pm 1°\text{C}$.

Thermoperiodicity in the first few days of an *LD 12*:12 regimen was
frequently 3 to 4° in amplitude; the wide range, indicative of periodic
torpor, then appeared. In extended DD (or in monochromatic light that
had no effect on the rhythm), duration of torpor tended to increase with
time. Occasionally a temperature cycle would be skipped; i.e., the body
temperature stayed low. However, arousal during the following cycle
would occur at the expected time (see Hayden, 1965).

Figure 2 is a graph of the data obtained from a mouse whose temperature was monitored during a test of phase shift by monochromatic light.
The plot shows the steep wave fronts in the cycle of body temperature
that are characteristic of periodic torpor (Lindberg, 1968). The overshoots
of body temperature on arousal were used as phase reference points in experimental series I. In series II, midpoint of activity was the phase reference point–i.e., the time halfway between midpoints of arousal from, and
re-entry into, torpor.

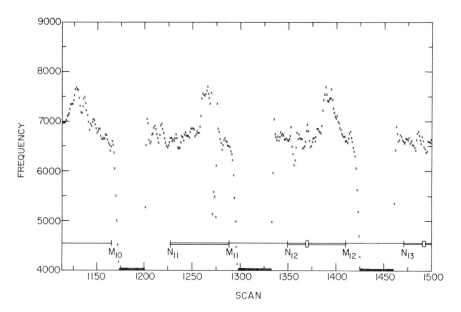

FIGURE 2 Transmitted frequencies (pulses/0.2 min) from a mouse at the 275-nm locus of the spectrograph. Each data point represents a scan; these scans occurred at 12-min intervals and are numbered sequentially on the abscissa. The line above the abscissa shows the exposure protocol: for the first 11 days the animal was exposed to the cycle LD 12:12 (white light), with dawn at time 0; at 1200 on the 11th day "continuous" darkness began, interrupted at Z.t. 1600 on the 12th and 13th days by a 1-hr monochromatic light pulse. The upper frequency range, ca. 7700 pulses/0.2 min, is equivalent to a body temperature of about 37°, and the arbitrary low-frequency cut-offs, at about 4000 pulses/0.2 min, a temperature of about 24°C. Room temperature 19°C. M=midnight; N=noon.

RESULTS AND DISCUSSION

We first determined the effectiveness of several irradiance levels of white light in shifting the phase of locomotor rhythms. Animals were entrained by an LD 12:12 cycle for 10 days, using "daylight" fluorescent lamps as the source of white light (WL), incident energy 600 $\mu W \cdot cm^{-2}$. Four hours after onset of darkness on the 11th day—i.e., at Z.t. 1600—exposure to 600, 12, or 0.3 $\mu W \cdot cm^{-2}$ WL, LD 12:12, began. Transients appeared by the second or third day of the new cycle at all three irradiance levels, and the animals were well-entrained by the fifth day at the 600 and 12 μW levels. The phase shift was not as well defined at the lowest irradiance level. When the rhythm was entrained to LD 12:12 (WL, irradiance ca. 10 $\mu W \cdot cm^{-2}$), a new regimen of LD 1:23 (WL, irradiance 6 $\mu W \cdot cm^{-2}$, first "dawn" at Z.t. 2100) initiated phase shifts within 2 days.

On the basis of these observations we designed an experiment (series I)
to indicate relative spectral efficiencies for phase shifting the body temperature rhythm. Our aim was to entrain mice by white light on an *LD
12*:12 regimen. They would then be exposed to a new cycle, *LD 1*:23,
first "dawn" at Z.t. 2100, with *L* consisting of different monochromatic
bands of light, in a protocol similar to that shown in Figure 2.

Caged mice were placed at various wavelength loci, from 275 to
750 nm, on the spectrograph focal curve, and kept under an *LD 12*:12
cycle for 11 days (WL, irradiance 10 μW·cm^{-2}). In the 12th cycle each
animal was exposed to monochromatic light, at an irradiance of 10
μW·cm^{-2}, for a period of 1 hr beginning at Z.t. 2100. This regimen continued for eight *LD 1*:23 cycles. Body temperature was recorded throughout the 20 days. The time in each cycle at which the arousal overshoot of
temperature occurred was determined from the plot of pulse frequency
versus time for each animal (Figure 3).

On the basis of the radiant energies of white light previously observed
to be effective in phasing motor activity, the consistent lack of entrainment of the temperature cycle is surprising, particularly under *LD 12*:12
(WL) at the 10 μW level. Of the 18 animals in this experiment for which
we have temperature records, 12 showed an apparently free-running
arousal rhythm in the 11 days of *LD 12*:12. The traces for animals number 5 and 11 (Figure 3), however, do show entrainment by the white light.
Probably several other animals (numbers 8, 9, and 10) entrained after
about a week, although this is not clearly demonstrated. Introduction of
a daily exposure to monochromatic light at Z.t. 2100 altered the period
significantly only at wavelengths 450, 500, and 525 nm. The mean arousal
time (± S.E.) of animal number 5 in the first 12 days is 4.5 ± 0.52; between days 13 to 20, it is 2.0 ± 0.42. The difference between these means
is highly significant ($p < 0.01$, Student's "t"). The response to the 450-nm
light pulse is a phase advance. The changes in τ following exposure to
monochromatic light for animals 7 and 8 are also statistically significant,
$p < 0.05$, suggesting that exposure to light at 500 and 525 nm is also effective in phasing. At the irradiance level used, the remaining monochromatic bands appear to be perceived as darkness: The arousal rhythm free
runs under those regimens.

In experimental series II, mice that had been under DD for three days
were placed on a regimen of *LD 12*:12 (WL), irradiance level about
60 μW·cm^{-2}. On day 7 the light was changed from white to monochromatic, irradiance about 12 μW·cm^{-2}; the new cycle was also *LD 12*:12,
but shifted 180°—i.e., monochromatic illumination began at Z.t. 1200.
Figure 4 shows the influence of this regimen on the time at which the

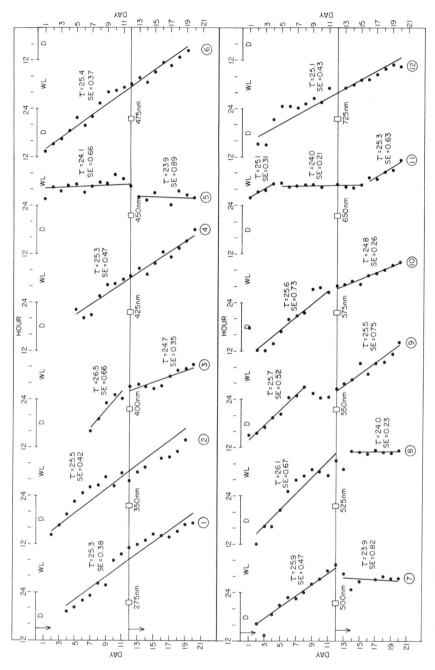

FIGURE 3 Arousal from torpor, as indexed by time of body temperature overshoot, under a regimen of *LD 12:12* (white light) for 11 days and then the cycle *LD 1:23* (monochromatic light) beginning on the 12th day at Z.t. 2100 (series I). Each circled number indicates the record of a separate animal.

368

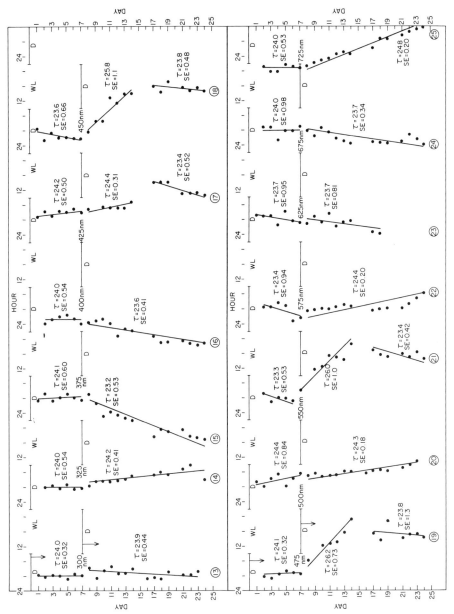

FIGURE 4 Effect of 180° reversal of photoperiod on the time of midpoint of the normothermic period. Animals were exposed to *LD 12:12* (white light); on day 7 the exposures were changed to *DL 12:12* (monochromatic light of the wavelengths shown) (series II).

369

midpoint of the normothermic period occurred. In no instance does the mean τ for the period before the 7th day differ significantly from 24 hr, and, perhaps more important, there is no significant monotonic drift of τ in this period. We infer that entrainment of the normothermic period took place in every animal under the LD *12*:12 (WL) photoperiod. With the exception of animal number 25, midpoints of the normothermic period occurred in the dark phase of the light cycle. Reversal of the photoperiodic sequence with monochromatic light at wavelengths 300, 325, 375, 400, 500, 575, 625, 675, and 725 nm allowed the previous trend in midpoint of normothermic temperature to free run with no shift in phase. On the other hand, shifts of this index to the new dark phase of the LD cycle took place at 450, 475, and 550 nm, and possibly also at 425 nm. Animal number 20, at the 500-nm locus, appeared recalcitrant.

These experimental data do not characterize adequately the spectral response for entrainment or for phase shift of the temperature rhythms of the pocket mouse. The tests are not replicated, and do not define energy-response relationships as a function of wavelength. From these limited observations, however, we do suggest that the most effective wavelengths for phase delay and perhaps for entrainment are not in the red spectral region, as we had anticipated, but are in the blue-green. In this respect, the photo-response of *P. penicillatus* appears to be closer to those of the invertebrate (Bruce and Minis, 1969; Frank and Zimmerman, 1969) than to those of the higher plant (Lörcher, 1958; Holdsworth, 1960; Wilkins, 1960; Bünning and Moser, 1966). A greater effectiveness of the blue-green light would be compatible with the proposition that the eye is the principal photoreceptive organ in the phasing of circadian rhythms of *Perognathus* by light, particularly in view of the predominant number (Karli, 1952) and function (Bonaventure, 1961) of rods (rhodopsin receptors) in the eye of the mouse.

ACKNOWLEDGMENTS

We are indebted to B. N. Jaroslow for the readout equipment, and to W. Eisler and D. LeBuis for their aid in its installation. We are also grateful to S. Tyler and M. Dipert for their assistance in programming and use of the computer, to E. Buess for data computation, and to R. Lindberg for his help and suggestions in the initiation of these experiments. This work was performed under the auspices of the United States Atomic Energy Commission.

REFERENCES

Bonaventure, N. 1961. Sur la sensibilité spectrale de l'appareil visuel chez la souris. Ct. Rd. Soc. Biol. 155:918–921.

Bruce, V. B., and D. G. Minis. 1969. Circadian clock action spectrum in a photoperiodic moth. Science 163:583–585.

Bünning, E., and I. Moser. 1966. Response-kurven bei der circadianen Rhythmic von *Phaseolus*. Planta 69:101–110.

Frank, K. D., and W. F. Zimmerman. 1969. Action spectra for phase shifts of a circadian rhythm in *Drosophila*. Science 163:688–689.

Gordon, S. A. 1964. Oxidative phosphorylation as a photomorphogenic control. Qt. Rev. Biol. 39:19–34.

Hayden, P. 1965. Free running period and phase shifts in two species of pocket mouse, p. 87 to 105. *In* R. G. Lindberg and J. J. Gambino [ed.], Investigation of *Perognathus* as an experimental organism for research in space biology. 1963 to 1964. Northrop Space Laboratories, Hawthorne, California.

Holdsworth, M. 1960. The spectral sensitivity of light-induced leaf movements. J. Exp. Bot. 11:40–44.

Jaroslow, B. N., and W. J. Eisler, Jr. 1968. Telemetry of hibernation, p. 103–105. *In* Argonne National Laboratory, BIM annual report. ANL-7535.

Karli, P. 1952. Retines sans cellules visuelles—recherches morphologiques, physiologiques, et physiopathologiques chez les rongeurs. Arch. Anat. Histol. Embryol. 35:1–76.

Kendall, M. G., and A. Stuart. 1966. The advanced theory of statistics. Vol. 3. Hafner, New York. 366 p.

Lindberg, R. G. [ed.]. 1968. Investigation of *Perognathus* as an experimental organism for research in space biology. Final Rep. Northrop Space Laboratories, Hawthorne, California.

Lindberg, R. G., G. J. De Buono, and M. M. Anderson. 1965. Animal temperature sensing for orbital studies on circadian rhythms. J. Spacecr. Rockets 2:986–988.

Lörcher, L. 1958. Die Wirkung verschiedener Lichtqualitäten auf die endogene Tagesrhythmic von *Phaseolus*. Z. Bot. 46:209–242.

Monk, G. S., and C. F. Ehret. 1956. Design and performance of a biological spectrograph. Radiat. Res. 5:88–106.

Sweeney, B. M. 1969. Rhythmic phenomena in plants. Academic Press, New York.

Wilkins, M. B. 1960. An endogenous rhythm in the rate of CO_2 output of *Bryophyllum*. II. The effects of light and darkness on the phase and period of the rhythm. J. Exp. Bot. 11:269–288.

A. D. LEES

The Relevance of Action
Spectra in the Study of
Insect Photoperiodism

In at least three species of insects—the aphid *Megoura viciae* (Lees, 1960, 1964), the silkmoth *Antheraea pernyi* (Williams and Adkisson, 1964), and the white cabbage butterfly *Pieris brassicae* (Claret, 1966)—photoperiodic responses are mediated by the direct action of light on the brain. Although this finding provides a useful pointer, it seems unlikely that spectroscopic examination of the light-absorbing pigments—either *in situ* or in brain extracts—will be feasible until the sites associated with the photoreceptive process have been identified more precisely. Since spectra for photoperiodic action cannot at present be matched against absorption curves, their usefulness in identifying the light acceptor is limited. Nevertheless, action spectra may well prove useful in another way. Photoperiodic time measurement in insects certainly involves a complex sequence of biochemical reactions, and changes in spectral sensitivity may be helpful in distinguishing and defining these steps. Before returning to this theme, I would like to comment on action spectra in general, with special reference to insect material.

The conditions that must be fulfilled if biological action spectra are to provide a reasonably accurate reflection of the properties of the absorbing system have often been enumerated (e.g., Loofbourow, 1948; Borthwick, Hendricks, and Parker, 1956; Blum, 1959). Even though the primary event is probably widely separated from the ultimate effect (hormonal control of diapause, polymorphism, etc.), it seems reasonable to assume provisionally that the end result is attributable to an initial photochemical change in a given number of acceptor molecules. On the other hand, it is impossible to decide whether the photochemical efficiency is independent of wavelength, as theory would demand. The extent of the attenuation of the radiation by scattering, reflection, or absorption by other nonacceptor substances is another imponderable. It should soon be possible, however, to correct for the differential filtering effect of certain structures overlying the chromophore (e.g., the cuticle), and it is doubtless a positive advantage that the path length is relatively short, especially in small insects.

If the primary photochemical events—as well as later processes concerned in the response—obey the Bunsen-Roscoe law of reciprocity, action spectra could be constructed from data that merely relate the total energy delivered to the biological effect. However, since it is now clear that considerable departures from reciprocity occur (except over very short time intervals), the duration of exposure for each test series of graded incident energies must be standardized. A valid action spectrum for one particular exposure time can then be obtained. Since the spectral sensitivity, as well as the intensity/time relationship, can change dramatically during the course of the cycle of illumination, the light exposure must be as short as possible. (The possibility of achieving this objective is, however, limited by the light energies available to the investigator, since reciprocity failures may be such that no response is given to short exposures even though the reciprocal energy levels are very high.) It follows also that the standard test periods of monochromatic radiation must be positioned at different points in the light–dark cycle to detect any differences in the action spectra. Preliminary work, using short exposures of white light, is obviously useful in defining the major differences in biological effect associated with such perturbations.

Some quantitative data on the wavelength sensitivity of diapause inhibition have been obtained in *Pieris rapae* (Barker, Cohen, and Mayer, 1964), in the leafhopper *Euscelis* (Müller, 1964), in the Colorado beetle *Leptinotarsa* (de Wilde and Bonga, 1958), and in the boll weevil *Anthonomus grandis* (Harris *et al.*, 1969). Also, approximate action spectra have been published for the red mite *Panonychus ulmi* (Lees, 1953) and much more detailed ones for the termination of diapause in the Lepidoptera *Antheraea pernyi* and *Laspeyresia pomonella* (Williams, Adkisson, and Walcott, 1965; Norris *et al.*, 1969). Unfortunately, in some instances the test exposures were unduly long and, if short, were applied only to one part of the light–dark cycle.

In recent work with the bean and vetch aphid *Megoura viciae*, I have attempted to avoid these limitations. In preliminary experiments, the periods of sensitivity relative to the light cycle as a whole have been charted by scanning the dark period with short interruptions of white light of moderate intensity (30 ft-c). Near-monochromatic light was then substituted for white light using exposures of like duration (usually 0.5 or 1.0 hr). Interference filters with an average band width at half peak transmission of 7 mμ were used, in conjunction with various tungsten filament sources. The photoperiodic response in this insect is maternal and controls the production of two types of female morph—the parthenogenetic viviparous form (virginoparae) and the egg-laying oviparae. The experimental regime was always applied during the first 8 days of the larval development of the mother, during which period approximately one third of the embryos contained in the developing parent become determined either as one morph or the other. The end-point of the assay is the energy level at which 50 percent of the parents produced virginoparous offspring at the beginning of their reproductive lives.

Previous work has shown that in *Megoura* the key factor in photoperiodic time measurement is the absolute length of the dark period (Lees, 1965, 1966). In any cycle comprising one light and one dark period, the former can be varied within wide limits without affecting the length of the critical dark period (close to 9.5 hr). Treatment with shorter night lengths yields parent aphids that produce only virginoparous daughters.

Night-interruption experiments, in which a dark period of rather more than critical length (10.5 hr) is scanned by 1-hr breaks of white light, show that there are two points of light sensitivity—one early (up to 3 hr after lights-off) and one much later (from 6 hr after lights-off up to the critical period). At hour 4, the insects appear to be totally insensitive to light, even of extremely high intensities (1 hr at 8,000 ft-c).

Although the overt effects of early and late night interruptions are the same—the promotion of virginopara-production—several lines of evidence indicate that the underlying biochemical pathways must be different. An early night break reverses the "dark reaction." This can be shown by interrupting the dark period after, say, 1.5 hr and following this treatment with a series of dark periods ranging from just below to just above the critical length. The dark period critical for the suppression of virginopara-production is, once again, 9.5 hr. Substantially longer dark periods have an identical effect. A cycle such as LD $8:1.5:1:12$ thus yields females that produce only oviparous daughters. The plot of response to monochromatic blue light against time of exposure also argues for the reversibility of the events taking place during the early part of the dark period. The irradiance required to produce a 50 percent response–approximately 0.2 μW cm^{-2} at 470 mμ for a 1-hr exposure–does not diminish when the duration of exposure is increased to 1.5 hr. This lack of reciprocity can be accounted for by assuming that after 1 hr the immediate back reaction is complete.

Interruptions in the late night produce entirely different effects. If a 1-hr light break, placed 7.5 hr after lights-off, is followed by a long dark period (e.g., LD 8:7.5:1:12), all the developing aphids will become virginopara-producers. In such regimes the terminal dark period (which can be made longer than 12 hr without influencing the result) is clearly in no sense "inductive" (of ovipara-producers). Moreover, the absence of a "critical length" shows that the 1-hr light break has not simply undone the events of the previous 7.5 hr, leaving the system ready to "measure" the subsequent dark period. At 7.5 hr therefore, the dark reaction appears to be irreversible. Other differences between early and late light breaks become evident when the energies required to produce the 50 percent response are measured for various exposure times. The response at 7.5 hr is more sensitive to short light interruptions than at 1.5 hr. The effectiveness of light when delivered at constant dose rates again increases disproportionately with time. The fact that it is still increasing after 1.5 hr indicates that the reaction associated with a late interruption is still incomplete after 1.5 hr. This is additional evidence for lack of reversibility.

The foregoing interpretation is reinforced by the action spectra for brief interruptions during the early and late night (Figures 1 and 2). Consider first a perturbation at hour 1.5: With the intensities available and a light exposure of 0.5 hr, the 50 percent response level was not achieved at any wavelength; with a 1-hr exposure, however, virginopara-production was strongly promoted, particularly in the blue region of the spectrum (Figure 1). Sensitivity is maximal at 450–470 mμ; there is a rapid fall in the green and very little sensitivity in the red. The precipitous decline in activity in the near UV is also noteworthy. The action spectrum for a 0.5-hr interruption placed 7.5 hr after the beginning of the dark period is entirely different (Figure 2). Maximum action is still in the 450–470 mμ region but the most striking feature is the considerable extension of sensitivity into the longer wavelengths. The action spectrum curve for a 1-hr exposure has much the same shape, although the departure from reciprocity is such that the threshold response level is reached with less than one fifth the incident energy.

These action spectra do not provide any decisive evidence on the nature of the chromophore, although the reversibility of the early light reaction could be interpreted as the dissociation of a protein-pigment complex. In view of the striking differences in the two spectra it also seems entirely possible that more than one pigment is involved. More definitely, the action spectra indicate that a sequence of at least four steps is involved in the dark timing reaction: (1) an early stage, reversible by blue light; (2) a light-insensitive stage; (3) a nonreversible, highly blue-red sensitive stage, that represents the natural long-day (or, more appropriately, the short-night) response; (4) a reaction that takes place when the uninterrupted dark period has reached critical length. At this point the product of the dark reaction again becomes light-insensitive.

The overall effect of the stage 4 reaction is, however, annulled (or perhaps concealed) if other cycles, either adjacent or more distant, include dark periods of 6–8 hr. A special case of this relationship is seen with

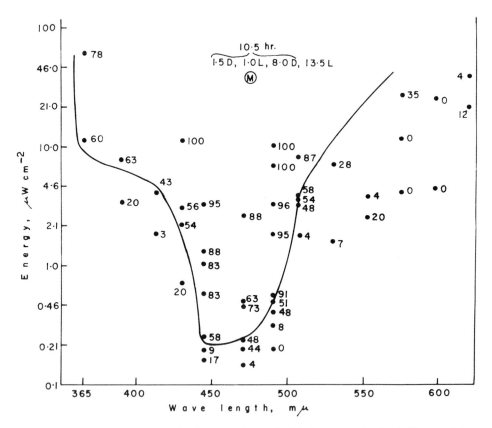

FIGURE 1 Action spectrum for the maternal control of virginopara-production in *Megoura viciae*. In this instance, near-monochromatic light (indicated by the symbol ⓜ) was applied in the early night, 1.5 hr after the beginning of a 10.5-hr dark phase. The freehand curve is drawn for incident energies at which approximately 50 percent of the parent aphids become virginopara-producers.

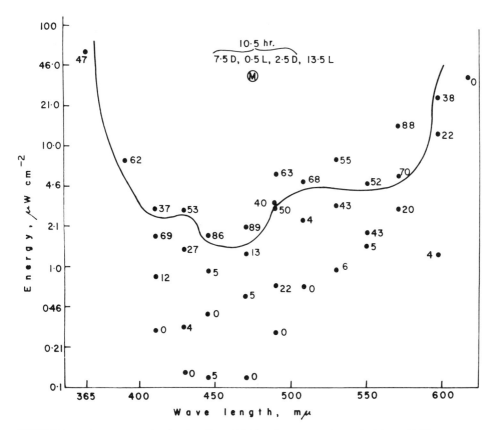

FIGURE 2   Action spectrum showing the effect of 0.5 hr of near-monochromatic light applied in the late night, 7.5 hr after the beginning of darkness.

sequences of "normal" short and long days, each consisting of only two components. It is very striking that when two or three short-night cycles (e.g., LD 16:8) are interpolated within a sequence of long-night cycles (e.g., LD 12:12) the effect is much greater than in the reverse situation. This suggests that the production of virginoparae is the positive event, perhaps mediated by the release of a morphogenetic hormone. If the interpretation outlined above is correct, the "dark reaction," when undisturbed by light, might be expected to result ultimately in the absence of the hormone.

It is understandable that cycles simulating the short night condition should also favor virginopara-production. These include regimes that incorporate a light break during the late night (Figure 3). Cycles involving an

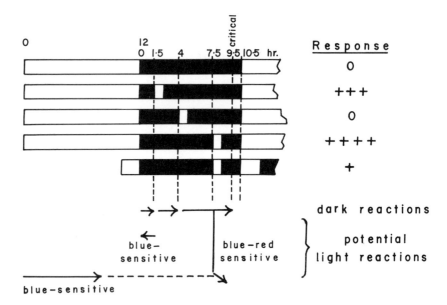

FIGURE 3 Diagram showing the minimum complement of light and dark reactions that appear to be involved in photoperiodic time measurement in *Megoura viciae*. Specific light cycles are shown as examples, together with the responses that reflect, in relative terms, the tendency of the parents to become virginopara-producers when exposed to these regimes during the early part of their larval development. The postulated sequences of dark and light reactions are shown below, the completion of each step being indicated by an arrowhead. The three photoresponses include a slow blue-sensitive reaction whose effects persist during the following dark period (interrupted line). This is required for the effective operation of a light break during the late night. [Added in proof: Recent work has also shown that ultra-short principal photoperiods cause the critical night length to increase.]

early light break are also promotive if the dark period following the break is reduced to less than the critical length (e.g., LD *12*:1.5:*1*:8). In this latter instance the 12-hr principal photoperiod also acts as a light break terminating the 8-hr dark period (see Figure 3). Nevertheless, the primary effect is clearly due to the 1-hr interruption, which re-sets the system so that dark period measurement starts from hour zero.

The light refractoriness that develops some 4 hr after the onset of darkness is puzzling. It is noteworthy, however, that if the night interruption is two or more hours long (e.g., LD *8*:4:*3*:12), a strong long-day response is induced. In other words the effect is the same as if a 1-hr night interruption is given 6 hr after lights-off, which suggests that the biochemical events that take place during this part of the "dark reaction" occur spontaneously, regardless of the light regime during hour 4.

Interruptions in the late night are relatively ineffective in causing virginopara-production unless they are preceded by a "main" photoperiod of at least 4–5 hr duration. The photoreaction of the "main" photoperiod also has an action spectrum. Maximum activity is again found at about 460 mμ with a sharp fall above 500 mμ. Indeed the curve resembles the 1-hr action spectrum for an interruption during the early night (Figure 1), although the time scale (4- or 8-hr exposures must be used) is greatly extended. The similarity of the curves may not be coincidental, since an early interruption of the dark period *becomes* a main photoperiod if it is prolonged.

The steps implicated in photoperiodic time measurement are summarized in Figure 3. It remains to be seen whether this type of hour-glass interval timer occurs in species other than *Megoura*. Although some features are frequently met with in insect photoperiodism (e.g., blue-sensitive photoresponses), it is becoming clear that considerable diversity is to be expected—possibly much more than in plants. Action spectrum studies will undoubtedly be of great assistance in elucidating the mechanisms involved.

REFERENCES

Barker, R. J., C. F. Cohen, and A. Mayer. 1964. Photoflashes: a potential new tool for control of insect populations. Science 145:1195–1197.

Blum, H. F. 1959. Carcinogenesis by ultraviolet light. Princeton Univ. Press, Princeton, New Jersey. 340 p.

Borthwick, H. A., S. B. Hendricks, and M. W. Parker. 1956. Photoperiodism, p. 479 to 517. *In* Radiation biology. Vol. 3. Academic Press, New York.

Claret, J. 1966. Mise en évidence du rôle photorécepteurs du cerveau dans l'induction de la diapause chez *Pieris brassicae* (Lepidoptera). Ann. Endocrinol. 27:311–320.

Harris, F. A., E. P. Lloyd, H. C. Lane, and E. C. Burt. 1969. Influence of light on diapause in the boll weevil. II. Dependence of diapause response on various bands of visible radiation and a broad band of infrared radiation used to extend the photoperiod. J. Econ. Entomol. 62:854–857.

Lees, A. D. 1953. Environmental factors controlling the evocation and termination of diapause in the fruit tree red spider mite *Metatetranychus ulmi* Koch (Acarina: Tetranychidae). Ann. Appl. Biol. 40:449–486.

Lees, A. D. 1960. Some aspects of animal photoperiodism. Cold Spring Harbor Symp. Quant. Biol. 25:261–268.

Lees, A. D. 1964. The location of the photoperiodic receptors in the aphid *Megoura viciae* Buckton. J. Exp. Biol. 41:119–133.

Lees, A. D. 1965. Is there a circadian component in the *Megoura* photoperiodic clock? p. 351 to 356. *In* J. Aschoff [ed.], Circadian clocks. North-Holland Publ. Co., Amsterdam.

Lees, A. D. 1966. Photoperiodic timing mechanisms in insects. Nature (Lond.) 210: 986–989.

Loofbourow, J. R. 1948. Effects of ultraviolet radiation on cells. VII. Growth. 12 (Suppl.): 77–149.

Müller, H. J. 1964. Über die Wirkung verschiedener Spektralbereiche bei der photo-periodischen Induktion der Saisonformen von *Euscelis plebejus* Fall. (*Homoptera*: *Jassidae*). Zool. Jahrb. Abt. Allg. Zool. Physiol. Tiere 70:411–426.

Norris, K. H., F. Howell, D. K. Hayes, V. E. Adler, W. N. Sullivan, and M. S. Schechter. 1969. The action spectrum for breaking diapause in the codling moth *Laspeyresia pomonella* (L.), and the oak silkworm, *Antheraea pernyi* Guer. Proc. Natl. Acad. Sci. U.S. 63:1120–1127.

Wilde, J. de, and H. Bonga. 1958. Observations on the threshold intensity and sensitiv-ity to different wavelengths of photoperiodic responses in the Colorado beetle (*Leptinotarsa decemlineata* Say.). Entomol. Exp. Appl. 1:301–307.

Williams, C. M., and P. L. Adkisson. 1964. Physiology of insect diapause. XIV. An endocrine mechanism for the photoperiodic control of pupal diapause in the oak silkworm, *Antheraea pernyi*. Biol. Bull. 127:511–525.

Williams, C. M., P. L. Adkisson, and C. Walcott. 1965. Physiology of insect diapause. XV. The transmission of photoperiodic signals to the brain of the oak silkworm, *Antheraea pernyi*. Biol. Bull. 128:497–507.

DISCUSSION

HILLMAN: Are you absolutely certain that none of your light effects is medi-ated by the plant tissue on which the aphids live? Do the aphids remain on the plants a long time after exposure?

LEES: Yes, they are there throughout the course of the treatment. However, I have also given the insect one photo-period and the leaves of the bean plant another. In this experiment, the host plants received short days but the aphids were temporarily removed from the plants at the end of the daily photoperiod and were given supple-mentary illumination. They then showed a long-day response. The effect is apparently directly and solely on the insects.

WILLIAM F. ZIMMERMAN
DONALD IVES

Some Photophysiological Aspects of Circadian Rhythmicity in *Drosophila*

Circadian rhythmicities in behavior and physiology are generally thought to reflect underlying circadian oscillations in biochemical processes. One approach to identification and localization of these oscillations is to trace the pathway of some stimulus known to affect the oscillator. Light is one such stimulus on which investigation seems particularly promising: It affects several outputs of the oscillation (presence of the rhythm and its phase, period, and amplitude), and all known photoreceptive phenomena involve specific conjugated molecules synthesized only by plants (carotenoids in vision, chlorophyll in photosynthesis, phytochrome in photoperiodism).

In this paper we report experiments whose purposes are (1) to elucidate further the phase-shifting effect of light on the circadian rhythm of adult emergence in *Drosophila* and (2) to determine the nature and location of the photopigment involved.

MATERIALS AND METHODS

Mass rearing techniques, fraction collectors, and monochromatic light projectors used in these experiments have been described previously

(Zimmerman, Pittendrigh, and Pavlidis, 1968; Frank and Zimmerman, 1969; Zimmerman, 1969). Two types of fraction collector were used: One type accepts all developmental stages surrounded by a puparium (prepupae, pupae, pharate adults) and allows them to be exposed on a flat surface to monochromatic light signals; the second type permits exposure of larvae to bright white light signals.

SPECTRAL SENSITIVITY OF THE *DROSOPHILA* EMERGENCE RHYTHM

The method of determining action spectra for phase-shifting the *Drosophila* rhythm is similar to that previously described (Figure 1; Frank and Zimmerman, 1969). Mixed populations of prepupae, pupae, and pharate adults are raised in *LD 12*:12 at 20°C, released into constant darkness (DD), and exposed to 15-min light signals 5 hr after the final dusk. In each experiment four populations were exposed to monochromatic light, one population was exposed to 1100 lumens/m² white fluorescent light (the "bright white" control), and one population was not exposed to a light signal (the "free-run" control). Action spectra were obtained by determining the relative number of quanta at wavelengths between 354 and 800 nm required to generate a phase shift ($\Delta\phi$) equal to about 50 percent of the saturating $\Delta\phi$ generated by the white light signal on the same day (see Figure 1).

Because the circadian oscillation is independent of developmental rate (Skopik and Pittendrigh, 1967), we assumed that the majority of flies in each emergence peak were within 24 hr of developmental synchrony when exposed to the light signal.

SPECTRAL SENSITIVITY OF THE PHOTORECEPTORS IN THE COMPOUND EYES OF *DROSOPHILA*

Spectral sensitivity of compound eye photoreceptors was determined by retinal action potential recordings. Recently emerged flies were mounted at the tip of a toothpick by immersing the lower head and thorax in a small drop of water-soluble casein glue. The corneal electrode consisted of a fine cotton wick protruding from a glass capillary tube filled with physiological saline. The reference (indifferent) electrode, another moist cotton wick, was in contact with the thorax. Reproducible recordings could be obtained for up to 5 hr (see Hotta and Benzer, 1969; Pak, Grossfield, and White, 1969). The equipment used to illuminate and record from insect eyes is described in Goldsmith (1965) and in Mote and Goldsmith (1970).

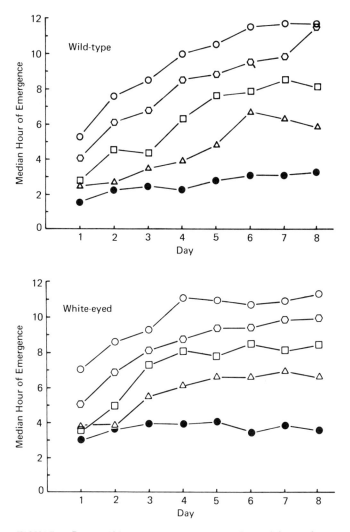

FIGURE 1   Phase-shifting the circadian rhythm in *Drosophila pseudo-obscura* by light. Shown are the median hours of adult emergence in fly populations grown in LD *12*:12 and released into DD; Day 1 is the first day of DD. One population was not given a light signal (closed circles), and the other populations were given 15-min light signals five hours after the final dusk (night of Day 0). Open circles indicate population exposed to white fluorescent light (1100 lumens/m^2). Remaining symbols indicate populations exposed to light of wavelength 456 nm; intensities in log quanta sec^{-1} cm^{-2} are 11.34 (triangles), 11.77 (squares), and 12.05 (hexagons).

Action spectra were obtained by determining the relative number of quanta at wavelengths between 370 and 700 nm required to elicit a corneal negative response of 5 millivolts in the dark-adapted eye; this sustained negative wave represents the electrical response of the photo-receptor (retinular) cells (Eichenbaum and Goldsmith, 1968). Light flashes of between 0.5 and 1.0 sec were sufficient to elicit and maintain the receptor response.

RESULTS AND DISCUSSION

LIGHT EFFECTS ON THE CIRCADIAN OSCILLATION IN *DROSOPHILA PSEUDOOBSCURA* LARVAE

Several previous publications on *Drosophila* emergence have for convenience referred to the experimental objects as "pupae" (Pittendrigh, 1960; Skopik and Pittendrigh, 1967; Zimmerman, Pittendrigh, and Pavlidis, 1968), but, of course, the rearing and flotation harvesting technique yields all developmental stages surrounded by a puparium—prepupae, pupae, and pharate adults. The responses to light of the circadian rhythm in these developmental stages are essentially identical (Pittendrigh, 1967; Skopik and Pittendrigh, 1967), but whether the circadian oscillation that gates adult emergence is also present in larvae, and whether its responses to light are the same as in later stages has not been reported.

The experiments summarized in Figure 2 demonstrate entrainment to an *L*D cycle in *Drosophila pseudoobscura* larvae and characterize the larval phase response curve. The upper graph shows the distribution of adult emergences in six populations of flies exposed to *LD 12*:12 cycles as larvae: Parent flies laid eggs for 8 days in an *LD 12*:12 cycle at constant 20°C; the parents were then removed and the larvae left in DD. Subsequent adult emergence exhibits a circadian rhythm phased by the *L*D cycle last "seen" by the larvae. Brett (1955) reached the same conclusion for *D. melanogaster* larvae from similar experiments.

The phase response curve of the larval circadian oscillation is shown in the bottom part of Figure 2. Parent flies laid eggs for 8 days in the fraction collectors, the *L*D *12*:12 cycle was discontinued on the seventh day, and at various times during the first free-running (DD) cycle, 15-min white fluorescent light signals were given. In each experimental series five populations were given light pulses (experimentals), and one population was not (free-run control). The resultant $\Delta\phi$'s in the rhythm of adult emergence (phase of experimental minus phase of control for each day) are plotted as

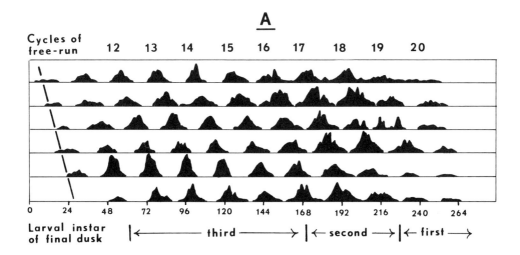

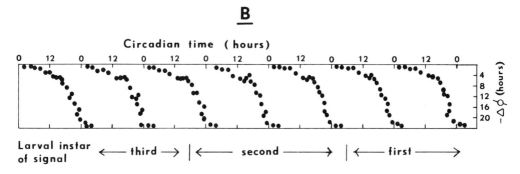

FIGURE 2   Light responses of the circadian rhythm in *Drosophila pseudoobscura* larvae. Upper figure (*A*) shows the distribution of adult emergences in populations of flies exposed as larvae to an *LD 12*:12 cycle, then transferred to DD; the slanted solid line is extrapolated dawn of the previous *LD* cycle. As τ for *Drosophila* is very close to 24 hr, "Cycles of free-run" refer to the number of days in DD. Lower figure (*B*) shows the Δφ generated by 15-min white fluorescent light signals applied to larvae at various phase points (circadian times) during the first free-running (DD) cycle, after previous entrainment to an *LD 12*:12 cycle. The resultant Δφ's are plotted as delays (see text). In both experiments, additional populations were sampled at 12-hr intervals for the distribution of developmental stages, and from this information the larval stage at which each emerging peak of adults was exposed to the light signal was estimated.

delays. The form and phase of the response curve are clearly the same in larvae as in later developmental stages (Pittendrigh, 1967; Skopik and Pittendrigh, 1967). This is strong evidence for invariance of the properties of the circadian oscillation throughout the life cycle of *Drosophila*.

SPECTRAL SENSITIVITY OF THE COMPOUND EYE PHOTORECEPTORS AND CIRCADIAN RHYTHM IN *DROSOPHILA*

A comparison of action spectra for a standard $\Delta\phi$ response of the *D. pseudoobscura* eclosion rhythm (Figures 2 and 3*A*) and a standard electrical response of the photoreceptor cells of the eye (Figure 3*B*) strongly suggests that the two light effects are mediated by different photoreceptive pigments: Whereas the rhythm is insensitive to light of wavelengths greater than 570 nm, the range of visual sensitivity extends up to 666 nm. The same comparison in *Drosophila melanogaster* yielded the same results (Zimmerman and Ives, 1969, unpublished experiments). Circadian rhythms

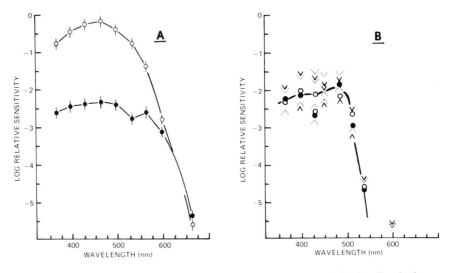

FIGURE 3   Spectral sensitivity of photoreceptor cells of the adult (*A*) and the circadian rhythm (*B*) in white-eyed (open circles, dotted arrows) and wild-type (closed circles, solid arrows) *D. pseudoobscura*. Ordinate: log reciprocal of the relative number of quanta required for a constant response (photoreceptive cells in compound eye: 5 millivolt corneal negative wave; rhythm: $\Delta\phi$ on Day 2). In *A*, the means (circles) and standard errors (vertical lines) are based on 9 white-eyed flies and 8 wild-type flies. In *B*, circles denote light energies giving 40 to 60 percent $\Delta\phi$ on Day 2, upward pointing arrows more than 60 percent $\Delta\phi$, and downward pointing arrows less than 40 percent $\Delta\phi$. The rhythm action spectra are based on flies that had fully pigmented eyes at the time of the light signal (Zimmerman and Ives, 1969, unpublished experiments).

in other organisms are also insensitive to wavelengths greater than 570 nm (*Neurospora*, Sargent and Briggs, 1967; *Pectinophora*, Bruce and Minis, 1969), while visual sensitivity in other organisms extends beyond 570 nm (insects: Goldsmith, 1964; Goldsmith and Fernandez, 1968). The action spectra of the *Drosophila* rhythm do not resemble action spectra for the separate UV and green receptor cells in insect eyes (Autrum and Kolb, 1968; Mote and Goldsmith, 1970).

The action spectra for pharate adults and early pupae are identical (Zimmerman and Ives, 1969, unpublished experiments), and this is further evidence against the involvement of visual pigments in mediating light-induced phase-shifts of the rhythm in any post-pupation stages. The argument for distinct photoreceptive pigments for these two light responses does not, however, imply that the photoreceptor of the circadian system could not also be a carotenoid-protein complex, as the visual pigment is known to be (Goldsmith and Fernandez, 1966).

LOCATION OF THE PHOTORECEPTOR OF THE *DROSOPHILA* RHYTHM

We have attempted to locate the rhythm photoreceptor in *Drosophila* by comparing the $\Delta\phi$ generated by dim monochromatic light applied to small areas of the pupae. Populations of pupae are glued, abdomen down, carefully painted with opaque black paint, and exposed to almost identical light signals. The only conclusion justified by our results to date is that during the first 5 days after pupation, the photoreceptor is located in the anterior end (Figure 4); light impinging on the anterior half generates as much $\Delta\phi$ as light impinging upon the entire organism, and no $\Delta\phi$ is generated by light impinging only upon the posterior half. Kalmus (1938) reached the same conclusion from experiments with cardboard masks, in which he showed that the eclosion rhythm in *Drosophila* entrains to *LD* cycles "seen" by the anterior half.

Although there is no conclusive evidence bearing on this question, there are several reasons for thinking that phase-shifts of the circadian rhythm in *Drosophila* might be mediated by direct light effects on nerve cells in the brain:

● The compound eyes do not seem to be involved, even when they have differentiated. Engelmann and Honegger (1966) found that the circadian rhythm of the *Drosophila melanogaster* mutant *sine oculis* (which lacks compound eyes and ocelli) entrains normally to an *LD 12*:12 (300 lux) cycle; we have confirmed this observation and have also shown that the response of the rhythm in *sine oculis* to dim monochromatic light

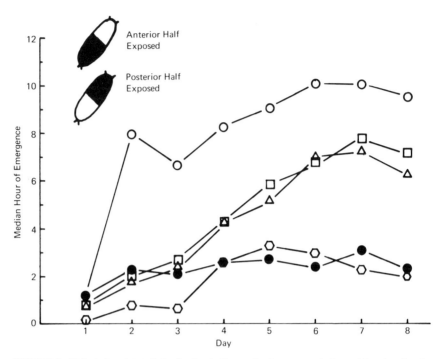

FIGURE 4   Light-induced $\Delta\phi$ of the rhythm in *D. pseudoobscura* populations with painted and unpainted puparia. Shown are the median hours of adult emergence in fly populations grown in *LD 12*:12 and released into DD; Day 1 is the first day of DD. One population was not given a light signal (closed circles) and the other populations were given 15-min light signals 17 hr after the time when dawn would have come on Day 1. Open circles: unpainted, white fluorescent light at 1100 lumens/m² ; open squares: anterior half exposed, light of 456 nm at an intensity of 11.34; open triangles: unpainted, 456 nm at 11.49; open hexagons: posterior half exposed, 456 nm at 11.33. Intensities are given in log quanta sec⁻¹ cm⁻². Control experiments showed no effect of the black paint on the phase or period of the free-running rhythm (Zimmerman and Ives, 1969, unpublished experiments).

signals is the same as that of wild type (Zimmerman and Ives, 1969, unpublished experiments). In addition, the data in Figure 3 argue against the involvement of a photoreceptor in the compound eyes: Whereas the compound eyes of the white-eyed mutant of *D. pseudoobscura* are 2 log units more sensitive to light than those of the wild type because of the absence of screening pigments (see Hengstenberg and Götz, 1967; Goldsmith and Fernandez, 1968), there is no such difference between white-eyed and wild type flies with respect to sensitivity of the circadian rhythm to light.

• Entrainment of circadian rhythms by light in two classes of vertebrates (Amphibia and Aves) is mediated by extraoptic pathways, and direct light effects on brain nerve cells are implicated (Menaker, 1968; Adler,

1969). Direct light effects on nerve cells are also implicated in photoperiodic control of differentiation in aphids (Lees, 1964) and testis growth in sparrows (Menaker and Keatts, 1968; Underwood and Menaker, 1970). A circadian rhythm photoreceptor need only detect changes in relative light intensity, and such changes can directly influence the electrical activity of nerve cells not obviously differentiated for photoreception (Kennedy, 1958a, b; Arvanataki and Chalazonitis, 1961).

• The 15-min white light response curve remains invariant throughout development, and we might thus expect the photoreceptor to be located in some structure that remains relatively invariant throughout development. The brain is one such structure.

• There is strong evidence that in insects the oscillation underlying circadian rhythmicity is located in the brain (Nishiitsutsuji-Uwo and Pittendrigh, 1968b; Truman, this volume; Truman and Riddiford, 1970); the latter two authors, using elegant brain transfer techniques, have also adduced evidence implicating the brain as the location of the photoreceptor. Proximity of the circadian oscillation and its photoreceptor would seem especially advantageous in an organism that undergoes complete metamorphosis.

Nishiitsutsuji-Uwo and Pittendrigh (1968a) concluded on the basis of painting experiments in the cockroach that the photoreceptor is located in the compound eyes; however, as Pittendrigh (personal cummunication) notes, they used bright white (saturating) light signals that might have penetrated through the eyes to a photoreceptor deeper in the head.

SUMMARY

In *Drosophila*, then, the response of the circadian system to white light signals is invariant with developmental stage, and the photoreceptive pigment involved is probably
  • The same in all postpupation stages,
  • Not the same one involved in vision,
  • Not located in the compound eyes or ocelli, and
  • Located in the brain.

ACKNOWLEDGMENTS

The experiments reported here were made possible by NSF Grant GB 8303 and by an NIH Special Fellowship (1969–1970) awarded to W. F. Z. We

thank Th. Dobzhansky (Rockefeller University) and R. Konopka and
S. Benzer (California Institute of Technology) for providing *D. pseudoobscura* and *D. melanogaster* mutants. The senior author is also indebted to
T. H. Goldsmith (Yale University) and Dr. M. Mote (Temple University)
for numerous informative discussions and instruction in the use of neurophysiological equipment at T. H. G.'s laboratory and to C. S. Pittendrigh
for several critical suggestions. Some of the experiments reported here
were carried out by L. Lander and illustrated by C. McKnight.

REFERENCES

Adler, K. 1969. Extraoptic phase shifting of circadian locomotor rhythm in salaman-
ders. Science 164:1290–1292.
Arvanataki, A., and N. Chalazonitis. 1961. Excitatory and inhibitory processes ini-
tiated by light and infrared radiations in single identifiable nerve cells (giant ganglion
cells of *Aplysia*), p. 194–234. *In* E. Florey [ed.], Nervous inhibition. Pergamon
Press, Oxford.
Autrum, H., and G. Kolb. 1968. Spektrale Empfindlichkeit einzelner Sehzellen der
Aeschnidan. Z. Vgl. Physiol. 60:450–477.
Brett, W. J. 1955. Persistent diurnal rhythmicity in *Drosophila* emergence. Ann.
Entomol. Soc. Am. 48:119–131.
Bruce, V. G., and D. H. Minis. 1969. Circadian clock action spectrum in a photoperi-
odic moth. Science 163:583–585.
Eichenbaum, D. M., and T. H. Goldsmith. 1968. Properties of intact photoreceptor
cells lacking synapses. J. Exp. Zool. 169:15–32.
Engelmann, W., and H. W. Honegger. 1966. Tagesperiodischer Schüpfrhythmik einer
augenlosen *Drosophila melanogaster*-Mutante. Z. Naturforsch. B. 22:1–2.
Frank, K. D., and W. F. Zimmerman. 1969. Action spectra for phase shifts of a circa-
dian rhythm in *Drosophila*. Science 163:688–689.
Goldsmith, T. H. 1964. The visual system in insects, p. 397 to 462. *In* M. Rockstein
[ed.], The physiology of insects. Vol. I. Academic Press, New York.
Goldsmith, T. H. 1965. Do flies have a red receptor? J. Gen. Physiol. 49:265–287.
Goldsmith, T. H., and H. Fernandez. 1966. Some photochemical and physiological
aspects of visual excitation in compound eyes, p. 125 to 143. *In* C. G. Bernard [ed.],
The functional organization of the compound eye. Pergamon Press, Oxford.
Goldsmith, T. H., and H. Fernandez. 1968. The sensitivity of housefly photoreceptors
in the mid-ultraviolet and the limits of the visible spectrum. J. Exp. Biol. 49:669–
677.
Hengstenberg, R., and K. G. Götz. 1967. Der Einfluss des Schirmpigmentgehalts auf
die Helligkeits- und Kontrastwarnehmung bei *Drosophila*-Augenmutanten. Kyber-
netik 3:276–285.
Hotta, Y., and S. Benzer. 1969. Abnormal electroretinograms in visual mutants of
Drosophila. Nature (Lond.) 222:347–351.
Kalmus, H. 1938. Die Lage des Aufnahmeorgans für die Schlupfperiodik von *Drosophila*.
Z. Vgl. Physiol. 26:362–365.
Kennedy, D. 1958a. Electrical activity of a "primitive" photoreceptor. A. N.Y. Acad.
Sci. 74:329–336.
Kennedy, D. 1958b. Nerve response to light in mollusks. Biol. Bull. 115:338.

Lees, A. D. 1964. The location of the photoperiodic receptors in the aphid *Megoura viciae* Buckton. J. Exp. Biol. 41:119–133.

Menaker, M. 1968. Extraretinal light perception in the sparrow. I. Entrainment of the biological clock. Proc. Natl. Acad. Sci. U.S. 59:414–421.

Menaker, M., and H. Keatts. 1968. Extraretinal light perception in the sparrow. II. Photoperiodic stimulation of testis growth. Proc. Natl. Acad. Sci. U.S. 60:146–151.

Mote, M., and T. H. Goldsmith. 1970. Spectral sensitivities of color receptors in the compound eye of the cockroach *Periplaneta*. J. Exp. Zool. 173:137–146.

Nishiitsutsuji-Uwo, J., and C. S. Pittendrigh. 1968a. Central nervous system control of circadian rhythmicity in the cockroach. II. The pathway of light signals that entrain the rhythm. Z. Vgl. Physiol. 58:1–13.

Nishiitsutsuji-Uwo, J., and C. S. Pittendrigh. 1968b. Central nervous system control of circadian rhythmicity in the cockroach. The optic lobes, locus of the oscillation? Z. Vgl. Physiol. 58:14–46.

Pak, W. L., J. Grossfield, and N. V. White. 1969. Nonphototactic mutants in a study of vision of *Drosophila*. Nature (Lond.) 222:351–354.

Pittendrigh, C. S. 1960. Circadian rhythms and the circadian organization of living systems. Cold Spring Harbor Symp. Quant. Biol. 25:159–184.

Pittendrigh, C. S. 1967. Circadian systems. I. The driving oscillation and its assay in *Drosophila pseudoobscura*. Proc. Natl. Acad. Sci. U.S. 58:1762–1767.

Sargent, M. L., and W. Briggs. 1967. The effects of light on the circadian rhythm of conidiation in *Neurospora*. Plant Physiol. 42:1504–1510.

Skopik, S. D., and C. S. Pittendrigh. 1967. Circadian systems. II. The oscillation in the individual *Drosophila* pupa; its independence of developmental stage. Proc. Natl. Acad. Sci. U.S. 58:1862–1869.

Truman, J. W., and L. M. Riddiford. 1970. Neuroendocrine control of ecdysis in silkmoths. Science 167:1624–1626.

Underwood, H., and M. Menaker. 1970. Photoperiodically significant photoreception in sparrows: Is the retina involved? Science 167:298–301.

Zimmerman, W. F. 1969. On the absence of circadian rhythmicity in *Drosophila pseudoobscura* pupae. Biol. Bull. 136:494–500.

Zimmerman, W. F., C. S. Pittendrigh, and T. Pavlidis. 1968. Temperature compensation of the circadian oscillation in *Drosophila pseudoobscura* and its entrainment by temperature cycles. J. Insect Physiol. 14:669–684.

DORA K. HAYES

Action Spectra for Breaking Diapause and Absorption Spectra of Insect Brain Tissue

Diapause can be terminated in pupae of the oak silkworm, *Antheraea pernyi* Guérin-Menéville, by exposing the pupae to a relatively long photoperiod (Williams and Adkisson, 1964). Williams, Adkisson, and Walcotts (1965) also found that the most effective wavelengths in the visible spectrum for termination of diapause were below 560 nm. We have determined the action spectra for breaking diapause in oak silkworm pupae and codling moth larvae, *Laspeyresia pomonella* (L.) (Hayes, Schechter, and Sullivan, 1968; Norris *et al.*, 1969) and the absorption spectra of whole brain tissues and brain slices. These investigations were undertaken to determine the mechanisms involved in the light-mediated breaking of diapause. Such information may provide new tools for insect control in the future as alternatives or supplements to pesticides.

ACTION SPECTRUM STUDIES

The spectrograph used to determine the action spectra and the techniques used for rearing, holding, and exposing the insects to different wavelengths of visible light have been described previously (Norris, 1968; Norris *et al.*, 1969). The spectrograph consisted of a quartz-iodine lamp with a wedge interference filter modified with two matched interference filters for one of the experiments; these filters gave a 10–13 nm band pass. A lens was used to project a 400–700 nm spectrum horizontally onto a holder containing the test insects. A graded neutral density filter was placed over the interference filter to produce a vertical intensity gradient. The entire holder was set in a constant temperature box [biological oxygen demand (BOD) type] in which the relative humidity was maintained between 55–70 percent. The beam of light was projected through a double window in a wooden door specially fitted to the box. Codling moths were maintained at $21 \pm 1.5°C$ and oak silkworms at $26 \pm 1.5°C$.

Three weeks before they were exposed to the spectrum, oak silkworm pupae, shipped from Japan, were removed from their cocoons and placed in the holder used in the experiment. The holder contained 136 insects, arranged with 8 insects in each of 17 vertical rows. The diapausing codling moth larvae (supplied by the Arid Areas Deciduous Fruit Insects Investigations, Entomology Res. Div., Yakima, Washington) were removed from the corrugated paper strips in which they had been shipped and placed in a holder (1,600 larvae in 32 vertical rows of 50 each). Two experiments were conducted on oak silkworm pupae and one on codling moth larvae. Before the insects were placed in the test chamber they were maintained in the holder under a regimen of 8 hr of light and 16 hr of darkness (LD 8:16) for at least 3 weeks at the same temperature as that of the experiment. During the experiment, the insects were exposed to 10 hr of white light (fluorescent) followed by 6 hr of monochromatic light; the number of insects breaking diapause at each wavelength were counted daily. Mortality was less than 10 percent in both species of insects. Two percent of short-day controls for codling moths emerged, while 92 percent of long-day controls emerged. Short-day controls for oak silkworms showed no emergence in both experiments, while long-day controls showed emergence frequencies of 100 percent and 96 percent in the two tests.

The most effective wavelengths for breaking diapause lie between 400 and 500 nm for both species of insects (Figures 1 and 2). Results from the two experiments on the oak silkworms are indicated by different symbols (Figure 1). We have not presented the confidence limits for action spectrum for the oak silkworm since only 8 insects were used to obtain each of the data points in these experiments. Since 50 insects were used for each data point obtained with the codling moth, confidence limits could be ascertained. The energy required to trigger the breaking of diapause of the oak silkworm decreases monotonously from 500 to 400 nm, while a peak at about 450 nm is observed in the codling moth response. Some emergence occurred at wavelengths above 600 nm, but never in as high a frequency as 50 percent.

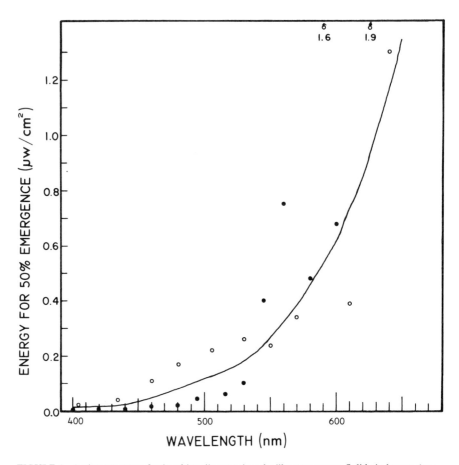

FIGURE 1 Action spectrum for breaking diapause in oak silkworm pupae. Solid circles are data from experiment 1 with a single filter spectrograph (20–25 nm band pass). Open circles are data from experiment 2 with a double filter spectrograph (10–13 nm band pass). (Reprinted with permission from Norris *et al.*, 1969.)

We have also determined that the amounts of sunlight that would reach each species of insect either in the cocoon or under a small piece of bark under natural conditions are adequate to break diapause (Norris *et al.*, 1969).

THE SITE OF PHOTORECEPTION

Williams and Adkisson (1964), Williams, Adkisson, and Walcotts (1965), and Williams (1969) have indicated that the primary target for light in the diapause response of the oak silkworm is the brain. Lees (1964) has shown that the brain is responsible for the photoreception involved in the regulation of sexual and asexual stages of the aphid, *Megoura viciae*. In *Antheraea pernyi* the dark purplish pigmented areas in the optic lobes located laterally in the pupal brain (Figure 3) are obvious candidates for the photoreceptive site, although further work by Williams (1969) indicates that the optic lobes as a whole are not as important as the central portion of the brain in the photoperiodic response in insects.

As a further check in our own laboratory, the dark spots were removed surgically from the brains of 38 pupae. Fifty control pupae were sham-

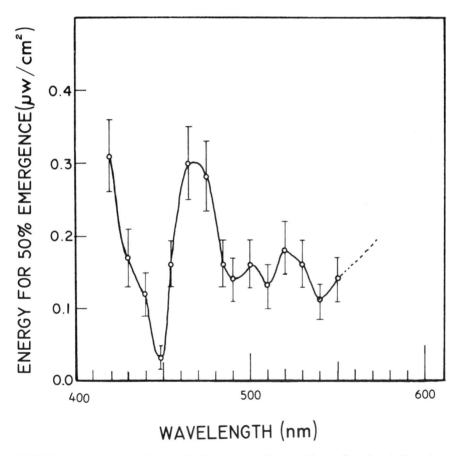

FIGURE 2   Action spectrum for breaking diapause in codling moth larvae. Error bars indicate the 95 percent confidence limits of the 50 percent response calculated by means of a moving average incorporating 5 intensity levels along the vertical intensity gradient. (Reprinted with permission from Norris *et al.*, 1969.)

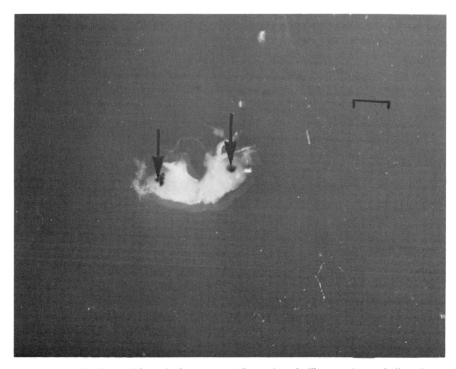

FIGURE 3   Brain dissected from *Antheraea pernyi* Guer., the oak silkworm. Arrows indicate location of dark spots. Bar is 1 mm in length.

operated. The rate of emergence was determined during exposure to *LD 16*:8; appropriate unoperated control animals were tested. Although a higher mortality was observed in the group from which the dark spots were removed, all insects that survived the surgery emerged. These results demonstrate that the dark spots are not necessary for the reception of light involved in the breaking of diapause.

We wished, then, to determine whether the ocelli of diapausing codling moth larvae were important in light reception during the breaking of diapause. We covered the ocelli of 19 larvae with wax mixed with lampblack; 22 larvae served as controls. We observed little difference in the rate of breaking diapause in the two groups of larvae exposed to *LD 16*:8, which breaks diapause in this species. This result is not final proof that the ocelli are not involved, since the semitransparent tissue of the insect could serve as a "light pipe" and receptors in the ocelli might thus be stimulated, but the results do indicate that the ocelli are probably not directly involved in the breaking of diapause.

FIGURE 4 Absorption spectrum of dark spots from lateral portion of oak silkworm brain.

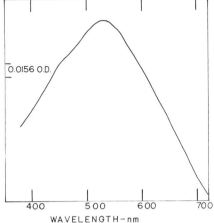

ABSORPTION SPECTRA OF POTENTIAL RECEPTOR TISSUE

A direct approach to the characterization of the receptor pigment is the determination of the absorption spectra of whole tissues or portions of tissue that are possible receptor sites. Brain tissues from both oak silkworms and codling moths were therefore examined. Absorption spectra of squashes of oak silkworm brains on microscope slides were obtained either with the microscope attachment on a Shimadzu QV50 Spectrophotometer, which was adjusted to measure the absorption of the material in a circular area 5 or 20μ in diameter, or with a microscope spectrophotometer designed by Norris (unpublished).

Using the microscope spectrophotometer we determined the absorption spectrum of a circular area 5μ in diameter of one dark spot from the optic lobe of the brain (Figure 4) of the oak silkworm. This spectrum, with the major peak at 535–545 and a small shoulder between 400 and 500 nm, does not correspond to the action spectrum obtained with these insects. From the work of Williams (1969), from our experiments, and from this absorption spectrum we conclude that these spots are not involved in the diapause-breaking response.

Small orange spots about 15–30μ in diameter were also observed distributed in a random fashion on the surface of many of the oak silkworm brains. The chemical nature and the function of these spots are unknown; their random arrangement suggests that their association with the brain may be fortuitous. The absorption spectrum of a circular area 20μ in diameter of one of these spots (Figure 5) was determined using the micro-

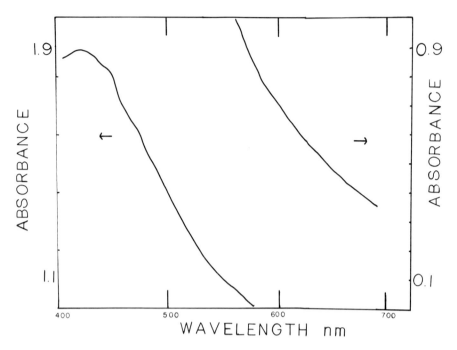

FIGURE 5   Absorption spectrum of orange spots randomly distributed on the surface of oak silk-worm brains.

scope spectrophotometer. Its absorption spectrum is somewhat similar to the action spectrum. No definite information is available concerning the possible role of these spots in photoreception.

The multipurpose spectrophotometer designed by Norris (Norris and Butler, 1961) was also used to obtain spectra of whole brains and brain slices. After dissection, the tissues were dipped into a 0.25 M solution of sucrose saturated with 1-phenyl-2-thiourea to prevent browning. Exposure to atmospheric oxygen was minimized by bubbling nitrogen into the solution in which the tissues were held after dissection. They were stored in dry ice until spectra were determined. At the time spectra were to be obtained, the tissues were thawed, placed in the well of a hanging drop slide, and kept moist. When necessary, blanks were used to correct for scatter effects. The spectrum obtained from larval codling moth brains is illustrated in Figure 6, and the spectrum from the intact midportion of diapausing oak silkworm pupal brains in Figure 7. Although there is undoubtedly present in these preparations absorbing material not involved in the photo-

FIGURE 6   Absorption spectrum of whole, isolated, washed codling moth brains. A, before reduction with sodium borohydride; B, after reduction with sodium borohydride.

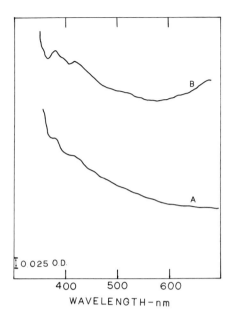

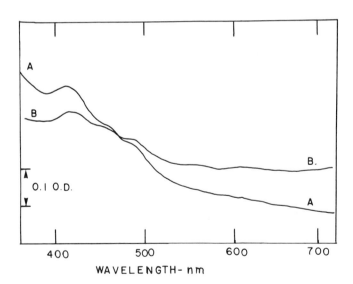

FIGURE 7   Absorption spectrum of the central portion of washed *Antheraea pernyi* brains. A, before reduction with sodium borohydride; B, after reduction with sodium borohydride.

periodic response, their absorbance between 400–500 nm is compatible with the action spectra obtained from the insects.

Biologically active pigments that have to be considered in light-mediated breaking of diapause include flavoproteins, hemoproteins, pteridines, corrinoids, bile-pigment proteins, and carotenoproteins and related rhodopsin-like materials (Hayes, Schechter, and Sullivan, 1968; Norris *et al.*, 1969). The maxima of the spectra illustrated in Figures 6 and 7 have not as yet been attributed to a particular pigment or class of pigments, but a combination of hemoprotein(s) and carotenoids or flavoprotein(s) could produce the peaks in the 400–500 nm region in the tissues before and after reduction with sodium borohydride. A broad absorption band is observed above 600 nm after treatment with sodium borohydride; there were also indications of diapause-breaking effects of light beyond 550 nm in the action spectra. More definitive identification awaits purification on a micro or submicro scale.

ACKNOWLEDGMENTS

The work reported in this paper is the result of the combined efforts of K. H. Norris, Market Quality Research Division, and M. S. Schechter, W. N. Sullivan, V. E. Adler, F. Howell, and the author, of the Entomology Research Division, Agricultural Research Service, U.S. Department of Agriculture. I should like to thank S. B. Hendricks for his advice during the course of this work.

REFERENCES

Hayes, D. K., M. S. Schechter, and W. N. Sullivan. 1968. A biochemical look at insect diapause. Bull. Entomol. Soc. Am. 14:108–111.
Lees, A. D. 1964. The location of the photoperiod receptors in the aphid *Megoura viciae* Buckton. J. Exp. Biol. 41:119–133.
Norris, K. H. 1968. A spectrograph for action-spectra studies in the 400- to 800-nm region. Trans. Am. Soc. Agric. Eng. 11:407–408.
Norris, K. H., and W. L. Butler. 1961. Techniques for obtaining absorption spectra on intact biological samples. IRE Trans. Bio-Med. Electron. 8:153–157.
Norris, K. H., F. Howell, D. K. Hayes, V. E. Adler, W. N. Sullivan, and M. S. Schechter. 1969. The action spectrum for breaking diapause in the codling moth, *Laspeyresia pomonella* (L.) and the oak silkworm, *Antheraea pernyi* Guer. Proc. Natl. Acad. Sci. U.S. 63(4):1120–1127.
Williams, C. M. 1969. Photoperiodism and the endocrine aspects of insect diapause. Symp. Soc. Exp. Biol. 23:285–300.

Williams, C. M., and P. L. Adkisson. 1964. Physiology of insect diapause. XIV. An endocrine mechanism for the photoperiodic control of pupal diapause in the oak silkworm, *Antheraea pernyi*. Biol. Bull. 127:511–525.
Williams, C. M., P. L. Adkisson, and C. Walcotts. 1965. Physiology of insect diapause. XV. The transmission of photoperiod signals to the brain of the oak silkworm, *Antheraea pernyi*. Biol. Bull. 128:297–507.

DISCUSSION

TRUMAN: I must admit that I am mystified by the orange pigment you found. I have done nearly a thousand operations on Pernyi brains, and I have never seen any pigment in the median area of the brain that might be part of the brain itself.

HAYES: The pigment dots are really quite small, and they may actually be on the membrane that surrounds the brain rather than on the brain surface itself. As I said, I don't know what they are, but I think they may be pteridines.

TRUMAN: I would tend to think that they are not actually part of the brain itself.

HAYES: But they are quite closely associated with it.

MENAKER: What do you see in section?

HAYES: We haven't done sections. I don't mean to give you the impression that I'm convinced that the orange pigment is responsible for photoperiodic reception. It is an interesting pigment with an interesting absorption spectrum that looks a bit like the action spectrum for the breaking of diapause. But that is where I'd like to leave it.

WILKINS: What is the pigment giving rise to the absorption spectrum in the brain itself? It could be a carotenoid, it could be a flavenoid—do you know what it is?

HAYES: No, we don't know. From the shape of the spectrum, I would say it is probably one of four things: a carotenoid, a flavoprotein, a pteridine, or one of the vitamin B_{12} enzymes. The absorption spectrum shows a tantalizing similarity to some of these known spectra, but there are differences, and the shape of the curve we get varies with the conditions under which we run the spectrum.

ROUSE: Do you have any information on fluorescence spectra of these pigments, at an activation wavelength of 285 nm, for example, and an emission peak of 350 nm, that might indicate the presence of substituted indoles?

HAYES: I don't know anything about the fluorescence of either of the two pigments I discussed. I think you are probably well aware, though, that insect tissue is fairly well loaded with pteridines, so that any tissue you look at is gorgeously fluorescent.

There is another hazard in interpreting any spectra from insect tissue; if you exclude oxygen when you are collecting your tissues, you get a very different absorption spectrum than if you expose the tissues to oxygen and subsequently try to reduce them. You can never really recover the reduced spectra, and I suspect that insect tissue doesn't ordinarily have much oxygen in it.

BRINKMANN: I have a general comment

on all papers dealing with action spectra in circadian rhythm research. I am surprised that no one considers that the quantitative interpretation of a photoprocess requires a direct relation between the input of energy (quanta at defined wavelengths) and the output of the photoprocess, commonly the concentration of a product. Furthermore, to identify pigments you have to compare stationary states of the photoprocess. There is good evidence, however, that phase shifts of the circadian rhythm are typical nonstationary processes; they do not simply depend on the energy of the light pulse. It is rather a rapid change of the light intensity—so-called "on" or "off" steps—that induces phase shifts, suggesting that the derivative of a flow of energy is more important than the flow itself. If this is true, the photoeffect is additionally affected by time-dependent parameters, called "adaptations" in the terminology of the old physiologists. In that situation it seems impossible to me to use action spectra of phase shifts to identify pigments involved in the circadian clock.

A photoperiodic stimulus may be an exception insofar as in many cases the photoinduction depends on the energy of the light pulse. But even here a clear photochemical interpretation is difficult because the photoinduction is essentially determined by the circadian phase at which the light pulse is applied—i.e., determined by the extent to which that pulse simultaneously shifts the circadian phase.

HAYES: I would like to clarify a point I perhaps didn't make clear in the presentation. A single long day is not sufficient to break diapause in these insects; you must give them at least 6 weeks of the inductive photoperiod at the proper wavelength to produce the emergence response.

IV

CIRCANNUAL RHYTHMS AND PHOTOPERIODIC CONTROL

EBERHARD GWINNER

A Comparative Study
of Circannual Rhythms
in Warblers

In recent years evidence has accumulated indicating that seasonal changes in biological, physiological and behavioral phenomena in various animal species are regulated, in part, by an endogenous rhythm with a period length of about 12 months. In animals kept for one or more years in seasonally constant conditions, biological functions have been shown to fluctuate or recur as they do in freeliving individuals of the same species. In some cases in which animals lived under such conditions for several years, the period length was observed to change from exactly 12 months (Stebbins, 1963; Pengelley and Fisher, 1963; Pengelley and Kelly, 1966; Gwinner, 1967, 1968a; Pengelley and Asmundson, 1969; Jegla and Poulson, 1970; Heller and Poulson, 1970). It seems justified, therefore, to refer to such periodicities as circannual rhythms.

Results of previous investigations indicate that circannual rhythms participate in the timing of migration and molt in the willow warbler, *Phylloscopus trochilus*, and in the wood warbler, *P. sibilatrix* (Gwinner, 1967, 1968a). These palaearctic species winter in central and southern Africa. Those wintering in the central region live in a tropical environment, deficient in regular seasonal changes, and the birds appear to utilize an endogenous calendar for the timing of winter molt and, especially, of spring migration. These findings support the old suspicion that animals inhabiting equatorial regions and other seasonally quasi-stable environments might compensate for the lack of reliable external information with the help of an internal timing device (e.g., Rowan, 1926; Chapin, 1932; Marshall, 1960a, 1960b). Whether circannual rhythms are restricted to such animals or whether they are involved more universally in the regulation of annual cycles in animals is an intriguing question raised by these data.

To illuminate this question, I began a comparative investigation of the annual cycles of migratory readiness, body weight, and molt in a long- and in a short-distance migrant, under seasonally constant environmental conditions. The willow warbler, *Phylloscopus trochilus*, and the chiffchaff, *P. collybita*, were the experimental animals. With respect to most features of their appearance and their general biology, these sibling species are nearly indistinguishable. They differ markedly, however, in their migratory behavior. The willow warbler leaves its breeding grounds in late summer for a long and speedy flight to tropical and southern Africa, where it lives until early March. The chiffchaff, on the other hand, starts migration in late autumn and travels in a leisurely way toward its Mediterranean and North African winter quarters, where it stays until late February. Thus, while the willow warbler inhabits the tropics or the southern hemisphere

TABLE 1   Composition and Treatment of the Experimental Groups

Group	Number of Birds (Year of Birth)	Procedure
I_W	2(1966), 4(1967)	Eni
I_C	7(1967)	Eni
II_W	3(1966)	Eni to 18 Sept. 1966, then Bno
	4(1966)	Eni to 18 Sept. 1966, then Bni
III_W	2+3(1966)	Eni to 15/20 Sept. 1966, then LD *12*:12
	4(1967)	Eni to 20 Sept. 1967, then LD *12*:12
III_C	8(1967)	Eni to 20 Sept. 1967, then LD *12*:12
IV_W	6(1967)	Eni to 20 Sept. 1967, then LD *18*:6
V_W	2(1964)	Eni to Feb. 1965, then LD *12*:12
V_C	2(1964)	Eni to Feb. 1965, then LD *12*:12
VI_W	8+6(1968)	LD *12*:12 as nestlings
VII_W	9+7(1968)	LD *18*:6 as nestlings

W = willow warbler, C = chiffchaff; Eni = Erling natural light conditions indoors, Bni = Bukavu natural light conditions indoors, Bno = Bukavu natural light conditions outdoors; LD *12*:12 (*18*:6) = artificial light–dark cycle with *12*(*18*) hr of light/24 hr (200:0.02 Lux in groups III, IV, VI, and VII; 200:0.2 Lux in group V). 6 birds of group VI and 7 birds of group VII originated from northern Sweden, the others from southwest Germany.

for about 6 months each year, the chiffchaff remains essentially a bird of the northern temperate zones. A starting point for this investigation was to determine whether these differences are correlated with differences in the endogenous control of annual rhythms.

Endogenous rhythms, normally synchronized with environmental cycles, demand *Zeitgebers*—i.e., periodically fluctuating factors in the environment, capable of influencing the phase of the rhythm (e.g., Aschoff, 1954, 1958; Hoffmann, 1969). It has been suggested that seasonal changes in daylength provide the dominant *Zeitgeber* for circannual rhythms; there is evidence that this is true for some species (e.g., Aschoff, 1955). Yet, although the effects of daylength are well known for many species in which circannual rhythms have not been demonstrated, very few investigations have been concerned with the problem of entrainment of circannual rhythms by changes in daylength. In only one instance has it been directly demonstrated that the annual cycle of daylength acts as a *Zeitgeber* (Goss, 1969a, b). The second aim of this study was, therefore, to determine how photoperiodic conditions affect the circannual rhythms of warblers, and which variables of the seasonal cycle of daylength might be involved in the process of entrainment.

MATERIALS AND METHODS

The experimental birds (with one exception) were taken from the nest in Southern Germany and in Northern Sweden at an average age of about 9 days and raised by hand. After they had reached independence, they were kept individually in activity registration cages (42 X 23 X 23 cm) in which one of the two perches was supported on electronic microswitches. The impulses were recorded on Esterline-Angus and Miniskript event recorders. This technique enabled me to collect information about migratory restlessness (*Zugunruhe*). *Zugunruhe* is nocturnal activity typically shown by nocturnal migrants during the migratory seasons (Wagner, 1930); it can (with some restrictions) be taken as an indication of the migratory readiness of a bird. In addition, body weights and molt data were usually taken twice a week (for details see Gwinner, 1968a, b; 1969).

The birds referred to in this paper belong to seven experimental groups (Table 1). The birds of groups I to IV were housed from the time they were taken from the nest to late September of the same year under natural photoperiodic conditions in a temperature-regulated room with large windows (21 ± 2° C) in Erling, Germany (48° N, 11° 11'E). Beginning in late September, however, they were subjected to the following experimental treatments:

Group I remained under the previous conditions until late December of the following year.

Group II was moved to a location in the central African winter quarters of the willow warbler (Bukavu, Congo, 2°14'S, 28°39'E), and kept there with a natural light cycle either outdoors (three birds) or indoors in controlled temperature conditions (23 ± 2°C) (four birds) until mid-June of the following year.

Groups III and IV were transferred to temperature-controlled chambers with artificial light–dark cycles (*LD 12*:12 in Group III; *LD 18*:6 in Group IV; light intensity during the light fraction was 200 lux, during the "dark" fraction 0.02 lux), and kept there for up to 16 months.

All birds of groups III and IV born in 1967 were housed in groups of four or five in chambers (1 m^3; temperature 20 ± 1°C) in Erling. Two willow warblers of Group III, hatched in 1966, lived under the same temperature conditions in Erling, but in larger chambers. The remaining three willow warblers of Group III were moved, together with Group II, to central Africa and kept there indoors in a chamber at constant temperature (23 ± 1°C).

The birds of *group V* lived from the time of fledging in 1964 to February 1965 in natural light conditions similar to those of the birds of the other groups during their first months of life. They were then transferred to a room (21 ± 2°C) with an artificial *LD 12*:12 cycle (200:0.2 lux) and kept there for 27 months until July 1967.

Groups VI and VII consisted of willow warblers taken from the nests at an age of about 9 days; they were immediately transferred to chambers under either a continuous *LD 12*:12 cycle (Group VI) or a continuous *LD 18*:6 cycle (Group VII), and kept there, in the same housing conditions as the birds of Groups III and IV, until January of the following year. Eight birds of Group VI and 9 birds of Group VII belonged to the same SW-German population as all the other experimental subjects. Six birds of Group VI and 7 birds of Group VII were taken from a Northern European population in Sweden (66°42'N, 20°25'E).

There were no obvious differences related to year of birth or sex between the birds of any of the groups in the parameters assayed.

RESULTS

WILLOW WARBLERS AND CHIFFCHAFFS IN NATURAL LIGHT CONDITIONS (GROUPS I AND II)

The performance of the birds kept throughout the whole period of observation in the natural lighting conditions of their breeding grounds (Group

I) is represented in Figures 1a and 2a. By and large, the temporal patterns of *Zugunruhe*, body weight changes and molt approximate the normal behavior of freeliving individuals of the two species.

Normally, both species carry out a postjuvenile body molt (see reviews by Ticehurst, 1938; Gwinner, 1968a). Adult willow warblers undergo a complete molt twice a year (postnuptial molt in midsummer; prenuptial molt in midwinter). Chiffchaffs replace their plumage completely only once a year (in midsummer) and later than the willow warbler's summer molt; in winter only a few individuals undergo a partial body molt. Differences between the two species in intensity and duration of fall *Zugunruhe* are correlated with differences in distances covered during migration: the willow warbler stays in migratory condition almost three times as long, and develops more than three times as much *Zugunruhe* as the chiffchaff (Gwinner, 1968b). In spring the situation is more complex in both species. As in various other species investigated under similar conditions, spring *Zugunruhe* is extended more or less to the onset of postnuptial molt. There is no quiescent phase during the natural breeding period. This may be because in migratory birds appropriate stimuli from the breeding grounds and from the sex-mate are necessary to inhibit migratory restlessness in spring (Merkel, 1956; Helms, 1963).

The largest discrepancies—as yet unexplained—between the behavior of the experimentals and of free-living conspecifics are found in the body weight cycle. The willow warblers show typical obesity only during the autumnal *Zugunruhe* period. In spring, some individuals increase their body weight slightly. In chiffchaffs, increase in body weight is slight or absent in some individuals in both spring and autumn. In contrast, wild birds of both species seem to be as fat in spring as in autumn.

The behavior of the willow warblers transferred in late September from the natural light conditions of their breeding grounds to those of central Africa (Group II) is in most respects indistinguishable from that of the willow warblers of Group I (Figure 3). The only important differences are in the timing of winter molt and spring *Zugunruhe*. Both events occur slightly earlier in the birds of Group II than in those of Group I, indicating modifying effects of the different photoperiodic conditions.

WILLOW WARBLERS AND CHIFFCHAFFS IN *LD 12*:12 (GROUPS III, V)

Willow Warblers The occurrence of appropriately timed events of the annual molt and migratory cycles in the birds of Group II living close to the equator, where seasonal changes in daylength are almost lacking, suggests the participation of an endogenous timing mechanism. This suggestion is confirmed by the behavior of those willow warblers kept for a long

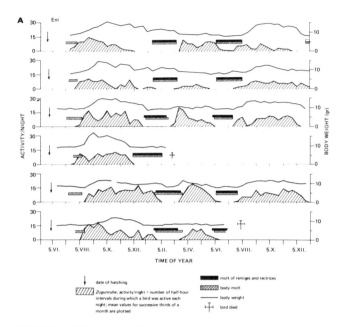

FIGURE 1a   Variations in *Zugunruhe*, body weight, and molt in 6 willow warblers kept for 19 months under natural light conditions (Eni) of
their breeding grounds (Group I$_W$).

time under *LD 12*:12 cycles (Group III). Figure 1b shows that, in the
birds of Group III, the characteristic patterns of molt, *Zugunruhe*, and
body weight persist over the whole 16 months of observation. The continuation of these annual rhythms for more than 1 year in the absence of
overt environmental stimuli suggests that the endogenous control mechanism is periodic.

The results obtained from the two willow warblers in Group V further
support this conclusion (Figure 4). While in both individuals fluctuations
in body weight apparently become arrhythmic, *Zugunruhe* and molt recur
at regular intervals twice a year over the whole 27-month period. Moreover, in at least one of the birds, the period length of these rhythms deviates considerably from 12 months: In bird *A* the phases of the cycles of
molt and *Zugunruhe* shift progressively forward, indicating a free-running
circannual periodicity of about 10 months.

Chiffchaffs   The behavior of the chiffchaffs kept in the continuous *LD
12*:12 cycle differs in many respects from that of the willow warblers. Figure 2b shows that in the chiffchaffs of Group III changes in body weight,
molt, and *Zugunruhe* become highly disorganized under such conditions.

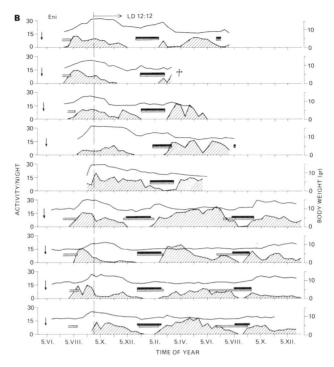

FIGURE 1b Variations in *Zugunruhe*, body weight, and molt in 9 willow warblers that were transferred in late September (vertical line) from natural light conditions of their breeding grounds to a continuous LD *12*:12 cycle (Group III$_W$). Bird No. 5 (second from bottom) was the only subject not raised by hand; it was caught during the autumn migration. Symbols as in Figure 1a.

Zugunruhe, in most birds initially patterned like that of the birds kept in natural light conditions, becomes increasingly irregular; interindividual variability is high. This is shown more quantitatively in Table 2, where the coefficients of variation of duration, total amount, and maximal value of autumn and spring *Zugunruhe* in the willow warblers and chiffchaffs are compared. Clearly, all three parameters are much more variable in the chiffchaffs than in the willow warblers. In addition, prenuptial molt is anomalous in most chiffchaffs under these conditions: It is extremely prolonged in some birds, and interrupted or reduced to a partial body molt in others. Changes in body weight, apparent in many of the chiffchaffs kept under natural lighting conditions and typical of all willow warblers, are almost entirely absent in these birds.

Rhythmicity disappears entirely in the two chiffchaffs of Group V

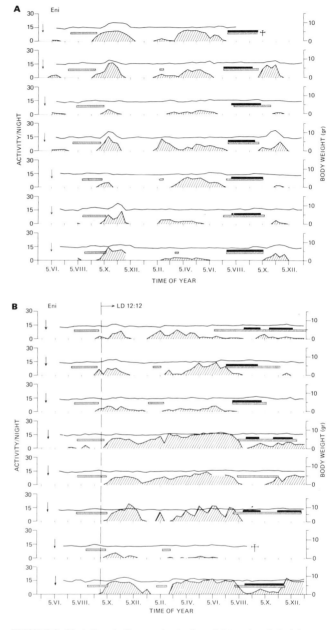

FIGURE 2   Variations in *Zugunruhe*, body weight, and molt in (a)
7 chiffchaffs kept for 19 months under natural light conditions of their
breeding grounds (Group I_C), and in (b) 8 chiffchaffs that were transferred in late September (vertical line) from these conditions to a continuous *LD 12*:12 cycle (Group III_C). Symbols as in Figure 1.

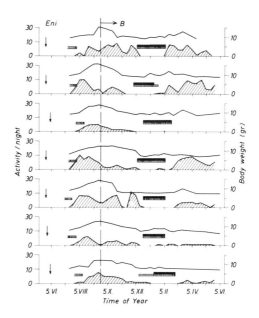

FIGURE 3 Variations in *Zugunruhe*, body weight, and molt in 7 willow warblers that were transferred in late September (dashed vertical line) from the natural photoperiodic conditions of their breeding grounds to central Africa (Group II$_W$). Symbols as in Figure 1.

kept for an extended time in *LD 12*:12. In one of the two birds, seasonal changes in all three functions are abolished by the end of the first year (Figure 4c). The other bird, which survived for only 17 months, behaved similarly. From these results we may conclude that the endogenous control of the annual cycles of *Zugunruhe*, body weight, and molt in birds kept in *LD 12*:12 is much less rigid in chiffchaffs than in willow warblers.

WILLOW WARBLERS IN *LD 18*:6 (GROUP IV)

The essentially normal performance of willow warblers kept from late September of their first year of life in *LD 12*:12 contrasts conspicuously with the behavior of birds transferred at the same time to a constant 18-hr day (Group IV, Figure 5). In the latter group, autumnal *Zugunruhe* ends relatively early, and winter molt starts more than 1 month before that of birds under *LD 12*:12 or natural lighting. The typical long-term alternations between *Zugunruhe* and molt are replaced by more or less irregular bouts of nocturnal activity, interrupted or accompanied by periods of molt of various durations. Body weight remains at a low and fairly stable

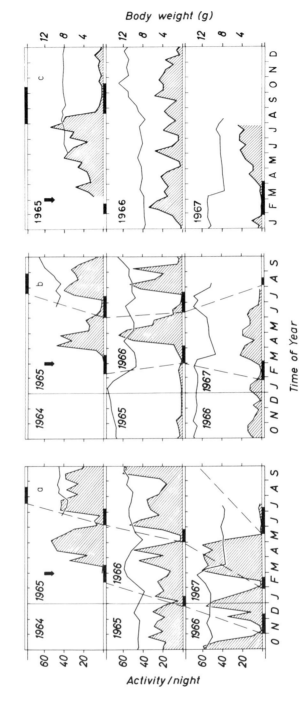

FIGURE 4 Variations in *Zugunruhe*, body weight, and molt in 2 willow warblers (a,b) and 1 chiffchaff (c) that were kept from late February (1965) (arrow) for 27 months in a continuous LD 12:12 cycle (Group V). Successive years are displayed one beneath the other. The dashed lines connect the onsets of corresponding molts in successive years. Activity/night = number of 10-min intervals during which a bird was active each night; mean values for successive thirds of a month are plotted. Solid bar indicates molt. Other symbols as in Figure 1. Records a and b from Gwinner (1968a).

414

TABLE 2 Coefficients of Variation of Various Parameters of Autumn and Spring *Zugunruhe* in Willow Warblers and Chiffchaffs kept in LD *12*:12

	Willow Warbler	Chiffchaff
Duration of autumn *Zugunruhe*	10.5	18.5
Amount of autumn *Zugunruhe*	21.2	54.7
Maximal value of autumn *Zugunruhe*	14.8	45.6
Duration of spring *Zugunruhe*	9.5	19.7
Amount of spring *Zugunruhe*	32.0	37.0
Maximal value of spring *Zugunruhe*	21.0	47.0

level. Thus, annual rhythmicity clearly expressed and retained in willow warblers kept under LD *12*:12 disappears under an 18-hr photoperiod.

GROUP DIFFERENCES IN THE TIMING OF EVENTS

Figure 6 shows the times at which the successive seasonal events occur in the willow warblers and chiffchaffs of the different groups. In both species, spring *Zugunruhe* tends to begin earlier and to end later in birds kept on a photoperiod of 12 hr than in birds kept in the natural lighting conditions of their breeding grounds. Correspondingly, in the former groups summer molt and the beginning of autumnal *Zugunruhe* are relatively delayed. Since in the birds kept in the natural lighting conditions of their breeding grounds, the photoperiod is shorter than 12 hr in winter but longer in summer, these differences may be explained on the assumption that longer photoperiods advance the time at which subsequent events occur. This assumption is consistent with the observation that in willow warblers kept on an 18-hr photoperiod, the end of autumnal *Zugunruhe* and the onset of winter molt and spring *Zugunruhe* occur earlier than in any of the other groups.

This, however, cannot be the whole story, since in chiffchaffs kept in LD *12*:12, autumnal *Zugunruhe* ends later than in chiffchaffs kept in the

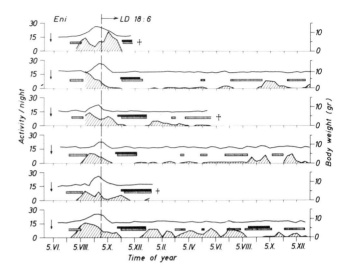

FIGURE 5 Variations in *Zugunruhe*, body weight, and molt in 6 willow warblers that were transferred in late September (vertical line) from
the photoperiodic conditions of their breeding grounds to a continuous
LD 18:6 cycle (Group IV). Symbols as in Figure 1a.

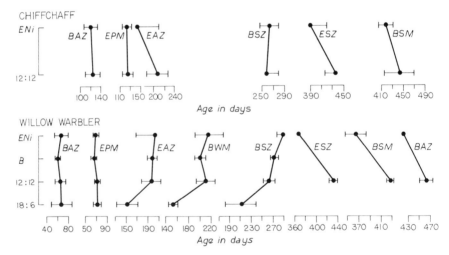

FIGURE 6 Comparison of the timing of successive annual events in willow warblers and in chiffchaffs kept in different photoperiodic conditions. Each circle represents the average age at which
an event occurred in that group. Horizontal bars are 99% confidence intervals. Corresponding mean
values of the different groups are connected by a line. BAZ = begin autumnal *Zugunruhe*; EPM =
end postjuvenile molt; EAZ = end autumnal *Zugunruhe*; BWM = begin winter molt; BSZ = begin
spring *Zugunruhe*; ESZ = end spring *Zugunruhe*; BSM = begin summer molt. B = birds of Group II_w
(Bno and Bni combined). Other abbreviations as in Figure 1.

natural photoperiod, which at this time of the year is shorter than 12 hr. The end of spring *Zugunruhe* and the onset of summer molt, on the other hand, occur earlier in the birds kept in the longer (natural) photoperiod. These differences might indicate that in this species a relatively longer day tends to delay the next event in late autumn and winter and to advance it in spring and summer. Such an argument is inconclusive, however, since the average duration of daylength is not the only difference in the experimental conditions. The birds living in natural light conditions are exposed to continuously changing photoperiods, and the rate of change is qualitatively and quantitatively different in spring and in autumn.

Since the way in which photoperiod affects different phases of the circannual rhythm is of theoretical interest with regard to problems of entrainment, another type of experiment, which allows firmer conclusions, is described in the following paragraphs.

SEASONAL DIFFERENCES IN THE RESPONSIVENESS TO PHOTOPERIOD

The willow warblers of Groups III and IV and those of Groups VI and VII were transferred from the natural lighting conditions of their breeding grounds to a 12-hr photoperiod and to an 18-hr photoperiod, respectively, at two different times of the year. The procedure is shown schematically in the upper graph of Figure 7. The birds transferred to either of the two conditions at time A (spring and early summer) were, on the average, 9 days old at the time of transfer; those transferred at time B (autumn) were, on the average, 110 days old. In the former groups, the age at which postjuvenile molt ended was determined, and in the latter, the age at which prenuptial molt started.

Results are given in the lower graph. The transfer of the birds to either long or short days had opposite effects at time A and at time B. At time A, molt ended relatively later in the birds transferred to the 18-hr day, whereas at time B, molt started relatively earlier in the birds transferred to the 18-hr day, and vice versa for transfer to the 12-hr day. The transfer to long days thus produced a relative delay of the subsequent event in spring, but a relative advance in autumn.

There is evidence that not only the event immediately following the transfer is affected by the change in photoperiod, but subsequent events as well. For instance, in the birds transferred to *LD 12*:12 at time A, winter molt started at an age of about 156 days, while in the birds transferred at the same time to *LD 18*:6, winter molt started at an age greater than 220 days. This suggests that the effects observed do, in fact, reflect responses of the oscillation itself and not only of its overt expression.

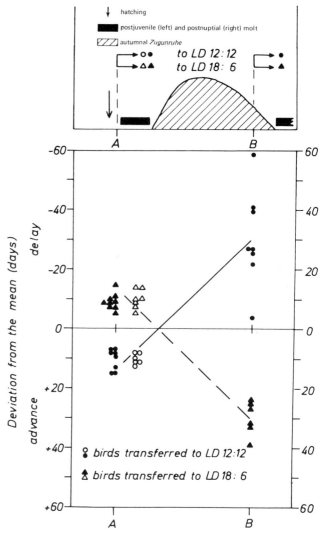

FIGURE 7 Phase-dependent differences in the effect of photoperiod on the annual cycle of willow warblers. The upper graph diagrams the procedure. Birds were kept in the natural lighting conditions of their breeding grounds and then transferred to either a 12-hr photoperiod (circles; groups III_W, VI_W) or an 18-hr photoperiod (triangles; groups IV_W, VII_W) at two phases of their annual cycle (A: at an age between 6 and 11 days in spring and early summer; B: at an age between 108 and 118 days in late September). At time A two subgroups of birds were tested: one consisted of birds from northern Sweden (open symbols), the other of birds from southwestern Germany (closed symbols). For the birds transferred to experimental lighting conditions at time A (groups VI_W, VII_W), the time postjuvenile molt ended was determined; for the birds treated the same way at time B (groups III_W, IV_W) the beginning of postnuptial molt was determined. The lower graph shows the results. Deviations of individual values from the mean of each group are plotted.

The results discussed above can be interpreted in two ways. One possibility is that the birds react to the continuous action of photoperiod. If this is true, we must conclude that the sensitivity of the birds to a given photoperiod changes with the time of year. The other possibility is that the birds react mainly to the transition from one daylength to another. In that case, the differences observed could be explained as a result of the differences in the stimulus situation: A bird transferred from natural lighting conditions to an 18-hr day in spring experiences a very different change in daylength from a bird treated the same way in autumn.

However, the following results render the latter alternative highly improbable. For the spring transfer, birds of two different populations were used: One subgroup came from latitude 48° and had experienced a photoperiod of about 15 hr prior to the transfer; the other subgroup came from latitude 66°42′ and had experienced approximately 24 hr of light per day previously. The magnitude of the change from natural lighting conditions to experimental conditions was therefore different in the two groups. Nevertheless, they reacted in the same way. The average age at which post-juvenile molt ended (days 45 and 46 respectively), as well as the differences in the termination of molt between the long- and the short-day group, were the same in birds from both populations. It appears highly probable, then, that these willow warblers reacted mainly to the continuous action of photoperiod. At the same time, the results support the assumption that the sensitivity to the same photoperiod is different at different phases of the annual cycle.

DISCUSSION

The results reported here demonstrate that an internal annual periodicity participates in the regulation of seasonal rhythms in warblers. The annual cycles of *Zugunruhe* and molt, and, to some extent, those of body weight, persist in willow warblers and chiffchaffs kept for 16 months in constant photoperiodic conditions of 12 hr of light and 12 hr of darkness. In two willow warblers that lived for 27 months in such conditions, the seasonal rhythmicity of *Zugunruhe* and molt continued for three periods. Moreover, the average period length of the *Zugunruhe*- and molt-cycle deviated so considerably from 12 months in at least one bird that it seems highly improbable that uncontrolled environmental factors caused the rhythm.

Such circannual rhythms have been postulated for a long time from observations of freeliving animals (e.g., Baker and Baker, 1936; Baker and Ranson, 1938; Aschoff, 1955; Marshall, 1959, 1960a, 1960b, 1961;

Immelmann, 1963, 1967). However, only within the last decade have convincing experimental demonstrations of endogenous annual cycles been published. Circannual rhythms have been found not only in willow warblers (Gwinner, 1967, 1968a), but in the cave crayfish, *Orconectes pellucidus* (Jegla and Poulson, 1969), the lizard *Sceloporus virgatus* (Stebbins, 1963), the ground squirrels, *Citellus lateralis, C. mohavensis* and *C. beldingi* (Pengelley and Fisher, 1963; Pengelley and Kelly, 1966; Pengelley and Asmundson, 1969; Heller and Poulson, 1970), the chipmunks, *Eutamias speciosus, E. alpinus, E. minimus,* and *E. amoenus* (Heller and Poulson, 1970), the sika deer, *Cervus nippon* (Goss, 1969a), and apparently in the slug, *Limax flavus* (Segal, 1960).

In other instances, the endogenous nature of the rhythms is uncertain since the period lengths observed are very close to or identical with 1 year. This has been the case in the fish, *Mystus vittatus* (Singh, 1968), the Whitethroat, *Sylvia communis* (Merkel, 1963), the weaver finch, *Quelea quelea* (Lofts, 1964), the dickcissel, *Spiza americana* (Zimmerman, 1966), the white-crowned sparrow, *Zonotrichia leucophrys gambelii* (King, 1968), and the woodchuck, *Marmota monax* (Davis, 1967).

A third group of reports comprises cases in which periodicities deviate to such an extent from 12 months that it is not clear whether they should be called circannual. This is the case in the testicular cycle of the domestic duck (Benoit, Assenmacher, and Brard, 1955, 1956, 1959; Benoit, 1961) and in body weight rhythms in robins, *Erithacus rubecula* (Merkel, 1963).

This brief review of the available evidence suggests that circannual rhythms are fairly widespread among long-lived animals. The results of comparative studies show, however, that the significance of endogenous rhythmicity for the control of annual cycles may vary considerably from species to species. The experiments discussed in this paper revealed differences in the rigidity of endogenous control and in the persistence of circannual rhythms in two closely related species kept in identical constant photoperiodic conditions—differences that are correlated with differences in the general biology of the species. The willow warbler is a very long-distance migrant, with winter quarters close to or beyond the equator. As has been pointed out repeatedly (e.g., Rowan, 1926; Chapin, 1932; Marshall, 1960b), rather complicated mechanisms would be necessary for the precise timing of seasonal events if such a bird relied exclusively on information provided by daylength and other cyclic environmental factors. This is especially true with respect to the initiation of winter molt and spring migration in those individuals that winter close to the equator where seasonal changes in the environment are either nearly lacking or highly variable from year to year. The adaptive significance for the willow warbler of an endogenous calendar is thus evident, but the need for such a mechanism is not as obvious in the chiffchaff. This species travels relatively short distances, wintering in southern Europe and in northern Africa, regions in which seasonal changes in daylength and other environmental factors are quite marked. In fact, the endogenous control of the annual rhythm of the chiffchaff seems to be less pronounced.

These results are clearly in agreement with the assumption that a rigid endogenous control of the annual cycle is a special adaptation of animals inhabiting (either permanently or during some part of the year) seasonally stable or unpredictable environments. Other data on birds are also compatible with this hypothesis. Endogenous annual rhythms are involved in the regulation of the testicular cycle of the tropical weaver finch *Quelea quelea* (Lofts, 1964). On the other hand, no indications of persistent circannual rhythms could be found in some classical photoperiodic species inhabiting lower latitudes (Miller, 1948, 1951; Wolfson, 1959, 1960; Weise, 1962; Farner and Follet, 1966).

Results obtained from mammals indicate that rigorous endogenous control of annually changing phenomena may have arisen under ecological situations other than tropical. In ground squirrels of the genus *Citellus* the annual rhythms of body weight and hibernation are endogenously controlled most rigidly in the obligatory hibernator *C. lateralis*, indigenous to the boreal life zone. Other species inhabiting generally more arid environments depend to a lesser degree on an endogenous rhythm (Pengelley and Kelly, 1966). These authors have pointed out that a circannual rhythm is an excellent adaptation for *C. lateralis*, which inhabits a severe and regularly changing environment, "where each event in the animal's life cycle must occur at an exact time in order for it to survive." Circannual rhythmicity enables the adnimals to prepare for a changing environment long in advance. For the other species of *Citellus* it seems to be more advantageous to be readily adaptable to the fluctuating conditions of their habitats, and endogenous rhythmicity is less marked in these animals. Similar relationships have been demonstrated between rigidity of the endogenous control and environmental conditions in four species of the genus *Eutamias* inhabiting different life zones (Heller and Poulson, 1970).

Circannual rhythmicity may have adaptive value quite apart from timing seasonal events. Results obtained from *Phylloscopus* warblers suggest that a time program, controlled by an endogenous annual rhythmicity, specifies the level of migratory activity that is just sufficient for the birds to reach the vicinity of their species specific winter quarters (Gwinner, 1968a, b). Such a temporal mechanism could assist a bird in finding its goal even if it were only capable of crude direction orientation. Circannual rhythms may thus be important not only for the initiation of migration at the appropriate time, but also for the termination of migration at the appropriate place.

It is doubtless premature to estimate how widespread circannual rhythms may be among animals. However, the suggestion that endogenous timing mechanisms control a variety of different adaptive functions speaks in favor of a fairly wide distribution. It should be emphasized that this view is not affected by results of experiments in which no endogenous annual rhythm could be found in animals kept in seasonally constant conditions. The results previously reported (see page 413) show how carefully such negative findings must be interpreted. Circannual rhythms, obvious in willow warblers kept in *LD 12*:12, cannot be observed in willow warblers kept in *LD 18*:6. Similarly, the expression of circannual rhythms in other animals depends on the photoperiodic conditions. In the sika deer, *Cervus nippon*, antlers are replaced on a circannual basis in animals exposed to constant light or to constant *LD 18*:6 or *6*:18; the rhythm disappears in *LD 12*:12, however (Goss, 1969b). The annual testicular cycle of the weaver finch, *Quelea quelea*, persists for at least 2 years in *LD 12*:12 but seems to disappear in *LD 8*:16 (Lofts, 1962, 1964). Comparable differences are indicated in the annual body-weight cycle of the whitethroat, *Sylvia communis* (Merkel, 1963). The most striking example of photoperiodic effect on circannual rhythmicity is reported recently by Schwab (this volume). Starlings (*Sturnus vulgaris*) show a well-defined annual rhythm of testicular size if kept in a continuous *LD 12*:12 cycle, but no periodicity is observed in birds kept in daily photofractions of 11 hr or less, or 13 hr or more. The range of permissive conditions for the expression of circannual rhythms is evidently held within narrow limits in these animals. Experiments in which no endogenous annual cycles could be detected in animals kept in only one or a few conditions are therefore comparatively meaningless with regard to the general question of the distribution of circannual rhythms among animals.

Circannual rhythms demand *Zeitgebers* which synchronize the endogenous periodicity with the natural year. The demonstration that photoperiodic conditions influence the rhythms of willow warblers and chiffchaffs suggests that in these species the annual daylength-cycle may have *Zeitgeber* properties.

Zeitgebers act by controlling the phase of the endogenous rhythm, and "one of the prerequisites for phase control is a periodically changing sensitivity of the organism to the stimuli of the *Zeitgeber*" (Aschoff, 1965). If seasonal changes in daylength act as a *Zeitgeber* for circannual rhythms, one would expect the sensitivity of an animal to the pertinent stimuli of daylength to be different at different phases of its circannual cycle. In other words, one should find a circannual response curve.

The results depicted in Figure 7 can be viewed as part of a circannual response curve. The timing of annual events in willow warblers transferred from natural light conditions to constant long or short days is affected differently depending on the phase at which the transfer took place. There are not yet enough data for the derivation of concrete ideas about the possible mechanisms of entrainment, but those available do throw some light on the question of which parameters of the annual daylength cycle might provide the effective stimuli. By analogy with the properties of *Zeitgebers* of circadian rhythms (e.g., Aschoff, 1960), at least two possibilities have to be considered: (1) The steady state of a given photoperiod may exhibit a continuous action effect on the angular velocity of the rhythm (proportional effect), and (2) the transition from one photoperiod to another may be of importance (differential effect).

The changes in the circannual cycle brought about by transferring willow warblers to constant photoperiodic conditions can hardly result from a differential effect, since birds exposed to different photoperiods prior to the transfer (and thus to steps of different magnitude) reacted alike. The opposite response to long- and short-day exposure at different times of the year is therefore most probably due to differences in response to daylength as such.

That photoperiod exerts continuous action effects on circannual rhythms is also suggested by experiments in which animals were kept in two different constant daylength conditions for more than one circannual cycle. In the fish *Mystus vittatus* (Singh, 1968), and in two bird species, (*Anas platyrhynchos domesticus*: Benoit, Assenmacher, and Brard, 1955, 1956, 1959; Benoit, 1961; and *Erithacus rubecula*: Merkel, 1963), the period length of the observed (possibly circannual) rhythms varies inversely with the photoperiod. While it is probably premature to draw an analogy between these relationships and those described by Aschoff's rule for the dependence of the circadian period-length on light intensity (e.g., Aschoff, 1960), these results demonstrate an influence of constant photoperiod on circannual rhythms that might be involved in the process of entrainment. The extent to which transitional stimuli derived from changes in daylength are also effective is unknown.

The increasing evidence for a wide distribution of circannual rhythms suggests that photoperiod might in many cases act as a *Zeitgeber* entraining an endogenous rhythm rather than as a causal stimulus. The early experimental literature supporting this suggestion has been reviewed by Aschoff (1955).

The demonstration that photoperiod is, in fact, capable of affecting the phase of circannual rhythms in warblers argues for such a view. In addition, the circannual response of warblers is strikingly similar to photoperiodic reactions of other animals. Phase-dependent differences in the reaction to photoperiodic conditions, as described here for circannual rhythms in willow warblers, are well documented for numerous vertebrates, especially for the gonadal responses of birds. Long days advance gonadal development between late autumn and spring, but delay gonadal recovery during the refractory phase in late summer. Short days generally have the opposite effect. Comparable results have been reported in studies of circadian rhythms (phase response-curves to light- and dark-steps; Aschoff, 1965). Since gonadal weight response-curves can easily be interpreted in terms of phase shifts, this similarity may reflect more than a formal analogy. Because changing sensitivity to environmental stimuli is typical of, and necessary for, the response of endogenous rhythms to a *Zeitgeber*, the discovery of such changing sensitivities suggests that endogenous circannual rhythms may be commonly involved in mediating the effects of photoperiod.

This study was supported, in part, by grants GB–5969X and GB–11905 from the National Science Foundation to Dr. D. S. Farner.

REFERENCES

Aschoff, J. 1954. Zeitgeber der tierischen Jahresperiodik. Naturwissenschaften 41: 49–56.

Aschoff, J. 1955. Jahresperiodik der Fortpflanzung bei Warmblütern. Stud. Gen. 8: 742–776.

Aschoff, J. 1958. Tierische Periodik unter dem Einfluss von Zeitgebern. Z. Tierpsychol. 15:1–30.

Aschoff, J. 1960. Exogenous and endogenous components in circadian rhythms. Cold Spring Harbor Symp. Quant. Biol. 25:11–28.

Aschoff, J. 1965. Response curves in circadian periodicity, p. 95 to 111. *In* J. Aschoff [ed.], Circadian clocks. North-Holland Publ. Co., Amsterdam.

Baker, J. R., and I. Baker. 1936. The seasons in a tropical rain forest. J. Linn. Soc. Lond. Bot. 41:248–258.

Baker, J. R., and R. M. Ranson. 1938. The breeding seasons of southern hemisphere birds in the northern hemisphere. Proc. Zool. Soc. Lond., A. p. 101–141.

Benoit, J. 1961. Opto-sexual reflex in the duck: physiological and histological aspects. Yale J. Biol. Med. 34: 97–116.

Benoit, J., I. Assenmacher, and E. Brard. 1955. Évolution testiculaire du Canard domestique maintenu à l'obscurité totale pendant une longue durée. C. R. Acad. Sci. 241:251–253.

Benoit, J., I. Assenmacher, and E. Brard. 1956. Étude de l'évolution testiculaire du

Canard domestique soumis très jeune à un éclairement artificiel permanent pendant deux ans. C. R. Acad. Sci. 242:3113–3115.

Benoit, J., I. Assenmacher, and E. Brard. 1959. Action d'un éclairement permanent prolongé sur l'évolution testiculaire du Canard pékin. Arch. Anat. Microsc. Morphol. Exp. 48:5–12.

Chapin, J. P. 1932. The birds of the Belgian Congo, I. Bull. Am. Mus. Nat. Hist. 65: 1–756.

Davis, D. E. 1967. The annual rhythm of fat deposition in woodchucks (*Marmota monax*). Physiol. Zool. 40:391–402.

Farner, D. S., and B. K. Follett. 1966. Light and other environmental factors affecting avian reproduction. J. Anim. Sci. 25(Suppl.):90–105.

Goss, R. J. 1969a. Photoperiodic control of antler cycles in deer. I. Phase shift and frequency changes. J. Exp. Zool. 170:311–324.

Goss, R. J. 1969b. Photoperiodic control of antler cycles in deer. II. Alterations in amplitude. J. Exp. Zool. 171:233–234.

Gwinner, E. 1967. Circannuale Periodik der Mauser und der Zugunruhe bei einem Vogel. Naturwissenschaften 54:447.

Gwinner, E. 1968a. Circannuale Periodik als Grundlage des jahreszeitlichen Funktionswandels bei Zugvögeln. Untersuchungen am Fitis (*Phylloscopus trochilus*) und am Waldlaubsänger (*P. sibilatrix*). J. Ornithol. 109:70–95.

Gwinner, E. 1968b. Artspezifische Muster der Zugunruhe bei Laubsängern und ihre mögliche Bedeutung für die Beendigung des Zuges im Winterquartier. Z. Tierpsychol. 25:843–853.

Gwinner, E. 1969. Untersuchungen zur Jahresperiodik von Laubsängern. Die Entwicklung des Gefieders, des Gewichts und der Zugunruhe bei Jungvögeln der Arten *Phylloscopus bonelli, Ph. sibilatrix, Ph. trochilus* und *Ph. collybita.* J. Ornithol. 110:1–21.

Heller, H. C., and T. L. Poulson. 1970. Circannian rhythms—II. Endogenous and exogenous factors controlling reproduction and hibernation in chipmunks (*Eutamias*) and ground squirrels (*Spermophilus*). Comp. Biochem. Physiol. 33:357–383.

Helms, C. W. 1963. The annual cycle and *Zugunruhe* in birds. Proc. XIIIth Int. Ornithol. Congr., Ithaca, New York, 1967, p. 925–939.

Hoffmann, K. 1969. Die relative Wirksamkeit von Zeitgebern. Oecologia (Berl.) 3: 184–206.

Immelmann, K. 1963. Tierische Jahresperiodik in ökologischer Sicht. Zool. Jahrb. Syst. 91:91–200.

Immelmann, K. 1967. Periodische Vorgänge in der Fortpflanzung tierischer Organismen. Stud. Gen. 20:15–33.

Jegla, T. C., and T. L. Poulson 1970. Circannian rhythms—I. Reproduction in the cave crayfish, *Orconectes pellucidus inermis.* Comp. Biochem. Physiol. 33:347–355.

King, J. 1968. Cycles of fat deposition and molt in white-crowned sparrows in constant environmental conditions. Comp. Biochem. Physiol. 24:827–837.

Lofts, B. 1962. Photoperiod and the refractory period of reproduction in an equatorial bird, *Quelea quelea.* Ibis 104:407–414.

Lofts, B. 1964. Evidence of an autonomous reproductive rhythm in an equatorial bird (*Quelea quelea*). Nature (Lond.) 201:523–524.

Marshall, A. J. 1959. Internal and environmental control of breeding. Ibis 101:456–478.

Marshall, A. J. 1960a. Annual periodicity in the migration and reproduction of birds. Cold Spring Harbor Symp. Quant. Biol. 25:499–505.

Marshall, A. J. 1960b. The role of the internal rhythm of reproduction in the timing of avian breeding seasons, including migration. Proc. XIIth Int. Ornithol. Congr., Helsinki, 1958, p. 475–482.

Marshall, A. J. 1961. Breeding seasons and migration. *In* A. J. Marshall [ed.], Biology and comparative physiology of birds. Vol. 2. Academic Press, New York.

Merkel, F. W. 1956. Untersuchungen über tages- und jahreszeitliche Aktivitätsänderungen bei gekäfigten Zugvögeln. Z. Tierpsychol. 13:278–301.

Merkel, F. W. 1963. Long-term effects of constant photoperiods on European robins and whitethroats. Proc. XIIIth Int. Ornithol. Congr., Ithaca, New York, 1962, p. 950–959.

Miller, A. H. 1948. The refractory period in light-induced development of the golden-crowned sparrow. J. Exp. Zool. 109:1–11.

Miller, A. H. 1951. Further evidence on the refractory period in the reproductive cycle of the golden-crowned sparrow, *Zonotrichia coronata.* Auk 68:380–383.

Pengelley, E. T., and S. M. Asmundson. 1969. Free-running periods of endogenous circannian rhythms in the golden-mantled ground squirrel, *Citellus lateralis.* Comp. Biochem. Physiol. 30:177–183.

Pengelley, E. T., and K. C. Fisher. 1963. The effect of temperature and photoperiod on the yearly hibernating behavior of captive golden-mantled ground squirrels (*Citellus lateralis tescorum*). Can. J. Zool. 41:1103–1120.

Pengelley, E. T., and K. H. Kelly. 1966. A "circannian" rhythm in hibernating species of the genus *Citellus* with observations on their physiological evolution. Comp. Biochem. Physiol. 19:603–617.

Rowan, W. R. 1926. On photoperiodism, reproductive periodicity, and the annual migrations of birds and certain fishes. Proc. Boston Soc. Nat. Hist. 38:147–189.

Segal, E. 1960. Discussion to A. J. Marshall. Cold Spring Harbor Symp. Quant. Biol. 25:504–505.

Singh, T. P. 1968. Effects of varied photoperiods on rhythmic activity of thyroid gland in a teleost, *Mystus vittatus* (Bloch). Experientia 24:93–94.

Stebbins, R. C. 1963. Activity changes in the striped plateau lizard with evidence on influence of the parietal eye. Copeia 1963:681–691.

Ticehurst, C. B. 1938. A systematic review of the genus *Phylloscopus.* British Museum of Natural History, London. 193 p.

Wagner, H. O. 1930. Über Jahres- und Tagesrhythmus bei Zugvögeln. Z. Vgl. Physiol. 12:703–724.

Weise, C. M. 1962. Migratory and gonadal responses of birds on long-continued short day-lengths. Auk 79:161–172.

Wolfson, A. 1959. Ecologic and physiologic factors in the regulation of spring migration and reproductive cycles in birds, p. 38 to 70. *In* Comparative endocrinology. John Wiley & Sons, New York.

Wolfson, A. 1960. Role of light and darkness in the regulation of the annual stimulus for spring migration and reproductive cycles. Proc. XIIth Int. Ornithol. Congr., Helsinki, 1958, p. 758–789.

Zimmerman, J. L. 1966. Effects of extended tropical photoperiod and temperature on the dickcissel. Condor 68:377–387.

DISCUSSION

FARNER: I would just like to say, for the record, something that I think we all realize but tend to forget. We often assume that because there are circadian periodicities with fairly uniform properties and because the earth has an annually cyclic environment, there should be circannual periodicities, hopefully with similarly uniform properties. But while it is possible to trace back to a very early stage in evolution the development of circadian periodicity, we should bear in mind that two thirds of the evolution of life was over before a circannual periodicity could have any adaptive significance because animals didn't live for as long as a year. And so I think that with circannual periodicities we are almost certainly dealing with a cluster of convergent phenomena—within certain physiological constraints, of course.

ROBERT G. SCHWAB

Circannian Testicular Periodicity in the European Starling in the Absence of Photoperiodic Change

The proper seasonal chronology of annual avian breeding cycles has considerable adaptive value: (1) utilization of optimal reproductive season; (2) minimization of the reproductive period; (3) synchronization of events within a population; and (4) temporal separation of incompatible events (for review see Farner, 1964). Some of these advantages would be more important to birds breeding at higher latitudes than to those breeding in tropical and equatorial regions. Even so, many equatorial avian species exhibit seasonal breeding cycles under the relatively slight annual fluctuations in daily light duration occurring there. The precise timing of breeding cycles in these birds is generally thought to be a function of such proximate environmental synchronizers as seasonal rains and associated changes in vegetation and food availability (Lofts and Murton, 1968). It seems likely that these synchronizers operate upon a broad chronological base established by a permissive physiological condition. That is, the animal must be physiologically capable of gonadal response when proximate environmental stimuli indicate the proper reproductive chronology.

The timing of annual avian breeding cycles in higher latitudes is closely correlated with the duration of daily photoperiods. The suggestion that reproductive cycles are photoperiodically controlled is supported by Rowan's (1925) discovery that artificially increased daily light periods in-

428

duce precocious gonadal growth in the slate-colored junco, *Junco hyemalis*. Many subsequent studies have shown that similar treatment can induce precocious gonadal growth in a large number of avian species, including the European starling, *Sturnus vulgaris* (Bissonnette, 1931; Burger, 1947, 1949, 1953).

Several hypotheses have been proposed to account for the influence of daily photoperiod on gonadal development. One hypothesis suggests that the photoperiodic effects result from a light-dependent reaction that produces an essential compound whose accumulation, duration of physiological activity, or release from site of production may be a function of daily photoperiod (Farner, 1959, 1964; Farner and Follett, 1966; Wolfson, 1959a, b, 1960).

A second scheme implicates a physiological circadian rhythm in the timing of gonadal growth in photoperiodic avian species (Farner, 1965; Farner and Follett, 1966; Hamner, 1963, 1964, 1966, 1968; Menaker, 1965; Wolfson, 1964, 1965). Hamner (1968) proposed that this circadian timing mechanism readjusts seasonally, and thus positions a photosensitive phase of the photoperiodic clock so that this phase is illuminated by increasing daily photoperiods after the winter solstice, thereby inducing gonadal growth. Hamner's scheme is certainly plausible and might well define the major time-measuring mechanism regulating the reproductive chronology in some birds. Such a scheme, however, in no way precludes the concurrent operation of a semiautonomous annual reproductive periodicity that is independent of seasonally changing daily photoperiods.

Such periodicities have been termed "quasi-annual" (Farner, 1964), "internal reproductive rhythms" (Marshall, 1959, 1960a, b, c), "innate reproductive rhythms" (Miller, 1955, 1959), and "circannian" or "circannual" rhythms (Pengelley, 1967; see also Lofts, 1962; Lofts and Murton, 1968; Marshall and Serventy, 1956, 1958, 1959).

Conclusive evidence for circannian reproductive periodicities in birds is scant, however. Benoit (1961) has found what appears to be an endogenous circannian reproductive cycle in domestic ducks (Pekin breed) under conditions of constant light or constant darkness. Lofts (1964) reported a circannian reproductive periodicity in the weaver finch, *Quelea quelea*, maintained under a chronic *LD 12*:12 regimen. Farner and Follett (1966) suggest that an endogenous circannian reproductive rhythm might account for the reproductive cycles reported in a number of avian species: short-tailed shearwater, *Puffinus tenuirostris* (Marshall and Serventy, 1959); the whitethroat, *Sylvia communis* (Merkel, 1961); and the greenfinch, *Chloris chloris* (Schildmacher, 1956).

Farner and Follett (1966), in their review of endogenous avian repro-

ductive periodicities, pointed out: "What is sorely needed at this point is
not further suggestive field observations, but rather careful observations
under constant conditions to ascertain whether gonadal cycles can occur
in an environment devoid of annually periodic environmental information.
Two consecutive, accurately timed cycles under such conditions would
constitute the minimum evidence for the existence of such periodicities;
more would be desirable." Furthermore, if such endogenous circannian
cycles are truly comparable to those under changing environmental conditions, they should exhibit *each* physiologically distinct phase that occurs
within the subject species during a calendar year.

This paper presents evidence that (a) properly timed testicular cycles
occur in the European starling under a chronic daily photoperiod and relatively stable temperature; and (b) such cycles exhibit the physiologically
distinct phases of the natural testicular cycle in this species. In addition,
expression of a circannian testicular periodicity in this species depends on
the duration of the fixed daily photoperiod under which the animals are
maintained.

MATERIALS AND METHODS

The starlings studied were juveniles captured in the San Joaquin Valley of
central California in June–August of 1966 and 1967. They were maintained on turkey pellets and fresh water *ad libitum* under natural daylengths in a large outdoor aviary at Davis, California. There the birds
reached maturity and progressed through their initial gonadal cycle. With
one exception (discussed later), the starlings has completed only one testicular cycle prior to exposure to programmed light regimens. As controls,
a large number of starlings were held in the outdoor aviary during their
second and third gonadal cycles, and the chronology of their testicular
cycles under natural photoperiods was monitored.

Starlings exposed to programmed light regimens were held in $2 \times 2 \times 4$-ft
wire cages enclosed in light-tight plywood "photochambers." The light intensity, from incandescent bulbs, was about 60 ft-c at the midpoint of the
wire cage. Daily temperatures within the photochambers fluctuated er-
ratically between 22 and 28° C.

All testicular measurements reported represent the width of the left
testis measured *in situ* following unilateral laparotomy. Testis widths of
less than 2 mm could not be measured accurately, and widths reported as
2 mm may actually have been up to 1 mm smaller. Certain experiments
necessitated a more precise evaluation of testicular condition. In such

cases, the testis was measured *in situ*, the animal sacrificed, and the functional condition of the testis evaluated by standard histological techniques. These histological assays indicate that testis width is a reliable index of development and that widths of 5.5 mm or greater characterize the spermatogenic stage of testicular metamorphosis in the European starling.

RESULTS

TESTICULAR CYCLES IN CAPTIVE STARLINGS UNDER NATURAL PHOTOPERIOD

The natural testicular growth phase begins in early January, and spermatogenesis is achieved by most individuals about the end of February. Spermatogenesis persists until the end of May or the first part of June, at which time testicular involution occurs. Most captive male starlings are reproductively quiescent by mid-June. A future report will present detailed analysis of the annual testicular cycle in captive starlings under natural daylengths at latitude 38° N. Here, only the approximate chronology (Figure 1) is given for comparison with the testicular condition of birds under the programmed light regimens.

EXPERIMENTS UNDER PROGRAMMED LIGHT REGIMENS

1. *Spontaneous Testicular Growth in Postinvolution Starlings under Fixed Daily Photoperiods*

On 21 June, groups of postinvolution starlings were exposed to fixed daily photoperiods of *LD 3*:21, *LD 9*:15, *LD 11*:13, *LD 12*:12, and *LD 13*:11. The testicular condition of the starlings under each light regimen was monitored monthly thereafter. Table 1 gives a statistical description of testicular growth and the duration of treatment under the *LD 3*:21, *LD 9*:15, and *LD 11*:13 regimens. In general, initial rate of testicular growth to the size associated with spermatogenesis is inversely related to daily light duration (Figure 2).

The average testicular size of birds under *LD 9*:15 reached a level indicative of spermatogenesis, involuted to nonspermatogenic levels, and then returned to a spermatogenic condition. The average size under *LD 3*:21, however, reached levels associated with spermatogenesis and remained there throughout the entire treatment.

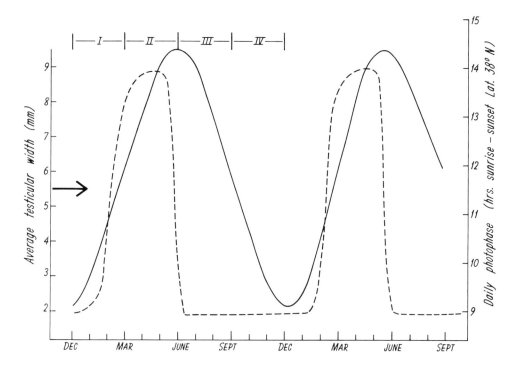

FIGURE 1 Temporal relationship of the annual testicular cycle in captive European starlings (broken line) to seasonally changing photoperiod (solid line). The testis width at or above which spermatogenesis can be expected in all individuals is indicated by the single horizontal arrow. Roman
numerals indicate seasonal chronology of physiologically distinct phases of the annual testicular
cycle; these are discussed in text.

TABLE 1 Statistical Description of Testis Size (mm Width) during Spontaneous Testicular Growth in Postinvolution Starlings under Fixed Photoperiods

	LD 3:21				LD 9:15				LD 11:13			
Month	Day	N	Mean ± SE	Range	Day	N	Mean ± SE	Range	Day	N	Mean ± SE	Range
June	21	7	<2.0		21	7	<2.0		21	10	<2.0	
July	13	7	<2.0		12	7	<2.0		29	10	<2.0	
Aug.	12	7	5.5 ± 0.8	2.1–8.3	11	7	4.2 ± 0.8	2.0–7.6	29	10	2.7 ± 0.3	2.0–5.2
Sept.	22	10	6.4 ± 0.6	3.5–10.6	1	10	5.6 ± 0.7	3.0–9.1	30	10	5.4 ± 0.7	2.3–8.2
Oct.	23	10	6.6 ± 0.5	4.0–9.5	11	10	4.4 ± 0.5	2.6–7.5	30	6	7.8 ± 0.3	7.1–8.9
Nov.	21	10	7.7 ± 0.5	5.2–10.5	21	9	5.6 ± 0.6	3.7–7.8	27	3	8.8 ± 0.4	8.1–9.3
Dec.	20	10	6.8 ± 0.4	4.9–9.0	19	9	5.9 ± 0.7	2.9–8.5	20	6	8.7 ± 0.6	8.2–9.5

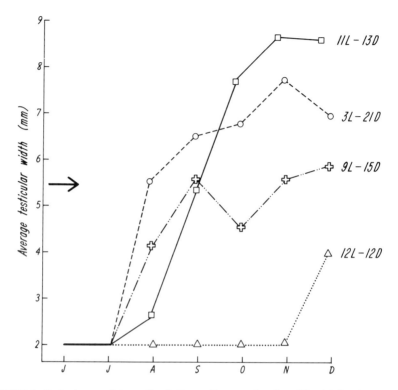

FIGURE 2   Testicular growth in postinvolution starlings as a function of fixed daily photoperiods. The testis width at or above which spermatogenesis can be expected is indicated by the single horizontal arrow. All photoperiodic treatments began on 21 June. (See also Tables 1 and 2.)

Unfortunately, failure of the watering system made it necessary to terminate the experiment under LD 11:13 prematurely, so we do not know whether these starlings would have remained in constant spermatogenesis. A subsequent experiment with juvenile birds suggests, however, that those birds might well have remained spermatogenic. In this experiment, groups of juvenile starlings (animals that had not previously experienced a testicular cycle) were exposed to a chronic LD 11:13 photoperiod, one group beginning in November and another in January. Most of these animals maintained constant spermatogenesis during 15 months of treatment (which would span two testicular growth-involution cycles in starlings under natural daylength). The particulars of that experiment are reported elsewhere (Schwab, 1970), but the pertinent average testicular size is shown in Figure 3.

Testicular growth occurred much later under LD 12:12 than with

shorter daily photoperiods. Even so, testicular growth began in December
and the animals achieved an average testicular size associated with spermatogenesis by mid-January. Monthly assays of testicular size during 27
months of treatment under LD *12*:12 (Table 2) document two testicular
cycles very nearly in phase with those observed in starlings under natural
daylengths.

No testicular growth was noted in starlings during 15 months of exposure to an LD *13*:11 regimen.

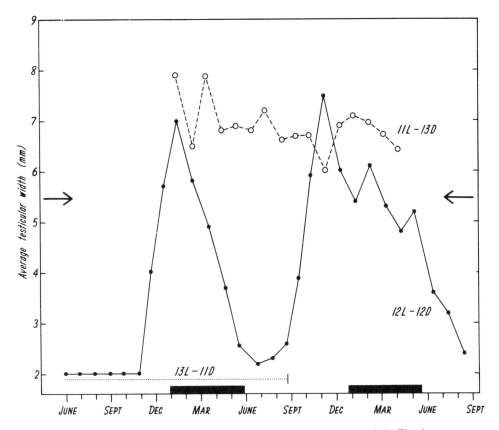

FIGURE 3   Testicular condition in European starlings under fixed daily photoperiods. The circannian testicular periodicity in starlings under LD *12*:12 is shown by the solid line (see Table 2 for
statistical description). Average testicular size in juvenile starlings under LD *11*:13 is indicated by
the broken line (see Schwab, 1970, for details). The duration of the LD *13*:11 treatment is shown
by the dotted line. Horizontal arrows indicate the testis width at or above which spermatogenesis
can be expected. For comparative purposes, the seasonal chronology of testicular growth-involution
in captive starlings under natural photoperiods is indicated by solid rectangles along the baseline.

TABLE 2 Statistical Description of Testicular Size (mm Width) in European Starlings Maintained under a Chronic LD $12:12$ Photoperiod

Month	1967				1968				1969			
	Day	N	Mean ± SE	Range	Day	N	Mean ± SE	Range	Day	N	Mean ± SE	Range
Jan.					19	18	5.7 ± 0.6	2.0–9.6	15	17	6.0 ± 0.5	2.3–9.7
Feb.					16	18	7.0 ± 0.5	2.0–10.0	17	13	5.4 ± 0.6	2.0–9.9
Mar.					18	18	5.8 ± 0.6	2.0–8.9	18	13	6.1 ± 0.6	2.0–9.7
Apr.					18	16	4.9 ± 0.6	2.0–8.3	29	11	5.3 ± 0.7	2.0–9.0
May					17	16	3.7 ± 0.5	2.0–6.8	28	9	4.8 ± 0.8	2.0–8.2
June	21	15	<2.0		19	16	2.5 ± 0.2	2.0–5.4	27	9	5.2 ± 1.0	2.0–10.3
July	19	15	<2.0		10	15	2.2 ± 0.2	2.0–5.1	28	9	3.6 ± 0.8	2.0–8.5
Aug.	21	14	<2.0		19	15	2.3 ± 0.3	2.0–5.6	20	9	3.2 ± 0.6	2.0–6.2
Sept.	18	15	<2.0		18	15	2.6 ± 0.4	2.0–7.2	24	7	2.4 ± 0.3	2.0–3.9
Oct.	20	15	<2.0		17	15	3.9 ± 0.6	2.0–7.3				
Nov.	16	15	<2.0		13	22	5.9 ± 0.6	2.0–9.8				
Dec.	18	13	4.0 ± 0.8	2.0–9.0	16	17	7.5 ± 0.3	4.6–10.0				

2. Resumption of Photosensitivity in Postinvolution Starlings

The period of quiescence following complete testicular involution in
European starlings consists of at least two physiologically distinct phases:
(a) the phase immediately following involution, during which no testicular
growth can be induced by exposing the animals to artificially increased
daily photoperiods; and (b) a subsequent phase characterized by gradually
increasing sensitivity to long daily photoperiods (that is, a photosensitivity
gradient).

The seasonal chronology of the two phases was determined by removing groups of postinvolution starlings from natural and from fixed *LD*
12:12 regimens each month from mid-June to mid-December. Each group
was exposed immediately to LL; testis size was assayed by laparotomy at
5-day intervals. These tests of photosensitivity showed that starlings taken
from both the *LD 12*:12 regimen and the natural photoperiod were refractory to long daylengths during June, July, and August. Thereafter, testicular growth in birds from both photoperiodic conditions was progressively
greater with LL treatments beginning in September, October, November,
and December (Figure 4).

There was one difference, however, in the light-induced testicular responses of birds taken from the two pretreatment light regimens. By
November, the birds taken from fixed *LD 12*:12 showed a maximal light-
induced testicular growth significantly greater ($P < 0.005$; Mann-Whitney
U test, one-tailed) than that of starlings taken from the natural photoperiod, even though the LL treatment began 11 days earlier for the
former group than for the latter group. This response suggests that the
birds of the *LD 12*:12 group emerged from the photorefractory condition
about 1 month sooner than birds under natural photoperiods. Moreover,
spontaneous testicular growth had already begun in birds from the chronic
LD 12:12 regimen (but not in birds from the natural photoperiod) at the
time the LL treatments were initiated in December (Table 2).

3. Testicular Responses in Postrefractory Starlings as a Function of Stable, Increasing, and Decreasing Daily Photoperiods

Beginning at the winter solstice, one group of European starlings was exposed to chronic *LD 9*:15 (the photoperiodic nadir at latitude 38° N) and
a second group to photoperiods decreasing 2 min each day from the nadir.
Starlings in both treatments achieved spermatogenesis in about 32 days
(Schwab and Darden, in press).

Starlings were also exposed to daily photoperiods increasing from the

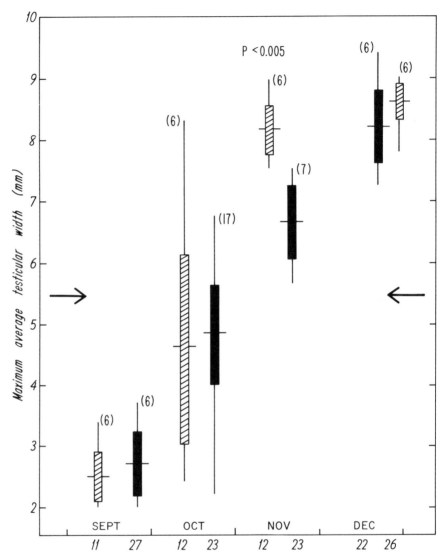

FIGURE 4 Photosensitivity gradient in starlings under chronic *L* D *12*:12 and under natural photoperiods. Indicated along the baseline is the date when each group of birds was removed from the natural photoperiod or the *LD 12*:12 artificial light cycle (prior to the initial testicular growth–involution cycle under this photoperiod) and exposed to continuous illumination. Testis width was measured by unilateral laparotomy at 5-day intervals. The maximum testis width achieved by each animal (regardless of amount of time under LL) is used to calculate the sample mean (horizontal line), the sample range (vertical line), and two standard errors above and below the mean (hatched rectangle for starlings from *LD 12*:12 and solid rectangle for starlings from natural photoperiod). Sample size is shown in parentheses, and level for spermatogenesis by the horizontal arrow. The P value (Mann-Whitney *U* test, one-tailed) of the statistically significant difference between the November samples is indicated. Values of the two groups for other months are not significantly different.

photoperiodic nadir at rates of 1, 2 (the natural average between the winter and summer solstice at this latitude), 3, 6, and 12 min each day. These increasing daily photoperiods induced spermatogenesis in about 79, 57, 47, 27, and 16 days, respectively. Thus, photoperiod increases must exceed 3 min per day to induce spermatogenesis in as short a time as does a fixed photoperiod of LD 9:15 or photoperiods decreasing therefrom at 2 min per day. Note that spermatogenesis will eventually occur in adult starlings exposed to any of the fixed or changing daily light regimens that we have used beginning after the winter solstice, and that only the rate of testicular growth is affected by the length of the photoperiod.

DISCUSSION

PHOTOPERIODIC EFFECTS ON ANNUAL TESTICULAR CYCLE

The annual testicular growth-involution cycle of the starling under natural photoperiods can be separated into at least four distinct phases (Figure 1). Results of exposing the birds to a variety of artificial photoperiods suggest that the photoperiodic effects are dependent on the natural phase the bird is in at the time of exposure.

Phase I spans the interval from the end of testicular quiescence to spermatogenesis. At latitude 38° N, Phase I occurs while the natural photoperiod increases in duration from about 9 to 12 hr per day, between the winter solstice and the vernal equinox. The conclusion that gonadal development occurs solely as a function of increase in photoperiod is not substantiated by experiment, however. Spermatogenesis will occur in starlings held in the laboratory under any stable or changing daily photoperiod beginning at the winter solstice, and photoperiodic manipulation affects only the rate at which spermatogenesis is achieved (Schwab and Darden, in press; Schwab, unpublished). This strongly suggests that daily photoperiods between the winter solstice and the vernal equinox regulate the rate but not the occurrence of testicular development in the European starling.

Phase II is characterized by continuing testicular growth, a period of spermatogenesis, and testicular involution to a quiescent condition. These stages normally occur between the vernal equinox and the summer solstice, while the natural daily photoperiod at this latitude is increasing from 12 to about 15 hr. The duration of spermatogenesis can be somewhat prolonged, although not extended indefinitely, by stabilizing the photoperiod at LD 12:12 (Schwab, unpublished) and by certain behavioral stimuli (Schwab and Lott, 1969). The duration of spermatogenesis can be shortened by exposing birds to daily photoperiods longer than natural daylengths during this phase (Schwab, unpublished). Involution of the testis does not occur in most individuals unless the daily photoperiod exceeds LD *11*:13 (Schwab, 1970).

The foregoing results indicate that testicular involution in the European starling is a function of the LD *12*:12 photoperiod at the beginning of this phase and that subsequent increases in daily photoperiods affect only the duration of spermatogenesis.

Phase III, the first half of the gonadal quiescent period, occurs between the summer solstice (about LD *15*:9 at this latitude) and the autumnal equinox. The testicular state of starlings during this period is characterized by the inability of daylengths from LD *13*:11 to LL to stimulate precocious testicular growth. Moreover, to remove starlings from such regimens, regardless of duration of treatment, and subsequently expose them to continuous light will not induce testicular growth. During this phase starlings are absolutely refractory to a chronic daily photoperiod of 13 hr or more. However, starlings under a chronic daily photoperiod of 11 hr or less, beginning at the summer solstice, will achieve spermatogenesis prior to the autumnal equinox, and the rate of testicular recrudescence to the spermatogenic stage is inversely related to the length of the daily photoperiod (Figure 2).

These results indicate that spontaneous testicular growth in postinvolution starlings is a function of daily photoperiods of less than 13 hr. But such photoperiods do not occur naturally until the end of Phase III. It seems likely, then, that the natural daylengths between the summer solstice and the autumnal equinox have little effect on the reproductive biology of the European starling other than extending the period of reproductive quiescence. That is, these daylengths neither initiate testicular growth nor permit the animals to regain photosensitivity.

Phase IV, the second half of the period of gonadal quiescence, is characterized by a photosensitivity gradient ranging from essentially zero, at or near the autumnal equinox, to maximum photosensitivity, at or near the winter solstice (about LD *9*:15 at this latitude).

The degree of photosensitivity appears to be chronologically somewhat more advanced in starlings from a fixed LD *12*:12 regimen beginning at the summer solstice than in starlings from natural photoperiods (Figure 4). Possibly the interval during which the natural photoperiod exceeds LD *12*:12 increases the length of time required before recovery of photosensitivity at shorter photoperiods (see Burger, 1947). It is equally plausible, however, that the starlings under LD *12*:12 began to increase in photosensitivity immediately after this treatment commenced and that this increase in photosensitivity reached measurable levels somewhat sooner than in birds under natural photoperiods. The chronological similarity in the photosensitivity gradient of birds from the fixed *LD 12*:12 regimen and of those under natural photoperiods is the result of the latter birds beginning later (after the duration of daily photophase reaches and subsequently falls below 12 hr), but subsequently having a faster rate of resumption of photosensitivity because they experience daily photoperiods of less than 12 hr. Recall that time to achieve spermatogenesis is inversely related to the duration of daily photoperiod (Figure 2).

Seasonally changing daily photoperiods do indeed influence the chronology of testicular cycles in starlings. These influences, beginning with postrefractory starlings (Phase I), are summarized below:

• The duration of daily photoperiod regulates the rate, though not the occurrence, of the testicular growth that ultimately results in spermatogenesis.

• Daily light durations of 12 hr or longer result in testicular involution. A photoperiod of *LD 11*:13 freezes the testicular cycle of most individuals in continuous spermatogenesis.

• Light durations longer than *LD 12*:12 do not permit resumption of photosensitivity, and the testicular cycle is frozen in continuous quiescence.

• Light durations of *LD 12*:12 or less facilitate resumption of photosensitivity, which is followed by spontaneous testicular development and spermatogenesis.

TESTICULAR CYCLES IN THE ABSENCE OF PHOTOPERIODIC CHANGE

The effects of natural daily photoperiods on the annual testicular cycle of the European starling suggest that this species has a functional photoperiodic threshold of *LD 12*:12. Furthermore, this single daily light regimen satisfies all regulatory photoperiodic functions necessary for a testicular cycle in this species. Exposure of starlings to fixed *LD 12*:12 at various times of the year has shown that this regimen facilitates testicular growth and associated spermatogenesis, testicular involution, photorefractoriness, and associated testicular quiescence. In addition, fixed photoperiods 1 hr longer or shorter (*LD 13*:11 and *LD 11*:13) result in diametrically opposed events—prolonged quiescence and prolonged spermatogenesis.

It seems reasonable, then, to conclude that the timing of testicular cycles in this species is a function of seasonal photoperiodic oscillations above and below a critical LD *12*:12 photoperiod threshold, and that these oscillations need not have an amplitude greater than 1 hr to facilitate synchronization of the photoperiodic timing mechanism with the annual environmental cycle. This, in turn, suggests the hypothesis that circannian testicular cycles would occur in European starlings in the absence of photoperiod change *if* the chronic daily regimen consisted of the operational photoperiod threshold, i.e., LD *12*:12. This was, in fact, the case. Starlings under the LD *12*:12 regimen underwent two testicular cycles (Figure 3, Table 2) which were essentially in phase with those observed in the control group, but with certain notable differences.

The testicular quiescence period in starlings under natural photoperiods is characterized by complete testicular involution in all individuals. Involution was complete in all but one starling under chronic LD *12*:12, but this individual did achieve testicular widths (5.4 and 5.1 mm) associated with a nonspermatogenic stage during June and July 1968. Certain individual starlings possibly possess a slightly higher photoperiod threshold for testicular involution than is "normal" for this species, accounting for the lack of complete involution in the one bird under the LD *12*:12 regimen. In a similar instance, Schwab (1970) found that continuous spermatogenesis could be prolonged for over a year in about 75 percent of a sample of European starlings under chronic LD *11*:13. The remainder of the sample exhibited involution to a testis size associated with cessation of spermatogenesis. In the former birds, the LD *11*:13 regimen possibly exceeded the photoperiodic threshold for continuous spermatogenesis.

Peaks of testicular size in starlings under natural daylengths are separated temporally by about 12 months. In contrast, the second peak of testicular size in birds under fixed LD *12*:12 was separated from the initial peak by about 9½ months. It is possible that the timing of the second testicular cycle under LD *12*:12 results in part from differential rates at which photorefractoriness is dispelled under this regimen and under natural photoperiods. This possibility, however, depends on the validity of the hypothesis proposed by Schwab and Darden (in press) that spontaneous testicular growth in European starlings under natural photoperiods is essentially a function of complete removal of photorefractoriness. While this may be the case under natural daylengths, such a process would not preclude a second, perhaps synergistic, mechanism for the timing of testicular cycles.

A more plausible explanation for the timing of the second testicular cycle under LD *12*:12 is that the chronology is a function of a free-running endogenous circannian rhythm. It seems likely that in our experiments this rhythm was not synchronized by any periodic exogenous stimulus, since the periodicity differed from that of the normal annual cycle.

The duration of spermatogenesis during the second cycle under *LD 12*:12 was considerably longer than that observed in starlings under natural daylengths. Furthermore, the involution phase was characterized by several minor fluctuations in average testicular size (Figure 3); no such phenomena were observed under natural daylengths. Possibly the differences in duration of spermatogenesis and in the involution phase under *LD 12*:12 and natural photoperiods result from (1) individual birds with operational photoperiodic thresholds for spermatogenesis and testicular involution slightly above or below the *LD 12*:12 level; (2) phase differences between individuals—i.e., the chronology of testicular cycles in some of the animals was advanced, in others delayed, and in others closely approximated the chronology of the natural cycle; and (3) limited persistence of the time-measurement mechanism in the absence of environmental cues, i.e., a damping of the endogenous circannian periodicity. Despite these differences, chronological similarity of testicular cycles in starlings under *LD 12*:12 and in those under natural daylengths suggests the existence of an endogenous circannian reproductive periodicity.

The meager data presently available on endogenous circannian reproductive rhythms in birds do not permit extensive speculation on the role of such periodicities in the control of reproductive cycles under natural conditions. Induction, with light cycles, of more than one testicular cycle in starlings during a single year (Burger, 1949) indicates that daily photoperiods can override the influence of an endogenous circannian reproductive periodicity. However, this result in no way precludes the possibility that such a periodicity results in an annual photosensitivity gradient facilitating timing of avian breeding seasons by natural daylengths and other proximate environmental synchronizers. Furthermore, what appears to be an endogenous circannian rhythm certainly can induce a breeding condition in male starlings maintained under a fixed *LD 12*:12 photoperiod. This circannian periodicity has considerable adaptive value with respect to the timing of breeding cycles in starlings inhabiting low latitudes, where the annual photoperiodic fluctuations are relatively slight. Such an endogenous circannian periodicity would in no way interfere with photoperiodically controlled breeding cycles in the higher latitudes. In fact, it seems reasonable to suppose that the typical photoperiodic control mechanism establishing the chronology of starling breeding cycles in higher latitudes evolved from the basic equatorial scheme, i.e., the endogenous circannian reproductive periodicity.

ACKNOWLEDGMENTS

I am especially indebted to S. T. Keating, T. R. Darden, and J. L. Osborne for technical and laboratory assistance during these investigations, and to the California State and County Departments of Agriculture for providing the starlings. The study was made possible by financial assistance from the State of California Starling Control Program and the Department of Animal Physiology, University of California at Davis.

REFERENCES

Benoit, J. 1961. Opto-sexual reflex in the duck: physiological and histological aspects. Yale J. Biol. Med. 34:97–116.

Bissonnette, T. H. 1931. Studies on the sexual cycle in birds. IV. Experimental modification of the sexual cycle in males of the European starling, *Sturnus vulgaris*, by changes in daily period of illumination and of muscular work. J. Exp. Zool. 58: 281–320.

Burger, W. J. 1947. On the relation of day length to the phases of testicular involution and inactivity of the spermatogenic cycle of the starling. J. Exp. Zool. 105:259–268.

Burger, W. J. 1949. A review of experimental investigations on seasonal reproduction in birds. Wilson Bull. 61:211–230.

Burger, W. J. 1953. The effect of photic and psychic stimuli on the reproductive cycle of the male starling, *Sturnus vulgaris*. J. Exp. Zool. 124:227–239.

Farner, D. S. 1959. Photoperiodic control of annual gonadal cycles in birds, p. 717 to 750. *In* R. B. Withrow [ed.], Photoperiodism. American Association for the Advancement of Science, Washington, D.C.

Farner, D. S. 1964. The photoperiodic control of reproductive cycles in birds. Am. Sci. 52:137–156.

Farner, D. S. 1965. Circadian systems in the photoperiodic responses of vertebrates, p. 357 to 369. *In* J. Aschoff [ed.], Circadian clocks. North-Holland Publ. Co., Amsterdam.

Farner, D. S., and B. K. Follett. 1966. Light and other environmental factors affecting avian reproduction, J. Anim. Sci. 25:90–115.

Hamner, W. M. 1963. Diurnal rhythm and photoperiodism in testicular recrudescence of the house finch. Science 142:1294–1295.

Hamner, W. M. 1964. Circadian control of photoperiodism in the house finch demonstrated by interrupted-night experiments. Nature (Lond.) 203:1400–1401.

Hamner, W. M. 1966. Photoperiodic control of the annual testicular cycle in the house finch, *Carpodacus mexicanus*. Gen. Comp. Endocrinol. 7:224–233.

Hamner, W. M. 1968. The photorefractory period of the house finch. Ecology 49: 211–227.

Lofts, B. 1962. Photoperiod and the refractory period of reproduction in an equatorial bird, *Quelea quelea*. Ibis 104:407–414.

Lofts, B. 1964. Evidence of an autonomous reproductive rhythm in an equatorial bird, *Quelea quelea*. Nature (Lond.) 201:523–524.

Lofts, B., and R. K. Murton. 1968. Photoperiodic and physiological adaptations regulating avian breeding cycles and their ecological significance. J. Zool. 155:327–394.

Marshall, A. J. 1959. Breeding biology and physiology: internal and environmental control of breeding. Ibis 101:456–478.

Marshall, A. J. 1960a. The role of the internal rhythm of reproduction in the timing of avian breeding seasons, including migration. Proc. XIIth Int. Ornithol. Congr., Helsinki, 1958, p. 475–482.

Marshall, A. J. 1960b. The environment, cyclical reproductive activity and behavior in birds. Symp. Zool. Soc. Lond. 1960(2):53–67.

Marshall, A. J. 1960c. Annual periodicity in the migration and reproduction of birds. Cold Spring Harbor Symp. Quant. Biol. 25:499–505.

Marshall, A. J., and D. L. Serventy. 1956. The breeding cycle of the short-tailed shearwater, *Puffinus tenuirostris* (Temminck), in relation to trans-equatorial migration and its environment. Proc. Zool. Soc. Lond. 127:489–510.

Marshall, A. J., and D. L. Serventy. 1958. The internal rhythm of reproduction in xerophilous birds under conditions of illumination and darkness. J. Exp. Biol. 35: 666–670.

Marshall, A. J., and D. L. Serventy. 1959. Experimental demonstration of an internal rhythm of reproduction in a transequatorial migrant (the short-tailed shearwater (*Puffinus tenuirostris*). Nature (Lond.) 184:1704–1705.

Menaker, M. 1965. Circadian rhythms and photoperiodism in *Passer domesticus*, p. 385 to 395. *In* J. Aschoff [ed.], Circadian clocks. North-Holland Publ. Co., Amsterdam.

Merkel, F. W. 1961. Der Einfluss eines kunstlich beibehaltenen Langtages auf Kleinvogel. Verh. Dsch. Zool. Ges. Saarbrucken 1961:357–363.

Miller, A. H. 1955. The expression of innate reproductive rhythm under conditions of winter lighting. Auk 72:260–264.

Miller, A. H. 1959. Responses to experimental light increments by Andean sparrows from an equatorial area. Condor 61:344–347.

Pengelley, E. T. 1967. The relation of external conditions to the onset and termination of hibernation and estivation, p. 1 to 29. *In* K. C. Fisher *et al.* [ed.], Mammalian hibernation. Vol. III. Oliver and Boyd, Edinburgh.

Rowan, W. 1925. Relation of light to bird migration and developmental changes. Nature (Lond.) 115:494–495.

Schildmacher, H. 1956. Physiologische Untersuchungen am Grunfinken, *Chloris chloris* (L.) in kunstlichen Kurztag und nach "hormonaler Sterilisierung." Biol. Zentralbl. 76:327.

Schwab, R. G. 1970. Light-induced prolongation of spermatogenesis in the European starling, *Sturnus vulgaris*. Condor 72:466–470.

Schwab, R. G., and T. R. Darden. 1970. Testicular metamorphosis in European starlings, *Sturnus vulgaris*, as a function of increasing, decreasing, and stable photoperiods. Condor.

Schwab, R. G., and D. F. Lott. 1969. Testis growth and regression in starlings, *Sturnus vulgaris*, as a function of the presence of females. J. Exp. Zool. 171:39–42.

Wolfson, A. 1959a. Role of light and darkness in the regulation of the refractory period in the gonadal and fat cycles of migratory birds. Physiol. Zool. 32:160–176.

Wolfson, A. 1959b. The role of light and darkness in the regulation of spring migration and reproduction in birds, p. 679 to 716. *In* R. B. Withrow [ed.], Photoperiodism. American Association for the Advancement of Science, Washington, D.C.

Wolfson, A. 1960. Regulation of annual periodicity in the migration and reproduction of birds. Cold Spring Harbor Symp. Quant. Biol. 25:507–514.

Wolfson, A. 1964. Role of day length and the hypothalamo-hypophysial system in the regulation of annual reproductive cycles. Proc. 2nd Int. Congr. Endocrinol., Lond., p. 183–187.

Wolfson, A. 1965. Circadian rhythm and the photoperiodic regulation of the annual reproductive cycle in birds, p. 370 to 387. *In* J. Aschoff [ed.], Circadian clocks. North-Holland Publ. Co., Amsterdam.

DISCUSSION

ENRIGHT: I can only marvel and applaud the patience of someone who will undertake experiments lasting two years or longer.

SCHWAB: One obvious advantage is that you don't have to work as fast.

ASCHOFF: I am extremely pleased by these data. The results you got in *LD 12*:12 are a beautiful example of what we would call a circannual rhythm, free-running with a period deviating from 12 months. I am especially struck by the difference in testicular response you found with stable and increasing photoperiods. The birds came to spermatogenesis in about 32 days under *LD 9*:15. It took them longer (about 57 days) when the photoperiod was increasing by 2 min a day, but not as long (only 16 days) when the photoperiod was increasing by 12 min a day. This is a most remarkable discovery. I would interpret this as a differential effect of the circannual *Zeitgeber.* It is also very remarkable that your circannual cycle oscillates only when the photoperiod is around *LD 12*:12. Now I had expected a rather small range of environmental conditions under which the circannual rhythm can oscillate, but not quite that small. When I met Goss in Hanover last year, we discussed this, and, if I remember correctly, he said that the circannual rhythms in deer antler growth do not oscillate under *LD 12*:12, but only when the photoperiod is longer or shorter. That makes your results even more fantastic. Now my last point is that if some circannual rhythms oscillate under only a very narrow range of conditions, I would take this to

indicate—in contrast to what Dr. Gwinner said—a more pendulum-like oscillation.

SCHWAB: I can do nothing but agree.

PAVLIDIS: It seemed that you emphasized the point that the rhythmicity you see is due primarily to the 12-hr photoperiod, while Dr. Gwinner interpreted his results as demonstrating some endogenous circannual oscillator. Am I right?

ASCHOFF: Schwab has a very clear-cut, beautiful endogenous circannual rhythm.

PAVLIDIS: Not necessarily.

SCHWAB: It is a question of whether the *LD 12*:12 cycle causes, or simply permits, the annual cycle.

ASCHOFF: An *LD 12*:12 cycle is just as constant an environment with respect to an annual rhythm as the heartbeat is for a circadian rhythm. The free-running cycles are from 9 to 11 months. It has nothing to do with season and nothing to do with multiples of *LD 12*:12.

PAVLIDIS: It is disturbing to me that you are assuming more and more clocks for the organism.

ASCHOFF: Oh, I like it.

FARNER: I have two somewhat general comments. First, these are of course extremely exciting data. I think they confirm something that has been suggested by the data of many investigations—namely, that we are dealing with three photoperiodically controlled phenomena, inseparable or nearly so in most species. One is the output of FSH-like hormone, another is the output of LH-like hormone, and the third—probably completely independent of these—is the development of the so-called photorefractoriness. It seems to me that you have here an ideal system to attack these independently, which is very, very difficult to do in the species with which I work. We can separate two of them by hypothalamic lesions, but so far that is all.

The other point that I would like to make—and I am on shaky ground here since I am not a traditional ornithologist—is that there are only a few species in the relatively large family (Sturnidae) that have become temperate-zone inhabitants; thus this may be a neat case of convergence with respect to photoperiodic control.

SCHWAB: Dr. Farner's points are well taken. I would predict that a similar system would be found in the house sparrow but not in all birds. We have been unable to duplicate the system—that is, to freeze the cycle in a state of continuous spermatogenesis—in the yellow-headed or tricolored blackbird.

W. M. HAMNER

On Seeking an Alternative
to the Endogenous Reproductive
Rhythm Hypothesis in Birds

Schwab (this volume) has clearly shown that male starlings, maintained continuously in the laboratory on a lighting regimen of LD *12*:12 for about 2 years, show periodic fluctuations in gonad size that correspond remarkably to the natural testicular cycle of wild starlings. Lofts (1964) reported similar observations for African weaver finches, *Quelea quelea*, also maintained on LD *12*:12. Some investigators regard these data as proof that the annual cycles of these birds are timed by an endogenous "circannual" clock that retains its seasonal accuracy by being reset anew each year by environmental or social stimuli (Lofts and Murton, 1968). There is a tendency among some investigators to regard this endogenous annual rhythm hypothesis as a fully sufficient mechanistic explanation of reproductive periodicity. For example, Lofts and Murton (1968) compare "photoperiodic" species to "autonomous" species and then attempt to develop evolutionary explanations of patterns of reproduction within these groups; Gwinner (this volume) has tried to extract general properties from all types of annual rhythms in order to generate a theoretical approach to the study of seasonality. It is possible, however, that some of the generalizations, particularly those about annual reproductive rhythms in birds, are premature. "Endogenous" species may be much rarer than has been assumed; perhaps they do not even exist. In this paper, I urge caution in dis-

cussing the "autonomous" nature of annual reproductive rhythms, a rec-
ommendation based on data on the testicular cycle of the European starling, *Sturnus vulgaris*. I propose as an alternative a hypothesis on control
of annual reproductive rhythms that makes specific predictions about the
testicular cycle of starlings.

Two general considerations originally prompted this investigation. The
first is related to the historical controversy about the endogenous or exogenous control of avian reproductive cycles [see Marler and Hamilton (1966)
on the endogenous–exogenous conflict throughout the field of animal
behavior]. Proponents of an endogenous mechanism for reproductive
periodicity have cited many observations that are difficult to interpret any
other way (Lofts and Murton, 1968; Gwinner, this volume). However,
regardless of whether endogenous annual rhythms exist or not, the re-
searcher who concludes that a particular breeding response is "endogenous"
often terminates his research on that subject and does something else
thereafter. After all, if the response is "endogenous," what else can one
do? Yet recent investigations with the house finch (Hamner, 1963, 1964,
1966, 1968), the starling (Schwab, this volume), and the bobolink
(Hamner and Stocking, 1970) indicate that photoperiodic mechanisms in
birds are much more complicated than had been envisioned. Since some
of these birds show responses similar or identical to the classic "endogenous" species, it is not unreasonable to suspect that there may be more to
learn about other "endogenous" species also. Although we have invoked
an endogenous component recently to explain some observations on
bobolinks (Hamner and Stocking, 1970), I regard the hypothesis as tentative and by no means suggest that it adequately substitutes for further
research. Much more photoperiodic research is required for bobolinks in
particular and for birds in general, and any conclusion that obscures this
need should be accepted reluctantly.

The second general consideration is not unrelated to the first. Popper
(1959) has pointed out that it is impossible to prove anything absolutely
in empirical science (see also Hempel, 1966), but that science advances by
elimination of alternative hypotheses, and, in particular, by disproof of a
null hypothesis. Yet, according to Ghent (1966)

the biological clock studies provide awkward situations in which to state a Null Hypothesis. The Null Hypothesis is supposed to be the conservative one that calls for a minimum of magic. But how is magic minimized by stating that the subtraction of every
conceivable directive stimulus will leave the rhythmicity unaffected? Such a statement
in reality is a Null Hypothesis in form only. With all known directive factors subtracted,
the Null Hypothesis becomes only a back-handed way of pointing very positively at the
organism itself as the originator of its own rhythmic activity through some internal
mechanism.

Thus we are in danger of attempting to prove that biological clocks exist when it is impossible for empirical science to prove anything at all, and, of course, if we do have "proof" we need no more data, and so we do no new experiments. This does not mean, of course, that biological clocks do not exist (indeed, I hope, they do; Hamner, 1963, 1964, 1966, 1968; Hamner and Enright, 1967), but it does suggest that we may be accepting a hypothesis in default of adequate alternatives. In the case of circadian rhythmicities, attempts have been made to test alternatives even though, as Ghent pointed out, it is not conceptually easy in this field to formulate alternatives. These attempts have generally failed, in the opinion of most students of rhythmicity (but see Brown, 1968). However, almost no attempt at all has been made to formulate alternative explanations for those data that are routinely accepted as proving the existence of endogenous annual rhythms.

These general considerations suggest that a hypothesis that utilized less of Ghent's "magic" might indeed be helpful; one such hypothesis is submitted below. This hypothesis makes predictions about the annual reproductive cycles of starlings that can be clearly tested and disproved. Other hypotheses are also possible but were not considered.

The customary test for the endogenous nature of a diurnal rhythm is to place the organism under constant conditions and observe whether the rhythm deviates regularly from 24 hr. Yet certain seemingly constant environmental conditions also entrain daily rhythms. For example, light cycles of LD 6:6:6:6 or LD 4:4:4:4 in some cases entrain a normally free-running rhythm to precisely 24 hr, a phenomenon called "frequency demultiplication"; the rhythm uses one of the light–dark transitions as a stable reference point. Frequency supermultiplication has also been observed (Hamner and Enright, 1967). Unfortunately, we have no reliable way of predicting which particular combinations will entrain a rhythm. Most tests for an endogenous annual reproductive rhythm in birds have been conducted under experimental conditions utilizing lighting regimens that either produce a daily light–dark cycle of exactly 24 hr (Lofts, 1964; Schwab, this volume) or occur in nature under conditions where daylengths do not change with season, as at the equator (Chapin, 1954; Chapin and Wing, 1959; Ashmole, 1962; Dorward, 1962; Snow, 1965). However, the fact that daylength does not change seasonally in the laboratory or the field does not mean that the experimental animals are exposed to "constant conditions." It is possible that the 24-hr light–dark cycle acts as a daily environmental cue to entrain the annual rhythm much as an alternating LD 4:4:4... cycle entrains a 24-hr rhythm by frequency demultiplication. In this case, one could seriously postulate an "annual frequency-demultiplication effect," in which the daily photoperiod provides the temporal cue for the annual gonadal cycle. Constant photoperiods could provide sufficient daily information both to synchronize and to drive the annual cycle. Synchronization by LD *12*:12 could produce the surprisingly accurate annual rhythms observed by Lofts and Schwab, heretofore interpreted as endogenous. The constant daily photoperiod also could drive the annual rhythm by providing the daily light stimulus that apparently is required for testicular growth and for termination of photorefractoriness. This requirement is a real and important difference between circadian rhythms and annual reproductive cycles of birds. Circadian systems are often entrained but are not driven by light; breeding cycles of birds may be synchronized by daylength and also physiologically driven by daylength through direct stimulation of the gonadotropic hormones required for gonadal growth. Available data are not sufficient for a detailed daily photoperiodic input hypothesis, but the above formulation suggests certain predictions.

If the annual rhythmicity seen under LD *12*:12 is indeed endogenous and the rhythm free-running, then LD *12*:12 must be seen as "constant" by the bird. One could then predict that in the absence of a light–dark cycle, under conditions of continuous illumination at specified light intensities, the annual rhythm would also free-run. If LD *12*:12 is "constant," then LL and DD would surely be "constant" also. Thus we should see a "circannual" rhythm under LL or DD if the mechansim is endogeneous, but if the rhythm is merely driven by constant daily LD *12*:12 ("annual frequency demultiplication"), then under LL or DD the rhythm would disappear. Benoit, Assenmacher, and Brard (1956) have done this experiment with domesticated ducks, but the testicular cycles of the ducks fluctuated so wildly that clear interpretation of their data is extremely difficult. No other similar tests have been reported.

One can make other predictions as well. If the daily light–dark regimen drives the entire seasonal reproductive cycle, then changing the photoperiod of a 24-hr LD cycle should alter the shape of the annual response curve. Long daylengths are known to hasten growth, maintenance, and regression of the testes, but apparently they delay photorefractory termination in most temperate-zone birds (Hamner, 1966); short daylengths delay growth, prolong gametogenic activity, and delay regression, but hasten termination of the refractory period (Hamner, 1966, 1968). Under the hypothesis outlined above, the shape of the normal annual testicular cycle would be expected to alter in birds exposed to continuous long days for over a year, so that the gametogenic phase is compressed and the photorefractory period prolonged; short-day birds should show a prolonged gametogenic phase and a shortened photorefractory period. Alternatively, if the rhythm is simply endogenous and if long days and short days are both interpreted as "constant conditions," then the annual cycle should persist with the same period, amplitude, and shape as seen under LD *12*:12.

Several kinds of predictions were thus stimulated by this "daily photoperiodic input" hypothesis and are amenable to disproof. Two experimental regimens were selected to test these predictions: constant illumination at four intensities: 10 lux, 1 lux, 0.1 lux, and 0.01 lux; continuous daily photoperiods of LD *15*:9, LD *13*:11, LD *11*:13, or LD *9*:15.

MATERIALS AND METHODS

Juvenile starlings (made available by Dr. Robert Schwab through the Starling Research Project) were placed in light-tight boxes, approximately 60×60×60 cm, inside wire mesh cages with large food-hoppers and automatic flow-tip watering devices. These boxes were lit by two 15-watt incandescent bulbs and were ventilated by individual fans. The boxes were constructed in banks of eight—four high and two wide. The birds could hear room noises. Temperature fluctuated between 24 and 30°C. Birds were moved into the experimental chambers from similar cages outdoors as soon as they had terminated their normal photorefractory period (Schwab, personal communication). A sample of 10 birds was laparotomized on November 6, 1967; in all these birds, the left testis was less than 2 mm wide. The birds on regimens of fixed daylength were exposed to light intensities of about 300 lux during the light phase of the cycle. Those birds exposed to continuous light of various intensities had the light from two 10-watt bulbs reduced by shielding the bulbs with layers of greenhouse cloth, which served as a neutral density filter. Fine adjustment of light intensity was accomplished by rheostats. Light intensity was measured at midcage. Testis width was measured in individually marked birds by laparotomy by means of vernier calipers inserted directly into the body cavity between the last two ribs. (Handling or operative procedures at intervals of 20 days does not alter testicular growth.) Birds maintained under continuous dim light were removed from the cage into a completely darkened room. Before each laparotomy the bird was hooded to prevent additional light stimulation. Birds were examined internally with the aid of a pencil-thin beam of light from a head lamp. Thus even during surgery these birds did not see any additional light. The birds on fixed lighting schedules were always laparotomized at the same relative time of day; this was usually also true for those receiving constant light but this factor was thought to be less important for those groups. Sample size in the very dim light conditions was small because the birds had difficulty adjusting to the darkened cages; some birds did not easily learn to drink from the automatic watering devices and therefore died. Birds newly placed in these cages were disoriented and were often killed by the original birds during the first few days.

The data presented show individual responses (solid circles), means of the groups (open circles), 95 percent confidence limits (vertical brackets), and a curve roughly fitted by eye to the means. The experiments were usually continued for 16 months.

RESULTS

FIXED DAYLENGTH REGIMENS (FIGURE 1)

The birds on the fixed daylengths tested in this experiment did not cycle, even though they do cycle on *LD 12*:12 (Schwab, this volume). However, there is a tendency toward rhythmicity under *LD 11*:13. Note the oscillation in testis width of the smallest gonad beginning in August 1968; these data were collected from the same individual. Schwab has also noticed this tendency to oscillate under *LD 11*:13, but in only a small proportion of his sample. Quite remarkably, and opposed to what is normally seen in temperate-zone photoperiodic species, the testes of starlings on *LD 11*:13 do not spontaneously regress. Whether this reflects the absence of an appropriate stimulus for regression or an actual inhibition of regression by *LD 11*:13 is not known. Except in birds under *LD 11*:13, the gametogenic phase of the cycle also corresponded to the predictions made in the Introduction. Under long daylengths (*LD 15*:9) the gametogenic phase is compressed, lasting only two months; under short daylengths (*LD 9*:15) it is extended, lasting almost 6½ months. The relationship of part of the gametogenic cycle to length of day is more complex than had been noted previously. With decreasing daylength, regression slows, until at *LD 11*:13, it does not occur at all. Under daylengths shorter than 11 hr, regression is apparently progressively accelerated. "Spontaneous" regression on *LD 13*:11 takes almost 2 months, but on *LD 9*:15 it takes only about 20 days. Schwab (personal communication) has found different effects of daylength on testicular regression in similar (but not identical) experiments; it is clear that more research is necessary to evaluate these data.

It is not surprising that no seasonal rhythmicity was observed in birds on *LD 15*:9. Very long days have effects identical to those of continuous bright light, and under these conditions photorefractoriness is not termi-

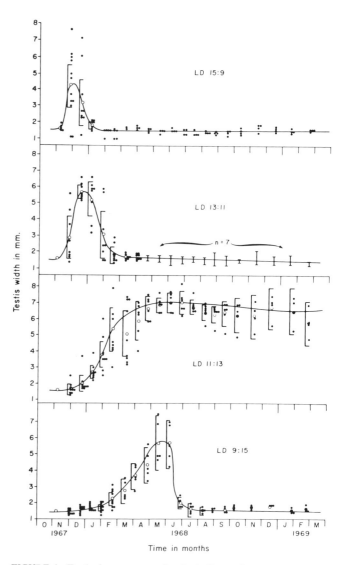

FIGURE 1   Testicular responses of male starlings to long-term constant daylengths. Solid circles are widths of left testes from individual birds; open circles are means. Vertical brackets show 95 percent confidence limits; the curve was drawn by eye.

nated. *LD 13*:11 also could be interpreted by the bird as a long day, with a similar effect. It is possible that these birds did terminate photorefractoriness but, like house finches under similar conditions (Hamner, 1968), were in a state of relative photorefractoriness such that they were sensitive only to daylengths longer than those to which they had been previously exposed. This possibility was not pursued. In any case, the testes of these birds did not recrudesce spontaneously, nor did the testes of birds on *LD 9*:15, although had the observations continued another 2 or 3 months, the birds on *LD 9*:15 might have shown a spontaneous testis cycle. Recrudescence should have begun in February 1969, however; if the period of the cycle is shorter than, or equal to, 1 year. If the cycle has a longer period, we would not have observed it.

Major differences in testicular response do occur in relation to relatively slight differences in daylength. I had thought that daylengths differing by 2 hr would be adequate to elicit the full range of photoperiodic response, yet the *LD 13*:11 and *LD 11*:13 schedules entirely missed the cycling seen by Schwab under *LD 12*:12.

CONSTANT LIGHT INTENSITY EXPERIMENTS

No rhythmicity occurred under continuous light varying in intensity over four orders of magnitude (Figure 2). Bright light (10 lux), as expected, induced responses similar to those seen under *LD 15*:9; in fact, the curves are almost identical. The testes of the birds under 10 lux enlarged rapidly and then regressed rapidly. They did not recrudesce spontaneously (but note that this experiment was too short to show an annual rhythm, if one should exist). Growth under 1.0 lux was just as rapid, but regression was very slow; 5 months passed before the testes of all the birds on this treatment had completely regressed. The 0.1 and 0.01 lux groups are statistically identical, with growth delayed until February (about when it starts normally); from that point, gametogenesis progressed gradually for the next year. No regression occurred in these two groups.[*] Contrary to what is generally believed, light at normal environmental levels apparently is not necessary for gametogenesis; gametogenesis will occur even at intensities below that of dim moonlight. Light can, however, trigger gametogenesis;

[*] Regressed testes are almost always less than 2 mm in width in the starling. The birds on these two regimens do not show the individual rhythmicity seen in Benoit's ducks, which would be masked by not plotting individual testicular growth curves. High points on the graph are from the same bird; low points from another individual. Desynchronization of "individual endogenous annual cycles" did not occur.

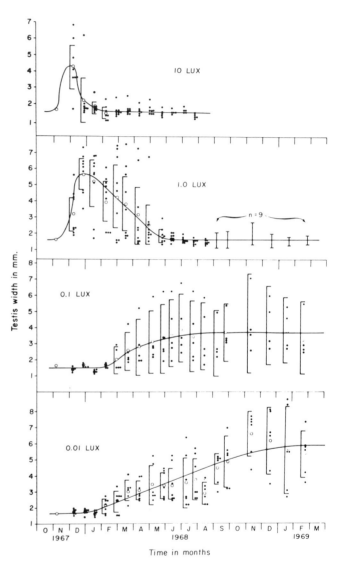

FIGURE 2 Testicular responses of male starlings to continuous illumination of fixed intensity.

thus, light is a sufficient, but not a necessary, stimulus for gonadal growth. On the other hand, gonadal regression requires light above 0.1 lux.

However, we saw previously that on *LD 11*:13 the testes did not regress even under bright light, so bright light *per se* is a necessary but not a sufficient stimulus for testicular regression. Obviously growth and regression of the testes are phenomena requiring different light intensities and different daylength conditions. The maintenance of enlarged testes in very dim light suggests that gametogenesis may well be a response to elimination of inhibition rather than the result of a direct light stimulus. In the absence of inhibition, gametogenesis proceeds spontaneously. This growth response may thus have several components: an intensity threshold (Figure 2), a daylength threshold, which may be flexible, and an internal timing component (birds exposed to *LD 9*:15, and LL at 0.1 lux and 0.01 lux, all showed gametogenic activity spontaneously in February).

DISCUSSION

One prediction of the "daily photoperiodic input hypothesis" was that in the absence of daily photoperiodic change (under continuous dim light) there would be no rhythmicity, and none was observed. If annual reproductive rhythms are similar to daily circadian rhythms, the endogenous nature of the rhythm should be more readily apparent if conditions are more constant. Since the gonads do not cycle under constant dim light of four different intensities, nor under any of four constant daylengths, but only under *LD 12*:12, it is difficult to conclude that this system is "endogenous." Apparently some historical bias is present in the timing mechanism. Under several conditions the gametogenic phase runs to completion, but under several different conditions gametogenesis begins about February. Parts of the cycle seem to be related both mechanistically and temporally to prior events. Schwab has shown that the prior reproductive cycle modifies the responses that occur thereafter in the starling (personal communication), but this modification by prior events is not endogenous rhythmicity in the usual circadian sense.

One might still wish to speak of the rhythmicity seen by Schwab and Lofts in birds exposed to *LD 12*:12 as having "limited ranges of entrainment," or to call these gonadal cycles "endogenous" or "autonomous," but such descriptions provide no new insight into the physiology or ecology of the birds. Our attempt to provide a testable alternative to the endogenous annual rhythm hypothesis, on the other hand, has led to specific experiments and new data that perhaps shed light on the physiological mechanisms controlling avian reproduction, and that, in any case, suggest that *LD 12*:12 is "peculiar," and perhaps unique, in evoking cyclic responses.

The data indicate clearly that the annual reproductive cycle of birds is not a cycle of degree, of change in the amplitude of one circulating hormone, of amount of physical activity, or of metabolic demand. It is, rather, a cycle of four distinct and qualitatively different reproductive phases that, in a particular sequence, produce an annual breeding rhythm. Each of these four phases of the reproductive cycle has separate daylength or light requirements and is thus physiologically distinct (even though the same gonadotropic hormones may be involved in the control of all four phases). The "endogenous" explanation tends to ignore these differences and leads to analogies with circadian systems (such as discussions of "alpha and rho") that have no physiological or ecological correlates. Use of circadian terminology for annual rhythmic events may permit model-building, but this model-building has not even told us anything that we did not already know about the physiology or ecology of circadian systems. In my opinion, extension of this terminology to the field of annual rhythms serves no useful purpose.

Reproductive timing mechanisms are under high selective pressures: Birds that make temporal errors do not leave offspring. The timing mechanisms for breeding are also undoubtedly plastic, since latitudinal changes in distribution of birds are common and since competition and predation force interspecific readjustments of breeding season as well. We do not search for one particular sweeping generalization to explain the diversity of characteristics such as feeding preferences or color patterns that are under similarly high selective pressures; we seek to explain these patterns in terms of evolutionary trends. Thus we commend the scholarly attempt by Lofts and Murton (1968) to examine the reproductive patterns of birds from an evolutionary perspective, even though we do not think their scheme is correct (see Hamner and Stocking, 1970).

I do not really intend to propose a serious alternative to the endogenous annual reproductive rhythm hypothesis for birds. The "annual frequency demultiplication hypothesis," or, in a somewhat less offensive form, the daily photoperiodic input hypothesis, is probably too simplistic an explanation for the complex breeding control system of starlings. Probably, as in plants and insects (Hamner, 1969), the timing mechanism is composed of several components that interact uniquely in each species to produce a synchronous and timely breeding effort. Nonetheless, the alternative I have proposed is clearly superior to the endogenous mechanism hypothesis, since it could be tested and since those tests produced data that provide new insights into the physiology of breeding cycles.

In its most extreme form the daily photoperiodic input hypothesis predicts that (1) under constant light of fixed intensity there would be no rhythmicity, whereas (2) under different photoperiodic combinations the system would cycle but the duration of the various phases of the rhythm would readjust to give an apparently endogenous annual cycle. The hypothesis, in this extreme form, was disproved by the experimental results. The former prediction was confirmed (there was no rhythmicity under LL) but the latter was not. The photoperiodic combinations tested did not result in reproductive cycling, but Schwab (this volume) observed rhythmicity under LD 12:12. Thus the daily photoperiodic input hypothesis is adequate for schedules approximating LD 12:12, but it is unsatisfactory for any other photoperiodic combination.

The hypothesis can, however, be expanded to account for the experimental results. Thus the reproductive cycle of the starling may be somewhat arbitrarily divided into four physiological stages (growth, maintenance, regression, and refractoriness) each with distinct daily light intensity and daylength requirements. These requirements are modified by the photoperiodic history of the bird and are probably internally flexible.

Some of the more important observations relevant to starling physiology should also be repeated and stressed.

Testicular growth is rapid when stimulated by light intensities above the 0.1 lux threshold, yet light *per se*, at normal environmental levels, is not a prerequisite for testicular development. Birds exposed to 0.1 lux and 0.01 lux show gonadal development, and these intensities are below that of full moon. This realization is new and of considerable physiological importance. Growth is stimulated by daylengths above LD 9:15 but will occur on LD 9:15 after a long time. Daylength modifies rate of growth, but does not determine whether growth will occur.

Maintenance of gonad size apparently can be modified by behavioral stimuli in multiply brooded birds, which suggests that maintenance is a separate and real phase of the cycle. The data presented herein for the starling do not confirm or deny this possibility. Maintenance of large testes seen in LD 11:13 or LL of 0.1 and 0.01 lux may only reflect the absence of stimuli for regression.

Regression of the gonads is not always spontaneous as had been thought. Threshold daylengths (LD 13:11 *versus* LD 11:13) and light intensity (1.0 lux *versus* 0.1 lux) components as related to the occurrence and rate of regression are identified for the first time.

Data pertaining to photorefractoriness were not obtained but could be analyzed in this system; the LD 9:15 and LL, 1.0 lux schedules would be most interesting to study in this regard.

Past history also modifies the system; birds under LD 9:14 and LL of 0.1 lux and 0.01 lux all show recrudescence in February. Schwab's starlings under LD 12:12 do the same. It is not clear how this is related to prior events in the cycle.

These data suggest at least one fairly novel approach to photoperiodic analysis with birds—modification of light intensity for birds pretreated under different lighting combinations. What happens when light intensity is raised to 1.0 lux in September in birds pretreated for one year with 0.01 lux? What is the intensity threshold for regression, and can it be modified? How does daylength modify a light intensity response? Will LL of 0.01 lux followed by LD 11:13 maintain enlarged testes? These and many more questions could be posed and resolved by further experiments, but we need more data, and I do not believe that continued adherence to the hypothesis of an endogenous annual rhythm will be as stimulating to further research as is the admission that the system is more complex than had been realized.

ACKNOWLEDGMENTS

I thank J. Stocking for his help in collecting the data and E. Callaway for occasional assistance. R. Schwab provided the birds and valuable discussion of the data. M. Ghislin, J. T. Enright, and P. Licht kindly read and improved the manuscript. This work was supported by the National Science Foundation through NSF GB 5739.

REFERENCES

Ashmole, N. P. 1962. The black noddy *Anous tenvirostris* on Ascension Island. Part I. General biology. Ibis 103b:235–273.

Benoit, J., I. Assenmacher, and E. Brard. 1956. Apparition et maintien de cycles sexuels non saisonniers chez le Canard domestique placé pendant plus de trois ans à l'obscurité totale. J. Physiol. (Paris) 48:388–391.

Brown, F. A., Jr. 1968. "Endogenous" biorhythmicity reviewed with new evidence. Scientia (Milan) 103:1–16.

Chapin, J. P. 1954. The calendar of wideawake fair. Auk 71:1–15.

Chapin, J. P., and L. W. Wing. 1959. The wideawake calendar 1953–1958. Auk 76:153–158.

Dorward, D. F. 1962. Comparative biology of the white booby and the brown booby *Sula* spp. at Ascension. Ibis 1036:174–220.

Ghent, A. W. 1966. The logic of experimental design in the biological sciences. BioScience 16:17–22.

Hamner, W. M. 1963. Diurnal rhythm and photoperiodism in testicular recrudescence of the house finch. Science 142:1294–1295.

Hamner, W. M. 1964. Circadian control of photoperiodism in the house finch demonstrated by interrupted night experiments. Nature (Lond.) 203:1400–1401.

Hamner, W. M. 1966. Photoperiodic control of the annual testicular cycle in the house finch, *Carpodacus mexicanus*. Gen. Comp. Endocrinol. 7:224–233.

Hamner, W. M. 1968. The photorefractory period of the house finch. Ecology 49:211–227.

Hamner, W. M. 1969. Hour-glass dusk and rhythmic dawn timers control diapause in the codling moth. J. Insect Physiol. 15:1499–1504.

Hamner, W. M., and J. T. Enright. 1967. Relationships between photoperiodism and circadian rhythms of activity in the house finch. J. Exp. Biol. 46:43–61.

Hamner, W. M., and J. Stocking. 1970. Why don't bobolinks breed in Brazil? Ecology 1970:743–751.

Hempel, K. G. 1966. Philosophy of natural science. Prentice-Hall, Englewood Cliffs, New Jersey.

Hinde, R. 1967. Aspects of the control of avian reproductive development within the breeding season. Proc. XIVth Int. Ornithol. Congr., p. 135–154.

Lehrman, D. S., P. M. Brady, and R. P. Wortis. 1961. The presence of the male and of nesting material as stimuli for the development of incubation behavior and for gonadotropin secretion in the ring dove *Streptopelia risoria*. Endocrinology 68:507–516.

Lofts, B. 1964. Evidence of an autonomous reproductive rhythm in an equatorial bird *Quelea quelea*. Nature (Lond.) 201:523–524.

Lofts, B., and R. K. Murton. 1968. Photoperiodic and physiological adaptations regulating avian breeding cycles and their ecological significance. J. Zool. 155:327–394.

Marler, P. R., and W. J. Hamilton, III. 1966. Mechanisms of animal behavior. John Wiley & Sons, New York.

Popper, K. R. 1959. The logic of scientific discovery. Basic Books, New York.

Snow, D. W. 1965. The breeding of Audubon's shearwater *Puffinus lherminieri* in the Galapagos. Auk 82:591–597.

DISCUSSION

GWINNER: My general reaction to your paper, Dr. Hamner, is that even if the experiments had turned out as predicted, your hypothesis would be no alternative to the concept of circannual rhythms. The term circannual rhythm is purely descriptive—it means that a phenomenon shows a free-running periodicity of about a year in annually constant conditions. Your hypothesis refers to a possible mechanism underlying circannual rhythms.

HAMNER: No; as I indicated in my paper, I really intended my "alternative" hypothesis to point out inadequacies in the assumption that the circannual rhythm is necessarily endogenous. The rhythm is clearly driven by the ambient photoperiod, and I think that the data presented show that different phases of the cycle are controlled by different daylengths and different light intensities. The fact that you can freeze the system in the maintenance phase indicates that you have got to have an appropriate lighting cycle for testes regression. I repeat that I believe the endogenous hypothesis to be sterile. The alternative I proposed most certainly was not; on the contrary, it led to new insights, and thus is pragmatically, if not theoretically, superior.

ASCHOFF: It is certainly clear that you need appropriate environmental conditions for the reproductive cycle to proceed normally. And as I mentioned before, with respect to Schwab's data, it is possible that the appropriate conditions are sharply restricted to around LD 12:12. But there is much evidence that the range can be wider, that the lighting regimen needn't be LD 12:12. It is even possible—and I did this in Hanover, before knowing Goss's data—to predict what the circannual period τ would be under different LD regimens. Now your proposition is that the annual system is driven by LD 12:12, but I don't understand how LD 12:12 can drive different circannual τ's between 9 and 14 months. As I said before, the LD 12:12 is to the circannual rhythm what the heartbeat is to a circadian rhythm.

HAMNER: Why then don't the birds cycle under constant light as all other circadian rhythms do?

ASCHOFF: You need an LD cycle to keep the circadian system entrained; otherwise the whole machinery doesn't work properly. The circadian system needs to be entrained—we learned this from Hillman's tomatoes.

HAMNER: I feel that use of concepts like "entrainment" lead to misleading, incorrect, and physiologically meaningless generalizations. I do not believe that extension of circadian terminology to encompass photoperiodic responses is at all useful.

GWINNER: If you look carefully at the data published by Chapin on sooty terns, you see that the average τ is very close to 10 lunar months. I wouldn't interpret that as free-running. My hypothesis is that the moon is synchronizing the rhythm in that case.

HAMNER: On the contrary, I think they are free-running.

PITTENDRIGH: Even if it isn't the moon that is doing it, the terns are clearly synchronized and are therefore being entrained by something. We don't know what the *Zeitgeber* is, but they are certainly not free-running.

FRANK H. HEPPNER
DONALD S. FARNER

Periodicity in Self-Selection
of Photoperiod

The accumulating evidence that many functions with diurnal periodicities are controlled by endogenous circadian oscillations entrained by the normal 24-hr light–dark cycle has led to the hypothesis of an analogous relationship between annual periodicities and endogenous circannual oscillations (Aschoff, 1955; Immelmann, 1963). Although this analogy should be pressed cautiously (Farner, 1970), there are now a few well-documented cases among mammals (Pengelley and Kelly, 1966; Davis, 1967; Pengelley, 1968; Goss, 1969a, b; Heller and Poulson, 1970) and birds (Merkel, 1963; Lofts, 1964; Zimmerman, 1966; Gwinner, 1968) that suggest that circannual periodicities do have an endogenous basis. For those species of birds in which gonadal cycles have been demonstrated to be controlled by daylength, information is limited and the picture is not clear. Male domestic mallards held in continuous light or continuous darkness showed fluctuations in testicular size (Benoit, Assenmacher, and Brard, 1956, 1959). However, the "cycles" were irregular and generally had periods of somewhat less than a year. With the white-crowned sparrow, *Zonotricia leucophrys gambelii*, we (Farner and Lewis, unpublished) have failed to detect any trace of a cycle in birds held on *LD 8*:16, although on *LD 20*:4 (in experiments still in progress) we have seen a few cases of limited, abortive testicular growth about 1 year after the last photoperiodically induced cycle. In the same species, however, King (1968) found evidence of an essentially normal, although phase-shifted, periodicity in molt and fat deposition under *LD 20*:4, and in fat deposition, but not in molt, under *LD 8*:16.

To investigate further the possibility of an endogenous component in the photoperiodically controlled annual testicular cycle, we began a study of the pattern of testicular growth under conditions in which isolated birds were permitted to select their own photoperiodic regimes. The results at present are incomplete but suggestive.

In *Zonotrichia leucophrys gambelii*, each interval of photoperiodically induced testicular development, whether natural or experimental, is followed by an interval of photorefractoriness and rapid regression of the testes to resting condition (Farner, 1964a). During this latter interval, testicular development cannot be induced by long days, but experimentally, as well as under natural conditions, photorefractoriness is eliminated by short days of 10 hr or less. Possibly an effective short day could be somewhat longer than 10 hr, since 12-hr days eliminate photorefractoriness in *Junco hyemalis* and apparently in *Zonotrichia albicollis* (Wolfson, 1960), and 14-hr days do so in *Carpodacus mexicanus* (Hamner, 1968).

METHODS

We first experimented with the night-perch scheme used so successfully by Wahlström (1964) with domestic canaries. Eleven birds were held under these conditions, some as long as 5 months; not a single bird ever chose to use the lights-off (night) perch for roosting although there was no tendency to avoid it during activity (Figure 1).

We then devised a scheme in which the bird was trained to go through a simple maze and hop on a perch in order to turn on the light. The circuit was so designed that after 30 min of light the lamps were turned off automatically. The small cage (22.5 × 22.5 × 40 cm) was placed inside a Hartshorne soundproof, lightproof box. Light was provided by a single 22-watt rapid-start Circline daylight fluorescent lamp and a 7-watt incandescent lamp wired in parallel and mounted above the cage. The intensity of light on the floor of the cage was 1500–1900 lux. In addition to the "lights-on" perch reached through the maze, the apparatus contained an activity-recording perch in the middle of the cage and a "dead" perch in the end opposite the maze; this arrangement appeared to produce the most uniform and intense patterns of activity. The light periods selected by the birds and the hops onto the activity perch were recorded with an Esterline-Angus event recorder. The training procedure and the experimental system are described in detail elsewhere (Heppner and Farner, 1971). The pattern of light selection by these birds is illustrated in Figures 2, 9, and 10.

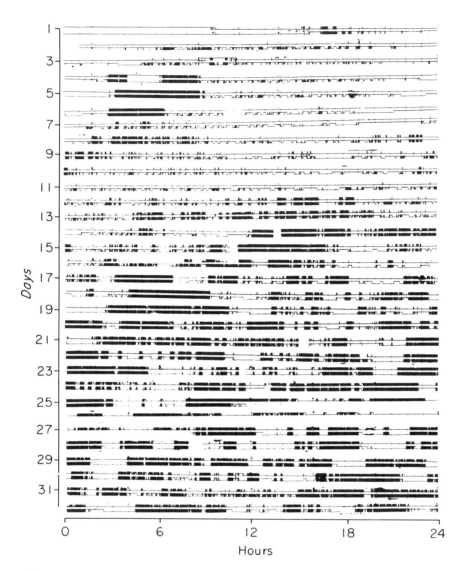

FIGURE 1 The behavior of a white-crowned sparrow in the night perch apparatus of Wahlström (1964). To keep the lights off, the bird had to sit continuously on the night perch. Each pair of traces across the record represents a day of recording, beginning at the left. The upper trace of each pair is a record of the position of the night perch. When the pen is in the "up" position the bird is on the night perch and the lights are off. The lower trace is a record of the position of the motor-activity perch in the middle of the cage: When the pen is in the "down" position the bird is on the perch. Note the absence of periodicity in motor activity and in "selection" of darkness. Similar results were obtained with all 11 birds tested by this method.

FIGURE 2 The performance of an initially photorefractory white-crowned sparrow (No. 4771) in the maze system (Heppner and Farner, 1971) in which hopping on the light perch automatically elicits 30 min of light. This is a portion of a much longer record; day 1 is therefore not the first day in the cage. Each pair of horizontal traces constitutes a day of recording (beginning from the left). The upper trace of each pair is a record of lighting in the cage; in the "down" position the lights are on. The lower trace records hopping on the activity perch in the middle of the cage; the pen is in "down" position when the bird is on the perch. Arrows indicate days on which laparotomies were performed.

This bird was exceptional in that it never selected long days (Table 1). Note conspicuous, "spontaneous" change in τ beginning on day 25.

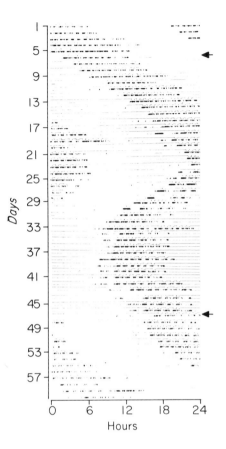

The experimental birds were caught in mist nets from migrating flocks in Kittitas County, Washington, during September. They were transported immediately to Seattle and held in large outdoor aviaries until mid-November, when they were moved into small indoor cages and kept there until the beginning of the training period. Half the birds were held on LD 20:4; they went through one cycle of testicular development and then became photorefractory. The remaining birds were held on LD 8:16 and as a result were photosensitive at the beginning of the training period. Drinking water and a vitamin- and mineral-enriched chick starter mash were available at all times during both the pre-experimental and experimental periods. Food and water were replenished weekly during a self-selected light period (with the exception of No. 2467, which was in the dark most of the time under self-selection conditions). Temperature was not controlled in the individual Hartshorne boxes; all were ventilated with room air of variable temperature, usually within the range of 18–25°C; the extremes were 10° and 30°C. We have been unable to identify any correlation between environmental temperature and the performance of the experimental birds.

To estimate the weight of the testes, laparotomies were performed initially on all birds and thereafter at approximately monthly intervals. Weights were estimated by comparison with a preserved series of testes of known weights.

The self-selection experiments in the maze apparatus were begun in June 1968. Since only one or two birds could be trained at a time, and since as birds died they were replaced, the dates for initiating self-selection of photoperiod vary over about 8 weeks. We obtained usable records for intervals of 4 to 11 months from 8 initially photorefractory birds and 10 initially photosensitive birds.

RESULTS AND DISCUSSION

As noted above, and as illustrated in Figure 1, the behavior of white-crowned sparrows in actively turning off lights is very different from that of canaries as described by Wahlström (1964). Of 11 birds not one chose dark periods of more than a few minutes and never in a regular pattern.

In designing the system in which the birds are given 30 min of light after pressing the lights-on perch, we failed to anticipate certain curious features of their behavior. Consequently it is possible, at least at this time, to present the results only in semiquantitative and qualitative terms.

With a single exception, the initially photosensitive birds (Group A, Table 1) tended to be cyclic in their selection of long days (often essentially continuous light) and short days. The period length for a complete long day–short day cycle varied from about 100 to 120 days (the period for No. 4501 is unclear). All of these birds induced apparently normal testicular cycles with the first interval of long days. However, only one (No. 4812), and possibly a second (No. 2673), induced another testicular cycle. The case of No. 2673 is unclear because a laparotomy, which revealed some increase in testicular size, apparently caused the bird to revert to short days. Within our current concept of the mechanism of the photoperiodic control of testicular growth, these were the only birds in this group that succeeded in eliminating photorefractoriness during the course of their cyclic selection of daylength. The single "noncyclic" bird (No. 2467) selected long days for a period sufficient to induce an apparently normal testicular cycle. Thereafter very little light was selected; frequently the brief light periods were separated by several days.

The behavior of the majority of the initially photorefractory birds

TABLE 1   Self-Selection of Photoperiod by White-crowned Sparrows

Group A—Initially Photosensitive		
(1)	4501 2673[a] 2414 2780	Initial long-day–induced testicular cycle; at least one subsequent "long-day" period without testicular growth. Four additional birds not carried through the entire experiment conformed to this pattern.
(2)	4812	Initial long-day–induced testicular cycle, followed by a period of short days. The ensuing long-day period failed to induce testicular growth. Another short-day period was followed by long days and induced gonadal growth.
(3)	2467	Initial long-day–induced testicular cycle; thereafter very little light was selected. Bird mostly in the dark.
Group B—Initially Photorefractory		
(1)	4714 4713 4711 4740 4781	No initial period of short days, or initial period of short days insufficient to eliminate photorefractoriness; at least two periods of long days; no testicular growth.
(2)	4865	Initial period of short days followed by long days and induced testicular cycle; three further long-day periods without testicular growth.
	4853	As in No. 4865, except that after the testicular cycle only short days were selected; no further testicular growth.
(3)	4771	Short days only; no testicular growth.

[a]May have begun a second long-day–induced testicular cycle.

(Group B, Table 1) resembled that of most of the initially photosensitive group *after* the initial testicular cycle. That is, five of the eight birds in Group B went through at least two cycles of photoperiodic selection, but failed to select a short-day regime that permitted the development of photosensitivity and hence the induction of testicular growth. Some birds may have gone through more than 2 daylength cycles, but interpretation is difficult because of the effects of laparotomies (see below). Two birds in Group B, however, did select an initial short-day regime adequate to induce photosensitivity; No. 4865 then went through one long-day induced testicular cycle. Three subsequent long-day regimes failed to induce testicular growth, presumably because the intervening short-day intervals were not adequate for elimination of photorefractoriness. The behavior of No.

4853 was similar except that after the testicular cycle only short days were selected; there was no further testicular growth. This bird might have become photosensitive by the end of the experiment but we did not test it. The remaining bird in Group B selected a short-day regime throughout the period of the experiment and showed no testicular growth. Among the six birds that selected daylengths in cycles, the periods were about 70–80 days long (No. 4713 is unclear).

Bird No. 2414 is characteristic of the majority of the initially photosensitive birds [Table 1, Group A(1)]. Figure 3 shows the number of hours of light (P) selected per solar day, the hours of photostimulation (S) per circadian period (τ), and the condition of the testes as assayed by laparotomy. On the basis of previous experiments (Farner, 1959, 1964b, and unpublished), any 2-hr or shorter period with at least 30 min of light was regarded as photostimulatory, in the calculation of S. Clearly the two light-fraction curves are functionally related; but the latter must be regarded as physiologically more significant since the circadian rhythms of the birds were free-running (not entrained to a solar day) and since the response system "sees" interrupted light, in patterns classed here as photostimulatory, as continuous light (Farner, 1964a).

Figure 4 shows for the same bird the total hours of photostimulation (S) per circadian period (τ) and the duration of τ, both as functions of time. This record and those for several other birds (see Figure 6) suggest the possibility that S/τ and τ are inversely related. Unfortunately the variables recorded (motor activity and selection of light) do not permit a measurement of τ during intense self-selection of light (see Figures 9 and 10). The exploration of this possible relation awaits further experiments with more sophisticated data-collection and analysis. Figures 5 and 6 illustrate the same analyses for another initially photosensitive bird (No. 4501).

No. 4781 (Figures 7 and 8) is representative of those initially photorefractory birds [Table 1, Group B(1)] that failed to select an adequate number of short days to become photosensitive. Again there appears to be an inverse relationship between τ and the total duration of photostimulation per period.

Except in intervals of continuous photostimulation, the birds generally displayed circadian periodicities in the selection of light and in motor activity. In no case, however, was τ stable throughout the entire experimental period; individual variation was of the order of 3–4 hr, and among all birds, values ranged from 21.7 to 26.5 hr. As noted above, τ itself may be cyclic and functionally related to the cycle in selection of light. Our records contain many cases of "spontaneous" changes in τ (Figure 2, for example).

When motor activity and selection of light are clearly circadian (Figures

2, 9, 10), there is a marked tendency toward a brief but variable period of motor activity before light is selected. This is especially true for the first light selected in each circadian period. The tendency appears to persist, although less conspicuously and less regularly, in periods of intense selection of light (effective LL) (Figures 9 and 10).

In most cases the effects of laparotomy on the patterns of motor activity and light selection appear to be ephemeral (see, for example, No. 4771, Figure 2, and upper left panel, Figure 10; and No. 2673, Figure 9, right

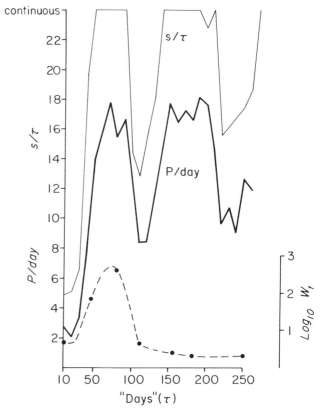

FIGURE 3   The history of an initially photosensitive bird (No. 2414) through approximately 8 months of self-selection of 30 min photoperiods. The heavy line shows the number of hours of light (P) selected per solar day (means for 10-day intervals); the abscissa units are solar days from the beginning of self-selection. The light line shows the number of hours of photostimulation (S) per circadian period (τ). The abscissa units are circadian periods (τ) from the beginning of self-selection. The broken line is the course of testicular weight ($\log_{10}$ mg) as estimated by visual inspection during laparotomy.

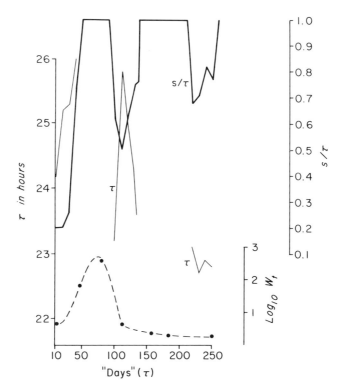

FIGURE 4 The same bird (No. 2414) as in Figure 3. The heavy line
shows the fraction of photostimulation (S) per circadian period τ (10-day
means) – right ordinate. The light line shows τ (10-day means) – left ordinate. No precise estimates of τ can be made for periods of continuous
photostimulation, i.e., when $S/\tau \leqslant 1$ (right ordinate). Abscissa units are
circadian periods (τ). Broken line as in Figure 3.

panel). On the other hand, laparotomy of No. 4781 on day 27 (Figure 9,
left panel) was followed by a persistent change in pattern, although this
was not the case for the laparotomy on day 69. The laparotomy of No.
2673 on day 29 (Figure 10, right panel) was followed by an abrupt change
from "continuous" photostimulation to a short-day regime with a circadian
cycle (τ = about 25 hr). In summary, we found that 49 laparotomies were
apparently followed by no change in pattern; in 18 cases laparotomies of
birds selecting continuous photostimulation were followed by a change to
a periodic pattern. Among 21 birds laparotomized while selecting circadian
patterns of light, τ increased in 20 and decreased in 1. It is clear that a
systematic study of the effects of handling, laparotomy, etc., as functions
of phase in circadian cycle and in the light-selection cycle is in order.

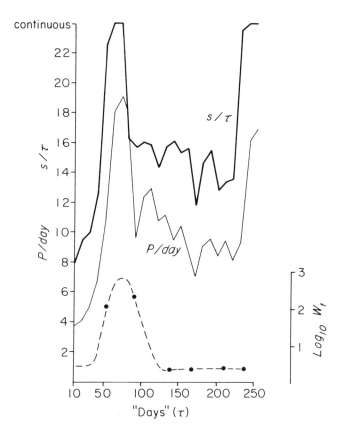

FIGURE 5 The history of an initially photosensitive bird (No. 4501) through approximately 8 months of self-selection of 30 min photoperiods. The light line shows (means for 10-day intervals) the number of hours of light (P) selected per solar day; the abscissa units are solar days from the beginning of self-selection. The heavy line shows the number of hours of photostimulation (S) per circadian period (τ). Other notation as in Figure 3.

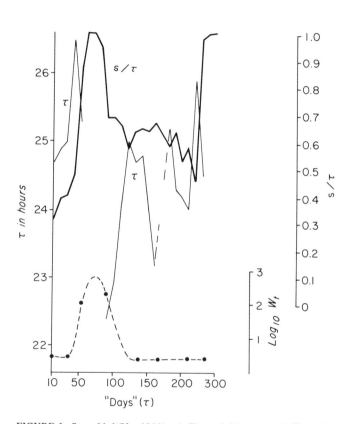

FIGURE 6   Same bird (No. 4501) as in Figure 5. Notation as in Figure 4.

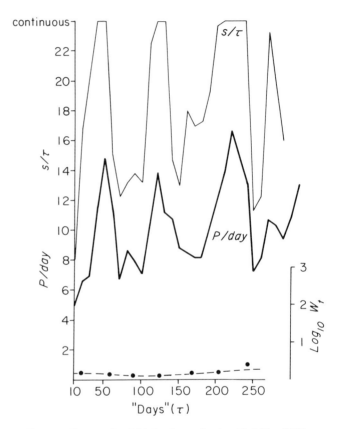

FIGURE 7 History of an initially photorefractory bird (No. 4781) through approximately 8 months of self-selection of 30-min photoperiods. Notation as in Figure 3. The low values at about day 80 follow a laparotomy; those at day 250 may have a similar cause.

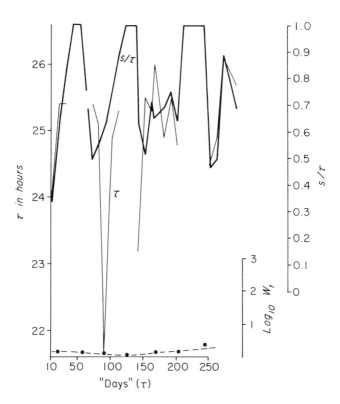

FIGURE 8   The same bird (No. 4781) as in Figure 7. Notation as in Figure 4. See legend for Figure 7 concerning the effects of laparotomies.

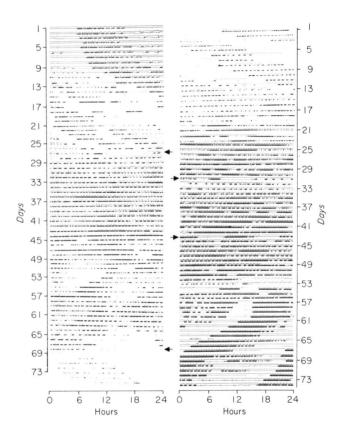

FIGURE 9 Left panel: An excerpt from the record of an initially photo-refractory bird (No. 4781) showing change from short days to continuous photostimulation (see text) and return to short days. Right panel: A similar excerpt from an initially photosensitive bird (No. 2673). Notation as in Figure 2.

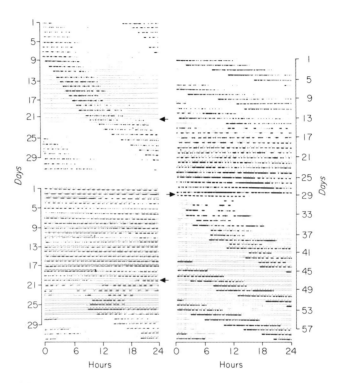

FIGURE 10   Effect of laparotomy on locomotor and self-selection behavior. For notation see Figure 2. Upper left panel is a portion of the record of an initially photorefractory bird (No. 4771). Note apparent lack
of effect of laparotomy on day 22. Lower left panel is from the record of
another initially photorefractory bird (No. 4781). In this case the laparotomy on day 21 is followed by a marked change in the pattern. Right
panel is from the record of an initially photosensitive bird (No. 2673).
The laparotomy on day 29 is clearly followed by a change in pattern.

ACKNOWLEDGMENTS

This investigation was supported by grant GB-5969X from the National
Science Foundation. We wish to acknowledge the invaluable assistance of
E. Gwinner in the analysis of our data and for many useful suggestions,
and R. Lewis for assistance with the laparotomies and for technical advice.

REFERENCES

Aschoff, J. 1955. Jahresperiodik der Fortpflanzung beim Warmblütern. Stud. Gen.
8:742–776.
Benoit, J., I. Assenmacher, and E. Brard. 1956. Apparition et maintien de cycles
sexuels non saisonniers chez le canard domestique placé pendant plus de trois ans
à l'obscurité totale. J. Physiol. (Paris) 48:388–391.
Benoit, J., I. Assenmacher, and E. Brard. 1959. Action d'un éclairement permanent
prolongé sur l'évolution testiculaire du canard pékin. Arch. Anat. Microsc. Morphol.
Exp. 48:5–12.
Davis, D. E. 1967. The annual rhythm of fat deposition in woodchucks (Marmota
monax). Physiol. Zool. 40:391–402.
Farner, D. S. 1959. Photoperiodic control of annual gonadal cycles in birds, p. 717 to
750. In R. B. Withrow [ed.], Photoperiodism and related phenomena in plants and
animals. Publ. No. 55. American Association for the Advancement of Science,
Washington, D.C.
Farner, D. S. 1964a. The photoperiodic control of reproductive cycles in birds. Am.
Sci. 52:137–156.
Farner, D. S. 1964b. Time measurement in vertebrate photoperiodism. Am. Nat.
98:375–386.
Farner, D. S. 1970. Predictive functions in the control of annual cycles. Environ. Res.
3:119–131.
Goss, R. J. 1969a. Photoperiodic control of antler cycles in deer. I. Phase shift and
frequency changes. J. Exp. Zool. 170:311–324.
Goss, R. J. 1969b. Photoperiodic control of antler cycles in deer. II. Alterations in
amplitude. J. Exp. Zool. 171:223–234.
Gwinner, E. 1968. Circannuale Periodik als Grundlage des jahreszeitlichen Funktionswandels bei Zugvögeln. Untersuchungen am Fitis (Phylloscopus trochilus) und am
Waldlaubsänger (P. sibilatrix). J. Ornithol. 109:70–95.
Hamner, W. M. 1968. The photorefractory period of the house finch. Ecology 49:212–
227.
Heller, H. C., and T. L. Poulson. 1970. Circannian rhythm: II. Endogenous and exoge-
nous factors controlling reproduction and hibernation in chipmunks (Eutamias) and
ground squirrels (Spermophilus). Comp. Biochem. Physiol. 33:357–383.
Heppner, F. H., and D. S. Farner. 1971. Training white-crowned sparrows, Zonotrichia
leucophrys gambelii, in self-selection of photoperiod. Z. Tierpsychol. 28:62–68.
Immelmann, K. 1963. Tierische Jahresperiodik in ökologischer Sicht. Zool. Jahrb.,
Abt. 1, Syst. Oekol. 91:91–200.
King, J. R. 1968. Cycles of fat deposition and molt in white-crowned sparrows in con-
stant environmental conditions. Comp. Biochem. Physiol. 24:827–837.
Lofts, B. 1964. Evidence of an autonomous reproductive rhythm in an equatorial bird
(Quelea quelea). Nature (Lond.) 201:523–524.

Merkel, F. W. 1963. Long-term effects of constant photoperiods on European robins and whitethroats. Proc. XIIIth Int. Ornithol. Congr. Ithaca, 1962. J. Hickey [ed.], 2:950–959.

Pengelley, E. T. 1968. Interrelationships of circannian rhythms in the ground squirrel, *Citellus lateralis*. Comp. Biochem. Physiol. 24:915–919.

Pengelley, E. T., and K. H. Kelly. 1966. A "circannian" rhythm in hibernation species of the genus *Citellus* with observations on their physiological evolution. Comp. Biochem. Physiol. 19:603–617.

Wahlström, G. 1964. The circadian rhythm in the canary studied by self-selection of photoperiod. Acta Soc. Med. Upsaliensis 69:241–271.

Wolfson, A. 1960. Role of light and darkness in the regulation of the annual stimulus for spring migration and reproductive cycles. Proc. XIIth Int. Ornithol. Congr., Helsinki, 1958. L. von Haartman [ed.], 2:758–789.

Zimmerman, J. L. 1966. Effects of extended tropical photoperiod and temperature on the dickcissel. Condor 68:377–387.

DISCUSSION

ASCHOFF: We have designed another apparatus for birds to self-select a photoperiod. Our birds have to hop a certain number of times to turn the lights on, and they learn quickly. But the lazy birds do just enough hopping to get the light—a nice case of learning.

FARNER: Yes, these finches learn the maze very quickly—in 5 days or less.

ASCHOFF: Now, these self-selection experiments are very clever, but also very complicated, because we know that under a self-imposed light–dark cycle, the circadian period usually lengthens. But your birds are in an unusual situation, because in order to get a long photoperiod in your conditions, activity time has to be long. Now, a long activity time usually goes with a high circadian frequency, but in the case of self-selected photoperiods, this cannot be so. So your birds are, from this point of view, in a "conflict" situation: They just can't do what they would like to do!

V

NEURAL AND ENDOCRINE CONTROLS

JAMES W. TRUMAN

The Role of the Brain in the Ecdysis Rhythm of Silkmoths: Comparison with the Photoperiodic Termination of Diapause

The giant silkworms of the family Saturniidae usually overwinter as diapausing pupae. Months later, in response to the warmth or lengthening days of spring, diapause terminates. Adult development then begins and the adult moth emerges a few weeks later. The ecdysis of these large moths from their pupal exuviae is a singularly impressive event, which prompted the series of investigations that are reported here.

The activity of the developing moths and the timing of ecdysis were recorded by an apparatus consisting of a series of heart levers that wrote on a smoked kymograph drum revolving once every 2 days. One to two days before ecdysis each animal was attached to a lever by means of a cotton thread that, several weeks earlier, had been attached to the tip of the pupal abdomen with melted wax. The time of ecdysis and the pattern of activity preceding and accompanying ecdysis were read directly from the records. All experiments were carried out at 26°C, the developing animals having been placed in their respective photoperiod regimens at least 2 weeks prior to ecdysis.

Each species consistently underwent ecdysis at a certain time of day. Ecdysis was preceded by a period of hyperactivity, the pattern of which was also species-specific. Figure 1 illustrates the ecdysis times and the pattern of emergence behavior of three species of saturniids, *Hyalophora*

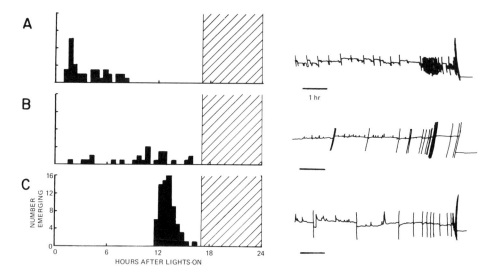

FIGURE 1   The distribution of ecdysis in an *LD 17*:7 photoperiod regimen (left), and tracings of the kymograph records of activity accompanying ecdysis (right). A, *Hyalophora cecropia*; B, *Antheraea polyphemus*; C, *A. pernyi.* Hatched area in all figures indicates dark period. (From Truman and Riddiford, 1970.)

cecropia, Antheraea polyphemus, and *A. pernyi*. Each species shows a predictable ecdysis "gate" under a regimen of 17 hr of light and 7 hr of darkness (*LD 17*:7). Moreover, the emergence behavior, while varying among species, is highly stereotyped for each species.

In this report the role of the brain in the photoperiodic control of ecdysis will be described and then compared with its role in the termination of pupal diapause—another photoperiodically controlled system in the pernyi silkworm (Williams and Adkisson, 1964).

NEUROENDOCRINE CONTROL OF ECDYSIS

There is convincing evidence that the mechanism that controls the time of emergence resides entirely within the brain (Truman and Riddiford, 1970). Although brainless moths are able to emerge, this event is no longer confined to a specific time of the day (Figure 2B). Ecdysis of brainless moths occurs randomly without respect to light or darkness, but sensitivity to photoperiod is restored when the excised brain is reimplanted into the tip of the abdomen. As shown in Figure 2C, the ecdysis of these "loose-brain" moths is clearly synchronized by photoperiod. Moreover, these moths demonstrate a free-running rhythm of ecdysis in continuous darkness (Figure 3).

In order to synchronize ecdysis with the photoperiod, a minimum of three components is necessary—a photosensitive system for the reception of light, a timing mechanism to measure the hours after lights-off and lights-on, and a mechanism to trigger the emergence behavior.

Williams and Adkisson (1964) have demonstrated that the pupal brain of the pernyi silkmoth is photosensitive. We find that the adult brain is also photosensitive. An experiment was performed in which the brain was removed from each of 20 cecropia pupae; in 10 pupae it was reimplanted into the head; in the other 10, it was reimplanted into the tip of the abdomen. Each pupa was then placed in a hole in an opaque diaphragm that

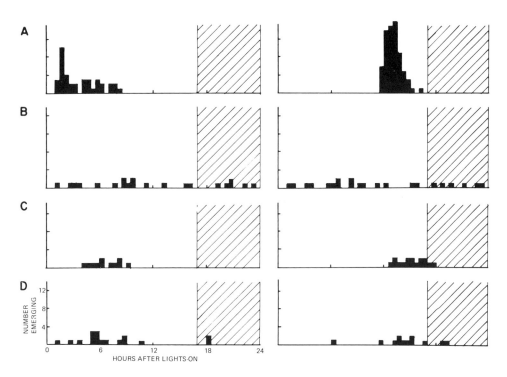

FIGURE 2   The distribution of ecdysis in an *LD 17:7* photoperiod regimen. A, unoperated moths: cecropia (left) and pernyi (right). B, brainless moths: cecropia (left) and pernyi (right). C, brainless moths, each with its own brain implanted in its abdomen: cecropia (left) and pernyi (right). D, brainless pernyi moths, each with a cecropia brain implanted in its abdomen (left); and brainless cecropia, each with a pernyi brain implanted in its abdomen (right). (Redrawn from Truman and Riddiford, 1970.)

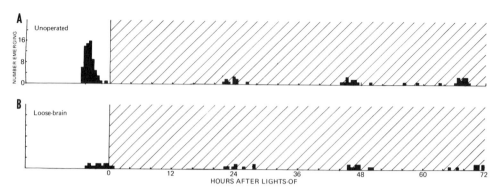

FIGURE 3 The ecdysis of pernyi moths that had developed in an *LD 17*:7 regimen and were placed in continuous darkness just prior to ecdysis. A, unoperated moths; B, moths with the brain removed from the head and reimplanted into the abdomen.

separated two photoperiod chambers. The anterior end was exposed to an *LD 12*:12 regimen and the posterior end was exposed to the reciprocal regimen. The photoperiod to which the brain was exposed determined the time of ecdysis. For example, all the moths that had their brains in their abdomens emerged early in the day to which the posterior end was exposed. Similarly, the moths that had their brains reimplanted into the anterior ends emerged early in the day to which the head end was exposed. These results demonstrate that the brain itself is photosensitive.

That the brain contains the emergence clock was shown by interchanging brains between cecropia and pernyi. As seen in Figure 2D, the time of ecdysis was characteristic, not of the host species, but of the species that donated the brain. The way in which this brain-centered timer interacts with the photoperiod will be considered in the next section. Suffice it to say here that by its interaction with the photoperiod, the timing mechanism determines a specific ecdysis gate. These dictates are then mediated through the release of a hormone from the brain. The neurotropic ecdysis hormone (Truman and Riddiford, 1970) is neither genus- nor species-specific and is a vehicle by which the brain communicates with the rest of the central nervous system.

When brainless animals received brain implants from another species, the moths emerged at the time dictated by the donor brain but showed the emergence behavior of the host. Thus, the brain-centered system serves only as a trigger of the ecdysis behavior. In response to the release of the hormone, the moth activates a preprogrammed behavior pattern that culminates in ecdysis about 1½ hr later.

THE ECDYSIS CLOCK

ITS COMPONENT PARTS: THE PHOTONON AND THE SCOTONON

To elucidate the brain-centered clock, ecdysis of the pernyi moth was studied in detail. The response of this insect to photoperiod is very similar to the well-documented eclosion system of *Drosophila* (Pittendrigh, Bruce, and Kaus, 1958; Pittendrigh, 1960, 1966). Under photoperiodic conditions, the time of ecdysis varied with the regimen (Figure 4). If developing moths were transferred into continuous darkness, the ecdysis rhythm free-ran with a period of 22 hr (Figure 5). By contrast, rhythmicity soon disappeared in conditions of continuous light, and the moths emerged randomly (Figure 5).

Data on ecdysis in pernyi can be most simply interpreted as reflecting a process that has two alternate pathways: the scotonon and the photonon (Truman, 1971). The scotonon is a dark-dependent process that has a duration of 22 hr, the period of the free-running rhythm. The photonon, by contrast, is initiated by a light interruption of the scotonon, and its duration is dependent on how much of the scotonon has been completed at the time of lights-on.

The Scotonon

In continuous darkness the primary timing process undergoes a periodic 22-hr oscillation. The dynamics of the timing process during one of these oscillations defines the scotonon whose duration is thus 22 hr. A central problem is to define the beginning or onset of the scotonon.

In his study of the ecdysis rhythm of *Drosophila*, Pittendrigh (1966) demonstrated that the timing process stops after 12 hr of light, but resumes in darkness. Moreover, the rhythm always resumed in a constant phase relationship to lights-off. Since moths placed in continuous darkness from photophases of either 12 or 17 hr assumed the same relationship to lights-off (Figure 5), the same is apparently true for the pernyi ecdysis rhythm. Under these conditions the point of the timing process that coincides with lights-off can be defined as the beginning of the scotonon (Truman, 1971).

The Photonon

Although the ecdysis clock describes a periodic oscillation in continuous darkness, under certain photoperiod regimens it behaves like an hourglass

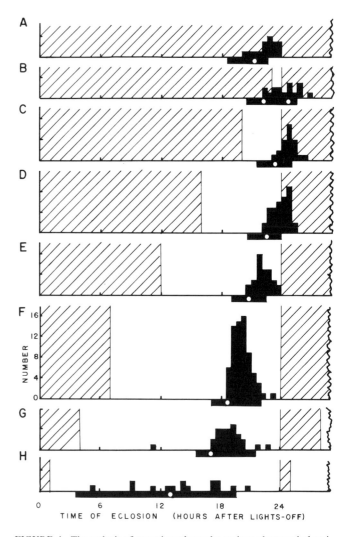

FIGURE 4   The ecdysis of pernyi moths under various photoperiod regimens. A, during the first day of continuous darkness after transfer from the *LD 12*:12 regimen; B, in an *LD 1*:23 regimen; C, *LD 4*:20; D, *LD 8*:16; E, *LD 12*:12; F, *LD 17*:7; G, *LD 20*:4; H, *LD 23*:1. The black bar gives the range for the onset of the emergence behavior in each regimen. The open circles are average times. In the *LD 1*:23 photoperiod, the average times for males and females are given separately.

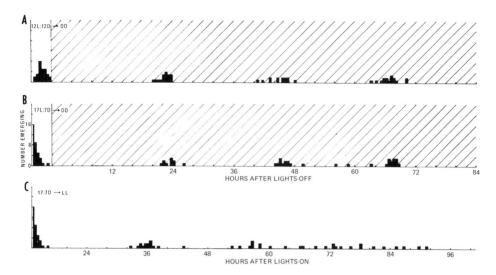

FIGURE 5   The ecdysis of unoperated pernyi moths exposed to a photoperiod regimen and subsequently placed in constant conditions just prior to ecdysis. Top, LD *12*:12 regimen to continuous darkness; middle, LD *17*:7 regimen to continuous darkness; bottom, LD *17*:7 regimen to continuous light. (From Truman, 1971.)

(Pittendrigh, 1966). Thus, when *Drosophila* were exposed to regimens having photophases of 12 hr or longer, the timing process stopped at some unknown point during the photophase of each cycle but resumed at the following lights-off. Apparently the lights-on signal alters the dynamics of the process so that it terminates during the light period. The pathway that the timing process follows after lights-on defines the photonon (Truman, 1971). Since under most regimens the termination of the photonon for *Drosophila* could not be determined, Pittendrigh was unable to formulate a satisfactory scheme to describe the behavior of the ecdysis clock under photoperiod regimens.

In the pernyi silkmoth the release of the ecdysis hormone and the subsequent onset of emergence behavior occurs as the timing process nears the end of a free-running period (21 and 22 hr after the beginning of the scotonon for males and females, respectively). On the assumption that ecdysis is triggered as the process nears the end of a cycle irrespective of which of the two alternate pathways is taken, I have used the time of initiation of emergence behavior as an approximation of the end of the photonon. Therefore, in regimens like those in Figure 4C–H, the duration of the photonon is taken as the time between lights-on and the initiation of emergence behavior.

But the photonon is not restricted to regimens having more than 12 hr of light. As the photophase is decreased to less than 12 hr, the termination of the photonon moves toward lights-off. Under extreme conditions, such as in an LD *1*:23 regimen (Figure 4B), the photonon terminates early in the scotophase of the next cycle. In this case, the scotonon begins about 1 hr after lights-off. Approximately 21 hr later, as in continuous darkness, the males begin their emergence behavior. But before the females can initiate emergence behavior, lights-on occurs, the photonon begins, and triggering of the behavior is delayed until 24½ hr after lights-off, i.e., ½ hr into the next scotophase. When the photonon ends, the next scotonon may then begin. Thus, the combination of scotonon plus photonon, which behaves like an hourglass under some regimens, produces a continuous oscillation if the photonon terminates in the dark.

Relationship between the Photonon and Scotonon

Since the initiation of the emergence behavior can be used to estimate the end of the photonon, it is simple to define the dependence of the photonon on the scotonon. For example, in an LD *17*:7 regimen, a cycle of the timing process begins at lights-off and continues according to the scotonon kinetics until lights-on. The process then finishes the cycle according to the photonon kinetics. The initiation of emergence behavior (1½ hr before ecdysis), which marks the end of the cycle, occurs about 12 hr after lights-on. Thus, if lights-on occurs 7 hr into the scotonon, the timing process completes the remainder of the cycle in 12 hr rather than the 15 hr that would have elapsed without the interruption. Figure 6 gives the relationship between the time of interruption of the scotonon and the length of the resulting photonon. When the light interruption occurs early in the scotonon, the process is completed in less than 22 hr. By contrast, when the interruption occurs late in the scotonon, the duration is greater than 22 hr.

GATE WIDTH AS A FUNCTION OF PHOTOPERIOD

In the pernyi silkmoth, the photoperiod determines not only the timing of the ecdysis gate but also the width of the gate (Truman, 1971). Under regimens having very short scotophases (LD *23*:1), the gate is very broad. As the length of the scotophase increases, gate width rapidly decreases until a minimum value of 3½ hr is attained. The gate width then remains essentially constant as the amount of darkness is further increased; the apparent exception of the LD *1*:23 regimen is due to the splitting of the

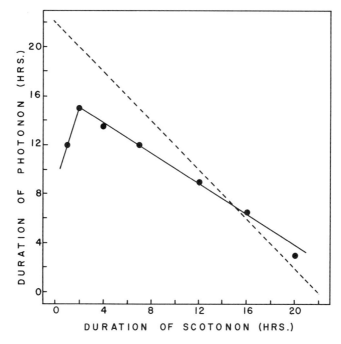

FIGURE 6 The effect of light on the dynamics of the clock. The timing
process always begins a cycle according to the scotonon kinetics, but a
light interruption switches the process to the photonon kinetics. The time
necesary to complete a cycle according to the photonon kinetics is
plotted as a function of the time at which the scotonon was interrupted
by light. The dashed line represents a combined duration of 22 hr (that
of the uninterrupted scotonon).

ecdysis peak by the lights-on signal, as discussed above. I have termed
3½ hr the "minimal gate." The number of moths emerging during 3½ hr
divided by the total number of moths provides a measure of the accuracy
of the timing process under each photoperiod regimen (Truman, 1971).

ACCURACY AND THE SCOTONON

Under conditions such as those shown in Figure 4C–H, the photonon ends
during the photophase, and a new scotonon is begun at each succeeding
lights-off. In the LD 23:1 regimen, the scotonon was utilized for only 1 hr
and the resulting accuracy was 0.38. The accuracy after various durations
of the scotonon is shown in Figure 7. Accuracy is obviously dependent on
the completion of an event that occurs early in the scotonon. The plot of

log (1−accuracy) versus duration of the scotonon shows that the likelihood of completion of this event falls off exponentially (Figure 7, inset).

These results show that the scotonon can be divided into two successive periods: the synchronization period and the dark-decay period. The synchronization period contains the events that lead to maximal accuracy and it extends through the first 2 hr of the scotonon—approximately the time necessary for 50 percent of the population to become maximally accurate. The dark-decay period comprises the events that occur during the remaining 20 hr of the scotonon.

SCHEMATIC REPRESENTATION OF THE PERNYI CLOCK

A schematic representation of the pernyi clock is shown in Figure 8. For simplicity, the pathways of the timing process have been represented as the fluctuation of a single substance [S] through time. The scotonon is represented as a simple saw-tooth curve. The steep 2-hr rising limb represents the synchronization period, whereas the gradual 20-hr falling limb is the dark-decay period. When [S] reaches S_0, a new scotonon can begin, but only in darkness. The shape of the photonon curve is dictated by the relationship given in Figure 6. The photonon begins at the concentration

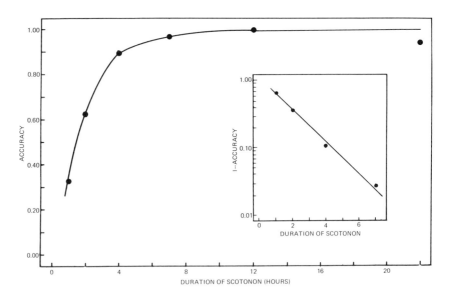

FIGURE 7   The relationship between the length of the scotonon completed prior to lights-on and the accuracy of the time measurement under that regimen. Inset: plot of log (1−accuracy) as a function of the duration of the scotonon.

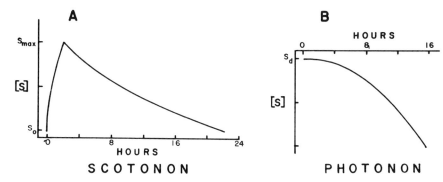

FIGURE 8 A schematic representation of the components of the pernyi ecdysis clock. The components are represented as the fluctuation of a hypothetical substance [S] versus time. The scotonon occurs only during darkness. The photonon is triggered by a light interruption of the scotonon; S_d is the concentration of S reached by the scotonon at the time of the interruption. The clock initiates emergence behavior as [S] approaches S_0. For further explanation see text. (From Truman, 1971.)

of S that the scotonon had reached at the time of the light interruption, S_d, and decays to S_0. Since the photonon is devoid of a rising limb, it follows that after S_0 is reached, the clock stops until the onset of darkness triggers a new scotonon. Since insects can entrain to skeleton photoperiods (Pittendrigh, 1965), it may be concluded that light is necessary for the onset of the photonon but unnecessary for its continuation. In all cases, emergence behavior is triggered as [S] approaches S_0.

An application of the pernyi scheme to regimens of LD *12*:12 and LD *23*:1 is shown in Figure 9. In the former case the lights-on signal occurs during the decay of the scotonon. For an interruption at any time during the decay, one would expect a relatively uniform variation in [S] at the time of the interruption and, consequently, a rather uniform gate width. However, when an interruption occurs during the rising limb, as in LD *23*:1, one would anticipate a much greater variation in the concentration of [S] and, consequently, a much broader peak. This relatively simple model can thus explain both the timing and the width of the ecdysis gate under the entire range of photoperiods.

ECDYSIS AND DIAPAUSE

SIMILARITIES OF THE SYSTEMS

Pupal diapause in the pernyi silkworm is both induced and terminated by photoperiod. The factors necessary for induction of diapause were exten-

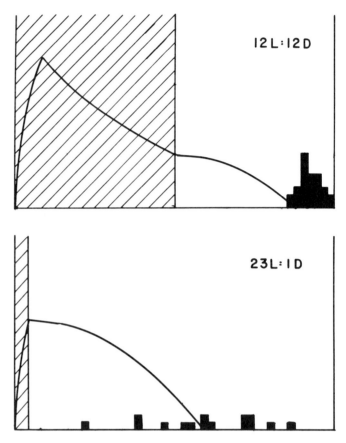

FIGURE 9   Application of the model of the pernyi ecdysis timer. A light interruption occurring during the gradual falling limb of the scotonon produces a relatively narrow gate (as in *LD 12*:12). An interruption of the steep rising phase produces a wide gate (as in *LD 23*:1).

sively studied 20 years ago by the Japanese investigator, Tanaka (1950a, 1950b, 1950c, 1951a, 1951b; see Lees, 1955, for English summary). Fifteen years later, Williams and Adkisson (1964) described the photoperiodic termination of diapause and demonstrated its control by a brain-centered neuroendocrine system.

The neuroendocrine control of ecdysis is strikingly similar to the control of diapause termination (Figure 10). Termination of pupal diapause and initiation of adult development are provoked by the release of the molting hormone (ecdysone) from the prothoracic glands. But these glands are in-

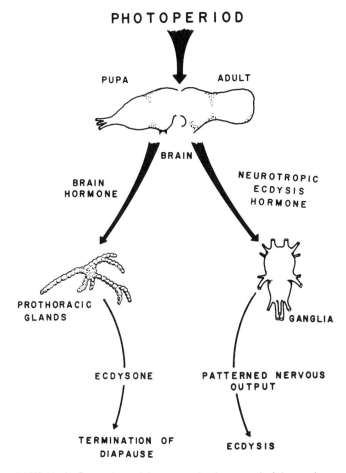

FIGURE 10   Comparison of the neuroendocrine control of the termination of diapause and of ecdysis.

capable of functioning without the tropic action of the brain (Williams, 1956). In the case of the photoperiodic termination of diapause in pernyi, the brain-centered mechanism reads the photoperiod, interprets the information, and thereby controls the release of brain hormone (Williams and Adkisson, 1964). In this respect the brain serves as the intermediary between the environment and the prothoracic glands.

The system that controls ecdysis is analogous. The actual ecdysis behavior is programmed into the neural circuitry of the thoracic or abdominal ganglia. This circuitry is unaffected by light but is linked to the photoperiod through the brain-centered ecdysis system. The outputs from both brain-centered mechanisms are neurosecretory hormones that provoke ecdysis or termination of diapause, as dictated by the reacting systems.

In controlling these events, the brain gives two different types of commands. In ecdysis a purely temporal command is given; the system controls when the hormone will be released. By contrast, in the termination of diapause, a yes or no command is given; the system controls whether or not the hormone will be released.

Except for this difference, the systems controlling ecdysis and diapause termination are similar in their response to photoperiod (Figure 11). The effects of photoperiod on diapause are clear-cut in regimens having from 6 to 20 hr of darkness. As more extreme regimens are approached, the response begins to break down (Williams and Adkisson, 1964). The accuracy of the ecdysis timer is also optimal in regimens having scotophases of 6 to 20 hr, and shorter or longer scotophases lead to inaccuracy. In the pernyi silkworm, the transition from regimens that produce 100 percent diapause to regimens that produce 100 percent development occurs over a span of less than 3 hr, as the photophase is lengthened from 13 to 16 hr. Similarly, if one corrects for the sexual dimorphism of the ecdysis times, under optimal conditions the entire ecdysis gate of pernyi is traversed in 2½ to 3½ hr.

MODEL OF THE DIAPAUSE CLOCK

It seems unlikely that the brain of the pernyi silkworm is equipped with two types of "clocks" that behave similarly through the range of photoperiod regimens (Figure 11). Therefore, one would suspect that both systems use the same type of photosensitive clock. Proceeding on this assumption, I have used the ecdysis clock as the basis of a hypothetical scheme to account for photoperiodic control of the secretion of the brain hormone that provokes termination of diapause.

However, the photosensitive clock described above determines only when the neurohormone is to be released, that is, the clock would determine the time of hormone release but would not provide the quantitative control a diapause response demands. In order to provide this "yes or no" control, an ancillary device has to be postulated. The type of device most consistent with the data requires that the day be divided into two periods; during the first period brain hormone release is inhibited and during the second period it is permitted. Thus, the proposed model for photoperiodic control of diapause has points in common with the more generalized models described by Pittendrigh and Minis (1964) and by Tyshchenko (1966) (as cited by Danilevsky, Goryshin, and Tyshchenko, 1970).

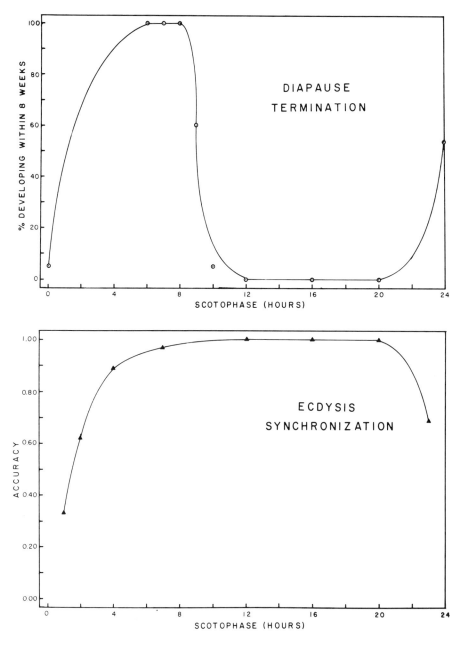

FIGURE 11   Comparison of diapause termination and ecdysis synchronization as a function of the
photoperiod. Above: the percent of pernyi pupae terminating diapause within 8 weeks (drawn
from data of Williams and Adkisson, 1964). Below: the accuracy of the ecdysis timing mechanism
as a function of the photoperiod.

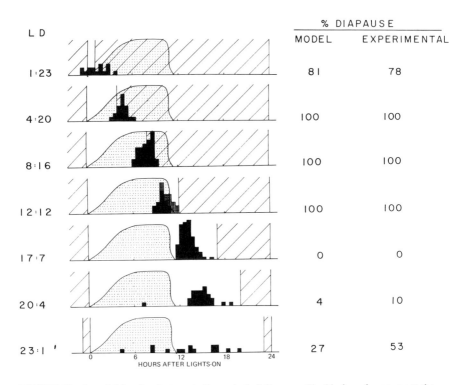

FIGURE 12 A model for the photoperiodic control of diapause. The black peaks represent the
times dictated by a photosensitive clock for brain hormone release. (To compensate for sexual
differences in ecdysis times, only the data for males are considered in the *LD 12*:12 regimen; see
text.) The stippled area represents the period during which brain hormone release is inhibited. The
experimentally determined values are interpolated from the data of Williams and Adkisson (1964)
as redrawn in Figure 11.

Figure 12 represents the interaction of the photosensitive clock and the
ancillary device. The solid peaks represent the times dictated by the photosensitive clock for the release of brain hormone. The stippled area represents the portion of the day during which the ancillary device inhibits the
release of brain hormone. (The shape of this area is arbitrary.) The inhibitory period is described in reference to the time the clock switches to the
photonon and covers the 11½ hours succeeding lights-on, regardless of the
ensuing photoperiod conditions.

As seen in Figure 12, from *LD 4*:20 to *LD 12*:12 the endocrine system
is inhibited at the time the clock signals the release of brain hormone, and
diapause is maintained. (To compensate for the fact that sexual differences
broaden the ecdysis peak from 2½ to 3½ hr in width, only the males are

considered in the *LD 12*:12 regimen.) As the photophase is lengthened to more than 12 hr, the signal for hormone release occurs outside the inhibitory period and diapause is terminated. At very long photophases the clock loses accuracy, and some of the release commands again fall in the inhibitory period. Under very short photophases, as in *LD 1*:23, the release of brain hormone partially anticipates the inhibition and some animals terminate diapause under this regimen. This scheme can thus accommodate the diapause response of pernyi over the entire range of photoperiod regimens.

DISSECTION OF THE DIAPAUSE CLOCK

Further evidence for the participation of a photosensitive clock of the type described above comes from the study of the effect of surgical manipulations on the photoperiodic clock of pernyi (Williams, 1969). In the pupa, the endocrine system necessary for termination of diapause can be uncoupled from its photoperiodic control by cutting between the median and lateral neurosecretory cells of the brain and reimplanting the small median piece into a brainless pupa. After this operation, the pupae responded in the same fashion to long- and short-day regimens—i.e., under both conditions, 25 percent initiated adult development. This percentage proved to be the same as that observed in continuous light. Since continuous light stops the clock and thus produces arrhythmia, one would expect a similar response from surgically rendering animals arrhythmic by removing the clock.

Williams further demonstrated that the dissociation of the optic lobes from the cerebral lobes consistently altered the response of the pupa to photoperiod. Of pupae from which the optic lobes were excised, only 80 percent terminated diapause under long-day conditions. Similarly, removal of the optic lobes had a marked effect on the ecdysis of pernyi in *LD 17*:7 regimen. The application of the ecdysis data from moths lacking optic lobes (Truman, unpublished) to the model for the control of diapause is shown in Figure 13. Here, again, the prediction of the model is in agreement with experimental results.

A UNIFIED SCHEME

Conceptual models of the processes underlying control of ecdysis and diapause in pernyi can be combined into a unified scheme. For both events, the brain interprets the photoperiod for the insect and probably uses the same type of photosensitive clock. If one must evaluate the relative importance of the light and dark periods for the functioning of the clock, the

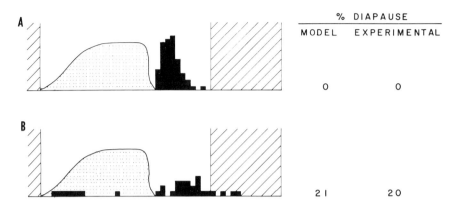

FIGURE 13   Application of the diapause model under an *LD 17*:7 regimen. Above: unoperated moths. Below: moths from which the entire brain was removed and the cerebral lobes reimplanted in the abdomen. Experimental data from Williams (1969).

scotophase would almost certainly have to be judged the more important. The length of the dark period determines the duration of the scotonon portion of the timing process. The latter, in turn, determines the length of the complementary photonon. Light initiates the photonon and also determines when the next scotonon will begin. It does the latter either directly by its presence, since the scotonon can occur only in the dark, or indirectly by triggering a photonon that extends into the next dark period, thereby delaying the beginning of the next scotonon (e.g., the *LD 1*:23 regimen). Upon completion of adult development, the endocrine system is "plugged into" the clock, which then determines the time of day that the ecdysis hormone is released.

A similar system can be envisioned for the termination of diapause. For a given photoperiod, the clock dictates a specific time for the release of brain hormone. The ancillary device then acts either to inhibit or to permit the release of brain hormone.

ACKNOWLEDGMENTS

I thank Professors C. M. Williams and L. M. Riddiford for helpful discussions during the investigation and the preparation of this manuscript. This work was supported in part by grant GB-7966 from the National Science Foundation to Professor L. M. Riddiford. The author was supported by a National Science Foundation predoctoral fellowship.

NOTE ADDED IN PROOF

Recent experiments, (Truman, in preparation) have shown that the length of the synchronization period in pernyi is on the order of 4 hr.

REFERENCES

Danilevsky, A. S., N. I. Goryshin, and V. P. Tyshchenko. 1970. Biological rhythms in terrestrial arthropods. Annu. Rev. Entomol. 15:201–244.

Lees, A. D. 1955. The physiology of diapause in arthropods. Cambridge Univ. Press, Cambridge. 151 p.

Pittendrigh, C. S. 1960. Circadian rhythms and the circadian organization of living systems. Cold Spring Harbor Symp. Quant. Biol. 25:159–184.

Pittendrigh, C. S. 1965. On the mechanism of the entrainment of a circadian rhythm by light cycles, p. 277 to 297. In J. Aschoff [ed.], Circadian clocks. North-Holland Publ. Co., Amsterdam.

Pittendrigh, C. S. 1966. The circadian oscillation in Drosophila pseudoobscura pupae: a model for the photoperiodic clock. Z. Pflanzenphysiol. 54:275–307.

Pittendrigh, C. S., V. G. Bruce, and P. Kaus. 1958. On the significance of transients in daily rhythms. Proc. Natl. Acad. Sci. U.S. 44:965–973.

Pittendrigh, C. S., and D. H. Minis. 1964. The entrainment of circadian oscillations by light and their role as photoperiodic clocks. Am. Nat. 98:261–294.

Tanaka, Y. 1950a. Studies on hibernation with special reference to photoperiodicity and breeding of the Chinese Tussar-silkworm. I [in Japanese]. Nippon Sanshigaku Zasshi 19:358–371.

Tanaka, Y. 1950b. Studies on hibernation with special reference to photoperiodicity and breeding of the Chinese Tussar-silkworm. II [in Japanese]. Nippon Sanshigaku Zasshi 19:429–446.

Tanaka, Y. 1950c. Studies on hibernation with special reference to photoperiodicity and breeding of the Chinese Tussar-silkworm. III [in Japanese]. Nippon Sanshigaku Zasshi 19:580–590.

Tanaka, Y. 1951a. Studies on hibernation with special reference to photoperiodicity and breeding of the Chinese Tussar-silkworm. V [in Japanese]. Nippon Sanshigaku Zasshi 20:132–138.

Tanaka, Y. 1951b. Studies on hibernation with special reference to photoperiodicity and breeding of the Chinese Tussar-silkworm. IV [in Japanese]. Nippon Sanshigaku Zasshi 20:191–201.

Truman, J. W. 1971. Hour-glass behavior of the circadian clock controlling eclosion of the slikmoth Antherala pernyi: Proc. Natl. Acad. Sci. U.S. 68:595–599.

Truman, J. W., and L. M. Riddiford. 1970. Neuroendocrine control of ecdysis in silkmoths. Science 167:1624–1626.

Tyshchenko, V. P. 1966. Two-oscillatory model of the physiological mechanism of insect photoperiodic reaction. Zh. Obshch. Biol. 27:209–222.

Williams, C. M. 1956. Physiology of insect diapause. X. An endocrine mechanism for the influence of temperature on the diapausing pupa of the Cecropia silkworm. Biol. Bull. 110:201–218.

Williams, C. M. 1969. Photoperiodism and the endocrine aspects of insect diapause. Symp. Soc. Exp. Biol. 23:285–300.

Williams, C. M., and P. L. Adkisson. 1964. Physiology of insect diapause. XIV. An endocrine mechanism for the photoperiodic control of pupal diapause in the oak silkworm, Antheraea pernyi. Biol. Bull. 127:511–525.

DISCUSSION

ZIMMERMAN: I think the transplantation experiments are beautiful. I can't see any loopholes in them at all; they are very nice. I'm still worried, however, about parallels between the behavior of photoperiods and circadian rhythm effects, especially after Pittendrigh's discussion of the new evidence in *Pectinophora*. I think you are right in not interpreting them as convincing evidence for a common mechanism. Also, what is the evidence for your assertion that the ventral ganglia are necessary for mediating the brain signals triggering emergence and that they are not photoreceptive?

TRUMAN: First, brainless moths can still emerge, but they do so randomly; they show no synchronization in an LD cycle. Second, cecropia and pernyi moths show a species-specific behavior pattern immediately preceding ecdysis. When the brains are interchanged between these two species, the moths emerge at a time characteristic of the species that donated the brain, but they still display the behavior typical of the host. By switching the brains, then, one does not switch the behavior, but only the synchronization of that behavior. Thus, the actual pattern must be programmed somewhere other than in the brain—most probably in the thoracic or abdominal ganglia. It is a system analogous to that seen in locusts. In that case, the flight behavior is prepatterned in the ventral ganglia and is driven by impulses coming from the cephalic ganglia. You have a pre-patterned center and all you have to do is push a button and start it going.

ZIMMERMAN: How do you distinguish the pernyi and cecropia behavior patterns?

TRUMAN: The differences in the behavior of the two species are very evident from the kymograph records of their emergence. Cecropia has a very stereotyped behavior with three distinct phases. It begins with a period of hyperactivity, followed by a comparable period of reduced activity and then a second hyperactive period, which culminates in eclosion. Pernyi emergence behavior is much simpler and less stereotyped. Typically one sees only a slow increase in activity during the hour and a half preceding ecdysis. The records are very different, and you can merely look at a kymograph trace and tell whether it was cecropia or pernyi that emerged.

ZIMMERMAN: The fact that this driving can be accomplished by brains implanted in abdomens surely leads to a comment about electrical versus hormone-mediated signals.

TRUMAN: I think that depends on the kind of rhythm you are looking at. In the emergence rhythm, the trigger is definitely hormonal. The cockroach activity rhythm could be triggered electrically. The triggering factor really has no bearing on the time-keeping mechanism; it's just the output of this mechanism.

MENAKER: I can't possibly say anything without first congratulating you on a beautiful piece of work. There are so many fascinating things about the story it is hard to focus on just one of them, but I think for many of us the transplant experiments are the most immediately interesting. An alternative explanation of your transplant results occurred to me as you were talking, and I would like to explore it with you. *A priori* it appears to me unlikely, but I can't exclude it; perhaps you can exclude it for me. It may be that the photoreceptor is not in the brain, but somewhere else, and that what is in the brain is the clock mechanism and the eclosion trigger. If that were true, you would expect to produce arrhythmicity by removing the brain and to restore rhythmicity by reimplanting the brain in the abdomen if, and only if, this coupling between the photoreceptor and the clock were hormonal.

TRUMAN: I agree that this possibility is not rigorously excluded by the data. However, it does not seem very likely to me either. [Author's note: The experiments that do exclude this possibility were subsequently performed and have been included in the text.]

STRUMWASSER: I think your idea of driving a set of neurons by a chemical release rhythm is very tenable. There are certainly nerve cells that produce neural hormones. For example, in *Aplysia* there is a specific set of nerve cells that produce a polypeptide that, 1 hr after injection, induces egg-laying behavior.

I have a brief question on the technical feasibility of pharmacologically manipulating the brain you implant. Would it perhaps be possible to culture it for a week and then implant it to see what changes are brought about in the timing of the ecdysis peak, for example? Specifically, I wondered if one could block protein synthesis for short periods of time, say 18 hr, and then implant the piece of tissue.

TRUMAN: It is extremely easy to do the operations on the pupae, but there is a 3-week span between the operation and the emergence of the moth. The problem is that any effects on the rhythm that pharmacological manipulations might produce could wear off during that time. I have performed operations on developing moths but it is a ghastly job because many of the developing tissues are in almost a gel-like condition. Again, the major problem is that there is just too much time between the manipulations and the subsequent emergence.

ASCHOFF: You presented data showing that width of gate is a function of photoperiod, and mentioned that darkness is necessary for the rhythm to continue. Now, what might happen if you could substitute another *Zeitgeber*, such as temperature, for light–dark, as has been done with tomatoes? Second, I am interested in when free-running starts. We know that the strength of the *Zeitgeber* depends on photoperiod, and several organisms have been shown to free-run in an LD cycle with only 1 hr of light or 1 hr of darkness. Now, you have only two cases where the gate becomes suddenly very wide, LD *1*:23 and LD *23*:1. Could it be that these animals are free-running? And would that interfere with your interpretation of your material?

TRUMAN: Concerning your first question, I haven't done any experiments along this line. Consequently, I don't know how the moths would react to a temperature cycle.

It isn't likely that the animals in the LD *23*:1 regimen are free-running. I have done some preliminary experiments with moths released into continuous darkness from the LD *23*:1

regimen. The resulting rhythm had the same phase relationships as seen in animals released from *LD 12*:12, and the emergence peaks were compact. One wouldn't see this pattern if free-running moths were released into DD.

MENAKER: Do you have any evidence on whether the clock runs during diapause, or could you suggest an experiment to determine that?

TRUMAN: Williams has some evidence on that, or at least on whether the clock runs during chilling of the diapausing pupae. He has taken two groups of pernyi and held them both at 5°C, one group under a 12-hr and the other under a 17-hr photophase. Then he terminated treatment and brought both groups to 25°C at a 17-hr photophase. Both groups broke diapause at the same time. So the pretreatment at either an inductive or noninductive photoperiod had no effect when the animals were returned to room temperature.

ADKISSON: I would like to answer that question. The clock certainly runs in pernyi during diapause, because both developing and diapausing insects respond to photoperiod. The critical transitional, or pivotal, photoperiod

separating the short-day from the long-day effect for pernyi has approximately 13.75 hr of light-time per day. Tanaka has shown that if pernyi larvae are grown in days shorter than this the resulting pupae enter diapause. If the larvae are grown in longer days, the pupae develop into moths without an arrest in growth. Williams and I showed that the pupae also are responsive to photoperiod. Days longer than 13.75 hr cause diapausing pupae to initiate adult development, while days shorter than this maintain them in the diapausing state. Thus, the evidence is clear that the clock is running during diapause. Regardless of whether the pernyi is a growing larva or a diapausing pupa, days shorter than 13.75 hr inhibit development—induce or maintain diapause—while days longer than this promote growth and development.

MENAKER: But which clock? We have yet to have any conclusive evidence on that.

ADKISSON: If diapause is controlled by the same endogenous mechanism as emergence, it is the same clock. Why should we postulate two clocks to do the job one can do?

COMMENT

DORA K. HAYES

We have found a rhythm in oxygen uptake by diapausing codling moth larvae (Hayes *et al.*, 1968). At 20° and 23°C there was a circadian periodicity of 24 hr; at 26°C we found an 8-hr periodicity. I would assume that this indicates that during diapause some sort of mechanism (the "clock" referred to by Adkisson)

exists for maintaining such a rhythm.

REFERENCE

Hayes, D. K., M. S. Schechter, E. Mensing, and J. Horton. 1968. Oxygen uptake of single insects determined with a polarographic oxygen electrode. Anal. Biochem. 26:51–60.

S. K. ROBERTS
S. D. SKOPIK
R. J. DRISKILL

Circadian Rhythms in Cockroaches: Does Brain Hormone Mediate the Locomotor Cycle?

A number of studies on circadian rhythms in cockroaches have considered possible hormonal links coupling different elements of the system. The earliest report (Harker, 1956) suggested a hormonal clock in the subesophageal ganglion. Subsequent failures to confirm these findings (Roberts, 1966; Nishiitsutsuji-Uwo and Pittendrigh, 1967, 1968; Brady, 1967) cast serious doubt on their reality. More recently, attention has focused on a different site in the nervous system, the neurosecretory cells in the brain. According to Nishiitsutsuji-Uwo and Pittendrigh (1968), a neural oscillator in the optic lobes elicits rhythmic hormonal secretion, which, via the circulatory system, modulates the locomotor activity sustained by the thoracic ganglia. The strongest evidence for such a hormonal link derives from their claim that severing the nerve tracts connecting the brain to the ventral nerve cords by cutting the circumesophageal connectives (CEC) fails to upset rhythmicity. In the absence of a direct neural connection between the driving oscillator in the optic lobes and locomotor centers along the ventral nerve cords, a continuance of rhythmicity would suggest hormonal involvement. In contrast to these findings, however, Brady (1967) reported that similar operations (performed on a different species) evoked apparent arrhythmicity, although the observed high level of activity precluded any strong conclusions.

In addition to their compelling evidence for an autonomous driving oscillator in the optic lobes, Nishiitsutsuji-Uwo and Pittendrigh propose a specific hormonal link that we now believe is untenable in light of the new
evidence. The purpose of this report is to present new evidence bearing on
their brain hormone hypothesis.

SURGICAL PROCEDURE

In a recent review of control links in the circadian system of the cockroach,
Brady (1969) presented evidence that clearly pointed to the protocerebrum
of the brain as the locus for the driving oscillator. He also considered the
issue of hormonal versus neural transmission of the oscillation to the
thoracic ganglia. It has been established that virtually all cephalic neuroendocrine tissues are not essential to the control of locomotor rhythmicity.
These tissues include the medial neurosecretory cells of the pars intercerebralis and the corpora allata–corpora cardiaca complex. However, the
lateral neurosecretory cells of the protocerebrum have not been ruled out
by any conclusive experiment. Indeed, Nishiitsutsuji-Uwo and Pittendrigh
have explicitly implicated them as the source of activity-modulating brain
hormone. We believe that careful and extensive repetition of experiments
is warranted since (1) Nishiitsutsuji-Uwo and Pittendrigh's conclusion is
based primarily on the observation that severing the CEC did not upset
rhythmicity (in five out of six animals tested), and (2) Brady's 1967 report
of conflicting observations was based on ambiguous records from four
animals.

The CEC have been cut in two ways:

• Brady's technique employs a head incision whereby the connectives
are exposed by removal of a small rectangle of cuticle from the frons overlying the anterior brain surface. We have found that this technique permits
clear exposure of the connectives as they descend bilaterally around the
esophagus. We have effected the CEC cuts with irridectomy scissors or fine
forceps and found that the cut ends generally retract from each other
enough to ensure complete separation.

• Nishiitsutsuji-Uwo and Pittendrigh made a U-shaped incision in the
ventral surface (mentum) of the neck. According to their detailed description of this operation, it affords ready access to the CEC at the point of
their junction with the subesophageal ganglion. Unfortunately, we have
found this technique more difficult than indicated. When the insect is

Circadian Rhythms in Cockroaches: Does Brain
Hormone Mediate the Locomotor Cycle?

507

taped down so as to expose the ventral surface of the neck, the connectives are not in fact "readily visible," but lie buried deep beneath the ganglion whence they pass through the tentorium to the brain. Unless the ganglion is forcibly reflected—a very destructive procedure—the connectives are not visible at all. It is, however, possible to pass a fine hooked wire or slender blade under the ganglion and blindly sever the connectives, but a postmortem inspection is clearly needed to ascertain whether the cuts were complete.

We have used both of these surgical techniques. Most of our experiments were carried out with adult male cockroaches: *Leucophaea madeirae* (used by Nishiitsutsuji-Uwo and Pittendrigh), and *Periplaneta americana* (used by Brady). Activity rhythms were recorded in standard running wheels (Roberts, 1960) maintained at a constant temperature of approximately 25°C. Except as noted below, all animals were maintained in *LD 12*:12 prior to and following surgical treatment.

EXPERIMENTAL RESULTS

CUTTING THE CEC THROUGH NECK INCISIONS

In Figure 1, A and B illustrate the apparent loss of rhythmicity following complete severance of the CEC. As indicated in Table 1, A, 15 out of 21 animals clearly appeared arrhythmic (−) during the first postoperative week. In one of the two animals judged rhythmic (+), we were able to confirm by postmortem that a single connective remained intact; the other roach was decomposed and postmortem was not possible. Dissection of the animal that became rhythmic during the fourth week disclosed a partially intact connective. In four cases tabulated as uncertain (?), the activity records did not permit any conclusion about rhythmicity. This was mainly due to radically low levels of activity—as little as one indication of movement in 2 or 3 days.

The depression of activity typically found after this kind of surgery was puzzling for two reasons. First, Nishiitsutsuji-Uwo and Pittendrigh had indicated that four of their six test animals showed normal or high activity levels—a finding consistent with Brady's observations following the head incision approach. Second, the subesophageal ganglion is generally considered to stimulate locomotor activity if uncoupled from inhibitory centers in the brain (Roeder, 1953). We must conclude that our technique

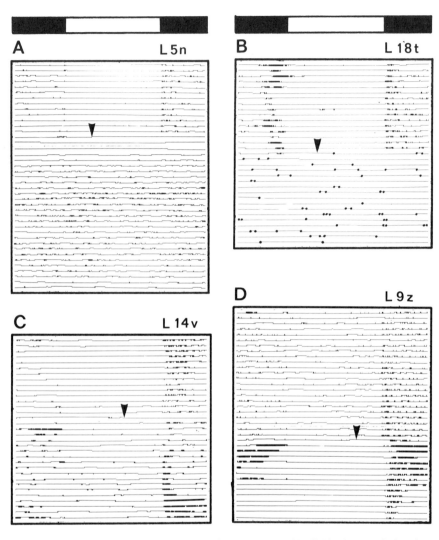

FIGURE 1   Locomotor activity of four *Leucophaea* recorded in *L*D *12*:12 prior to and after the
following surgical procedures: A and B, both circumesophageal connectives cut by neck incision—
the low level of postoperative activity in L 18t is made visible by dotting each recorder pen stroke;
C, only one circumesophageal connective cut by neck incision; D, circumesophageal connectives
left intact but mouthpart nerves cut by neck incision. Time of operations is indicated by arrows;
baseline is 24 hr.

TABLE 1 Effects of Surgical Lesions on the Circadian Locomotor Rhythms in Cockroaches

Operation Procedure	Number of Animals	Postoperative Observations												Activity Level		
		State of Rhythm														
		First Week			Second Week			Third Week			Fourth Week					
		+	–	(?)	+	–	(?)	+	–	(?)	+	–	(?)	Low	Normal	High
A. Cut both CEC (neck incision)	21	2	15	4	0	11	0	0	5	0	1	3	0	17	2	2
B. Cut both CEC (head incision)	43	0	38	5	1	21	1	0	9	0	6	1	0	11	21	11
C. Cut single CEC (head and neck)	15	13	1	1	6	1	3	5	1	0	4	0	0	1	9	5
D. Sham (head and neck)	5	5	0	0	5	0	0	4	0	0	Discontinued			0	5	0
E. Cut maxillary and mandibular nerves	5	5	0	0	3	0	1	2	0	1	2	0	1	0	5	0

509

damages the ganglion, a view seemingly justified in light of the generally high levels of activity found in later experiments in which the CEC were cut without exposing the subesophageal ganglia to surgical trauma.

CUTTING THE CEC THROUGH HEAD INCISIONS

Because of the relative ease with which the CEC can be cut by an approach through head incisions, this technique has provided the bulk of our experimental data. The operation was performed on a total of 43 animals and the postoperative observations include more than 600 days of recordings. In Figure 2, A and C clearly demonstrate the typical loss of a circadian rhythm in two of these roaches. A total of 38 animals showned an arrhythmic pattern during the first week after surgery. The intense postoperative activity displayed by *Periplaneta* (Figure 2, B) was scored as questionable (?) since such incessant activity might mask an underlying rhythm. We entirely agree with Brady that in this species the operation generally evokes such high levels of activity—much higher than in *Leucophaea*—that interpretation is difficult. It should be emphasized, however, that the test group as a whole (see Table 1, B) includes 32 animals with either low or normal activity levels.

CONTROL PROCEDURES

We adopted three different operative procedures to ensure that general trauma does not induce loss of rhythmic behavior. Sham operations were carried out on five animals: In four cases the brain was exposed and manipulated, and in one case the neck area and subesophageal ganglion were similarly treated. As recorded in Table 1, D, all five animals showed normal rhythms for at least 2 weeks, with no significant change of activity level.

In 15 additional animals, incisions were made in the head (11 cases) or neck (4 cases), and a single connective was cut. Because the operation through the neck is so difficult, each of these animals was subsequently killed and, on examination, was found to possess only one intact connective. It may therefore be concluded, on the basis of the 13 animals that remained rhythmic (e.g., Figure 1, C and D), that the operation leaves the circadian system essentially undisturbed. This conclusion is further supported by the fact that free-running rhythms were observed in three animals maintained in constant darkness. Thus the single connective roaches are not simply responding in a photokinetic fashion to alternations of light and dark.

Circadian Rhythms in Cockroaches: Does Brain
Hormone Mediate the Locomotor Cycle?

511

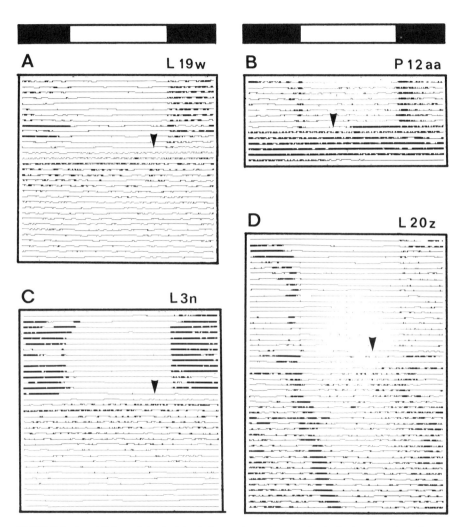

FIGURE 2   A and C, loss of rhythmicity in *Leucophaea* after severing both circumesophageal connectives by the head incision technique; B, development of extreme hyperactivity in *Periplaneta*
following severance of both circumesophageal connectives–this record was scored (?) due to possible masking of rhythmicity; D, *Leucophaea* 20z is initially entrained in LD. At the time indicated
by the arrow, one circumesophageal connective was cut and the roach was placed in DD.

The third procedure was adopted after it was discovered that CEC cuts through the neck invariably necessitated severing the maxillary and mandibular nerves as well. These two pairs of nerves emerge from the subesophageal ganglion and run anteriorly through the area that permits access to the connectives. Without attempting to sever the CEC, we simply cut these two pairs of mouthpart nerves (in five animals). Figure 1, D, illustrates one of these; inspection of Table 1, D, indicates that the operation leaves the rhythm intact.

DISCUSSION

The proposition that a brain hormone controls the expression of the circadian locomotor rhythm in the roach must be reviewed in the light of the foregoing experiments. The hypothesis depends heavily on Nishiitsutsuji-Uwo's and Pittendrigh's observation that rhythmicity persists after cutting the nervous connections between the driving oscillator in the optic lobes and the locomotor centers along the ventral nerve cord. The extensive experiments described here indicate that the overwhelming majority of the experimental animals lost their rhythms following the CEC operation. We believe, as the control procedures attest, that the persistent arrhythmicity we observe after this operation precludes the interpretation that activity resulting from surgical trauma simply masks an underlying rhythm. Surprisingly, the CEC cutting operation in the hands of Nishiitsutsuji-Uwo and Pittendrigh (1968) failed to upset the rhythm even temporarily—this suggests that a blood-borne hormone, rather than a neural link, is involved. It is difficult to reconcile this interpretation with the fact that removal of the corpora cardiaca and corpora allata does not upset rhythmicity (Roberts, 1966). Surely it is via these neurohaemal structures that the presumed brain hormone would enter the blood (Scharrer, 1952; Willey, 1961). Even if one assumes that the hormone could enter the blood by another route, a temporary disruption of the rhythm would seem likely.

We re-emphasize here the difficulty in carrying out the neck incision procedure and of ensuring complete bilateral severance of the CEC. Whereas we have verified complete severance in a number of instances by postmortem dissection, such confirmation was apparently not done in the earlier investigation. As one of our control groups shows, incompletely cutting the connectives permits normal expression of rhythmicity.

It should also be noted that rhythmicity in LD following "cut-and-see" surgery does not, of itself, constitute proof of an intact circadian system. Unless the rhythm can be demonstrated to persist and to free-run in constant darkness, the less interesting alternative previously mentioned cannot be ruled out. Since the earlier observations, supporting the claim of persistence of rhythms after CEC cuts, were made in LD, we hesitate to conclude that an intact hormonal link persists.

The failure to confirm the earlier CEC cutting experiments is difficult to explain except in terms of differences in surgical techniques. All other aspects of the experiments were virtually identical: type of insect, method of anaesthesia, activity recording methods, temperature control, and light intensities. Brady (1969) has suggested that the original description of the neck incision technique does not rule out the possibility that mouthpart nerves, rather than CEC, were cut. This idea had independently occurred to one of us (S.R.) and prompted the appropriate experiment. Clearly, cutting the mouthpart nerves leaves the rhythm (in LD) intact. The fact that no mention was made of these nerves in the earlier report strengthens Brady's suggestion. It is also impossible to exclude incomplete cutting of the CEC as a basis for the conflicting results. On the strength of these considerations, we think the weight of evidence does not at present justify the conclusion that hormones are involved. Indeed, we now concur with Brady's (1969) assertion that the control pathway most likely involves neural (electrical) channels through the CEC and ventral nerve cord.

ACKNOWLEDGMENTS

The authors wish to acknowledge support from the National Science Foundation (GB-3716), from the University of Delaware Research Foundation, and from a grant in aid of research from Temple University.

REFERENCES

Brady, J. 1967. Control of the circadian rhythm of activity in the cockroach. II. The role of the sub-oesophageal ganglion and ventral nerve chord. J. Exp. Biol. 47:165–178.

Brady, J. 1969. How are insect circadian rhythms controlled? Nature (Lond.) 223:781–784.

Harker, J. E. 1956. Factors controlling the diurnal rhythm of activity in *Periplaneta americana*. J. Exp. Biol. 33:224–234.

Nishiitsutsuji-Uwo, J., and C. S. Pittendrigh. 1967. The neuroendocrine basis of midgut tumor induction in cockroaches. J. Insect Physiol. 13:851–859.

Nishiitsutsuji-Uwo, J., and C. S. Pittendrigh. 1968. Central nervous system control of circadian rhythmicity in the cockroach. III. The optic lobes, locus of the driving oscillation? Z. Vgl. Physiol. 58:14–46.

Roberts, S. K. deF. 1960. Circadian activity rhythms in cockroaches. I. The free-running rhythm in steady-state. J. Cell. Comp. Physiol. 55:99–110.

Roberts, S. K. deF. 1966. Circadian activity rhythms in cockroaches. III. The role of endocrine and neural factors. J. Cell. Comp. Physiol. 67:473–486.

Roeder, K. D. 1953. Reflex activity and ganglion function, p. 423–462. *In* K. D. Roeder [ed.], Insect physiology. John Wiley & Sons, New York.

Scharrer, B. 1952. Neurosecretion. XI. The effects of nerve section on the intercerebralis-cardiacum-allatum system of the insect *Leucophaea madeirae*. Biol. Bull. 102:261–272.

Willey, R. B. 1961. The morphology of the stomodeal nervous system in *Periplaneta americana* (L.) and other Blattaria. J. Morphol. 108:219–262.

COMMENT

S. K. ROBERTS

I should like to report briefly some additional evidence in support of Nishiitsutsuji-Uwo's and Pittendrigh's conclusion (1968) that the optic lobes play a crucial role in maintaining the locomotor rhythm in cockroaches. This conclusion principally derived from two types of experiments: (1) bilateral severance of the optic tracts, which isolates the optic lobes from the remaining protocerebrum; and (2) bilateral severance of the optic nerves, which leaves the optic lobes attached to the protocerebrum, but separated from the ommatidia of the compound eyes. This last procedure was reported to leave the rhythm intact and free-running (in LD) in the majority of animals tested. However, cutting both optic tracts in most cases evoked a loss of rhythms.

I have been able to confirm both of these important findings in the roach *Periplaneta americana*; the earlier experiments mainly employed *Leucophaea madeirae*. In Table 1, A summarizes the results following bilateral cuts through the optic tracts. In each of the 36 tested animals the normal (entrained) rhythm in

LD disappeared immediately after surgery. In the majority of cases (21), arrhythmicity continued for many weeks until either the animal died or the experiment was discontinued. The remainder of the animals displayed either ambiguous running patterns (3 cases) or a resumption of rhythmicity after initial loss (12 cases). The average time for reappearance of a rhythm was 5 weeks. Although there are no supporting histological data available, this suggests successful regeneration of some optic tracts.

While separation of the optic lobes from the brain clearly upset the activity rhythms, cutting the optic nerves did not. Although the latter operation necessitated more extensive surgical trauma, the majority of animals recovered (11 out of 20; Table 1, B) and showed free-running rhythms in LD. Of the remaining 9 animals, 5 resumed rhythmicity within about a week and, with one exception, all free-ran in LD. The exceptional case remained entrained to the LD cycle—most likely the nerves to the compound eyes were incompletely cut. Another

TABLE 1   Effects of Surgical Lesions on the Circadian Locomotor Rhythms in
Periplaneta americana

Type of Operation	Number of Animals	Postoperative Observations				
		Weeks of Observation (Average)	State of Rhythm			
			+	– to +	–	(?)
A. Cut both optic tracts	36	2–34 (11)	0	12	21	3
B. Cut all optic nerves	20	1–34 (6)	11	5	1	3
C. Cut single optic tract	11	2–14 (8)	8	2	0	1

procedure that left the rhythm intact was to cut the optic tracts on one side only. Nishiitsutsuji-Uwo and Pittendrigh demonstrated that the remaining optic lobe, with tracts intact to the protocerebrum, was fully competent to sustain rhythmicity. My findings confirm this observation as well (Table 1, C).

Finally, I have completed some preliminary experiments to determine whether the whole optic lobe, or only part of it, serves to maintain a rhythm. The optic lobe contains three major neuropile masses (Bullock and Horridge, 1965). These masses, which appear as separately distinct areas in histological preparations, are (1) the outer *lamina* below the ommatidia; (2) the intermediate *medulla*; and (3) the inner *lobula*. My results indicate that at least part of the optic lobe is dispensible; only the area containing the *lobula* and *medulla* is needed to express a rhythm. The majority of animals (8 out of 9) with partially

ablated lobes either remained rhythmic and free-running in LD, or regained their rhythms after initial loss. Histological examinations indicated that only the *lamina* had been removed in this group. It is not clear at this time whether these inner neuropile areas, or other surrounding neural tissues, constitute the crucial circadian element.

REFERENCES

Nishiitsutsuji-Uwo, J., and C. S. Pittendrigh. 1968. Central nervous system control of circadian rhythmicity in the cockroach. III. The optic lobes, locus of the driving oscillation? Z. Vgl. Physiol. 58:14–46.
Bullock, T. H., and G. A. Horridge. 1965. *In* Structure and function in the nervous system of invertebrates. W. H. Freeman and Company, San Francisco, p. 1079–1080.

DISCUSSION

PITTENDRIGH: I would just like to say that this is the second time I have heard Roberts make an elegant presentation clearing up confusion in the literature on cockroach clocks and that I consider the circumesophageal connectives of Dr. Uwo and myself to have been properly cut only this morning!

JACKLET: Considering the rate at which cockroaches regenerate their neurons, is it possible to keep a cockroach long enough after severing the nerves for rhythmicity to return?

ROBERTS: Yes; for example, in some older experiments where I made lesions through the pars intercerebralis with an apparent loss of rhythmicity, rhythmicity sometimes reappeared after 5 or 6 weeks. Also, there are cases (unpublished experiments of mine) where the optic tracts have been cut, and, after a period of several weeks, rhythmicity is resumed. Attempts to correlate the reappearance of the rhythm with specific neural regeneration have not been made, however.

PITTENDRIGH: On this point, in almost all the operations that were published by Uwo and myself concerning the optic lobes, there was a tendency for the rhythm to regenerate in animals that lived long enough.

J. BRADY

The Search for an Insect Clock

Ever since the original report that rhythms could be transferred between cockroaches by parabiosis (Harker, 1954), research into the source of insect periodicities has concentrated on hormonal mechanisms to the complete exclusion of other possibilities. A review of the evidence for such hormonal clocks has recently been published (Brady, 1969) and, insofar as it concerns the control of the cockroach locomotor rhythm, is summarized again here in Figure 1. The apparent conclusions to be drawn from this work are that:

- No rhythms are transferable by gland transplant;
- Virtually all the cephalic endocrine organs can be removed without stopping the rhythm;
- The only operations that do stop the rhythm are those that interrupt the neural pathways between the optic lobes and the brain, or between the brain and the thorax.

This scarcely adds up to a case for humoral timing of the rhythm; a much simpler interpretation is that the driving oscillation is transmitted neurally.

Although it seems likely that the primary clocks of other insects may prove to function similarly, this does not preclude hormonal periodicities participating at lower levels. Indeed, there is much evidence that implies

517

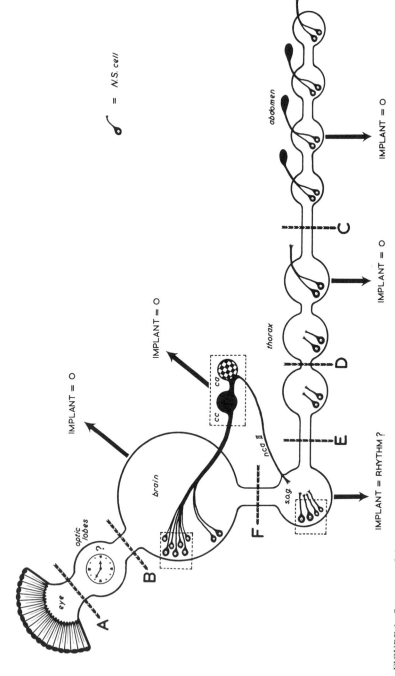

FIGURE 1 Summary of the search for hormonal clocks controlling the circadian rhythm of locomotor activity in the cockroach. The ganglia of the central nervous system are represented by the linked spheres. Dotted boxes indicate endocrine tissue that can be removed without stopping the rhythm (Roberts, 1966; Brady, 1967a, b). Arrows indicate organs transplanted from rhythmic donors to headless, arrhythmic recipients: *O* means that the recipient shows no detectable rhythm (Harker, 1956; Roberts, 1966; Brady, 1967b). Heavy broken lines indicate where the nerve trunks have been severed: cuts *B, E, F,* or splitting the protocerebral lobes bilaterally, apparently stop the rhythm; cuts *A, D, C,* splitting the pars intercerebralis midsagittally, or removing one protocerebral lobe, do not (Brady, 1967b; Nishiitsutsuji-Uwo and Pittendrigh, 1968a, b; Roberts, Skopik, and Driskill, this volume; see also Azaryan & Tyshchenko, 1969); cut *A* may be made between the medulla and lamina of the optic lobe without stopping the rhythm (Roberts, this volume). *N. S.,* neurosecretory; *s.o.g.,* subesophageal ganglion; *cc ca,* corpora cardiaca–allata complex.

518

such participation. On the one hand, there are histologically and pharma-cologically detectable cycles (e.g., Rensing, Thach, and Bruce, 1965; Fowler and Goodnight, 1966; Rao and Gropalakrishnareddy, 1967; Hinks, 1967); on the other hand, there is evidence for endocrine control of the level of locomotor activity as such (e.g., Ozbas and Hodgson, 1958; Milburn, Weiant, and Roeder, 1960; Haskell and Moorhouse, 1963; Haskell *et al.*, 1965; Brady, 1967a). In the first case, however, it is impossible to tell from such observations alone whether the rhythms are driving or being driven, and in the second, the results imply hormonal control of the amount of activity only; nothing is revealed about its timing. So far, no experimental evidence other than Harker's work (1956) supports the con-tention of a primary endocrine clock timing daily repeated functions (see, however, Truman, this volume).

As a result of preoccupation with hormones, the literature contains little information suggestive of alternatives. The only relevant work seems to be that of Nishiitsutsuji-Uwo and Pittendrigh (1968b) and Roberts (this volume), which shows that cutting neural pathways to the eyes does not stop the cockroach rhythm, whereas cutting nerves to the optic lobes does. The work of Jahn and Wulff (1943) demonstrating a circadian rhythm in the electroretinogram of beetles may also prove to be relevant. Since a role for the connections between the optic lobes and the protocerebrum is indicated by these reports, one may start the new search for a clock by looking at what is known of the neuroanatomy of this area. Figure 2 dia-grams the known tracts from the optic lobes. Each lobe has direct links with the optic tubercle, mushroom body, protocerebral neuropile, ventral nerve cord, and contralateral lobe. Most of these connections are presum-ably concerned with visual interpretation.

Surgically isolating the optic lobes must sever all these tracts, and one may ask what effect this is likely to have on systems other than those con-cerned solely with time-keeping. How can one tell that the absence of a detectable locomotor rhythm after such an operation is not just the result of interrupting more general aspects of locomotor coordination? As re-vealed by Roeder (1967) and Huber (1965), the central control of locomo-tion in insects mainly involves the system outlined in Figure 3. This shows that the subesophageal ganglion has an excitatory influence on thoracic motor activity, but is itself inhibited by the mushroom bodies. The central body exerts an excitatory influence similar to that of the subesophageal ganglion and is likewise held in check by the mushroom bodies. Sensory input from the compound eyes and, presumably, proprioceptive feedback dictate the direction of turning locomotion, which is regulated by a balance of mutual inhibition between the two mushroom bodies.

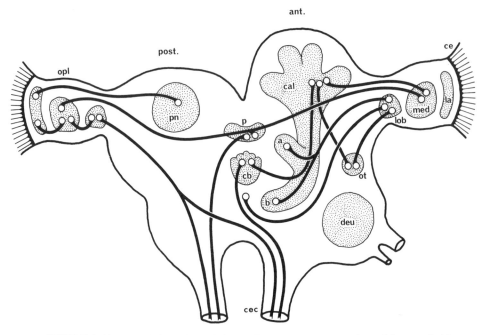

FIGURE 2   Nerve tracts from the optic lobes of insects. A diagrammatic pastiche compiled from
the works of Kenyon (1896, bee), Jawlowski (1958, bee), Bretschneider (1913, *Periplaneta*), and
Hanström (1928, *Periplaneta*). Other tracts, not included, mainly duplicate the general links shown
here (see Horridge, 1965). No directionality is implied in the pathways as shown; stippled areas represent main neuropile structures. Left portion of figure represents a more posterior aspect, and
the right a more anterior aspect. *ant*, anterior; *post*, posterior; *a*, α lobe; *b*, β lobe; and *cal*, calyces
of mushroom body; *cb*, central body; *cec*, circumesophageal connectives; *ce*, compound eye; *deu*,
antennal glomeruli of deutocerebrum; *la*, lamina; *lo*, lobula; *med*, medulla; *opl*, optic lobe; *ot*, optic
tubercle; *p*, pons; *pn*, general protocerebral neuropile.

If the system controlling the integration of locomotion (as opposed to
its coordination, which is thoracic) is not much more complex than this
scheme implies, it would seem safe to infer that disconnecting the optic
lobes should not particularly upset it. The only obvious interference with
the system is the removal of visual input, and in any case the rhythm will
survive the much more destructive operation of splitting the protocerebrum
midsagittally. Thus, other operations that stop the rhythm without grossly
distorting locomotor behavior may perhaps also be assumed to have done
so directly, by cutting the link between the clock and the thorax, rather
than indirectly by upsetting central integrating circuits. On this assumption, one may examine such experimental evidence as there is concerning
the location of the clock.

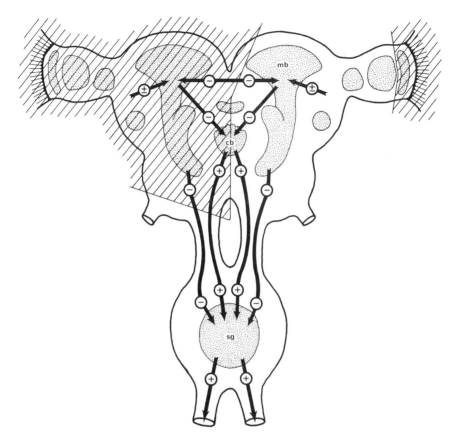

FIGURE 3 Schematic representation of the central control of locomotor activity (after Huber, 1965). + indicates excitatory pathway; – indicates inhibitory pathway; these pathways are not yet anatomically identified. Cross-hatching represents areas that can be removed without stopping the rhythm (Brady, 1967a; Nishiitsutsuji-Uwo and Pittendrigh, 1968a, b; Roberts, this volume). *mb*, mushroom body; *sg*, subesophageal ganglion; other conventions as in Figure 2.

All that is necessary for the cockroach locomotor rhythm to be maintained is for the bilaterally duplicated parts of one protocerebral ganglion, plus its ipsilateral lobula and medulla, to remain in contact with the thorax. The medial protocerebral structures, the distal parts of the optic lobe, the deutocerebrum, and all the contralateral parts of the brain can be dispensed with (Figure 1, and cross-hatching, Figure 3). Even so, this is little enough to go on, and the suggestions that follow are therefore speculative.

It is not clear from the published data whether the central body ever survived undamaged in the split protocerebral preparations, but some of the cuts were apparently off-center (Nishiitsutsuji-Uwo and Pittendrigh, 1968b) so that it is certainly possible that it could have. In view of Huber's work (Figure 3) it would therefore be very interesting if retention of a rhythm after this operation proved to correlate with retention of an intact central body. Without more specific data, however, it is simpler, for the moment, to assume that splitting the brain does effectively destroy the central body.

If, then, the primary time-keeping mechanism resides in the bilaterally paired structures in the protocerebrum, there must clearly be two clocks. Normally, these could be presumed to keep each other in phase, but what happens when the brain is split and they are separated? The two clocks must then oscillate independently, unless synchronized via the subeso-phageal ganglion. Alternatively, in the absence of such feedback, will they free-run at different periods so that the locomotor rhythm then gradually develops bimodality? And if this occurs, will the cockroach develop patho-logical symptoms as Harker's work would imply (Harker, 1958, 1961)?

Of the neural tracts illustrated in Figure 2, the only ones likely to be left intact by brain operations that do not disturb the rhythm are those connecting the lobula and medulla to the mushroom body. In the Roeder/ Huber system of locomotor control (Figure 3), it is the mushroom bodies that inhibit the excitatory centers in the central body and subesophageal ganglion. It is tempting to suggest that the clock may control activity by acting directly on the mushroom bodies to disinhibit the subesophageal center, possibly with a loop incorporating the optic tubercles.

As to the nature of the clock, nothing is revealed by any of this work— the experiments are concerned only with the transmission of the oscilla-tion. However, it appears that circadian time is still measured by a cock-roach after it has been injected with tetrodotoxin (Truman, personal communication). This implies that the clock can function without action potentials, so that reverberatory axon circuits or other related phenomena are presumably not involved.

Whether these suggestions will survive future experiments remains to be seen, but it must be remembered in any case that the whole of this paper has been constructed on the assumption that some sort of primary circa-dian pacemaker does exist in the insect. It would not be fair to close the discussion without raising some pertinent doubts on this central point. First, the neuropile structures and tracts that have been mentioned lend themselves to consideration mainly because their presence is obvious; in reality there are also countless other pathways that could equally well be involved. Moreover, electrical stimulation of a great many widely differing sites in the insect brain releases locomotor activity (see Huber, 1967).

Secondly, the rhythm under consideration here is a motor one, which may call for a neural control mechanism; other tissues may well use other control systems (Truman, this volume). Thirdly, circadian periodicity seems to be a fundamental property of all cells, and some insect tissues can apparently exhibit this periodicity *in vitro* (Rensing, this volume; and perhaps Neville, 1967). Thus, although there is a certain logical tidiness in expecting to find a single primary clock, this may prove to be an oversimplification.

REFERENCES

Azaryan, A. G., and V. P. Tyshchenko. 1969. The role of brain neurons in the control of the circadian rhythm of behaviour in the cricket *Gryllus domesticus* L. [in Russian]. Dokl. Akad. Nauk SSSR 186:484–487.

Brady, J. 1967a. Control of the circadian rhythm of activity in the cockroach. I. The role of the corpora cardiaca, brain and stress. J. Exp. Biol. 47:153–163.

Brady, J. 1967b. Control of the circadian rhythm of activity in the cockroach. II. The role of the sub-oesophageal ganglion and ventral nerve cord. J. Exp. Biol. 47:165–178.

Brady, J. 1969. How are insect circadian rhythms controlled? Nature (Lond.) 223:781–784.

Bretschneider, F. 1913. Der Centralkörper und die pilzförmigen Körper im Gehirn der Insekten. Zool. Anz. 41:560–569.

Fowler, D. J., and C. J. Goodnight. 1966. The cyclic production of 5-hydroxytryptamine in the opilionid. Am. Zool. 6:187–193.

Hanström, B. 1928. Vergleichende Anatomie des Nervensystems der wirbellosen Tiere unter Berücksichtigung seiner Funktion. Springer, Berlin.

Harker, J. E. 1954. Diurnal rhythms in *Periplaneta americana* L. Nature (Lond.) 173:689–690.

Harker, J. E. 1956. Factors controlling the diurnal rhythm of activity of *Periplaneta americana* L. J. Exp. Biol. 33:224–234.

Harker, J. E. 1958. Experimental production of midgut tumours in *Periplaneta americana* L. J. Exp. Biol. 35:251–259.

Harker, J. E. 1961. Diurnal rhythms. A. Rev. Ent. 6, 131–146 (see p. 136).

Haskell, P. T., D. B. Carlisle, P. E. Ellis, and J. E. Moorhouse. 1965. Hormonal influences in locust marching behaviour. Proc. XIIth Int. Congr. Entomol., London, 1964, p. 290–291.

Haskell, P. T., and J. E. Moorhouse. 1963. A blood borne factor influencing the activity of the central nervous system of the desert locust. Nature (Lond.) 197:56–58.

Hinks, C. F. 1967. Relationship between serotonin and the circadian rhythm in some nocturnal moths. Nature (Lond.) 214:386–387.

Horridge, G. A. 1965. Arthropoda: receptors for light, and optic lobe, p. 1077 et seq. *In* T. H. Bullock and G. A. Horridge [ed.], Structure and function in the nervous systems of invertebrates. W. H. Freeman Co., San Francisco.

Huber, F. 1965. Neural integration (central nervous system), p. 333 to 406. *In* M. Rockstein [ed.], The physiology of insecta. Vol. 2. Academic Press, New York.

Huber, F. 1967. Central control of movements and behavior of invertebrates, p. 333 to 351. *In* C. A. G. Wiersma [ed.], Invertebrate nervous systems—Their significance for mammalian neurophysiology.

Jahn, T. L., and V. J. Wulff. 1943. Electrical aspects of a diurnal rhythm in the eye of *Dytiscus fasciventris*. Physiol. Zool. 16:101–109.

Jawlowski, H. 1958. Nerve tracts in bee (*Apis mellifica*) running from the sight and antennal organs to the brain. Ann. Univ. Mariae Curie–Sklodowska, Sect. D. 12:307–323.

Kenyon, F. C. 1896. The brain of the bee. A preliminary contribution to the morphology of the nervous system of the arthropoda. J. Comp. Neurol. 6:133–210.

Milburn, N., E. A. Weiant, and K. D. Roeder. 1960. The release of efferent nerve activity in the roach, *Periplaneta americana*, by extracts of the corpus cardiacum. Biol. Bull. 118:111–119.

Neville, A. C. 1967. A dermal light sense influencing skeletal structure in locusts. J. Insect Physiol. 13:933–939.

Nishiitsutsuji-Uwo, J., and C. S. Pittendrigh. 1968a. Central nervous system control of circadian rhythmicity in the cockroach. II. The pathway of light signals that entrain the rhythm. Z. Vgl. Physiol. 58:1–13.

Nishiitsutsuji-Uwo, J., and C. S. Pittendrigh. 1968b. Central nervous system control of circadian rhythmicity in the cockroach. III. The optic lobes, locus of the driving oscillation? Z. Vgl. Physiol. 58:14–46.

Ozbas, S., and E. S. Hodgson. 1958. Action of insect neurosecretion upon central nervous system *in vitro* and upon behavior. Proc. Natl. Acad. Sci. U.S. 44:825–830.

Rao, K. P., and T. Gropalakrishnareddy. 1967. Blood borne factors in circadian rhythms of activity. Nature (Lond.) 213:1047–1048.

Rensing, L., B. Thach, and V. Bruce. 1965. Daily rhythms in the endocrine glands of *Drosophila* larvae. Experientia 21:103.

Roberts, S. K. 1966. Circadian activity rhythms in cockroaches. III. The role of endocrine and neural factors. J. Cell. Physiol. 67:473–486.

Roeder, K. D. 1967. Nerve cells and insect behavior. 2nd ed. Harvard Univ. Press, Cambridge, Massachusetts.

DISCUSSION

ZIMMERMAN: By an extension of your argument about connectives, could it be the case that the clock is in the subesophageal ganglion and requires a connective circuit that goes to the protocerebrum?

BRADY: I doubt it, because Pittendrigh and Uwo showed that you can cut off nearly all the subesophageal ganglion except the through tracts without disturbing rhythmicity.

PITTENDRIGH: I am a little disturbed by your implication that Roberts' clarification of the circumesophageal cutting showed that the clock really was not hormonal. There was never any suggestion in the Princeton papers that the clock was hormonal.

BRADY: I am sorry; I did not mean to imply that your suggestion was that the clock was hormonal, but only that the couplings were hormonal.

PITTENDRIGH: Those are very different statements. The other thing I wish to say, without retracting the strength of my comment about the character of the results that Roberts, Skopik, and Driskill presented, is that there is something in the published data that neither you nor Roberts has alluded to and that I think deserves some comment. That is the paper we published on the pars intercerebralis. Those data still remain to be explained if we are to dismiss a hormonal coupling altogether. We removed, or attempted to remove, you recall, the pars intercerebralis from 47 animals. There was a statistically significant difference in the histological data gathered after the experiment was over, showing that in those few animals in which a rhythm did persist, some of the neurosecretory cells were left. There is, in other words, some correlational evidence remaining for a hormonal coupling.

BRADY: If I remember correctly, you looked at 22 brains histologically after this operation and found that there were neurosecretory cells left in many of the animals that had remained rhythmic. On the other hand, you also found no neurosecretory cells in eight animals that remained equally rhythmic. Much depends on how you interpret these results, as I have discussed elsewhere (Brady, 1969).

HALBERG: It is a statistical matter. That is, if you find neurosecretory cells in sizable numbers in some insects, the fact that you could not find them in some others would not be conclusive proof that such cells are not involved in maintaining the rhythm.

BRADY: But you can also argue that the other way around.

PITTENDRIGH: There is a statistical difference between the two groups, and the question is in the interpretation of the statistics.

BRADY: No, I believe the question is in the interpretation of the observations: It depends on whether, when you cannot see a neurosecretory cell, you think that it is gone, or you think that it is still there, but invisible.

STRUMWASSER: It should be pointed out—and I think this is your point, Dr. Halberg—that it is very difficult to be sure, histologically, of the presence of neurosecretory cells. Electron microscopic criteria are generally accepted, but the work of Pittendrigh and Uwo was done by light microscopy. They employed some esoteric stains that are reasonable, but certainly not perfect, indicators of neurosecretion.

BRADY: I also had seven animals that remained rhythmic after microcautery of the pars intercerebralis. Not all the neurosecretory cells were removed, it is true, but there was a massive reduction in their numbers (in two animals to only one or two cells each) without apparently affecting the rhythm.

DeCOURSEY: Would it be possible to work with some rhythm other than locomotor activity in these animals and get away from all these difficulties? Could you look at oxygen consumption or perhaps some cellular rhythm, not as the main method, perhaps, but at least for some of the critical experiments?

BRADY: Certainly; oxygen consumption or some other rhythm would be a very good thing to measure.

STRUMWASSER: Your point is that the clock may be still intact, but when you get rid of the coupling factor, locomotion becomes arrhythmic.

BRADY: Quite so; in animals made arrhythmic by surgical manipulation of the brain, the only output of the clock that is being measured is locomotion, which may well be disturbed, while other rhythms may be left intact and show, for instance, in oxygen consumption.

HALBERG: As you might recall, Janet Harker had, at the time of your work with her, a few slides of midgut tumors that presumably developed upon transfer of the subesophageal gland from a cockroach on a given light–dark cycle to another insect that had been kept on a reversed lighting schedule. These are most sensational studies of cancer induction. I would like to know if there has been any follow-up; inferential statistical information would be particularly interesting.

BRADY: I know even less about those slides than you do; I never saw them. But using Harker's data, I worked out the statistical differences between the controls and the experimental transplants; they were significant. There has been no follow-up that I know of.

PITTENDRIGH: Uwo and I attempted to repeat Harker's work on tumors, with negative results. Our work was reported in 1967 in the *Journal of Insect Physiology*, Volume 13, pages 851–859.

BRADY: The significance of Harker's results derives from her claim that the transplanted ganglia were secreting rhythmically, whereas you have no information that your own transplants were; indeed, I imagine that your own feeling is that they were not. Without positive information on the activity, or lack of it, of implanted ganglia, it seems to me to be difficult to make meaningful comparisons. However, I readily accept the serious implications of your results.

LUDGER RENSING

Hormonal Control of Circadian Rhythms in *Drosophila*

A descriptive approach to the study of circadian rhythms in *Drosophila* has yielded a more or less complete morphology of the different rhythmic functions. The temporal relations among some of these functions, to be described in this paper, suggest such underlying causal mechanisms as the effect of light, neurosecretory cells, and ecdysone production on nuclear volume and chromosomal puff formation. Only the experimental approach, however, can reveal the true mechanism. Cell or organ cultures can provide information on whether there is a group of pacemaker cells in the organisms or whether each cell of an organism has its own oscillator. Data obtained from *Drosophila* salivary glands *in vitro* and other data in the literature (see review by Rensing, 1970) indicate that cellular circadian rhythms persist in culture. This fact raises the question of how a population of cellular oscillators in the organism is maintained in synchrony and how it is synchronized with external *Zeitgeber* stimuli. Since light cannot reach every single cell in most animal organisms, mediation by hormones seems a likely mechanism. This hypothesis can also be tested *in vitro*. In this report, experiments with the hormone ecdysone will be discussed in relation to possible generalizations regarding synchronizing mechanisms and the "circadian rule," and to the molecular processes involved in cellular rhythms.

CIRCADIAN RHYTHMS *IN VIVO*

Drosophila larvae on the seventh day of their development (20°C) show rhythmicity in LD in the following functions (Figure 1): nuclear volumes of neurosecretory cells, corpus allatum, and prothoracic glands (Rensing, unpublished); nuclear and nucleolar volumes of salivary gland cells (Rensing, 1969c). The same rhythms are observable on the sixth day, but with different phases (Rensing, Thach, and Bruce, 1965; Rensing, 1966b, 1969c). All rhythms are bimodal, with maxima 12 hr apart. The transverse profiles of the nuclear volumes of the salivary glands are almost the same after fixation as in living cells. The ecdysone-inducible puff 74 EF/75B (Becker, 1959, 1962) appears at the end of larval development and, in the majority of cases, at a certain time of day, followed 6–8 hr later by a maximum of puparium formation (Rensing and Hardeland, 1967).

From these experiments and from data in the literature, one can arrive at an approximate temporal sequence of events as follows: Stimulated by light or following the endogenous circadian rhythm, neurosecretory cells synthesize RNA (Figure 1c). The maximum of RNA synthesis may occur 0.5–1.0 hr earlier than, or at the same time as, the maximum of nuclear and nucleolar size (Figure 1a), as the data on neurosecretory cells of *Acheta* (Cymborowski and Dutkowski, 1969) indicate. Neurohormone is probably synthesized about 4–6 hr later (Figure 1c; Cymborowki and Dutkowski, 1970) and is subsequently transported to the corpus cardiacum sometime over the next 6–9 hr in adult *Drosophila* (Rensing, 1964, 1966b). The release of neurohormone induces the release of ecdysone (see the review by Scharrer and Scharrer, 1963), which in turn induces certain gene activities within 1 hr (Clever, 1961). It might be concluded that ecdysone is released about 10–12 hr after the maximum of RNA synthesis or, in any case, some hours after the dark–light transition (Figure 1c); it is at this time that the maximum of ecdysone-inducible puffs is observed (Figure 1b), followed 6–8 hr later by the maximum of puparium formation. [This latter interval has also been described by other workers (Becker, 1962).]

It is not known, however, how much time must elapse from the release of neurohormone and ecdysone to the stimulation of a new maximum of RNA synthesis. It is possible that this interval is rather short and that a new cycle in all tissues, including the neurosecretory cells, starts immediately. It is also possible that this interval is longer, perhaps up to 12 hr more. There is a close connection between this problem and the question whether the observed bimodal rhythm of nuclear size and oxygen consumption (Rensing, 1966a) is a 12-hr (demicircadian) rhythm or whether it is the result of two 24-hr (circadian) rhythms with a 12-hr phase difference.

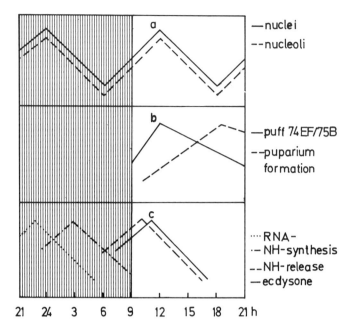

FIGURE 1 Circadian functions in *Drosophila* larvae at the seventh day of development (20°C). (a) Nuclear volumes of neurosecretory cells, corpus allatum, prothoracic glands, and fat body; and nuclear and nucleolar volumes of salivary gland cells (from Rensing, unpublished and 1969c). (b) Number of ecdysone-inducible puffs (74 EF/75B) and number of newly built prepupae (from Rensing and Hardeland, 1967). (c) Indirect evidence for RNA and neurohormone (NH) synthesis, NH release, and ecdysone release; see text. Abscissa: time of day.

CIRCADIAN RHYTHMS *IN VITRO*

The findings described in Figure 1 do not clearly indicate whether a rhythm of neurohormone and ecdysone concentration induces rhythms in all the other tissues or whether each cell has an autonomous rhythm, which is merely synchronized by the action of hormones. To answer this question we kept *Drosophila* salivary glands in a defined medium, on LD *12*:12, for at least 4 to 5 days and measured the nuclear size. The experiments showed that a bimodal rhythm of nuclear volume persists under these conditions and matches that measured *in vivo* (Figure 2; Rensing, 1969c). This rhythm can be observed also in constant light (100 lux) and with different K^+ and Na^+ concentrations in the medium. It can be phase-shifted by phase shifts of the light–dark regime during larval life.

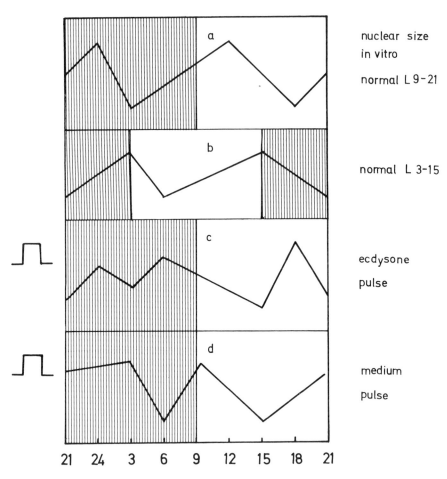

FIGURE 2 The temporal distribution of the maxima of nuclear size in *Drosophila* Salivary glands kept *in vitro*. (a) Undisturbed conditions with light from hr 9–21. (b) Undisturbed conditions with light shifted to hr 3–15. (c) Ecdysone pulse of one hour at different times between hr 13 and hr 17 during the previous light period. (d) Medium pulse of 1 hr as control. Ordinate: number of maxima of nuclear size; absicssa: time of day. (From Rensing, 1969c.)

If this phase-shift were due to the action of hormones mediating the light–dark stimuli, addition of ecdysone to the salivary glands *in vitro* might phase-shift the rhythm as well. Such an effect can, in fact, be observed after adding an ecdysone-containing medium (80/μg/ml) for 1 hr during the phase of relatively low nuclear size. In most cases, the effect cannot be observed immediately. The next "normal" maximum of nuclear size still appears between hr 21 and hr 24. The following maxima, how-

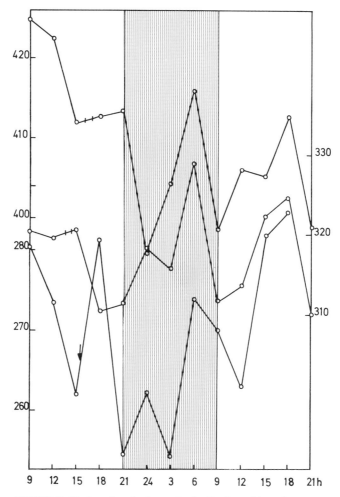

FIGURE 3 Nuclear size of salivary gland cells after a 1-hr ecdysone pulse or ecdysone step ↓. Points of the upper curves: mean values of 5–6 nuclei; lower curve: a single nucleus. Ordinates: product of the longest and shortest diameter of the nuclei; abscissa: time of day. (Reprinted with permission from Rensing, 1969c.)

ever, are advanced 3–6 hr compared to the controls (Figure 2c, d; Figure 3). The difference in the distribution of maxima between untreated (Figure 2a) and control salivary glands (Figure 2d) may be attributed to small changes in oxygen tension (Ritossa, 1964) or to other changes in the microconditions of the glands. The observed differences, however, occur within the normal range of temporal variance: maxima from hr 21 to hr 3 and hr 9

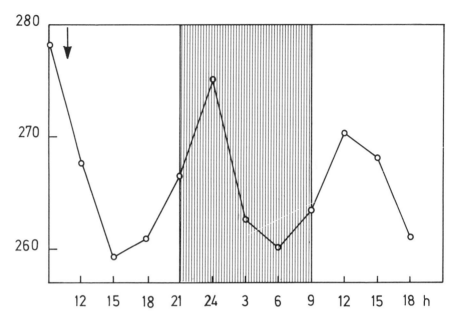

FIGURE 4   Nuclear size of salivary gland cells after an ecdysone step ↓ at a phase of relatively high nuclear size. Mean values of five nuclei. Ordinate and abscissa: see Figure 3 (Rensing, unpublished).

to hr 15, minima from hr 24 to hr 6 and from hr 12 to hr 18. Addition of ecdysone during the phase of maximal nuclear size had no effect on the phase of the rhythm (Figure 4), a fact that may be related to the hypothesis that ecdysone is released at about the same time *in vivo*.

HORMONES AND SYNCHRONIZATION

From the above experiments it can be concluded that hormones act on the cellular oscillator. A review of the available data in the literature on hormones and their effects on circadian rhythms (Table 1; Rensing, 1969b) suggests the following generalizations:

 • Changes in hormone level (caused by the removal of endocrine glands) (see Table 1), different light intensities (Tilgner, 1967), or changes in the course of ontogeny (see review by Gray and Bacharach, 1967) are correlated with changes in frequency and phase of the circadian rhythm (see Table 1).
 • Abrupt changes in the hormone level caused by light pulses, addition

or injection of hormones (see Table 1), or by stress (Rensing, 1969b) induce changes in the phase of the rhythm (see Table 1).

These generalizations lead to the hypothesis that there may be two effects of hormones on circadian rhythms: one depending on different hormone levels (proportional effect), the other depending on a sudden change in the hormone level (differential effect). Both effects can also be observed by applying certain light programs to the organisms—either different constant light intensities or light pulses or steps (Aschoff, 1960). These observations, together with the fact that the direct action of light has similar molecular effects on nucleic acid and protein metabolism as have some hormones (see Rensing, 1969b), suggest the following hypothesis (Figure 5): Organisms have developed mechanisms that transform external stimuli (especially light) into hormonal signals. It is this system that synchronizes the population of cellular clocks to *Zeitgeber* cycles and that synchronizes the cellular oscillators with each other and modifies the free-running frequency according to the level of hormone.

The transformation of external stimuli into internal hormonal signals is almost certainly different in different species. The "circadian rule" proposed by Aschoff and Wever is derived from species in which this transformation follows certain rules for a certain range of conditions; other groups—insects, for example— have other transformation rules (Rensing and Brunken, 1967). Possible mechanisms involved are discussed in the following section.

HORMONES AND MOLECULAR PROCESSES IN CIRCADIAN RHYTHMS

If I may simplify the present knowledge on hormonal actions at the molecular level (see Karlson, 1967), three main effects (taking ecdysone as a representative example) can be distinguished:

- Induction of a certain gene sequence (Clever, 1961, 1964; Berendes, 1967);
- Stimulation of RNA and protein synthesis rate; and
- A change in the permeability and/or potential of the cellular membranes (Ito and Loewenstein, 1965; Kroeger, 1966).

Since ecdysone influences circadian rhythms on the cellular level, as shown in the experiments discussed above, it seems most probable that

TABLE 1 Effects of Changes in the Hormonal System on Circadian Rhythms

Function	Organism	Conditions	Parameter of Circadian Rhythm Affected	Author
Body temperature	Mouse	Hypophysectomy	Period length	Ferguson et al., 1957
Locomotor activity	Gecarcinus	Eye stalk removal	Period length	Bliss, 1962
Blood glucose	Rat	Hypophysectomy, adrenal medullectomy	Phase	Pauly and Scheving, 1967
Blood amino acids	Mouse	Adrenalectomy	Phase	Feigin, Dangerfield, and Beisel, 1969
Blood glucose	Man	Oral administration of percorten	Phase	Menzel et al., 1948
Number of erythrocytes	Man	Administration of doryl, ephetonin	Phase	Goldeck and Siegel, 1948
Kidney function	Man	Injection of cortisone	Phase	Rosenbaum et al., 1952
Kidney function	Man	Cirrhosis, ascites	Phase	Jones, McDonald, and Last, 1952
Adrenal secretion	Hamster	Administration of ACTH	Phase	Andrews, 1968
Locomotor activity	Canary	Progesteron, medroxyprogesteron	Phase	Wahlström, 1968
Kynurenine transaminase	Rat	Injection of cortisol	Phase	Hardeland, 1969a
Locomotor activity	Pyrrhula	Stress	Phase	Rensing, 1969b
Nuclear volume	Drosophila	Administration of ecdysone	Phase	Rensing, 1969c
Body temperature	Man	Ontogeny	Phase	Review: Hellbrügge, 1960
Electrical skin resistance	Man	Ontogeny	Phase	Review: Hellbrügge, 1960
Locomotor activity	Man	Ontogeny	Phase	Review: Hellbrügge, 1960
Amylophosphorylase	Mouse	Ontogeny	Phase	Egli, 1967
Oxygen consumption	Drosophila	Ontogeny	Phase	Rensing, 1966a
Nuclear and nucleolar volume	Drosophila	Ontogeny	Phase	Rensing, 1969c

FIGURE 5   The possible pathway of certain
light programs influencing the basic cellular
oscillators in animals with light impenetrable integuments (A) and in organisms without such
integuments (B).

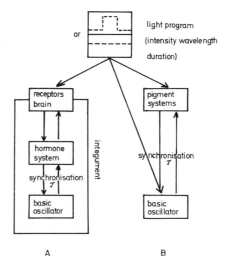

one or all of these effects of ecdysone are involved. On the other hand, we
have to consider the following facts and hypotheses regarding molecular
processes of the cellular oscillators:

• Different proteins are synthesized at different times of day
(Hardeland, 1969a, b), which could be due to sequential gene activity—
an hypothesis extensively worked out by Ehret and Trucco (1967) (the
"chronon" concept). There are also circadian periodicities in R N A and
protein synthesis (Barnum, Jardetzky, and Halberg, 1957; Sestan, 1964;
Vanden Driessche and Bonotto, 1968; Cymborowski and Dutkowski,
1969, 1970).
• The rate of R N A and protein synthesis may play a role in determining parameters of the circadian rhythms. The experiments with antibiotics
that provide the evidence for this proposition, however, are subject to
various interpretations; the point needs further investigation.
• Penetration of molecules through cellular and nuclear membranes
seems to be a factor in circadian processes, as suggested, for example, by
the experiments with ethanol (Bünning and Baltes, 1962). In any case, induction and repression of gene activities, together with feedback mechanisms, are probably essential features of a circadian system (Jacob and
Monod, 1963; Goodwin, 1963). An indication of changing repressor concentrations during 24 hr was found in our own experiments (Hardeland
and Rensing, 1968).

A PRELIMINARY MODEL

The hormone experiments described above and other data on hormone action and cellular circadian rhythms serve as a basis for the following preliminary model for molecular processes in circadian rhythms of *Drosophila*.

There is a sequence of gene activities, A–B–C, which repeats every 12 (or 24) hr: A stimulates B, B stimulates C, and C stimulates A. Each step involves RNA and protein synthesis, membrane permeations, and certain precursor pool sizes. A, B, and C regulate the synthetic activity of different numbers of genes, accounting for the quantitative differences in overall RNA and protein synthesis. There is a feedback system built in between each of A, B, and C that is an important part of the self-sustained oscillation. No hormone is needed to maintain this oscillation.

A pulse of ecdysone induces several puffs, which in turn stimulate A. [This indirect action has to be assumed, because ecdysone-inducible puffs cannot be observed *in vitro* (Nagel and Rensing, 1971).] If the pulse is given at a time when A is already induced by C (at maximum nuclear size) this additional stimulation by ecdysone does not change the cyclic sequence—no phase-shift is observed.

If the ecdysone pulse, however, is given at a time when B is stimulated and A repressed (at relatively low nuclear size), the hormone will derepress A and start a new cycle with a certain phase difference to the original cycle. Both cycles may at first be superimposed, giving rise to the observed rhythm. This part of the model would account for the phase-shifts in the circadian rhythm (differential effect) of pulsed addition of hormones.

The effects of different hormonal levels (proportional effect) could be explained in terms of influences on the rate of RNA and protein synthesis or influences on the penetration rate of these molecules through membranes. Critical pool sizes would be reached earlier or later and thus the period length of the gene sequence would be shorter or longer. This, in turn, would lead to different phase relationships to the *Zeitgeber* cycle. The proportional effect may be observed after removal of hormone glands, during inhibition of RNA and protein synthesis, or under constant light conditions.

If the hormone titer is high enough to override feedback control, the gene sequence should stop oscillating: At any time of the 24-hr period all the genes involved would be active. A drop in the hormonal level should always start the cycling at a certain phase—in this case, at C, because the assumed feedback mechanism is between B–A and C–B.

This model may well prove inadequate as a description of the mechanism of cellular oscillation and synchronization, but it does incorporate recent discoveries on the molecular mechanisms of hormone action and on gene activation in salivary gland chromosomes.

REFERENCES

Andrews, R. V. 1968. Temporal secretory responses of cultured hamster adrenals. Comp. Biochem. Physiol. 26:179–193.

Aschoff, J. 1960. Exogenous and endogenous components in circadian rhythms. Cold Spring Harbor Symp. Quant. Biol. 25:11–28.

Barnum, C. P., C. D. Jardetzky, and F. Halberg. 1957. Nucleic acid synthesis in regenerating liver. Tex. Rep. Biol. Med. 15:134–147.

Becker, H. J. 1959. Die Puffs der Speicheldrüsenchromosomen von *Drosophila melanogaster*. I. Mitt. Beobachtungen zum Verhalten des Puffmusters im Normalstamm und bei zwei Mutanten, giant und lethal-giant larvae. Chromosoma 10:654–678.

Becker, H. J. 1962. Die Puffs der Speicheldrüsenchromosomen von *Drosophila melanogaster*. II. Mitt. Die Auslösung der Puffbildung, ihre Spezifität und ihre Beziehung zur Funktion der Ringdrüse. Chromosoma 13:341–384.

Berendes, H. D. 1967. The hormone ecdysone as effector of specific changes in the pattern of gene activities of *Drosophila hydei*. Chromosoma 22:274–293.

Bliss, D. E. 1962. Neuroendocrine control of locomotor activity crab, *Gecarcinus lateralis*. In H. Heller and R. B. Clark, Neurosecretion, 391–410, Academic Press, New York.

Bünning, E., and J. Baltes. 1962. Wirkung von Äthylalkohol auf die physiologische Uhr. Naturwissenschaften 49:19.

Clever, U. 1961. Genaktivitäten in den Riesenchromosomen von *Chironomus tentans* und ihre Beziehungen zur Entwicklung. I. Genaktivierung durch Ecdyson. Chromosoma 12:607–675.

Clever, U. 1964. Actinomycin and puromycin: effects on sequential gene activation by ecdysone. Science 146:794–795.

Cymborowski, B., and A. Dutkowski. 1969. Circadian changes in RNA synthesis in the neurosecretory cells of the brain and suboesophageal ganglion of the house cricket. J. Insect Physiol. 15:1187–1198.

Cymborowski, B., and A. Dutkowski. 1970. Circadian changes in protein synthesis in the neurosecretory cells of the central nervous system of *Acheta domesticus*. J. Insect Physiol. 16:341–348.

Egli, K. 1967. Histochemische Untersuchungen in der Mäuseleber. Physiologische Tagesschwankungen und postnatale Entwicklung. Histochemie 8:164–174.

Ehret, C. F., and E. Trucco. 1967. Molecular models for the circadian clock. I. The chronon concept. J. Theor. Biol. 15:240–262.

Feigin, R. D., H. G. Dangerfield, and W. R. Beisel. 1969. Circadian periodicity of blood amino-acids in normal and adrenalectomized mice. Nature (Lond.) 221:94–95.

Ferguson, D. J., M. B. Visscher, F. Halberg, and L. M. Levy. 1957. Effects of hypophysectomy on daily temperature variation in C3H mice. Am. J. Physiol. 190:235–238.

Goldeck, H., and P. Siegel. 1948. Die 24-Stunden-Periodik der Blutreticulocyten unter vegetativen Pharmaka. Aerztl. Forsch. 2:245–248.

Goodwin, B. C. 1963. Temporal organization in cells. Academic Press, New York.

Gray, C. H., and A. L. Bacharach. [ed.]. 1967. Hormones in blood. Academic Press, New York.

Hardeland, R. 1969a. Circadiane Rhythmik und Regulation von Enzymen des Tryptophan-Stoffwechsels in Rattenleber und -niere. Z. Vgl. Physiol. 63:119–136.

Hardeland, R. 1969b. Zur Regulation der tagesperiodischen Proteinsynthese, Verh. Dtsch. Zool. Ges., Innsbruck 1968, Zool. Anz. Suppl. 32:307–317.

Hardeland, R., and L. Rensing. 1968. Circadian oscillation in rat liver tryptophan pyrrolase and its analysis by substrate and hormone induction. Nature (Lond.) 219:619–621.

Hellbrügge, T. 1960. The development of circadian rhythms in infants. Cold Spring Harbor Symp. Quant. Biol. 25:311–324.

Ito, S., and W. R. Loewenstein. 1965. Permeability of a nuclear membrane: changes during normal development and changes induced by growth hormone. Science 150:909–910.

Jacob, F., and J. Monod. 1963. Genetic repression, allosteric inhibition and cellular differentiation, p. 30–64. In M. Locke [ed.], Cytodifferentiation and macromolecular synthesis. Academic Press, New York.

Jones, R. A., G. O. McDonald, and J. H. Last. 1952. Reversal of diurnal variation in renal function in cases of cirrhosis with ascites. J. Clin. Invest. 31:326–334.

Karlson, P. [ed.]. 1967. Wirkungsmechanismen der Hormone. Springer, Berlin.

Kroeger, H. 1966. Potentialdifferenz und Puff-Muster. Exp. Cell Res. 41:64–80.

Nagel, G., and L. Rensing. 1971. The puffing pattern and puff size of Drosophila salivary gland chromosomes in vitro. Cytobiologia 3:288–292.

Menzel, W., and I. Othlinghaus. 1948. Inversion des Blutzuckertagesrhythmus durch Percorten. Dtsch. med. Wschr. 73:326.

Pauly, J. E., and L. E. Scheving. 1967. Circadian rhythms in blood glucose and the effect of different lighting schedules, hypophysectomy, adrenal medullectomy and starvation. Am. J. Anat. 120:627–636.

Rensing, L. 1964. Daily rhythmicity of corpus allatum and neurosecretory cells in Drosophila melanogaster (Meig.). Science 144:1586–1587.

Rensing, L. 1966a. Zur circadianen Rhythmik des Sauerstoffverbrauchs von Drosophila. Z. Vgl. Physiol. 53:62–83.

Rensing, L. 1966b. Zur circadianen Rhythmik des Hormonsystems von Drosophila. Z. Zellforsch. 74:539–558.

Rensing, L. 1969a. Genetische Untersuchungen über den circadianen Rhythmus des Sauerstoffverbrauches von Drosophila. Verh. Dtsch. Zool. Ges. Innsbruck 1968, Zool. Anz. Suppl. 32:298–307.

Rensing, L. 1969b. Zur Ontogonese und hormonellen Steuerung circadianer Rhythmen. Nachr. Akad. Wiss. Goettingen 1969 (8):57–70.

Rensing, L. 1969c. Die circadiane Rhythmik der Speicheldrüsen von Drosophila in vivo, in vitro und unter dem Einfluss von Ecdyson. J. Insect Physiol. 15:2285–2303.

Rensing, L. 1970. Circadiane Rhythmik von Zellen in vitro. Verh. Dtsch. Zool. Ges. Wuerzburg 1969. Zool. Anz. Suppl. 33:166–171.

Rensing, L., and W. Brunken. 1967. Zur Frage der Allgemeingültigkeit circadianer Gesetzmässigkeiten. Biol. Zentralbl. 86:545–565.

Rensing, L., and R. Hardeland. 1967. Zur Wirkung der circadianen Rhythmik auf die Entwicklung von Drosophila. J. Insect Physiol. 13:1547–1568.

Rensing, L., B. Thach, and V. Bruce. 1965. Daily rhythms in the endocrine glands of Drosophila larvae. Experientia 21:103.

Ritossa, F. M. 1964. Behaviour of RNA and DNA at the puff level in salivary gland chromosomes of Drosophila. Exp. Cell Res. 36:515–523.

Rosenbaum, J. D., B. C. Ferguson, R. K. Davis, and E. C. Rossmeisl. 1952. The influence of cortisone upon the diurnal rhythm of renal excretory function. J. Clin. Invest. 31:507.

Scharrer, E., and B. Scharrer. 1963. Neuroendocrinology. Columbia Univ. Press, New York.

Sestan, N. 1964. Diurnal variations of ^{14}C-leucin incorporation into protein of isolated rat liver nuclei. Naturwissenschaften 51:371.

Tilgner, S. 1967. Beziehungen zwischen Licht, Auge und Nebennierenrindenaktivität (Demonstriert am Verhalten der Eosinophilenzahl im peripheren Blut). Biol. Rundsch. 5:267–277.

Vanden Driessche, T., and S. Bonotto. 1968. The circadian rhythm in RNA synthesis in *Acetabularia mediterranea*. Biochim. Biophys. Acta 179:58–66.

Wahlström, G. 1968. Sleep induction by progesteron and medroxy-progesteron in the canary. Acta Pharmacol. Toxicol. 26:583–596.

DISCUSSION

HALABAN: All of the data that you have presented are from light–dark conditions. Do you have data also from constant conditions?

RENSING: Yes, I did experiments in LL with a light intensity of 100 lux, and it turned out that the rhythm of nuclear size persisted under these conditions for at least 5 days.

ENRIGHT: I am a bit concerned about the reproducibility of your curve for nuclear volume. The amplitude of the rhythm you showed there was on the order of 5–10 percent change in nuclear volume. That would mean changes of 2–3 percent, maximum to minimum, in nuclear diameter. Just how reproducible is a measurement of nuclear diameter?

RENSING: I worried about that, too, and when I first saw it I thought it could be an artifact. So I had three of my students make the same measurements without their knowing that I had found anything, so their measurements would not be influenced by expectation. I did this also for another experiment in which I changed the rhythm by ecdysone. I have not yet tackled the problem of how to deal with the statistics of the *in vitro* method. I measured, I think, a total of about 500 nuclei, but I did statistics on only a very few.

HALBERG: There is a need, from a technical viewpoint, to distinguish various cellular bioperiodicities. On the one hand, there are several extensively documented circadian rhythms persisting, e.g. after 4 days in LL, in RNA-labeling, DNA-labeling, and so forth, in mouse liver cells (Physiologic 24-hr periodicity in human beings and mice, the lighting regimen and daily routine. *In*: Photoperiodism and related phenomena in plants and animals, R. B. Withrow, ed., Washington, D.C., AAAS, 1959, pp. 803–878). Moreover, in the liver of mice kept in LL (Continuous light or darkness and circadian periodic mitosis and metabolism in C and D_8 mice. Am. J. Physiol. 201:277–230, 1961) the rhythm in RNA-labeling desynchronizes in DNA-formation while both these rhythms free-run from that in the incorporation of p^{32} into phospholipid, and so on. These findings then cannot be attributed to the very real changes in the water content of the preparation used. However, rhythms in cell volume or cell constituent volume are gravely complicated by changes in the water content. Was the water content the same throughout the span of your measurements?

RENSING: I did not say that the rhythm ceased after 3 days; we have observed these rhythms as long as we have kept the glands *in vitro*; up to 10 days. As to your second point, I have no information about water content. But in response to your first point, we measured individual nuclei, so if there is a tendency to desynchronize it won't matter.

HALBERG: But I think Dr. Enright's point was a good one. You could still do a simple analysis of variance or somehow estimate how many of the cells you examined exhibited statistically significant changes.

RENSING: Well, I would be very grateful for help with these statistics because the size of the nuclei are quite different. I don't see how you can get good statistics with these data. Longitudinal measurements on arrhythm of a single individual raise statistical problems which have—to my knowledge—not been solved unequivocally.

EDMUNDS: If I read your graph correctly, the nuclear size rhythm shows a bimodal peak at hr 3 and hr 15, where hr 0 is the beginning of the light period. Does this bimodality persist in constant conditions?

RENSING: Yes, I think it persists at least until day 10 under these conditions.

BRINKMANN: Is this bimodal state observed under free-running conditions temperature-compensated too?

RENSING: I can't really say anything about that. I have measured too few cycles, and in only one series of experiments. And then I was just determining whether the rhythm persisted under constant conditions; I wasn't really looking at free-running frequency. I think another interesting point to look at would be to see if you could phase-shift the rhythm just by changing the light–dark cycle.

GORDON: Are the changes in nuclear and nucleolar volume with time measured in a single cell or separate cells?

RENSING: I have shown results from both kinds of measurement. Results in the first profile I showed, when I fixed the nuclei of salivary glands, were from independent cells; measurements *in vitro* were of individual cells. So the *in vitro* results represent changes in volume of the nucleus within a single cell.

GORDON: I ask, in part, because the tips of secondary roots of the broad bean, *Vicia faba*, growing in continuous darkness, show a roughly 24-hr cyclic variation in mean nuclear volumes of the cell populations. Correlated with these volume changes is a periodicity of the mitotic index. Thus the variations in volume may be as simply a reflection of an approximately 24-hr, but not necessarily circadian, periodicity of the cell cycle.

RENSING: Yes, your argument has already been raised by Bünning. And there are indications that these changes in the nuclei might be due to cell division, but actually in the salivary glands you have only endomitosis, and even that doesn't occur *in vitro*. So I think one can exclude the division factor in this case.

ZIMMERMAN: Did you say anything about the developmental synchrony of the pupae or larvae?

RENSING: No, I said nothing about that. We have a very rough measure of synchrony in that we prepare salivary glands from larvae that crawl up the glass wall of the vials. It's not a very exact measure, I agree, but it is in the range of 24 hr.

The Influence of the Pineal
Organ on the Circadian
Activity Rhythm in Birds

Removal of the pineal organ of the house sparrow, *Passer domesticus*, results in loss of the overt circadian activity rhythm in constant darkness (Gaston and Menaker, 1968; Gaston, 1969). This response to pinealectomy has been observed in about 100 house sparrows (Figure 1A), as well as in two white-crowned sparrows (*Zonotrichia leucophrys gambelii*) (Figure 1B).

In addition to abolishing the free-running rhythm, pinealectomy affects the pattern of entrainment in the house sparrow. Pinealectomized birds can be entrained by light cycles. If a bird is pinealectomized while entrained, an increase in the value of the phase angle is consistently obtained (Figure 2; Table 1). Figure 2A is the record of a bird entrained to LD 3:21. Before pinealectomy this bird's activity began 1–2 hr prior to the onset of the light. After pinealectomy the phase angle increased to 6 hr. Figure 2B is the record of a bird entrained to LD 6:20. As a result of pinealectomy, the phase angle increased from 8 hr to 11 hr in advance of the signal. Development of arrhythmicity can be seen in the locomotor patterns of both the birds in the subsequent DD regimen.

Controls were carried out to determine if a change in the phase angle could be produced by manipulations accompanying pineal removal. In these procedures the bird was removed from its cage, anesthetized, allowed to recover in the surgery room, and then returned to its cage. Controls

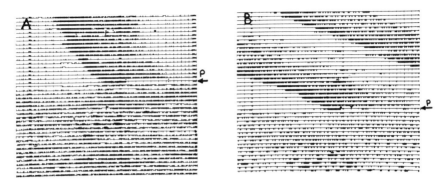

FIGURE 1 Activity records before and after pinealectomy of *Passer domesticus* (A), and *Zono-trichia leucophrys gambelii* (B). Both birds are in constant darkness throughout the records shown. P indicates the day of pinealectomy.

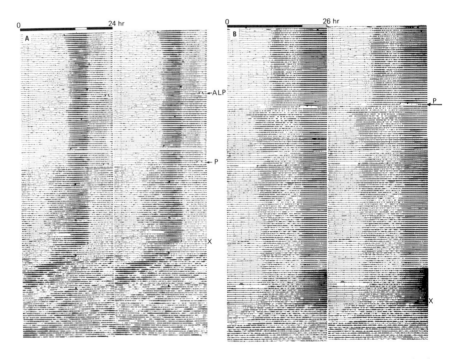

FIGURE 2 Activity records of two house sparrows that were pinealectomized while entrained to *LD 3*:21 (A), and *LD 6*:20 (B). Each record is duplicated, the right side displaced one day above the left. The light and dark bands at the top of the left side of each record designate the positions of the daily light period. ALP—administration of a control procedure (see text); P—day of pineal-ectomy; X—first day of constant darkness.

TABLE 1 Birds That Were Penealectomized While Entrained to a Light Cycle

Bird	LD	BPX	APX	BPA	APA	Δφ	Control	Rhythmicity	Hist
2079	3:21	16	4	1-5	9	4-8	+	A	neg
2112	3:21	19	4	1 1/2	6	4 1/2	+	A	neg
2115	3:21	11	4	1	-1/2-2	-1/2-+1	+	R	scat
2709	6:18	12	6	2	2-5	0-3	-	A	-
3424	6:18	7	6	2-2 1/2	3 1/2-4	1-2	-	A	scat
3699	6:18	12	8	3	5	2	-	A	-
3819	6:18	15	5		0	0	+	R	lopc
3103	1/2:23 1/2	20	12	1/2	3	2 1/2	-	R	scat
3861	1/2:23 1/2	20	13	3	13	10	-	E	-
3849	1/2:23 1/2	34	8	2	11-12	9-10	-	E	-
3934	6:20	32	16	8	11	3	-	A	neg
3558	6:20	48	8	8 1/2	9	1/2	S	A	neg
3828	6:20	32	16	8 1/2-9	10	1-1 1/2	-	E	-
3233	6:20	48	8	7-8	8 1/2	1/2-1 1/2	S	E	-

LD—photoperiod to which the bird was entrained; BPX—number of weeks entrainment before pinealectomy; APX—number of weeks entrainment after pinealectomy; BPA—phase angle in hours of the bird before pinealectomy; APA—phase angle of the bird after pinealectomy; Δφ—phase shift (APA–BPA); Control: +, administration of a control procedure to this individual prior to pinealectomy (see text); -, no control was performed; S, sham pinealectomy provided the control procedure; Rhythmicity—rhythmicity of the bird at the time of sacrifice or death: A, arrhythmic in DD; R, rhythmic in DD; E, died while entrained, no DD data; Hist—histological evaluation of the brain: neg, no pineal parenchymal cells were found; scat, scattered parenchymal cells were found in the pineal site; lopc, luminally organized pineal tissue was present in the pineal site; minus, histology was not possible.

543

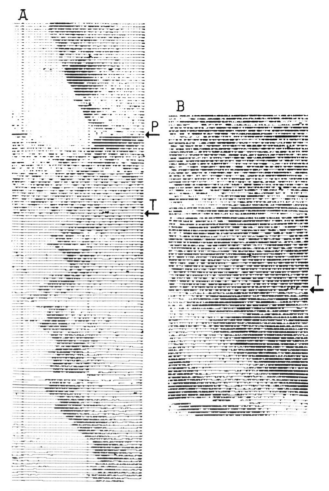

FIGURE 3   Two persistently rhythmic implanted birds. Bird A was pinealectomized as shown by P; B was pinealectomized prior to the portion of
the record shown. T indicates the day on which the bird received a pineal
implant. Both birds were in DD throughout.

were always performed at the same subjective time relative to activity onset as was the subsequent pinealectomy. Such a procedure should simulate
any effect of the light shock that necessarily accompanies surgery. No
phase angle changes were observed in the 5 birds that were subjected to
this procedure (Table 1). Two of the birds entrained to LD 6:20 were
sham-pinealectomized as a control. In neither case did the sham operation
affect the bird's phase angle.

Table 1 summarizes the results obtained on 14 birds that were pinealectomized while entrained to light cycles. In most cases the pinealectomized
birds were put into DD after entrainment, to check for arrhythmicity. In
all the birds that were subsequently arrhythmic in DD, an increase in phase
angle resulted from pinealectomy. The 3 birds that were rhythmic in DD
at the time of sacrifice or death are the ones that, among birds on the same
photoperiod, showed least phase angle increase after pinealectomy. From
3 to 8 months of data in DD were collected on these 3 rhythmic birds to
verify persistence of rhythmicity. Histological examination of the brains
of 4 out of 5 of the arrhythmic birds showed no pineal parenchymal cells,
whereas all 3 of the rhythmic pinealectomized birds showed evidence of
pineal parenchymal tissue at the pineal site.

Thirty attempts were made to implant pineals from intact birds into the
pineal site of arrhythmic pinealectomized birds. Two out of 30 resulted in
persistently rhythmic activity in DD of the implanted birds (Figure 3).
Histology revealed that the bird whose record is shown in Figure 3A had
lumenally organized pineal parenchymal tissue present in the pineal site.
However, the brain of the bird in Figure 3B showed no evidence of any
pineal cells.

In 8 of the 30 implanted birds, transient rhythmicities persisting for
about 7 days accompanied the implant. Histology revealed lumenally organized pineal tissue in 3 of these 8 birds. Further studies have shown, however, that such a transient rhythmicity can be obtained simply from the
light shock accompanying surgery. In these studies 15 arrhythmic pinealectomized birds were subjected to 8-hr light shocks. In 5 of these birds
transient rhythmicities were produced and persisted for 6–7 days (Figure 4).

Of the 20 birds remaining arrhythmic after pineal implantation, 1 showed
lumenally organized pineal tissue in the pineal site (Gaston, 1969).

In summary, the effect of pinealectomy on the free-running circadian
activity rhythm of the house sparrow was repeated in the white-crowned
sparrow, suggesting that the effect may be general in birds. In the house
sparrow pinealectomy affects entrainment to light cycles as well as the
free-running state. An increase in phase angle results when entrained birds
are pinealectomized and returned to the entraining light cycle. Since light
pulses are capable of producing transient rhythmicities. Two out of 30 attempts to implant pineals into arrhythmic pinealectomized birds resulted
in persistently rhythmic birds. In 8 other cases transient rhythmicities were
produced by the implant surgery or alternatively by the light shock accompanying it. It is clear that the pineal organ is essential for the expression of
normal activity rhythms in some birds, and its further study should contribute to our understanding of the physiology of circadian rhythmicity.

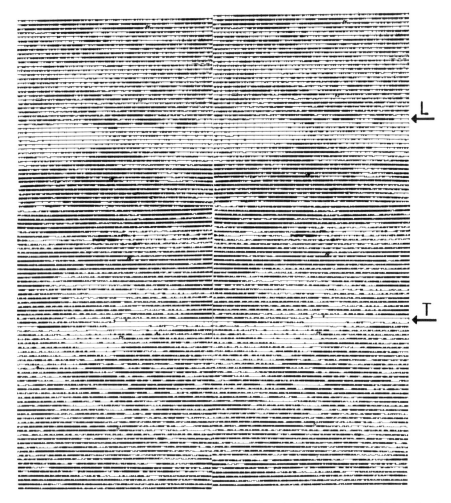

FIGURE 4 Effect of an 8-hr light shock (L) and an implanted pineal (T), both of which produced transient rhythmicity in a pinealectomized bird held in DD throughout the record.

REFERENCES

Gaston, S. 1969. The effect of pinealectomy on the circadian activity rhythm of the house sparrow, *Passer domesticus*. Ph.D. Thesis. University of Texas at Austin.
Gaston, S., and M. Menaker. 1968. Pineal function: the biological clock in the sparrow? Science 160:1125–1127.

DISCUSSION

CROWLEY: I have just one comment on this very interesting work. I think you may be missing something in your data, Dr. Gaston. You refer to the shift that occurs after pinealectomy as a shift in phasing. It is hard to tell from this record, but it appears to me to be a shift in wave form, wherein the wave develops earlier. There may well be a phase angle shift too, but in any case the wave form is different. Thirty years ago Kleitman demonstrated that in humans activity begins to develop before the human being awakens and begins daytime activity. I think it very interesting that you have perhaps demonstrated the same phenomenon in pinealectomized birds.

GASTON: But intact birds also have positive phase angles, that is, they too anticipate lights-on.

CROWLEY: Yes, but your pinealectomized birds seem to do so much earlier. Also, there is no sharp onset, but rather a gradual development of activity. Blips begin appearing on the record, then get more and more dense.

STRUMWASSER: Are you referring to the light–dark entrained, pinealectomized bird?

CROWLEY: Yes.

ZIMMERMAN: Isn't the issue whether there has been a change in phase or an increase in activity? It appears to me to be the latter.

GASTON: Although that may appear to be the case from looking exclusively at the data I have just now presented, increased amount of activity is not a general consequence of pinealectomiz-

ing either free-running or entrained birds. Besides, many times normal sparrows will abruptly change their "blip density." Since a bird's activity is not limited to perch-hopping, we don't know whether change in recorded activity represents a real change in the bird's total activity.

SWEENEY: You might be losing information by measuring the phase angle difference only by onset of activity, rather than including offset in your calculations. Cessation of activity appears to be changed very little or not at all in pinealectomized birds as compared with intact entrained birds. The end point appears to be very sharp also.

ESKIN: The offset of activity is generally masked in sparrows, being very much influenced by lights-out, so that it is not a very good measure of phase.

WAHLSTRÖM: To raise another question, have you tried any treatments other than the transplants to restore the circadian pattern? You haven't tried drugs or anything like that in the pinealectomized animals?

GASTON: No, we haven't.

ENRIGHT: I am not sure whether this was published or whether I heard it from you in conversation, but I know that at one time you were considering the possibility that, after pinealectomy, the bird interprets constant darkness as constant bright light. This would explain the arrhythmia and would also be consistent with the phase angle shift.

GASTON: If the pinealectomized bird could not distinguish between light and dark, we wouldn't expect consistently to find the major portion of the activity during lights-on when the bird is entrained. We may postulate that pinealectomy in constant conditions somehow changes the level of the "activity–rest oscillation" with respect to some threshold. But we have no idea of what it might mean physiologically for a bird to "interpret" some set of conditions as constant bright light.

HALBERG: It might be worthwhile in some of these studies to control food as the synchronizer. Synchronization by feeding time has been reported to override even the effects of lighting (e.g., in the case of the circadian rhythm in bloodeosinophil counts of mice restricted in carbohydrate and fat intake).

GASTON: Food and water are constantly available to all our birds and are replaced only at 3-week intervals. We haven't attempted to use food as a *Zeitgeber*.

MARVIN E. LICKEY
SHELDON ZACK
PAMELA BIRRELL

Some Factors Governing Entrainment of a Circadian Rhythm in a Single Neuron

One of the large identifiable neurons in the abdominal ganglion of the sea hare, *Aplysia californica*, has a circadian rhythm of activity in the frequency of its autorhythmic spikes. Strumwasser (1963) originally discovered the rhythm in long-term intracellular recordings obtained from preparations of the isolated ganglion. In these experiments, wherein the ganglion was maintained at constant temperature and illumination and isolated from all sources of periodic sensory input, the neuron typically showed a maximum spike frequency at the time of projected dawn in the aquarium environment of the intact animal prior to dissection. The neuron is a neurosecretory cell (Coggeshall, 1967) and is labeled R15 in the nomenclature of Frazier *et al.* (1967). Because of its characteristic bursting spike pattern, Strumwasser (1965) has named it the "parabolic burster."

By means of this circadian rhythm, neuron R15 can acquire and store, for extended periods, information about certain aspects of the external environment. Indeed, this and similar phenomena (Aréchiga and Wiersma, 1969) provide some of the very few cases where information arising in the environment can, by physiological measurement, be recognized in the long-term coded output of a neuron. Because of the very great general importance of neural mechanisms capable of providing long-enduring coded copies of an individual's past history, this rhythm has captured our interest in the most compelling way.

549

Accordingly, we have undertaken several studies in an attempt to further define the domain of environmental information that may be encoded by the rhythm and to explore possible pathways over which such information is read into the neuron during entrainment. In one set of experiments we have tested the neuron for entrainment after exposure of the animal to light cycles having periods of either 21 hr or 27 hr (Lickey, 1969), and in a second we have tested for entrainment in eyeless *Aplysia*. During the course of these studies, we have found the rhythm to be much more complex than we had naively supposed at the outset.

Figure 1 is an activity record from R15 that illustrates our technique and replicates some of the findings originally reported by Strumwasser (1965). The abdominal ganglion was removed from an *Aplysia* that had been exposed to continuous light for 13 days and then transferred to *LD 12*:12 for 5 additional days to entrain the rhythm. R15 was then impaled during the early portion of the first projected night. The pipette remained in place and the cell remained active for 55 hr 19 min, during which time the temperature in the recording chamber remained within the range of 12.0–12.5°C. The number of spikes occurring during successive 5-min intervals was automatically recorded on line for subsequent analysis and plotting by computer. The activity record shown is a moving average of spike frequency as a function of clock time; the averaging window was 30 min wide and plotted points are 5 min apart.

Three principal properties of the model that we have thus far applied in the interpretation of our data are illustrated in Figure 1. First, the major activity peak is interpreted as the expression of a circadian oscillator that can free-run for at least a few cycles. In this example the major peak during the first 24 hr occurred at 22:55 PDT, and on the second day another peak, although not the major one, occurred at 22:50. Second, the rhythm can be entrained by *LD 12*:12, and, as would be predicted from Strumwasser's original report, the peak spike frequency at steady state occurred near the time of projected dawn. Third, and this point is critical for our subsequent analyses, the neuron is potentially capable of assuming a second stable phase angle difference that is about 180° different from that shown during any given cycle. During the second day of the illustrated experiment, there was a pronounced peak well synchronized with the projected dusk. There was no peak at dusk during the preceding cycle, and, indeed, most preparations give only a single peak during the first 24 hr. Among any group of five to ten similar preparations, however, each potential phase relation will be represented at least once during the first 24 hr.

In addition to the circadian peaks, R15 has a shorter-term oscillation that, in this preparation, had a period of approximately 1.25 hr. Although

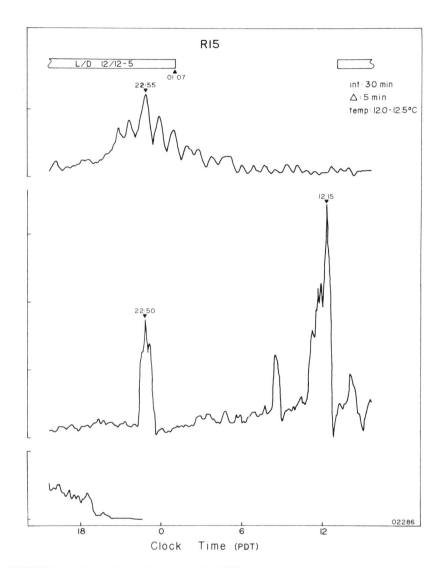

FIGURE 1   Activity in R15 following *L*D *12*:12 (170:0). Ordinate: moving average of spike frequency. Abscissa: Pacific Daylight Time. Projected dark indicated by rectangle at top. Projected dawn indicated by triangle at 01:07 PDT. The symbol "5" enclosed in the rectangle indicates that the intact *Aplysia* had experienced the light cycle for 5 days. Activity peaks indicated by triangles at 22:55, 22:50, and 12:15 PDT. Feb. 28–March 2.

we have not made a detailed study, the two oscillations appear to be independent of one another. Note that the hourly oscillation was present only during the first day, while the circadian peak was expressed on both days.

Customarily we do not interpret the activity records beyond the first 24 hr following impalement; the entire run has been included in Figure 1 only for illustration. The fact that the cell will remain active for considerably longer than 1 day, however, suggests that activity peaks observed during the first day are not the result of cellular degeneration in the recording chamber.

SEASONAL MODULATION

Contrary to our expectations and Strumwasser's original findings, many of our runs failed to give their peak near the time of projected dawn or dusk. Indeed, in the majority of cases the activity peak has occurred either near the projected midday or the projected midnight. Though this result was distressing when first observed, consideration of a large number of runs obtained at various times of year has convinced us that this discrepancy is due to a seasonal modulation of the steady state phase angle difference when the rhythm is entrained to LD *12*:12.

Although we have fewer data than we would like for much of the year, and none for August, presently available evidence indicates that the activity peak occurs near the projected midday or projected midnight during the months of April through October (20 out of 24 successful runs), but near the projected dawn or dusk during the remainder of the year (8 out of 8 successful runs). The segregation of phase angle differences according to season is statistically significant [$p < .025$, Runs test (Siegel, 1956)]. This finding is corroborated by a series of 9 runs obtained during the month of March by Strumwasser (1965), in which all the activity peaks occurred near the projected dawn.

NON-24-HOUR ENTRAINMENT

From November through March the rhythm in R15 can be entrained to light cycles having a 27-hr period. As an example of such entrainment, Figure 2 shows an activity record obtained from R15 of a ganglion drawn from LD *13.5*:13.5 (top), which was run simultaneously in the same chamber with R15 from another ganglion drawn from constant light (bottom). Following 27-hr light cycles the peak occurred at 15:05 PDT and was well

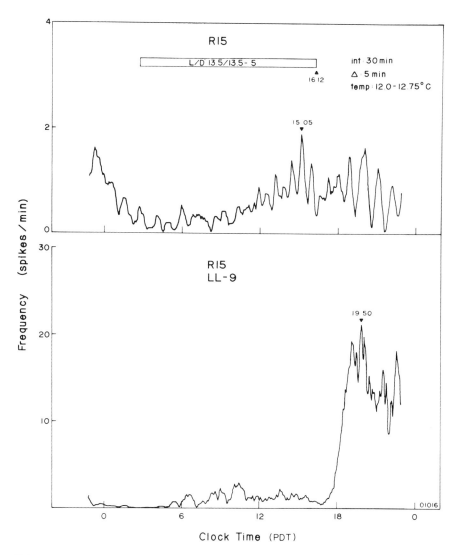

FIGURE 2   Activity in R15 following *LD 13.5*:13.5 (170:0) and LL (170 lux). Conventions as in Figure 1. Jan. 1-2. (Reprinted with permission from Lickey, 1969.)

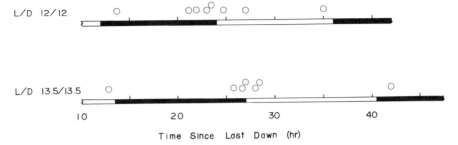

FIGURE 3   Distributions of phase angle differences following *LD 12*:12 and *LD 13.5*:13.5 (170:0). Circles give time of peaks since last experienced dawn. Nov.–March. N(*LD 12*:12)=8; N(*LD 13.5*:13.5)=7.

synchronized with the projected dawn; following constant light, the companion peak occurred 4 hr 45 min later at 19:50 PDT. Although one cannot predict the time of the peak following a history of constant light, the phase angle difference between the two oscillating neurons suggests that the timing was not determined by factors common to both neurons in the recording chamber or in the aquarium environment. In addition, the occurrence of a pronounced peak following constant light provides further evidence that R15 continues to oscillate in the absence of a recent history of periodic light cues (Strumwasser, 1965).

Figure 3 is a compilation of all our data on peak timing following *LD 12*:12 and *13.5*:13.5 (170:0) during November–March. In all cases, the temperature in the recording chamber was held within a one-degree tolerance in the range of 10.0–13.5°C, and no cell showed significant signs of injury or degeneration throughout the first 24 hr of the run. The plotted points, which show the elapsed time between the last experienced dawn and the major activity peak for each preparation, are precisely clustered. None of the activity peaks in either distribution fell closer to the projected midday or midnight than to projected dawn or dusk (p <.01 for *LD 12*:12, p <.02 for *LD 13.5*:13.5; two tail, binomial expansion), and we conclude that R15 can be entrained as readily by *LD 13.5*:13.5 as by *LD 12*:12 during the winter season.

In contrast, entrainment to 27-hr light cycles is quite unreliable during the remainder of the year. Figure 4, which gives the distributions of phase angle differences following three different light cycles, summarizes all our data on non-24-hr entrainment during April through October. Following *LD 13.5*:13.5, peaks occurred at widely scattered times throughout the projected light cycle (Figure 4, bottom), and we doubt that many of the preparations were entrained. Thus it appears that there is a seasonal modulation in the range of entrainment such that long periods are more easily established in the winter than in other seasons.

Consistent with this interpretation is the finding that individual cells drawn from 21-hr light cycles showed marked synchronization among themselves with a modal phase angle difference of about −2 hr (Figure 4, top). These points deviate significantly from a rectangular distribution [p <.05, Kolmogorov-Smirnov test for goodness of fit (Siegel, 1956)], and we conclude that many of the preparations were entrained by *LD 10.5*:10.5. Furthermore, comparison of the 21- and 24-hr distributions strongly suggests that the phase angle difference typically assumed by the neuron can be dependent on the length of the entrained period. No such dependence is detectable, however, when comparing the winter distributions of 24- and 27-hr preparations shown in Figure 3.

In addition to affecting the period, non-24-hr light cycles suppress the mean level of the oscillation in R15. The median of mean level, expressed as spikes per min, in preparations drawn from *LD 12*:12 (*170*:0) is 11.58 (N=51); from *LD 10.5*:10.5 (*170*:0) it is 4.21 (N=12), and from *LD 13.5*: 13.5 it is 3.04 (N=21) [p <.01, Kurskal-Wallis one-way analysis of variance, Siegel (1956)]. Thus far we have been unable to detect any seasonal changes in the mean level, nor have we found it to depend on the success of entrainment to the preceding light cycles.

The seasonal change in both the phase angle difference following *LD*

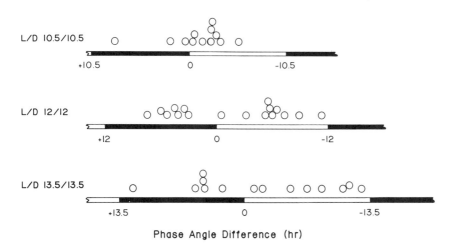

Phase Angle Difference (hr)

FIGURE 4   Distributions of phase angle differences following *LD 10.5*:10.5, *LD 12*:12, and *LD 13.5*:13.5 (170:0). Circles give phase angle difference with respect to nearest projected dawn. Apr.–Oct. N(*LD 10.5*:10.5)=12; N(*LD 12*:12)=17; N(*LD 13.5*:13.5)=14.

12:12 and the entrainability to *LD 13.5*:13.5 might be explained by assuming an appropriate seasonal change in the phase response curve or in the free-running period of the endogenous oscillator (Aschoff, 1965; Pittendrigh, 1965). A change of the latter type has been observed empirically by Menaker (1961) for the free-running period of the body temperature rhythm in bats. Data currently available to us on R15, however, do not permit a critical test of these hypotheses.

IMPORTANCE OF EYES FOR ENTRAINMENT

As a preface to our experiments on the importance of the eyes for entrainment of R15, we may recall the report of Jacklet (1969; also, this volume) that the isolated eye and optic nerve have a circadian rhythm in the frequency of spontaneous compound action potentials, and that *Aplysia* retinal cells contain granules of a neurosecretory type (Jacklet, 1968). These facts lead one to suspect that entrainment of the oscillation in R15 might be accomplished through a neural or endocrinological coupling between the oscillating eye and the oscillating neuron. Present evidence, however, indicates that this is not the case; some of our most striking examples of entrainment to *LD 12*:12 have been obtained in eyeless *Aplysia*.

Figure 5 shows an activity record from R15 obtained during the month of June from an *Aplysia* from which the eyes had been surgically removed prior to exposure to the entraining light cycles. The peak frequency occurred at 21.7 PDT, 4.3 hr prior to projected dawn. Similar peak times occurred in control preparations drawn from either intact or sham-operated animals. In the latter, a small patch of skin just anterior to the eyes had been removed just prior to entrainment. Such records suggest that the eyes cannot be the sole source of information responsible for entrainment in R15, and consideration of all our data from eyeless *Aplysia* confirms this interpretation.

Figure 6 gives the distributions of phase angle differences observed in intact, sham-operated, and eyeless *Aplysia* drawn from *LD 12*:12 (*170*:0). All the eyeless runs were obtained in April–June 1969, and during this time most of the control preparations gave activity peaks during the projected night (10 out of 14; p = .18, two tail, binomial expansion). Substantially the same result was obtained from the eyeless group, where 15 of the 19 peaks occurred during the projected night (p = .02, two tail, binomial expansion).

On the other hand, there is more variance among the eyeless preparations than among the controls. Although this difference is not statistically

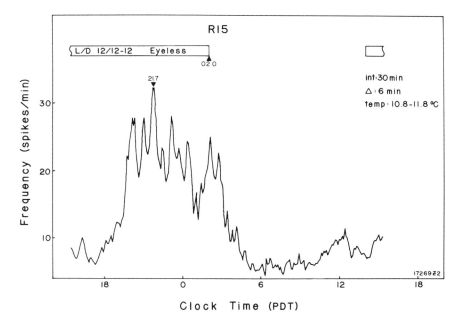

FIGURE 5   Activity in R15 from eyeless *Aplysia* following *LD 12*:12 (170:0). Conventions as in Figure 1. June 21–22.

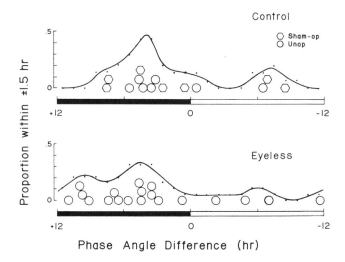

FIGURE 6   Distributions of phase angle differences in eyeless and control *Aplysia*. Circles and hexagons give phase angle difference with respect to nearest projected dawn. For explanation of smooth curve see text. Apr.–July. N(unop)=7; N(sham-op)=7; N(eyeless)=19.

significant, we have attempted to emphasize it by constructing continuous functions above the distributions whose ordinates represent the degree of clustering to be found at various times during the projected light cycle. Each point gives the proportion of peaks that occurred within ± 1.5 hr of the time of the point; the lines have been fitted by eye. Thus about 45 percent of all activity peaks in the control group can be found in a 3-hr window centered at 4 hr prior to projected dawn; another 20 percent can be found in a 3-hr window centered at 4.5 hr prior to midnight. By contrast, the curve from the eyeless distribution is flatter, and there is a prominent extra bump centered at 2.5 hr following projected dusk. The significance of this extra bump, consisting of four cases, as evidence for a possible role for the eyes, hinges upon whether it is considered reproducible. We believe that, at most, the eyes play only a minor role in the entrainment of R15.

In this respect the circadian rhythm in R15 may be similar to the locomotor rhythms of sparrows (Menaker, 1968; also, this volume) and salamanders (Adler, 1969; also, this volume), where the light reception critical for entrainment appears to occur directly in the brain. Indeed, some *Aplysia* neurons in the abdominal ganglion are known to be directly responsive to light (Arvanitaki and Chalazonitis, 1961); R15, however, does not appear to be one of these (but see Frazier *et al.*, 1967). Alternatively, entrainment might be accomplished via extraoptic peripheral photoreceptors such as are presumed to exist in other species of gastropods (Bullock and Horridge, 1965, pp. 1336–1337).

Some of the properties of the rhythm in R15 that are probably relevant to theories of entrainment and that illustrate some of the ways in which information about temporal properties of the external world can be encoded in the activity of this neuron are cited below. The summary is drawn from the previous work of Strumwasser (1963, 1965) as well as our own.

• The spike frequency oscillates with a circadian period that can free-run for at least a few cycles under constant conditions (Strumwasser, 1965).

• The rhythm can be entrained by LD *12*:12 light cycles (Strumwasser, 1965) as well as by LD *10.5*:10.5 and LD *13.5*:13.5 (Lickey, 1969).

• The mean level of the oscillation is reduced by non-24-hr light cycles (Lickey, 1969, and this report).

• There are two potential steady-state phase relations between exogenous LD *12*:12 light cycles and the endogenous rhythm; these are about 180° apart (Lickey, 1969).

- In response to an advancing phase shift in the exogenous *LD 12*:12 light cycle, the rhythm recovers its steady-state phase angle difference through a series of overshooting transients lasting about 5 days (Strumwasser, 1965).

- The susceptibility of the rhythm to perturbation by heat shock is dependent upon the phase of the rhythm at the time the shock is applied; similar susceptibility cycles have been indicated in response to pharmacological agents, especially actinomycin D (Strumwasser, 1965).

- The phase angle difference can be dependent upon the length of the entrained period, especially as demonstrated with 21-hr entrainment (this report).

- There is a seasonal modulation of the steady-state phase angle difference such that activity peaks occur at either projected dawn or dusk in the winter but at projected midday or midnight in other seasons (Lickey, 1969).

- There is a seasonal modulation in the range of entrainment such that entrainment to cycles with long periods is more easily obtained in the winter than in other seasons (this report).

- The eyes are not essential for entrainment, although they may play a minor role (this report).

ACKNOWLEDGMENTS

We gratefully acknowledge the assistance of J. Wilson for technical assistance, M. Franek for programming, and D. Holtan for typing the manuscript. The work was supported in part by PHS Grant 07458 and PHS Predoctoral Fellowship 5 F01 MH34335.

REFERENCES

Adler, K. 1969. Extraoptic phase shifting of circadian locomotor rhythm in salamanders. Science 164:1290–1291.

Aréchiga, H., and C. A. G. Wiersma. 1969. Circadian rhythm of responsiveness in crayfish visual units. J. Neurobiol. 1:71–85.

Arvanitaki, A., and N. Chalazonitis. 1961. Excitatory and inhibitory processes initiated by light and infra-red radiation in single identifiable nerve cells (giant ganglion cells of *Aplysia*). *In* E. Florey [ed.], Nervous inhibition. Pergamon Press, New York.

Aschoff, J. 1965. The phase-angle difference in circadian periodicity. *In* J. Aschoff [ed.], Circadian clocks. North-Holland Publishing Co., Amsterdam.

Bullock, T. H., and G. A. Horridge. 1965. Structure and function in the nervous system of invertebrates. W. H. Freeman Co., San Francisco.

Coggeshall, R. E. 1967. A light and electromicroscopic study of the abdominal ganglion of *Aplysia californica.* J. Neurophysiol. 30:1263–1287.

Frazier, W. T., E. R. Kandel, I. Kupfermann, R. Wazieri, and R. E. Coggeshall. 1967. Morphological and functional properties of identified neurons in the abdominal ganglion of *Aplysia californica.* J. Neurophysiol. 30:1288–1351.

Jacklet, J. W. 1968. Synchronized neuronal activity and neurosecretory function of the eye of *Aplysia*. Proc. Int. Union. Physiol. Sci. 7:213.

Jacklet, J. W. 1969. Circadian rhythm of optic nerve impulses recorded in darkness from isolated eye of *Aplysia*. Science 164:562–563.

Lickey, M. E. 1966. Further studies of a circadian rhythm in a single neuron: seasonal modulation and adiurnal entrainment. Physiologist 9:230.

Lickey, M. E. 1969. Seasonal modulation and non-24-hour entrainment of a circadian rhythm in a single neuron. J. Comp. Physiol. Psychol. 68:9–17.

Menaker, M. 1961. The free running period of the bat clock; seasonal variations at low body temperature. J. Cell Comp. Physiol. 57:81–86.

Menaker, M. 1968. Extra-retinal light perception in the sparrow. I. Entrainment of the biological clock. Proc. Natl. Acad. Sci. U.S. 59:414–421.

Pittendrigh, C. S. On the mechanisms of the entrainment of a circadian rhythm by light cycles. *In* J. Aschoff [ed.], Circadian clocks. North-Holland Publ. Co., Amsterdam.

Siegel, S. 1956. Nonparametric statistics for the behavioral sciences. McGraw-Hill Book Co., New York.

Strumwasser, F. 1963. A circadian rhythm of activity and its endogenous origin in a neuron. Fed. Proc. Fed. Am. Soc. Exp. Biol. 22:220.

Strumwasser, F. 1965. The demonstration and manipulation of a circadian rhythm in a single neuron. *In* J. Aschoff [ed.], Circadian clocks, North-Holland Publ. Co., Amsterdam.

DISCUSSION

STRUMWASSER: Are there any differences in mean level of oscillation of the cells between summer and winter?

LICKEY: No.

STRUMWASSER: Also, are there any seasonal differences in free-running period?

LICKEY: We would like to know that. Unfortunately, our conclusions about the free-running period are limited by the fact that the activity records cannot be confidently interpreted for more than about 24 to 36 hr after impalement, so we can't determine precisely for a single preparation. As you have suggested before, organ culture techniques might permit longer recording sessions, but we have not tried this. Another approach to the problem would be to apply our present technique in an exhaustive examination of the range of entrainment at various times of year. If, as we might expect, the free-running period is shorter in summer than in winter, the range of entrainment should be correspondingly displaced toward shorter values in the summer than in the winter.

GWINNER: Could the seasonal changes in phase angle be related to seasonal changes in reproductive activity?

LICKEY: I think it is entirely possible. We don't yet have a full year's data on reproductive tract weights. The animals are freshly collected from different places along the southern California coast and are maintained in the laboratory for only 2 to 5 weeks before dissection. Reproductive tract weights vary depending on where the animals are collected as well as the time of year. Last spring, however, reproductive tract weight did show a dramatic increase beginning about April 1. At about the same time the activity cycle in R15 changed its phase angle by about 90°.

JACKLET: I would like to bring up the matter of the intensity of the entraining light cycle. It is quite possible that at lower intensities of light the eyes may assume more importance than they seem to have at this intensity.

LICKEY: Yes, I agree wholeheartedly. We cannot yet rule out the possibility that the eye is necessary, and even sufficient, under some conditions of illumination.

ZIMMERMAN: You say all these properties exist in a single cell. Is the possibility excluded that interactions take place between R15 and other neurons with which it is synapsed?

LICKEY: I don't know of hard evidence that completely excludes the possibility that other neurons participate in generating the circadian rhythm recorded in R15. Neither has it been proved that the rhythm in R15 originates endogenously in this cell. In fact, there exist other identifiable neurons in the ganglion that also appear to have circadian activity cycles, so the phenomenon is not confined to a single neuron. On the other hand, it seems unlikely that the timing of the activity peak in R15 could be governed solely by periodic reverberatory activity in a network of synaptic connections. It is more likely that there are one or more cells in the ganglion that have an endogenous circadian rhythm. At the moment we cannot say for sure whether R15 is one of these. Perhaps Dr. Strumwasser has better data than we on this point.

STRUMWASSER: I'm very glad you asked that. Zimmerman's question is a good one. There are two pharmacological tricks for stopping synaptic transmission, one of which was only discovered 6 years ago. There is a magical natural agent called tetrodotoxin that abolishes nerve impulses in all sodium-requiring membrane systems. Also, one can use calcium-free artificial seawater, and in this way abolish any synaptic activity; it is very clear that calcium is required for transmitter release. I have two figures. The first figure was obtained from an animal kept in LL and shows an intracellular recording from R15, the parabolic burster. You can see that impulses have been abolished in the presence of tetrodotoxin. One can prove, by recording from a number of nerves that emerge from the ganglion, that in calcium-free artificial seawater with tetrodotoxin added, all the impulses in the ganglion recordable from the various nerve trunks or by penetrating cells are abolished. One is then left with just the driving oscillator in that cell itself (expressed in Figure 1 as periodic membrane oscillations). One cannot exclude, however, the possibility of glial–neural interaction, of which we know little at this stage. Typically, under these conditions, LL animals have parabolic bursters that show rather high oscillatory rates; these oscillations run down slowly in just ordinary artificial seawater.

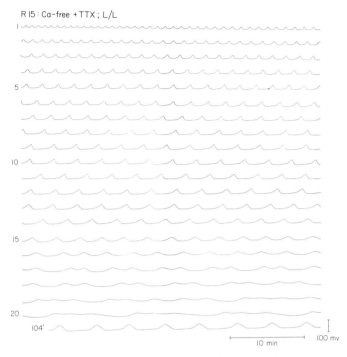

R I5 : Ca-free + TTX ; L/L

FIGURE 1   Continuous intracellular recording from the parabolic
burster in the presence of calcium-free artificial seawater and tetrodotoxin (25 μg/ml). The parieto–visceral ganglion, in which the parabolic
burster is located, was obtained from a sea hare, *Aplysia californica*, that
had been kept in constant light for about 1 week. Record reads from top
to bottom; numbers 1–20 on left are line numbers; each line is 40 min
long. 104 minutes are skipped between line 20 and the last line. The line
shows that the oscillations can be restored in amplitude by moderately
hyperpolarizing the membrane. 100 mV calibration applies to all lines.
Chamber temperature was 14°C. (From Strumwasser, 1969, unpublished.)

Figure 2, which was obtained from
a light–dark entrained animal, shows
that the cell is remarkably different
in that it is completely quiet for long
periods of time and then starts oscillating. Now in the case I show here, the
circadian oscillation is phase-shifted
180°. This is a summer animal from
which the ganglion was obtained, but
interpretation of the phase shift is
complicated by the presence of tetrodotoxin and lack of calcium. So the cycle
of the circadian rhythm, measured
now not in impulse rate but in terms
of the driving oscillation in the membrane, can appear when all classical
impulses and classical transmitter release, as we understand it, have been
abolished in the ganglion. A quantitative model of the membrane oscilla-

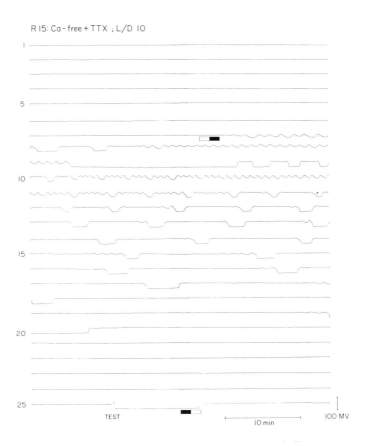

FIGURE 2   Continuous intracellular recording from the parabolic
burster in the presence of calcium-free artificial seawater and tetrodotoxin (25 μg/ml). The ganglion was obtained from a sea hare that had
been kept in LD cycles (12:12) for 10 days. Projected LD cycle indicated
on lines 7 (lights-out) and 25 (lights-on). Record reads from top to bottom; numbers 1–25 on left are line numbers; each line is 40 min long.
The last line shows that although oscillations of the membrane potential
have stopped, initiation of a step to a hyperpolarized state can still be induced by application of a short depolarizing current at "test." (See
Strumwasser, F. *The Physiologist* 10:318, 1967.) 100 mV calibration applies to all lines. Chamber temperature was 14°C. (From Strumwasser,
1969, unpublished.)

tions (not the circadian cycle) has been developed by Strumwasser and Kim (*The Physiologist* 12:367, 1969).

JACKLET: I notice that you have flip-flops only in the LD animal. Is that right?

STRUMWASSER: You can also initiate flip-flops by manipulating the membrane potential in the constant light condition, but that is another complicated story, a separate neurophysiological question. What is the nature of the driving force producing these membrane oscillations? The response to that question would take some time to develop.

WILLOWS: Clearly, Dr. Strumwasser, you have produced a rhythm in that *L*D entrained animal. What Dr. Lickey has said previously indicates that light to the eyes probably doesn't have much of a direct effect on entrainment. What about the obvious possibility of light having a direct effect on the cell?

LICKEY: Yes, there is that possibility. There are light-sensitive neurons in the ganglion (Arvanitaki and Chalazonitis, 1961). It is, in fact, recorded in the literature that light inhibits R15 (Frazier *et al.*, 1967), but we have never observed this and are inclined to believe that R15 is not directly responsive to light. The preparations are run under constant illumination of about 200 lux.

WILLOWS: You mentioned Dr. Strumwasser's result with the heat pulse. Have you ever tried a light pulse?

STRUMWASSER: I have never tried it myself, but I think someone really should do that experiment. If we do it and it fails, though, I don't think it will mean very much.

VI

CELLULAR AND
BIOCHEMICAL
MECHANISMS

KLAUS BRINKMANN

Metabolic Control of Temperature Compensation in the Circadian Rhythm of *Euglena gracilis*

In considering biochemical mechanisms possibly underlying circadian rhythmicity, it has become commonplace to point out that liquid enzyme systems (such as glycolysis; Chance, Hess, and Betz, 1964; Chance, Schoener, and Elsasser, 1964), structure-bound enzyme systems (such as protein synthesis; Monod and Jacob, 1961), or active transport systems (e.g., ion transport through mitochondrial membranes; Chance and Yoshioka, 1966; Höfer and Pressmann, 1966) are all capable of producing stable oscillations. Since these oscillations may be self-sustained and may even exhibit phase-dependent phase-shifts upon the application of external synchronizers (Betz and Becker, in press), they are reasonably proposed as models for circadian oscillations (Pye, this volume). In addition, periodic chemical systems can be described by a series of equations (Higgins, 1967) that may even account for circadian frequencies running some 300 times more slowly then biochemical oscillations.

At least one important difference remains, however: All these short-term oscillations are strongly temperature-dependent, with Q_{10} values for

Portions of this paper have been reported at the First International Symposium on Biochemical Oscillations, Prague, 1968, and at the Fifth International Congress on Photobiology, Hanover, N.H., August 26–31, 1968.

frequency of 2 or more, whereas typical circadian rhythms are almost in-sensitive to temperature. The Q_{10} of frequency of the glycolytic oscillator, for instance, is between 2 and 4 (Betz and Chance, 1965), while the Q_{10} of frequency of most "biological clocks" is in the range of 0.9 to 1.1 (Wilkins, 1965). It has been shown that in feedback-induced biochemical oscillators, frequency could become independent of temperature if the temperature coefficients for the reactions involved had the appropriate ratios (Spangler and Snell, 1961). There are indications, however, that circadian rhythms are neither temperature-independent nor temperature-compensated in that simple way. Rather, they are equipped with special regulatory mechanisms that can compensate for temperature-induced disturbances in a manner dependent on the phase of the oscillator. Regu-latory processes are suggested by both types of temperature reactions that can occur in *Euglena gracilis*. The one type is characterized by transients of increased frequency following rapid increases in temperature. After the transients, the former frequency is exactly re-established but with the final phase angle shifted relative to an undisturbed control (Brinkmann, 1966). This type is restricted to cultures in which respiration is reduced in favor of a lactic acid fermentation.

In the other type of temperature compensation, which occurs in young, strongly respiring cultures, the frequency is slightly dependent on the abso-lute level of temperature, and the phase angle is unaffected by temperature steps. Compensating processes are evident in that case, too, since the tem-perature coefficient (Q_{10}) of frequency is less than 1 (Figure 2). [A Q_{10} of frequency less than 1 is not exceptional. It has been repeatedly reported for other algae; see, for instance, Bühnemann (1955).] In *Euglena gracilis*, both types of temperature compensation can operate alternatively in the same cell, depending only on the metabolic situation. It is the purpose of the work reported here to look for metabolic processes that determine the type of compensation.

THE PHYSIOLOGICAL BACKGROUND

TESTING THE RHYTHM

Our experimental subject is the strain 1224-5/9 of *Euglena gracilis* (green, all pigments, eye spot, flagella), from the Algensammlung Pringsheim, Göttingen, West Germany. They were grown in an illuminated, thermo-statically controlled chamber, either autotrophically or mixotrophically, in media as described previously (Brinkmann, 1966).

To test for circadian rhythmicity, we measure the daily fluctuation of random motility in a large population (100 ml with 1×10^6 cells/ml), as indicated by differences in the sedimentation equilibrium. These differences are measured by monitoring the optical density at a fixed level of the cell suspension (photo-element 18×18 mm with lower edge 1.5 cm above the bottom of a 12-cm high cuvette, mounted in a temperature-controlled water bath). Since the equilibrium of sedimentation is reached in less than 100 min of darkness, we used a test-light rhythm of 20 min light (white, approximately 1000 lux) and 100 min darkness throughout. Test-light rhythms of 2-hr period length or less do not entrain the circadian rhythm either directly or by frequency demultiplication (Schnabel, 1968).

When biochemical assays were to be performed, we thoroughly mixed the culture during the illumination in order to provide all cells with equal exposure to light. In experiments of a merely descriptive nature, the cells were not stirred at all. Regardless of whether or not the cell suspensions have been stirred during illumination, the rhythm of sedimentation continues for as long as several months, exhibiting a good sinusoidal wave form.

As an example, Figure 1 shows the sedimentation test in an autotrophic culture in which growth is just slowing down. It illustrates, in addition, that an increase in cell number can be recorded simultaneously if a magnetic stirrer is switched on during the light pulses. Most of our experiments employed cells that were not dividing.

THE TWO TYPES OF TEMPERATURE COMPENSATION

The circadian rhythm of random motility in *Euglena* exhibits two alternative types of temperature compensation: one frequency-sensitive and one phase-sensitive. The two types of temperature compensation are complementary: either the phase is insensitive to temperature steps but the frequency is sensitive to different constant temperatures (frequency-sensitive type), or the phase is sensitive to temperature steps but at the end of the transient cycles the original frequency is re-established, regardless of the absolute level of temperature within 15–32°C (phase-sensitive type).

Although, as I will show later, it is not the culture medium itself that determines the type of compensation used, the two types can most easily be demonstrated in two different media (Figure 2). In young mixotrophic cell cultures the circadian period is longer at higher constant temperatures, but it is not affected by temperature steps (glucose–peptone medium, pH 4.7, vitamins B_6, B_{12} and trace elements). In old autotrophic cultures (Ca, Na, K, Mg, nitrate, sulfate, phosphate, Cl, pH less than 4; no growth because of the absence of vitamins and trace elements), the period length is constant at approximately 23.3 hr at different temperature levels, but the phase is shifted by a sudden increase in temperature and the former frequency is re-established via short transient cycles that approach the former "steady-state" frequency logarithmically. With one exception, a negative temperature step does not affect the phase of the rhythms. If it is applied exactly at the point of positive inflection, it quickly reverses the phase without inducing transients (Brinkmann, 1966).

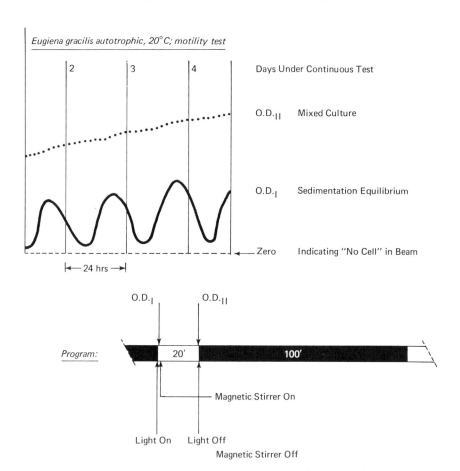

FIGURE 1   Test-light program for motility measurement and the time course of optical density in a suspension of slowly growing *Euglena* cells. OD I is representative of the sedimentation equilibrium reached during 100 min of darkness. OD II is representative of the overall cell number in the cuvette (stirred culture). OD I increases with decreasing sedimentation. Maximum of OD I indicates highest random motility. If entrained by a *12*:12 hr *L*D cycle, a maximum of OD I would coincide with the middle of day time. The slope of OD II is similar to the increase of the amplitude in OD I, indicating that an increase in population size is superimposed on the motility rhythm and that the new cells are in phase with the whole population.

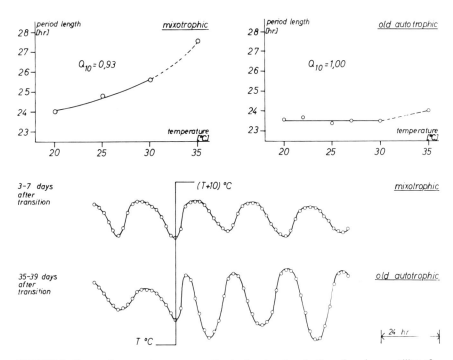

FIGURE 2 Types of temperature compensation in the circadian rhythm of random motility of *Euglena gracilis*. In the two upper graphs the free-running frequencies in young mixotrophic and old autotrophic cultures are compared as a function of different constant temperatures. The lower graphs illustrate the phase reaction of the two cultures after a single temperature jump of $+10^\circ$C given at a minimum of the cycle. The mixotrophic culture represents the frequency-sensitive type of temperature compensation, and the old autotrophic culture represents the phase-sensitive type of temperature compensation (after Brinkmann, 1966). "Transition" means transition from *L*D *12*:12 hr to the test-light program of *L*D *20*:100 min.

Since the reaction following a stepwise increase of temperature was the main test for temperature compensation in the biochemical work, this reaction warrants further discussion.

A single positive temperature step completely synchronizes the rhythm. Regardless of the time at which the step is applied, the peak occurs approximately 23.3 hr later. The phase response curve is of the "slope one" type (Figure 3). Above a low threshold (as yet undefined because of interfering population effects in this range), the reaction is independent of the height of the temperature step (Figure 4), which suggests that the actual input to the oscillator is the differential quotient of a temperature regimen. This hypothesis can be tested by applying a sinusoidal temperature program with a period length of $\Upsilon = \tau$. If the length of the entraining period Υ equals the free-running period τ, the biological rhythm is entrained at resonance condition, and no phase shift occurs between the entraining oscillator and the entrained one. Under this condition, and if the first derivative of the temperature program is really what entrains the biological rhythm, the rhythm should follow the cosine instead of the sine of the temperature program. The result is complicated, however, by the fact that cell motility also depends directly on temperature. If a sinusoidal temperature program is applied, we expect to see direct temperature effects superimposed on the circadian rhythm.

As Figure 5 shows, the two components can be identified in the superposition. Curve a is a control; curve b illustrates again the phase inversion following a single positive temperature jump applied at the minimum of the curve. The two solid curves below (c and d) show the results of entrainment by a temperature program; the program itself is indicated by the dotted curve at the bottom (e). Arrows downward indicate the positive

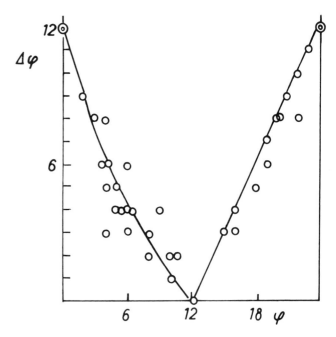

FIGURE 3   Phase response curve for the temperature step reaction (+10°C) of old autotrophic cultures. Ordinate: phase shift, in hours, measured after reestablishment of the previous frequency; abscissa: phase at the beginning of the temperature step after the preceding minimum, in hours. The final phase difference between the shifted rhythm and the control is plotted, neglecting the sign of the phase shift (from Brinkmann, 1966).

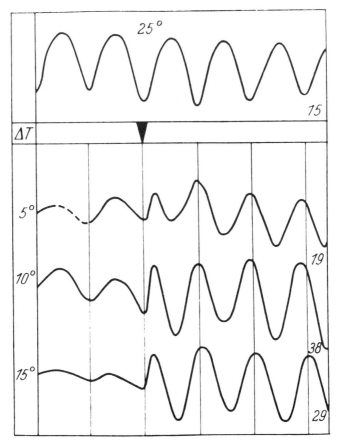

FIGURE 4 Phase inversion in old autotrophic cultures following single
positive temperature steps of different heights applied at the minimum of
motility. Numbers at left: height of temperature step in °C; numbers on
the right: days after transition into test-light rhythm; arrow: application
of temperature step; vertical lines represent 24-hr intervals (from Brinkmann, 1966).

points of inflection in the temperature program (where the circadian maximum should be); arrows upward indicate the position of these points regardless of whether or not the temperature program is in operation. In
curve c and curve d, at least two components are superimposed. After the
rhythm is released from entrainment, only one component persists, with a
maximum corresponding to the position of the point of positive inflection
in the former temperature program. That means the circadian component

has followed the cosine of the sinusoidal temperature program. This phase relationship can be expected, of course, only if the period length of the entraining program is very close to the characteristic period length of the rhythm (Υ). If the entraining period differs only slightly from Υ, characteristic phase angles will result (a typical feature of self-sustained rhythms). At the limits of entrainment ($\Upsilon = 12$ hr and $\Upsilon = 48$ hr) the phase angle can become as high as 180° in *Euglena*, as has been demonstrated with entraining light–dark cycles (Schnabel, 1968).

RESTRICTING THE PROBLEM

Figure 2 shows that phase shifts following a temperature step are restricted to old autotrophic cultures; young mixotrophic cells are insensitive to temperature steps. It is not the growth medium, however, that determines the type of temperature compensation.

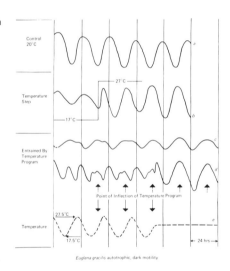

FIGURE 5 Entrainment of the motility rhythm in old autotrophic cultures by a sinusoidal temperature program with a period of 24 hr and an amplitude of 10°C; (a) free-running motility rhythm with period τ, (b) influence of single temperature step. A second temperature step applied 24 hr later (Υ=24) would coincide with maximal motility and would not further influence the phase. Prediction from curve b: a rhythm of repeated temperature steps with a period of $\Upsilon = \tau$ would entrain the biological rhythm, and the maximum of motility would coincide with the step up in temperature; in the case of a sinusoidal temperature program the maximum of motility would coincide with the point of positive inflection of the temperature program. c–d results from a motility rhythm under entrainment by a rhythmic temperature program as indicated by the dotted line below (e). The arrows indicate points of positive inflection of the temperature program. Both rhythms (c and d) show a superimposition of different components, including thermokinesis, during entrainment; after being released from the temperature program only one component persists and the remaining maxima coincide with the points of inflection in the preceding program (dotted arrows). That is, the circadian component has followed the cosine of a sinusoidal temperature program of $\Upsilon = \tau$ as predicted. Curve c has been measured with the photoele-

ment 1.5 cm above the bottom of the cuvette, as in the case of all other records reported in this paper. Curve d was measured with the photoelement 3.5 cm above the bottom of the cuvette in order to show that the relative amplitudes during the entrainment (but not the final circadian phase under free-running condition) depend also on the height at which the OD of the culture was taken.

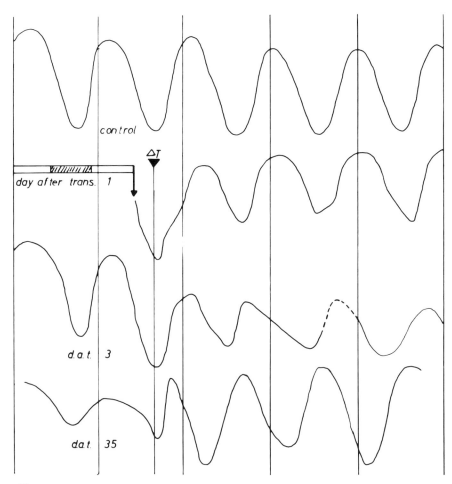

Euglena gracilis , autotrophic

FIGURE 6   Phase shift of the circadian motility rhythm in autotrophic cultures. The phase shift is
induced by single temperature step of +10°C given at different numbers of days after transition
(d.a.t.) from *LD 12*:12 hr to *LD 20*:100 min and simultaneously stopping aeration when beginning
the *LD 20*:100 min.

Figure 6 shows the results of experiments in which a positive temperature step at the minimum of motility was applied to cells of different ages
(that is, different lengths of time after transfer to the test light rhythm and
stopping aeration). At the first day no phase shift can be induced. On the
third day an intermediate phase shift results, and in older cell suspensions
full inversion takes place. The decreasing amplitude of the response in the

cuvette stirred at the third day after transfer may indicate that the results from fully reacting cells are being superimposed on those from cells not reacting at all. The fact that both types of temperature compensation are possible in the same media allows us to concentrate our attention on autotrophic cultures.

THE BIOCHEMICAL APPROACH

THE PROBLEM

The problem to be investigated herein is, What is the metabolic factor—a substance, a regulatory condition, or whatever—that actually determines in *Euglena gracilis* whether its temperature compensation follows the frequency-sensitive or the phase-sensitive type? Since the alternative compensation mechanisms are found associated with either young, still-growing cells (frequency-sensitive type) or an old, stagnating-cell suspension (phase-sensitive type), two directions of approach are open:

- The metabolic situation in the two states can be compared under steady-state conditions; or
- We can look at what happens while the cells are changing their type of temperature compensation.

The following section summarizes current experiments employing both approaches.

METHODS

In these experiments, concentrations of metabolites were measured enzymatically. Cell suspensions were taken from the circadian test situation described above (test-light rhythm, 20 min L:100 min D; constant temperature, 20°C or as indicated) 90 min after the preceding lights-off step. Still in darkness, the cells were quickly injected into liquid nitrogen and lyophilized. The dry weight was taken from the lyophilized sample, including substances from the medium. The cells were extracted with 5 percent perchloric acid and washed once in 1 percent perchloric acid. The extracts were combined and neutralized with KOH to remove the perchloric ions.

To compare the concentrations of metabolites inside the cells with their concentrations in the medium, samples from the circadian test situation were filtered by suction with the aid of "super cel." The cells were extracted with perchloric acid and prepared as described above; the clear filtrate was concentrated by lyophilization. The cell volumes were calculated from hematocrit centrifugation.

Oxygen consumption was measured with a Clark-type electrode under temperature-controlled conditions and corrected for the number of cells. To estimate the rate of protein synthesis, *Euglena* cells were incubated with labeled amino acids as indicated. Incorporation was terminated by injection of perchloric acid (5 percent final concentration), and the cells were washed twice. Radioactivity was measured with a liquid scintillation counter either as total activity in the acid-insoluble pellet, or after fractionation into acetone extract ("lipids"), material solubilized by tryptic digestion ("protein"), and insoluble residue.

COMPARISON OF STEADY-STATE CONDITIONS IN YOUNG, STILL-GROWING CELLS AND OLD, STAGNATING CELLS

Ample information from different laboratories is now available on synchronization of metabolic processes with the cycle of cell division. In green or colorless strains of *Euglena*, nucleic acid metabolism, protein synthesis, and synthesis of cell-wall material are known to oscillate with well-defined phase angles to one another (Blum and Padilla, 1962; Edmunds, 1965; Walther and Edmunds, 1970). These rhythms are expressed in both the turnover rates and the accompanying pool sizes of intermediate compounds. Although cell division cycles are probably coupled to the hypothetical circadian timer rather than being capable of independent circadian oscillation (Bruce, 1965), a cycle in cell division forces almost all metabolic activities to fluctuate. Additionally, in *Euglena* a circadian rhythm of protein synthesis may continue in young cultures for a few cycles even if cell division has already terminated (Feldman, 1968).

In contrast, old autotrophic cultures that have shown no mitotic activity for at least a week do not show daily fluctuations in most of their metabolic activities, although they exhibit the most stable rhythm of random motility. The following paragraphs describe processes in which we could detect no oscillations:

1. The pools of the glycolytic pathway compounds and the pools of the high-energy compounds ATP and ADP are maintained at a constant level at all circadian phases.

As metabolism in photoautotrophic cells is strongly influenced by light, we had to check metabolite concentrations first with respect to the test-light cycle. As Figure 7 shows, the concentrations of ATP and G-6-P in-

crease after the test light has been switched on, but return almost to the
initial level in the dark. Samples were taken from the minimum and from
the maximum of the circadian rhythm; they differ slightly in the height of
the peak and in the descending slope during darkness. These differences
suggest that the photosynthetic capacity and the turnover rate of photosynthetic products are higher at the maximum phase than at the minimum;
this hypothesis has not yet been checked in detail.

In this context it is important that at the end of darkness in the test-
light cycles the concentration of ATP and G-6-P are the same throughout
the circadian period, although the motilities differ considerably. Figure 8
shows that the pool sizes, in samples taken 90 min after the light of each
test-light cycle has been turned off, do not oscillate significantly with the
circadian rhythm. Similar results have been found in *Gonyaulax* (Sweeney,
1969). The constant level of ATP is especially surprising, since an oscillation in dark motility must be characterized at least by an oscillating turnover rate in phosphorylation. Obviously the regulation of pool sizes is independent of the circadian regulation of flow rates. That suggests that pool
sizes have a different significance with regard to short-term metabolic regulation and to the long-term regulation of the biological clock. In this sense,

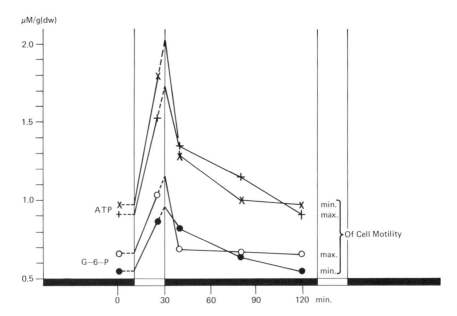

FIGURE 7 Fluctuations of ATP and G-6-P during one test-light cycle in an old autotrophic culture. Light cycle (20 min light, 100 min darkness) indicated below. Samples were taken at both the
minimum and the maximum of the circadian rhythm of random motility.

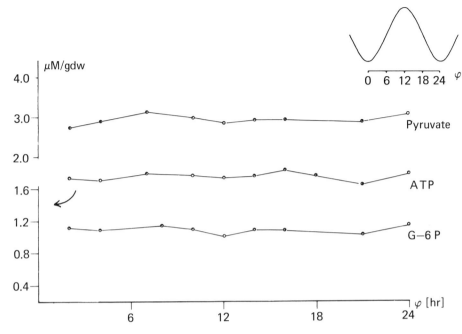

FIGURE 8 Pools of ATP, G-6-P, and pyruvate at different phases of one circadian cycle in an old autotrophic culture. Each sample was taken 90 min after the test was switched off (see Figure 7). Upper right: definition of the phase scale in terms of the circadian rhythm of motility.

Figure 7 represents a short-term regulation and Figure 8 the simultaneous events of a long-term regulation; the difference between the two time constants is large enough to indicate different processes.

2. Old autotrophic cultures of *Euglena* do not show a rhythm in respiration; at very low pH they do not show measurable respiration at all. Their metabolism is characterized by lactic acid fermentation at the expense of photosynthetic products. However, the concentrations of pyruvate and lactate do not oscillate either.

3. We could not detect any trace of oscillation in protein synthesis (incorporation of ^{14}C histidine and ^{3}H glutamine), although incorporation persists at a rate of approximately one tenth of that in young cells. From Figure 13 it is evident that, after a rise in temperature, old cells show an increase in incorporation similar to that in young cells but without a lag phase.

In general, metabolic events are uncoupled to a remarkable extent from the persistent circadian rhythm in old cultures. Whether or not this un-

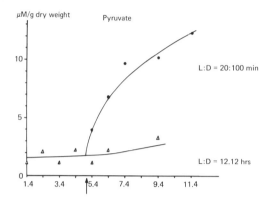

FIGURE 9 Left: accumulation of pyruvate in an autotrophic culture after transfer into test conditions of *LD 20*:100 min (arrow). Control remained under *LD 12*:12 hr. Abscissa: time in days.
The values indicate the total pyruvate, including that in the medium. (Note: the abscissa is an absolute time scale. Read April 1, April 3, April 5. . . .) Right: actual concentrations of pyruvate and
lactate inside and outside the cells in an old autotrophic culture (under test conditions for almost a
month).

coupling is responsible for the different type of temperature compensation
is still unknown. More information is needed with respect to direct influences of temperature steps.

TRANSIENTS IN METABOLISM THAT ACCOMPANY THE CHANGE IN
TEMPERATURE COMPENSATION

During the first days in the test conditions, *Euglena* shifts from respiration
to lactic acid fermentation; pyruvate accumulates and pH in the medium
decreases to 1.9. The accumulation of lactate and pyruvate starts immediately after the transfer of an autotrophic culture into the test-light rhythm
(Figure 9). The redox equilibrium is shifted far to the reduced state, toward
lactate. The ratio of lactate to pyruvate is higher inside the cells than outside, indicating that the leakage of lactate is approximately a thousand
times less than that of pyruvate.

The final level of pyruvate is shown in Figure 8; pyruvate concentration
rises to 50 percent of the final level within the first week. Of the same
order of magnitude is the time needed for the change in the type of temperature compensation (about 6 days), and the time in which the curve
shape of the motility rhythm changes from an asymmetric form to a symmetric, almost sinusoidal, one (Brinkmann, in press, a).

This induced lactic acid fermentation suggests that respiration is inhibited. We compared the respiration of young and old autotrophic *Euglena* cells at two different temperatures and measured dependence of respiration on pH. Figure 10 shows that respiration differs drastically in the two cases. Young, still-growing cells exhibit a typical pH dependence of respiration with a temperature coefficient of approximately 1.9. The old cells, however, show almost no respiration at 20°C and pH 2.0; these conditions are typical of standard conditions in an extended motility test. But after a rise in temperature, respiration resumes. The cell suspension was maintained at 20°C during the preceding test, and the cells were removed for measurement of respiration during the dark time of the test-light rhythm without a change in temperature. After the endogenous respiration had been recorded at 20°C, the temperature was quickly raised to 31°C. After some transients, oxygen consumption became constant within 10 min. The respiration rate between 10 and 20 min was taken as representative. The original pH of the cell suspension was 2.0 in the old culture and 4.7 in the young culture. Adjusted pH values had to be maintained for 30 min

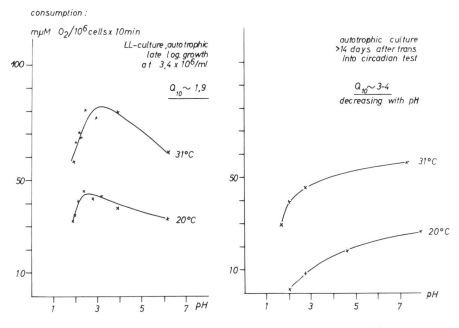

FIGURE 10 Endogenous respiration in a growing (left) and an old (right) autotrophic culture as a function of pH and temperature. The original pH of the growing culture is 4.6; that of the old culture 14 days after transition into test conditions is about 2.0. Other pH values were adjusted immediately prior to the polarographic measurement of the O_2-consumption.

before respiration could be estimated. Oxygen measurements were performed in complete darkness.

Figure 10 shows that both the pH dependence and the effect of temperature are different in young and old cultures. The Q_{10} of respiration in old cultures varies between 3 and 4, depending on the pH. Below pH = 3 it is extremely high. At present, we have no idea what factor inhibits pyruvate utilization. Whatever it may be, the unusually great increase in respiration after a rise in temperature is indicative of a drastically changing metabolism. We do not yet know whether or not this increase in oxygen consumption corresponds to a similarly increased oxidative phosphorylation.

We do know, however, that correlated with radically altering the oxygen consumption, temperature steps induce a shift of the circadian phase. Temperature steps induce only advancing phase shifts in *Euglena*; that is, they temporarily increase the circadian frequency (Brinkmann, 1966).

Comparison of the two types of temperature compensation and their correlated respiratory behavior reveals that the phase sensitivity of old *Euglena* cells is exceptionally high. The large Q_{10} of respiration in old cultures is itself unusual. Furthermore, the temperature step advances the phase by 12 hr, which could be considered a change of 12 hr in frequency of the oscillation for a single cycle. In most other organisms, the circadian frequency depends only slightly on temperature, and temperature steps are only weak *Zeitgeber*.

ACTIDIONE AS A SPECIFIC POISON OF THE TEMPERATURE STEP REACTION

As yet very little is known about the mechanism of phase-shifting in circadian rhythms. Nobody knows what is really oscillating—whether there is a single master clock, a small number of processes coupled in an oscillating network, or a large number of different interacting oscillators that together make up the circadian oscillator. But two questions can be asked even now, Do phase-shifting signals coming from different sources—e.g., temperature steps and light steps—enter the oscillating system via the same channel? and How far—if at all—is it possible to distinguish different paths of information? Of relevance to these questions is our discovery that it is possible to inhibit the temperature-induced phase shift without affecting phase setting by light. The specific inhibitor is Actidione(=cycloheximide). Figure 11 shows different phase responses following a positive temperature step in the presence and in the absence of Actidione. It can be shown that Actidione is not merely masking a phase shift during 1 or 2 days of transients; the phase remains unaffected for more than a week after the signal (Brinkmann, in press b). On the other hand, Actidione does not inhibit the reinduction or phase setting of a circadian rhythm by light steps

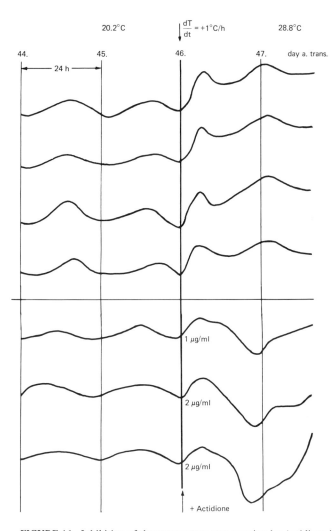

FIGURE 11 Inhibition of the temperature step reaction by Actidione in old autotrophic cultures (44–47 days after transition into test conditions). The arrow above indicates the beginning of the rise in temperature. Upper four curves: control; lower three curves: Actidione is added simultaneously with the temperature step (the numbers give final concentration of Actidione). In the controls the phases have been almost inverted by day 47 as compared with the original phase on day 45; no inversion occurs in the presence of Actidione, however. The first peak after the increase in temperature and the high amplitude later on are due to thermokinesis (see Figure 5).

(Figure 12), which means that these two different *Zeitgeber* must act on different mechanisms.

In contrast to the frequency-decreasing effect of Actidione on younger *Euglena* cultures described by Feldman (1968), in our experiments Actidione seems not to act via an inhibition of protein synthesis. Figure 13 shows the incorporation of an amino acid during a rise in temperature in the presence and in the absence of Actidione in young cells (left) and old cells (right). The upper curves show the motility rhythms up to the date of the incorporation experiment. The scale of the curve at right is one tenth of the scale at left, indicating that the incorporation rate in old cells is reduced to approximately one tenth of that in young cells. But in contrast to the young cells, the incorporation rate in old cells is only slightly depressed by Actidione. This slight decrease is similar in the protein fraction, the acetone-soluble fraction, and the cell-wall material, indicating that the poison might have caused an unspecific decrease in the permeation of the amino acid (Brinkmann, in press, b).

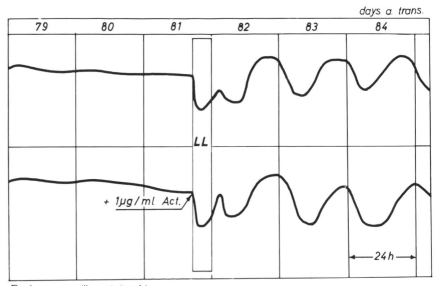

Euglena   gracilis autotrophic

FIGURE 12 Reinduction and phase setting of the circadian rhythm of random motility by means of a 6-hr light pulse in very old autotrophic cultures. Upper curve: control; lower curve: the effect of Actidione, which was added simultaneously with the beginning of the light pulse. The downward slope during the light pulse is a superimposed phototactic response. Both curves show an identical initial phase setting after the light-off step, which is typical for autotrophic cultures (Schnabel, 1968); in the presence of Actidione, however, the stable frequency is decreased (Feldmann, 1968).

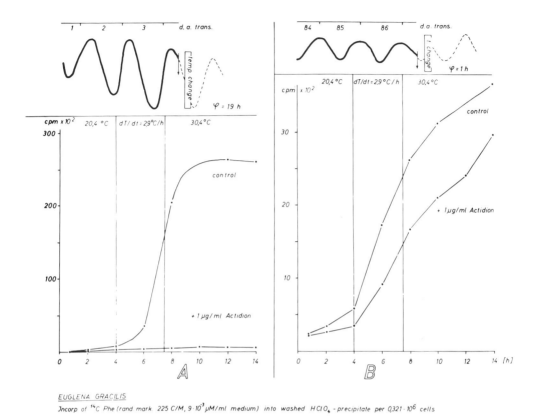

EUGLENA GRACILIS
Jncorp. of ¹⁴C Phe (rand. mark. 225 C/M, 9·10³ μM/ml medium) into washed HClO₄ -precipitate per 0,321·10⁶ cells

FIGURE 13   Influence of Actidione on the incorporation of ^{14}C-Phenylalanine into the acid-insoluble fraction of *Euglena gracilis*. A: Young autotrophic cells; B: autotrophic cells nearly 3 months old; both with and without Actidione. Time scale in hours. The curves above show the original rhythm before the start of the incorporation experiment. During the incorporation the temperature was linearly increased between the vertical lines. Note: the scale cpm in B is 1/10 of the scale in A. The circadian phases at the time of starting the incorporation are not identical. Other experiments have shown, however, that at least in the old autotrophic cultures the effect of Actidione is independent of the circadian phase.

We conclude, therefore, that Actidione specifically inhibits the temperature step reaction of old autotrophic cultures without inhibiting protein synthesis. Preliminary experiments have failed to show an influence of Actidione on the respiration rate of *Euglena* cells. If this finding is confirmed, we must conclude that the phase-shifting temperature signal passes through at least one other, Actidione-sensitive process located between the step affecting respiration and the final step influencing the oscillator.

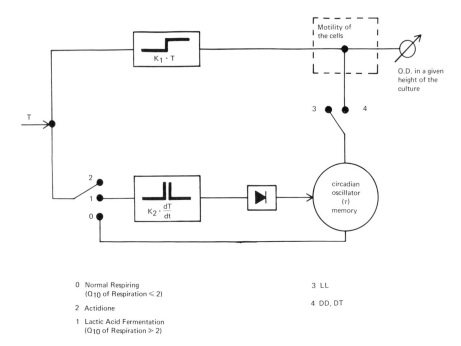

FIGURE 14   Scheme summarizing different effects of temperature on the random motility of *Euglena gracilis*. For details see text.

DISCUSSION

Our results may be summarized by a schema showing the effects of temperature on the random motility of *Euglena* (Figure 14). Motility, indicated by the broken line at top right, is controlled by a temperature signal, T, entering from left via two channels: One, shown at the top, represents direct sensitivity to temperature (thermokinesis); the other one, shown at the bottom, represents control via the circadian rhythm. This channel has a differentiating operator and a gate that allows only positive values to affect the rhythm. Obviously the two different control effects are additively combined to determine overt motility.

The flip-flop 0–1–2 is to be interpreted as follows: Position 0 is that of young, normally respiring cells in which the phase is not shifted by temperature steps (only the stable frequency is slightly dependent on temperature); position 1 occurs under lactic-acid fermentation conditions, where full entrainment by positive temperature steps is possible; and position 2 is found when Actidione is applied to old, lactic-acid fermenting cultures.

We do not yet know whether position 2 is identical with position 0. When the flip-flop 3–4 is in position 4 [under continuous test, DT (period of test-light rhythm between 12 min and 4 hr) or continuous darkness], motility is controlled by the circadian oscillator. Position 3 represents continuous constant bright light under which random motility is not controlled by a circadian rhythm. This preliminary diagram could be extended by a similar arrangement symmetrically at right, showing different ways in which light affects motility either directly via photokinesis or via entrainment by light steps similar to the entrainment by temperature.

Our main conclusion, that the mechanism of temperature compensation in the circadian rhythm of *Euglena* may be controlled by the temperature coefficient of respiration, is based on the correlation between the induction of a lactic-acid fermentation (accompanied by an increase in Q_{10} for the rate of endogenous oxygen consumption) and the transformation from frequency sensitivity to phase sensitivity in temperature compensation of the circadian rhythm. A correlation, however, does not demonstrate causality. However, the general idea that control mechanisms associated with mitochondria may determine the type of temperature compensation of circadian rhythms is supported by other general observations. Bacteria, which have no mitochondria, also display no circadian rhythmicity. In fungi, where mitochondria are supposed to be of a less efficient type than those of mammals and higher plants, circadian rhythms (and other long-term rhythms as well) vary considerably with respect to their sensitvity to external conditions (Jerebzoff, 1965). Finally, isolated mitochondria from different species of poikilothermic and homoiothermic animals are able to compensate for a drift in temperature by adjusting their succinate-induced oxygen consumption (Newell and Walkey, 1966; Newell, 1967).

ACKNOWLEDGMENTS

This work was supported by the Deutsche Forschungsgemeinschaft and the Stiftung Volkswagenwerk. The author would like to express his gratitude to Dr. A. Betz for his helpful criticism and to Mrs. R. Hinrichs, Miss P. Dahle, and Mrs. S. Berking for their expert technical assistance.

REFERENCES

Betz, A., and J. U. Becker. In press. Phase shift induced by pyruvate and acetaldehyde is oscillating NADH of yeast cells. *In* B. Chance [ed.], Symposium on biochemical oscillations.

Betz, A., and B. Chance. 1965. Influence of inhibitors and temperature on the oscillation of reduced pyridine nucleotides in yeast cells. Arch. Biochem. Biophys.
109:579–584.

Blum, J. J., and G. M. Padilla. 1962. Studies on synchronized cells: The time course of
DNA, RNA and protein synthesis in *Astasia longa*. Exp. Cell Res. 28:512–523.

Brinkmann, K. 1966. Temperatureinflüsse auf die Circadiane Rhythmik von *Euglena
gracilis* bei Mixotrophie und Autotrophie. Planta 70:344–389.

Brinkmann, K. In press, a. Respiration depending types of temperature compensation
in the circadian rhythm of *Euglena gracilis*. *In* B. Chance [ed.], Symposium on
biochemical oscillations.

Brinkmann, K. 1970, b. The role of actidion in the temperature jump response of the
circadian rhythm in *Euglena gracilis*. *In* B. Chance [ed.], Symposium on biochemical oscillations.

Bruce, V. G. 1965. Cell division rhythm and the circadian clock, p. 125 to 138. *In*
J. Aschoff [ed.], Circadian clocks. North-Holland Publ. Co., Amsterdam.

Bühnemann, F. 1955. Das endodiurnale System der Oedogonium Zelle. III.: Über
den Temperatureinfluss. Z. Naturforsch. B10, 305–310.

Chance, B., B. Hess, and A. Betz. 1964a. DPNH-oscillations in a cell-free extract of
Saccharomyces carlsbergensis. Biochem. Biophys. Res. Commun. 16:182.

Chance, B., B. Schoener, and S. Elsasser. 1964b. Control of the wave form of oscillations of the reduced pyridine nucleotide level in a cell-free extract. Proc. Natl. Acad.
Sci. U.S. 52:337–341.

Chance, B., and T. Yoshioka. 1966. Sustained oscillations of ionic constitutes of
mitochrondria. Arch. Biochem. Biophys. 117:451–465.

Edmunds, L. N., Jr. 1965. Studies on synchronously dividing cultures of *Euglena
gracilis* Klebs (strain Z). II. Patterns of biosynthesis during the cell cycle. J. Cell.
Comp. Physiol. 66:159–182.

Feldman, J. F. 1967. Lengthening the period of a biological clock in *Euglena* by
cycloheximide, an inhibitor of protein synthesis. Proc. Natl. Acad. Sci. U.S.
57:1080–1087.

Feldman, J. F. 1968. Circadian rhythmicity in amino acid incorporation in *Euglena
gracilis*. Science 160:1454–1456.

Higgins, J. 1967. The theory of oscillating reactions. Ind. Eng. Chem. 59:19–62.

Hofer, M., and B. Pressmann. 1966. Stimulation of oxidative phosphorylation in
mitochondria by potassium in the presence of valinomycin. Biochemistry 5:3919–
3925.

Jerebzoff, S. 1965. Manipulation of some oscillating systems in fungi by chemicals,
p. 183 to 189. *In* J. Aschoff [ed.], Circadian clocks. North-Holland Publ. Co.,
Amsterdam.

Monod, J., and F. Jacob. 1961. General conclusions: Telenomic mechanisms in cellular
metabolism, growth and differentiation. Cold Spring Harbor Symp. Quant. Biol.
26:389–401.

Newell, R. C. 1967. Oxidative activity of poikilotherm mitochrondria as a function of
temperature. J. Zool. 151:299–311.

Newell, R. C., and M. Walkey. 1966. Oxidative activity of mammalian liver mitochondria
as a function of temperature. Nature (Lond.) 212:428–429.

Schnabel, G. 1968. Der Einfluss von Licht auf die Circadiane Rhythmik von *Euglena
gracilis* bei Autotrophie und Mixotrophie. Planta 81:49–63.

Spangler, R. A., and F. M. Snell. 1961. Sustained oscillations in a catalytic chemical
system. Nature (Lond.) 191:457–458.

Sweeney, B. 1969. Transducing mechanisms between circadian clock and overt rhythms
in *Gonyaulax*. Can. J. Bot. 47:299–308.

Walther, W. G., and L. N. Edmunds, Jr. 1970. Periodic increase in deoxyribonuclease
    activity during the cell cycle in synchronized *Euglena*. J. Cell Biol. 46:613–617.
Wilkins, M. B. 1965. The influence of temperature and temperature changes on biologi-
    cal clocks, p. 146 to 163. *In* J. Aschoff [ed.], Circadian clocks. North Holland Publ.
    Co., Amsterdam.

DISCUSSION

VANDEN DRIESSCHE: I wonder if one cannot explain your results with Actidione—that is, the small effect of Actidione on amino acid incorporation in old cells—by the fact that Actidione acts fairly specifically on cytoplasmic ribosomes, which are very active in protein synthesis in young cells and less active in older cells. That is interesting also in relation to the finding that circadian rhythms seem to depend on nuclear genes rather than on genetic information carried in organelles. It is also true, of course, that mitochondrial and chloroplastic functions are integrated in the cell by information coded in the nucleus.

BRINKMANN: That may be so.

SWEENEY: I would just like to call attention to the neat way Dr. Brinkmann has used studies of temperature compensation as a way of looking at the mechanisms underlying the oscillations, a mechanism so difficult to get at directly. I am also going to take this opportunity to put in a plug for using motility or phototaxis as an essay of rhythmicity and to tell you of another study of motility that has been carried out at Santa Barbara by Dr. Davenport and Dr. Forward. They found that the dinoflagellate *Gyrodinium dorsum* is phototactic only when it has been exposed to red light just prior to being tested. This response to red light is not constant throughout time, but rather shows a very nice rhythm in otherwise constant conditions. That is, if you leave your culture in DD without any red preillumination, you never see any phototaxis at all. If you give a short exposure to red light, you find that phototaxis can be seen, but only during a small fraction of each 24 hr, corresponding to the region around dawn of the previous LD cycle. Furthermore, the effect of this short exposure to red light can be negated by exposure to far red. So we think we are seeing a phytochrome response, and what we see as a rhythm in phototaxis is really a rhythm in phytochrome action.

EHRET: I wonder, Dr. Brinkmann, if you have done experiments something like those I will describe now. Take a culture that grows through 10 generations, say, of rapid growth at 19°C, another at 25°C, and another at 35°C, until you have had prolonged rapid growth. Now compare the young culture and the old culture from each of the three sets with respect to free-running circadian output, and with respect to some assay that would be proportional to the activation energy for an enzyme—an Arrhenius plot of flagellar motility, perhaps. Have you done anything like this? Your ordinary experiments, as I understand them, generally go from a temperature T_n, to another temperature T_{n+5} or T_{n-5}. And then you observe an old culture and a young culture, both circadian, after that shift. What I am concerned with are the previous growth conditions. If you allow your cells to divide under prolonged ultradian conditions at the temperatures in question—instead of simply applying the temperature switch to the infradian or nondividing cells—and then test the derivative cultures for their circadian outputs, your results might be quite different. The low-temperature organisms might respond quite differently when given a chance to synthesize populations of new enzymes appropriate to and characteristic of that temperature.

BRINKMANN: I haven't done that; all cultures were originally grown at one temperature, 25°C. But I think the previous growth conditions—different growth rates at different temperatures, as you proposed, for instance—do not play any role in determining the type of temperature compensation. Once grown (either mixotrophically at high velocity or autotrophically and more slowly), the cultures can display either type of compensation depending on the subsequent conditions, that is. Without any cell division we can induce the cells to switch from the frequency-sensitive to the phase-sensitive type of temperature compensation by starting the test light program. Only the velocity of that transition is affected by the medium—it is much higher in autotrophic than in mixotrophic cultures. In every case the transition is correlated with the

reduction of mitochondrial respiration and the induction of lactic-acid fermentation. Thus, I believe that the type of temperature compensation is determined by the way the intracellular energy turnover is connected with the mitochondrial respiration, rather than by the selection of temperature-specific generations of enzymes as you proposed.

PAVLIDIS: How have you determined the time constant of the compensation mechanism? The way you describe it here, you simply fix the derivative so that the time constant of compensation is zero. Dr. Zimmerman's experiments with *Drosophila pseudoobscura* indicate that the time constant there is something like 2 hr.

BRINKMANN: If you entrain in *Euglena* a self-sustained biological rhythm of the characteristic period length of τ with a sinusoidal temperature rhythm of a period length of $T = \tau$ (that is, resonance entrainment), you will get under steady-state conditions a cosine of the biological rhythm. The fact that the biological rhythm follows the first derivative of the entraining temperature sine wave with the phase angle $\Delta\phi = 0$ may have misled you to the suggestion that the time constant of the phase shifting reaction is zero. But you cannot conclude this from the steady state of a resonance entrainment.

The time constant of the phase shifting reaction was demonstrated in another experiment (Figure 4) in which the biological rhythm was forced to shift phase by the application of a single positive temperature step during the minimum of the random motility of an old autotrophic culture. In Figure 4 you can see that the phase is not instantaneously shifted. Instead of that it takes up to 24 hr (1–2 transient cycles) until finally adjusting the phase. If you now plot the time between two succeeding extremes of the transient cycles versus the number of these half-cycles after input of the temperature step, you get an expression of the aperiodic logarithmic transient function of frequency which was induced by the temperature step (Brinkmann, 1966, Figure 31). This plot suggests that in *Euglena gracilis* the time constant of the phase shifting reaction by far exceeds the time constant of 2 hr measured by Dr. Zimmerman with *Drosophila pseudoobscura*.

PAVLIDIS: Are the transients that you observe due to the secondary system?

BRINKMANN: I do not know how to test this and would like to avoid such a differentiation at all.

FELDMAN: I have a couple of questions. First, would you clarify for me the point you made about mitochondria in the fungi?

BRINKMANN: I hit on this idea first by just looking to see what kind of circadian rhythms could be found in fungi. Dr. Jerebzoff has shown that in fungi one finds a variety of different temperature compensations depending on the metabolic situation. On the other hand there are no circadian rhythms in bacteria lacking mitochondria. Now I am speculating about the function of mitochondria. There seems to be a trend in the evolution of the circadian clock beginning with bacteria lacking both the mitochondria and circadian rhythms, continuing via fungi with a variety of mitochondria (variable in the compartmentation as well as in the

types of carbon acid cycles) and a variety of partially temperature-compensated circadian and ultradian rhythms, and finally reaching the standard cellular organization of plants and animals, which all exhibit a similar type of mitochondria and which are all equipped with a temperature-compensated circadian rhythm. But as we have learned from *Euglena*, even in the latter case we can manipulate the type of temperature compensation by manipulating the intracellular activity of mitochondria. It was the speculation about the evolution of a temperature-compensated circadian clock and its correlation with the evolution of "standard" mitochondria that I had in mind in my remark about fungi.

FELDMAN: I think the results of the Actidione experiments are exceedingly interesting. I wondered, since there was no inhibition of protein synthesis in the old cells, whether you checked with labeled Actidione to make sure the Actidione was actually getting into the cells.

BRINKMANN: No, but of course we assume it did get into the cells because it poisoned the temperature step reaction.

FELDMAN: But that could have been an effect on the plasma membrane.

BRINKMANN: I tend to think Actidione is having its effect on the mitochondrial membrane. But I cannot exclude the possibility of an effect on the plasma membrane; the only thing we can say for sure is that Actidione is not acting by inhibiting protein synthesis.

EHRET: I feel that you missed the thrust of the point I made earlier, so I will raise it again. A much more direct attack at the temperature compensation problem would be to identify enzyme half-lives and other properties as a function of temperature. Your indirect measures derived from antibiotics can be very deceiving, especially in algae, where different intracellular targets have different thresholds. I see the problem as a question of whether, in temperature compensation, we are observing cellular behavior as a function of the same enzyme for several cycles after a temperature step, or whether we are seeing a new enzyme that appears because of some genetic switch mechanism. Such induction is a phenomenon well known at these temperature ranges, whether it be in protozoan serotypes or in trout enzymes.

BRINKMANN: The fact that it takes at least two days to switch from one type of temperature compensation to the other may suggest that the synthesis of a new generation of enzymes is involved. On the other hand it has been shown that this transfer takes place under a highly reduced state of protein synthesis that is unaffected by Actidione. Thus, I suppose that physical aging (why not the oxidation of some lipid components of the membrane?) is involved rather than the selection of a new generation of enzymes.

EDMUNDS: It might be interesting to examine a paralyzed mutant of *Euglena* if you get a chance. It would be nice to have as a parallel system a mutant— hopefully involving a small mutation— that gives you a nonmotile system versus the existing motile system. There is at least one report of such a paralyzed mutant having been isolated about 8 years ago, but it hasn't been studied further. It is not known just how extensive the mutation is.

HOCHACHKA: If, under the physiological conditions in which you are running your experiments, you could show that enzyme activities as a whole were functioning in a temperature-independent way, would you still have a problem?

BRINKMANN: No, if my model consisted of an enzymic action.

HOCHACHKA: I didn't mean any specific enzyme. If all the enzymes in any cell, as a whole, were—under your experimental conditions—essentially temperature-independent catalysts, would you still have a temperature compensation problem? I wonder if you may not be to some extent looking for a problem that doesn't exist.

HASTINGS: In the cases where it has been looked at, the component enzymes are temperature-dependent.

HOCHACHKA: Are they temperature-dependent under the physiological conditions?

HASTINGS: How do you know how to set up those conditions? You want to make an *in vitro* clock.

HOCHACHKA: That's right; it would be nice.

There is a standing observation in the literature, which we recently pointed out, that enzyme-substrate affinity is inversely related to temperature. So under conditions that occur in most cells, with substrate concentrations at nonsaturating levels, reaction velocities are determined more by enzyme-substrate affinity than by catalytic activity *per se*. And, at least in a number of systems that we have examined, this inverse relationship between temperature and enzyme-substrate affinities is such that enzyme reaction rates are temperature-independent as long as substrate concentrations are at K_m values or lower. In these *Euglena* conditions you have described to me, I would suspect that your enzymes will never be saturated with substrate. Hence these enzyme activities would all be temperature-independent.

BRINKMANN: It depends on what enzymes you mean. Lactic acid dehydrogenase of the old autotrophic cells, for instance, should be highly saturated with substrate.

HOCHACHKA: The statement I'm making is a general one for enzymes with a high affinity constant. For example, in the case of citrate synthetase in the Krebs cycle, the enzyme concentration is so high that the concentration of substrate is considered to be negligible, if free substrate is there at all. So in this case, the catalytic rate is going to be determined strongly by enzyme-substrate affinity. And that, as I pointed out, is inversely related to temperature; so you have a built-in temperature stability mechanism in the enzyme. I know of only two examples of enzymes where this generalization does not hold. For a clock mechanism, it seems to me that enzymes showing this characteristic would be at a tremendous selective advantage through evolutionary time. I'm not certain I understand all the experimental conditions you are using, but again I wonder if you are not looking for a problem that, to *Euglena*, doesn't exist.

BRINKMANN: Well, there is a reason for our not having calculated activity rates of enzymes in *Euglena*. There is no theoretical base defining the activity of an enzyme. We have shown in some of our studies that when you are testing so-called enzyme activities in a crude extract, what you are really measuring is a masking effect of the enzyme-substrate complex, including the action of allosteric affectors. There seems to be no way of showing what activity of an enzyme really is without knowing that. It may even be that such complete understanding is not required because I expect that biological clocks are specialized to act independently of the actual size of intermediary metabolic pools. Your built-in temperature compensation would render the clock operation too sensitive to random fluctuations in the basic metabolism.

In any case, if the enzymes would catalyze in a temperature-compensated way such as you are proposing, it would not yet explain the metabolic control of different types of temperature compensation we are dealing with in the case of *Euglena*.

PAVLIDIS: In a temperature-independent system, how would you explain temperature entrainment?

HASTINGS: That's a question about the oscillator, not about temperature dependence.

BRINKMANN: In a temperature-independent system I would not expect temperature entrainment. Temperature compensation, on the other hand, is a matter of basically temperature-sensitive mechanisms. There are many things I don't know. We have only looked at the correlation between respiration and temperature compensation. I have no suggestions about the underlying system at this point.

LELAND N. EDMUNDS, JR.

Persistent Circadian Rhythm of Cell Division in *Euglena*: Some Theoretical Considerations and the Problem of Intercellular Communication

Cell division in autotrophically grown cultures of *Euglena*, as in numerous other microorganisms, can be synchronized by appropriate 24-hr cycles of light and darkness (Cook and James, 1960; Petropulos, 1963; Edmunds, 1964, 1965a) or of temperature (Pogo and Arce, 1964; Neal *et al.*, 1968; Terry and Edmunds, 1969, 1970a), so that the population approximately doubles every 24 hr. The mechanism by which cell division is synchronized remains obscure.

As a working hypothesis, we have assumed (Edmunds, 1966; Edmunds and Funch, 1969a, b) that an endogenous circadian clock underlies this persistent rhythmicity, although other cyclic mechanisms probably supplement or supplant circadian oscillations in the control of many cell division and mitotic rhythms (see Bruce, 1965). Appropriate *Zeitgeber* would thus entrain the cell division rhythm of individual cells of the population, and under free-running conditions divisions would be timed or "gated" by an endogenous clock operating through biochemical cell cycle controls.

The results reported here (including experiments with a photosynthetic mutant of *Euglena*) substantiate and extend our initial hypothesis and raise three theoretical problems:

Persistent Circadian Rhythm of Cell Division in *Euglena*: Some Theoretical
Considerations and the Problem of Intercellular Communication

595

- What is the significance of a "free-run" under non-constant conditions?
- What is the meaning of variability in the free-running period (τ_{fr})?
- What are the implications of a strongly persistent free-running rhythm with little, if any, decay of synchrony in a population of *Euglena*?

MATERIAL AND METHODS

ORGANISM

The experimental organism used in these studies was *Euglena gracilis* Klebs (strain Z), which has been maintained in this laboratory for over 5 years, and *E. gracilis* var. *bacillaris* strain Z (Pringsheim) P_4 ZUL (furnished by Dr. H. Lyman of the Department of Biological Sciences, State University of New York at Stony Brook). The latter mutant has been isolated and described by Russell and Lyman (1968); it contains somewhat reduced levels of both chlorophyll and carotenoids and normal amounts of the reductive pentose phosphate cycle enzymes, but is unable to carry out the Hill reaction because of a block at or near photosystem II in the photosynthetic electron transport chain. Consequently, it is unable to grow autotrophically on minimal salt medium and must be maintained on heterotrophic medium (see Schiff, Lyman, and Russell, in press).

CULTURE AND ASSAY OF RHYTHM

Wild-type *Euglena* cultures were grown axenically and autotrophically in either 4-liter or 8-liter serum bottles at $25 \pm 0.5°C$ on an aerated inorganic salt medium containing vitamins B_1 and B_{12} (described in detail by Edmunds, 1965a, 1965b; Edmunds and Funch, 1969b). The P_4 ZUL mutant was grown in 4-liter serum bottles at either 25 or 19°C on a defined low pH (3.5) medium containing malic and glutamic acids, as described by Jarrett and Edmunds (1970); thioglycolic acid (5×10^{-5} *M*) was added to improve synchrony. The magnetically stirred cultures were maintained in environmental chambers furnished with clock-controlled, cool-white fluorescent light banks. A miniature fraction collector, a Brewer automatic pipetting machine, and a Coulter electronic cell-counter were used routinely throughout all experiments to assay cell number (Edmunds, 1965a; Edmunds and Funch, 1969b); the onset of cell number increase in the population was taken as the phase reference point (ϕ_r).

RESULTS AND DISCUSSION

BACKGROUND INFORMATION

Substantial evidence supporting our hypothesis of an endogenous circadian oscillation underlying cell division in *Euglena* has already been gathered. A summary of the general cell division patterns in cultures of *Euglena* grown under various light regimes is presented in Table 1. These findings demonstrate that

- Synchronous cell division in a population can be directly and precisely entrained to a 24-hr period by a series of *L*D cycles in which T = 24 hr; *e.g.*, *LD 4*:20; *LD 8*:16; *LD 10*:14; and *LD 14*:10; however, not all cells necessarily divide during any given cycle—as reflected by step-sizes (ss)* less than 2.0 (Edmunds, 1965a; Edmunds and Funch, 1969b);
- Direct entrainment by *L*D cycles of T ≠ 24 hr may also occur; the limits of entrainment for cycles consisting of equal *L* and D fractions are about 20 hr and 28 hr (*LD 10*:10 and *LD 14*:14) (Edmunds, unpublished; Edmunds and Funch, 1969b);
- "Skeletons" of T = 24 hr photoperiods (see Table 1) will entrain the population rhythm to a precise 24-hr period, indicating that the continuous action of light is not required for synchrony (Edmunds and Funch, 1969b);
- Rhythmic cell division will persist for many days under continuous dim illumination (800 lux) and constant temperature (25 ± 0.5°C) with a mean free-running period of 24.2 hr (Edmunds, 1966);
- Under high-frequency LD cycles, such as *LD 2*:6, *LD 1*:3, or *LD 1/4*:1/2, the division rhythm "free runs" with a period of 25–30 hr (Edmunds and Funch, 1969b);
- The division rhythm also "free runs" under "random" lighting regimes in which the duration and sequence of relatively short L and D fractions has been randomly ordered (Edmunds and Funch, 1969a, b).

LOW-FREQUENCY CYCLES

The foregoing analysis has been expanded to include various low-frequency cycles (Table 1). If an *LD 12*:24 regime is imposed on a culture of wild-type *Euglena*, direct entrainment of the division rhythm results, with fission bursts in the population occurring every 36 hr, at the onset of the

---

*Ratio of number of cells per milliliter of culture after the cell division burst to the number of cells per milliliter before the onset of the burst.

Persistent Circadian Rhythm of Cell Division in *Euglena*: Some Theoretical
Considerations and the Problem of Intercellular Communication

597

TABLE 1 Summary of the General Cell Division Patterns Obtained in Cultures of
Euglena gracilis (Z) Grown under Various Illumination Regimes

Type of Regime[a]	Example[a,b]	Results[a]
Complete photoperiod (T = 24 hr)	LD 10:14	Entrainment (τ = 24 hr)
Complete photoperiod (T < 24 hr)	LD 10:10	Entrainment (τ = 20 hr, lower limit)
Complete photoperiod (T > 24 hr)	LD 14:14	Entrainment (τ = 28 hr, upper limit)
Skeleton photoperiod (LD 12:12)	LD 3:6:3:12	Entrainment (τ = 24 hr)
Constant conditions	LL (800 lux), following LD 10:14	Free-run ($\bar{\tau}$ = 24.2 hr)
High frequency	LD 1:3 following LL or LD 10:14	"Free-run" (τ = ~ 24 hr, variable)
Low frequency	LD 12:24 LD 12:36	Entrainment (τ = 36 hr) (τ = 48 hr)
Random[c]	Variable L/variable D following LL or LD	"Free-run" (τ = ~ 24 hr, variable)

[a]It is important to distinguish an LD cycle from the division cycle as observed in the population of cells (some or all of which may divide at about the same time), and from the individual cell cycle, whose period can only be inferred from the size of the division bursts and from their period (τ).

[b]Unless otherwise stated, cool white fluorescent illumination (8,000 lux) was used, and the temperature of the culture was maintained at 25°C.

[c]For practical reasons the following arbitrary protocol was adopted: total duration of L during any given 24-hr time span = 8 hr; total duration of D during any given 24-hr time span = 16 hr; individual L and D fractions ranged between 0.25 hr and 1.0 hr, and 0.5 hr and 1.5 hr, respectively; and the values for the L and D fractions for a given 24-hr time span were written on small slips of paper and were drawn alternately in random fashion from separate containers (Edmunds and Funch, 1969a).

long dark period. The average step-size of these bursts is 1.74 over the concentration range of 5×10^3–2×10^4 cells/ml. Similarly, under an *LD 12*:36 cycle, the rhythm is entrained with a period of 48 hr and the ϕ_r still occurs at the onset of the dark period, with $\overline{ss}$ = 1.79 (Figure 1). In both of the low-frequency experiments, the cultures had been previously synchronized by an *LD 12*:12 cycle.

These driving cycles having long periods are particularly interesting since each provides exactly 12 hr of light per cycle, which is approximately the amount of light needed to furnish enough energy for growth so that the population can double in cell number. (That $\overline{ss}$ < 2 is probably due to maintenance and turnover requirements during the extended dark fractions.) Consequently, it is not clear whether the observed phasing of cell division under these particular regimens is due merely to the imposed regime itself (i.e., a passive oscillation) or whether actual entrainment of an endogenous self-sustaining oscillation has occurred. Indeed, it is even possible that in

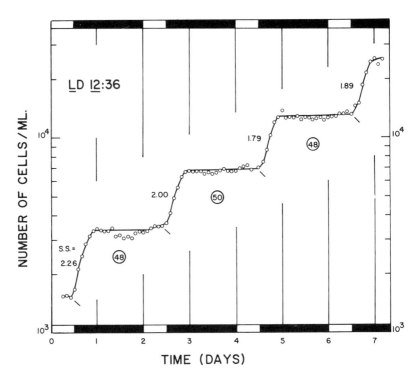

FIGURE 1   Entrainment of the cell division rhythm in a population of *Euglena gracilis* (Z) grown phototrophically on minimal salt medium at 25°C in *LD 12*:36. Ordinate: cell concentration (cells per ml); abscissa: elapsed time (days). Step-sizes (ratio of number of cells per milliliter following a division burst to that just before the onset of divisions) are indicated for the successive division bursts. The period (τ) of the rhythm is also given in hours (circled just to the right of each burst). Note that the average period of the rhythm in the culture is nearly identical to that of the synchronizing cycle and that cell number usually occurs every 48 hr.

the *LD 12*:36 regime (but not in *LD 12*:24) the putative circadian gating oscillation has been entrained to a 24-hr period: a "gate" would open during the middle of the long 36-hr plateau periods of the cell number curve (Figure 1) but since most of the cells would have already divided 24-hr earlier during the preceding gate and no more energy is available, no cells would be able to proceed through this gate.

RESULTS WITH THE PHOTOSYNTHETIC MUTANT

The requirement of light for photosynthesis in the *E. gracilis* (Z) strain in autotrophic culture conditions imposes limitations on the study of synchronization by light. Results are often difficult to interpret since they are

Persistent Circadian Rhythm of Cell Division in *Euglena*: Some Theoretical
Considerations and the Problem of Intercellular Communication

599

confounded by the dual use of light signals—for nutrition and for the phase-shifting and entrainment of a light-sensitive clock. For this reason we used the P_4ZUL mutant, an obligate heterotroph incapable of photosynthesis.

The effects of an *LD 10*:14 cycle on cultures of the mutant grown at two different temperatures are diagrammed in Figure 2. At 25°C exponential growth occurs with a generation time (G.T.) of about 10 hr (i.e., "ultradian" in the terminology of Wille and Ehret, 1968). If the temperature is lowered to 19°C, however, the results are strikingly different: Division in the population becomes synchronous and the fission bursts are apparently entrained to the 24-hr period of the driving cycle. Since $\overline{ss}$ = 1.96 the population approximately doubled during each burst, thereby yielding a G.T. of about 24 hr (i.e., the "circadian–infradian" growth phase discussed by Wille and Ehret).

But synchronization by LD cycles does not of itself provide conclusive evidence that the division cycle is controlled by an endogenous clock. It is possible that the alternating light intervals enhance the photoassimilation of organic molecules from the medium or that they reversibly inhibit cell division, or both, thereby synchronizing the culture. Figure 3 suggests, however, that this is not the case. When a culture grown at 19°C is synchronized by an *LD 10*:14 cycle and subsequently placed in continuous darkness, the cell division rhythm persists for at least 6 days before it damps out because of the onset of the stationary growth phase. Furthermore, the rhythm free-runs with a $\overline{\tau}$ significantly shorter than that of the prior entraining cycle. The period of the individual cell cycles can be inferred to be considerably longer on the basis of the reduced step-sizes, and is given by the expression $n\tau$, where n is an integer and τ is the free-running period observed in the population (Edmunds, 1966; Edmunds and Funch, 1969b).

The effects of a single change in illumination (Figure 4) also support the gating hypothesis. If a mutant culture growing exponentially (G.T. = 26 hr) in DD is suddenly subjected to a "switch-up" in irradiance (5,000 lux) by a single D to L transition, synchronous cell division results with $\overline{\tau}$ = 23 hr. Once again, not all the cells in the population divide during any one circadian cycle ($\overline{ss}$ = 1.88). Apparently either (a) the hypothesized endogenous oscillations in the individual cells were running but asynchronous in DD and were "reset" and synchronized by the D to L transition; or (b) the circadian oscillations were absent in DD and were initiated by the D to L transition. It is surprising that a free-running rhythm was observed in continuous bright illumination of an intensity that quite frequently causes rhythms in other circadian systems (and in autotrophically grown wild-type *Euglena*) to damp out.

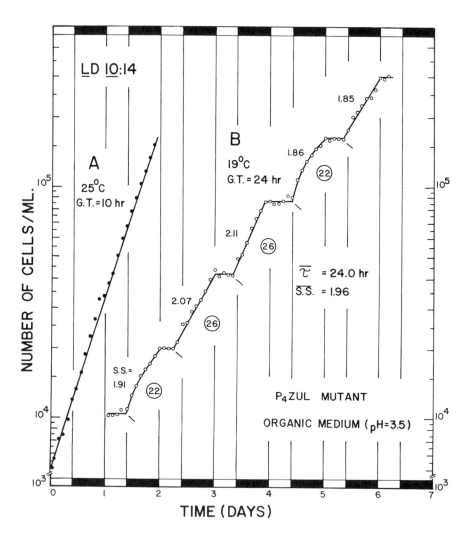

FIGURE 2 Population growth of a photosynthetic mutant (P$_4$ZUL) of *Euglena gracilis* var. *bacillaris* strain Z (Pringsheim) grown on a defined heterotrophic medium in *LD 10*:14. Curve A: exponential increase in cell number (generation time of 10 hr) at 25°C. Curve B: entrainment of the cell division rhythm at 19°C. Other notation as in Figure 1.

Persistent Circadian Rhythm of Cell Division in *Euglena*: Some Theoretical
Considerations and the Problem of Intercellular Communication

601

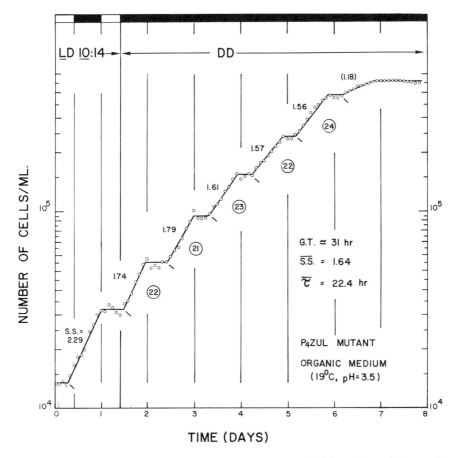

FIGURE 3 Persistent free-running circadian rhythmicity in the cell division rhythm of cultures of
the P_4ZUL mutant of *Euglena* grown heterotrophically at 19°C in continuous darkness. The culture had been previously entrained to a precise 24-hr period by an *LD 10*:14 regime. Other notation
as in Figure 1.

"MIXING" EXPERIMENTS

Preliminary experiments were conducted to test for the possibility of intercellular communication via chemicals released into the medium by the
cells. (See following section.) Two identical cultures of *E. gracilis* (Z) were
grown 10 hr out of phase on minimal medium at 25°C in *LD 10*:14; each
culture was entrained by the imposed cycle as shown in Figure 5 (Panels A
and B). One-liter aliquots were then taken from each culture at 1100 hr—
the beginning of the light phase in Chamber A (Panel A, Figure 5), and the

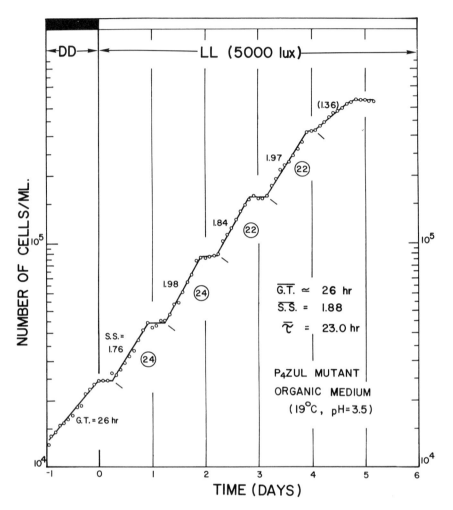

FIGURE 4   Initiation of a persistent free-running circadian rhythm of cell division by a single "step-up" transition (DD to LL) in cultures of the P_4ZUL mutant of *Euglena*. Notation as in Figure 1.

end of the light phase in Chamber B (Panel B, Figure 5)—when the cell concentration in each was about 1.5×10^4 cells/ml. The aliquots were then mixed in a third vessel, which was immediately placed into a third chamber at the onset of the dark phase of an imposed *LD 10*:14 regime; the results are illustrated in Panel C. Cell division occurred at once, just as it did in the control culture (shown in Panel B). The step-size was reduced (ss = 1.39), but this was to be expected since the aliquot from culture A con-

Persistent Circadian Rhythm of Cell Division in *Euglena*: Some Theoretical
Considerations and the Problem of Intercellular Communication

603

tained cells that had just finished dividing. The dotted line represents the
calculated step-size (1.42) based on the performance of the control cultures and on the assumption that no communication occurred; it does not
differ significantly from the step actually observed in the mixed cultures.
Similar results were obtained if the Millipore filtrate of one culture were
mixed with an equal aliquot of whole cells at the same concentration from
a second culture out of phase with the first.

FINAL CONSIDERATIONS AND CONCLUSIONS

Entrainment of the cell division rhythm in *Euglena* by low-frequency LD
cycles and the results previously obtained (Table 1) with skeleton photoperiods (Edmunds and Funch, 1969b), high-frequency cycles (Edmunds
and Funch, 1969b), random lighting regimes (Edmunds and Funch, 1969a),
and constant dim light (Edmunds, 1966) are consistent with the concept
of a circadian oscillation that controls the timing of cell division. This gating oscillation would be similar to that envisioned for the *Drosophila* eclosion system (Pittendrigh, 1966; Skopik and Pittendrigh, 1967) and is assumed to be conceptually distinct from the other events of the developmental sequence making up the cell cycle (Edmunds and Funch, 1969b).

All of this work, however, dealt with photoautotrophically grown
Euglena, and was thus complicated by the possible interaction of nutritional factors with the clock-phasing machinery. Consequently, the discovery of a high-amplitude, persistent circadian rhythm of cell division in
the P_4ZUL photosynthetic mutant of *Euglena* grown on organic medium
in DD was of great importance (Jarrett and Edmunds, 1970). We now will
be able to derive a phase response curve for the mutant without the complication of the cells' utilization of the assaying light signals for growth and
division.

Similar persistent rhythms of cell division have been described in only a
very few other heterotrophic unicells and in somewhat less detail; these include *Tetrahymena pyriformis* (Wille and Ehret, 1968), *Paramecium
bursaria* (Volm, 1964), and *P. multimicronucleatum* (Barnett, 1969). In
addition, Mitchell (in press) reports that he has been able to synchronize
heterotrophically grown, UV-bleached *Euglena* with LD cycles. Finally,
we (Jarrett and Edmunds, 1970) have recently synchronized the hetero-
trophically-grown wild-type *Euglena* (Z strain) using an *LD 10*:14 cycle
under appropriate temperature conditions.

The fact that the P_4ZUL mutant can be synchronized by LD cycles
when grown at 19°C (Figure 2B), but not when grown at 25°C (Figure 2A)

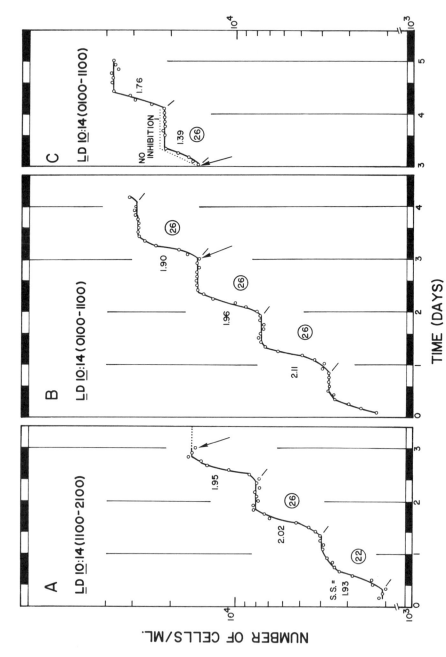

FIGURE 5 Mixing experiment to test for intercellular communication in *E. gracilis* (Z). Numbers in parentheses indicate the times during which the lights are on. Other notation as in Figure 1. See text for experimental procedure.

604

Persistent Circadian Rhythm of Cell Division in *Euglena*: Some Theoretical
Considerations and the Problem of Intercellular Communication

605

seems significant. One of the consequences of the chronon model for circadian clocks (Ehret and Trucco, 1967; Ehret and Wille, 1970) is that
although eukaryotic cells (as well as higher organisms) may possess regulatory capacities for circadian time-keeping, these capacities can be expressed only when the cells are in the circadian–infradian mode of growth
(G.T. $\geqslant$ 24 hr). The ability of exponentially growing cells to enter into a
synchronous circadian rhythm of cell division following asynchronous
growth at an ultradian G.T. ($<$ 24 hr) has been called the "*G-E-T*-effect"
(Wille and Ehret, 1968) after *Gonyaulax* (Sweeney and Hastings, 1958),
Euglena (Edmunds, 1966), and *Tetrahymena* (Wille and Ehret, 1968), all
of which exhibit the capacity for such entrainment.

The data from the P_4 ZUL mutant further support this hypothesis
(Figure 2): as predicted, entrainment did not occur at 25°C when the G.T.
was only 10 hr but was observed at 19°C when the overall G.T. was increased to about 24 hr. Furthermore, a single "switch-up" transition from
DD to LL (5,000 lux) was sufficient to elicit a persistent circadian rhythm
in the mutant, which had been previously growing exponentially (G.T. $\sim$
26 hr) at 19°C. It may well be that the paucity of reports of light-induced
synchrony in heterotrophically grown unicells may be due to the fact that
at the laboratory temperatures commonly used (23 to 25°C), cell division
may take place too rapidly. Finally, evidence (Edmunds, 1965a) of the
breakdown of synchrony in cultures of wild-type *Euglena* (growing photoautotrophically in LD at 25°C with a G.T. of 24 hr) by exposure to bright
light (LL = 3,500 lux), or by the addition of ethanol, acetate, or glutamic
acid to the medium, may also be interpreted by the circadian–infradian
"rule": upon introduction of a utilizable, exogenous carbon source the
G.T. was decreased to 13–15 hr (i.e., a switch from infradian to ultradian
growth mode), and asynchronous exponential growth resulted.

All our observations on the patterns of cell division in cultures of *Euglena*
grown under various illumination regimes have raised three important theoretical questions:

1. The Significance of a "Free-Run" under Nonconstant Conditions

We have shown that numerous high frequency LD cycles (Edmunds and
Funch, 1969b) or random illumination regimes (Edmunds and Funch,
1969a) elicit a persistent circadian rhythm of cell division when imposed
on photoautotrophically grown *E. gracilis* (Z). One might designate this as
a "free-run"; according to the "circadian vocabulary" compiled by Aschoff,
Klotter, and Wever (1965), however, a free-running rhythm is a self-
sustaining oscillation under constant conditions—and here the oscillation

occurs in a fluctuating (random or ordered) LD regime. How are these observations then to be explained?

Under high-frequency cycles, a light-sensitive A-oscillator of the type envisioned by Pittendrigh (1965, 1966) presumably would be immediately phase-shifted by the successive light "signals." But would this result in stable equilibrium and entrainment? The behavior of the *Drosophila pseudoobscura* eclosion rhythm* in *LD 1:3* was calculated from its phase response curve and the accompanying primary assumptions in Pittendrigh's (1965, 1966) theory of entrainment of a circadian rhythm by LD cycles (see also Ottesen, 1965). No matter at what subjective circadian time (CT) the first light pulse, P_1, falls on the phase response curve, the system soon reaches a state in which alternate pairs of successive L pulses (e.g., P_n, P_{n+2}) generate identical phase delays ($-\Delta\phi$); thus, P_n, P_{n+2}, ... each generate $\Delta\phi = -3.3$ hr, while P_{n+1}, P_{n+3}, ... each generate $\Delta\phi = -4.7$ hr. The ultimate result is that after 24 hr of real time has elapsed (equivalent to six repeated *LD 1:3* cycles), a total phase delay of 24 hr has been generated. Of course, the phase response curve is normalized to 24 hr for convenience, and for a non-normalized curve (reflecting a circadian τ), the total $-\Delta\phi$ only approximates 24 hr.

But what is the meaning of this phase delay of approximately 24 hr? If the analysis applies to the cell division system in *Euglena*, it would appear that although high-frequency L signals do indeed affect the putative underlying light-sensitive (A) oscillation, no steady-state analytical equilibrium solution is possible, and no entrainment, as such, can occur; this theoretical analysis is borne out by the empirical findings. One can postulate that the A-oscillation is unable to remain coupled to the hypothetical B-oscillation that more directly underlies the cell division rhythm; the latter oscillation (and the overt rhythm) would then "free run" under these particular nonconstant conditions in a manner similar to that observed in dim LL (Edmunds, 1966). Alternatively, perhaps the two oscillators *are* coupled, and the phase delay of approximately 24 hr generated by the high-frequency LD cycle during a 24-hr time span in some way drives the overt rhythm with a circadian periodicity. One problem, of course, is that the $-\Delta\phi$ is calculated on the basis of a phase response curve derived from the effects of single light perturbations on a true DD free-run. And finally, perhaps

---

*Unfortunately, a phase response curve for the *Euglena* cell division rhythm has not yet been derived because of technical limitations inherent in the autotrophic system. Examination of the phase response curve for the circadian rhythm of phototactic response in *Euglena* that Feldman (1967) has obtained, however, yields results comparable to those described here for the *Drosophila* system.

Persistent Circadian Rhythm of Cell Division in *Euglena*: Some Theoretical
Considerations and the Problem of Intercellular Communication

607

the light-sensitive oscillation is coupled to the B-oscillation but is unable
to respond to the *LD 1*:3 cycle, rendering it effectively free-running.

A similar analysis has been made of the predicted effects on the
Drosophila system of the "random" lighting regimen used with *Euglena*
(Edmunds and Funch, 1969a). Again, regardless of the CT of the first L
pulse, the system eventually reaches a state where each successive signal
impinges on the relatively "dead" portion of the phase response curve and
generates a small phase delay, which is then offset by the progress of real
time. Of course, since the L and D periods were not usually of equal duration, there was some variability in the phase delay engendered with each
L pulse. On the average, however, each L pulse generated $\Delta\phi \simeq -1.3$ hr;
and since there were 18 L pulses in a given 24 hr in the particular regime
presented to the *Euglena* cultures (see Edmunds and Funch, 1969a), the
total $-\Delta\phi$ per 24 hr amounted to approximately 24 hr on the basis of calculations made using the *Drosophila* response curve. It is plausible, therefore, to expect a circadian periodicity under a random lighting regime, and
one is indeed observed in the *Euglena* system.

2. The Meaning of Variability in the "Free-Running Period" (τ)

The data for the circadian cell division rhythm in *Euglena* exposed to high
frequency and random illumination regimes (Edmunds and Funch, 1969b),
have a surprisingly high degree of variability in the successive τ values in
any one experiment. These values range, for example, from 22 to 28 hr for
an *LD 1*:3 cycle and from 24 to 32 hr for an *LD 2*:4 cycle. Similarly, τ may
range from 26 to 30 hr in a random regime. Although there is usually no
discernible trend in the changing values of τ, in one case (*LD 4*:4) we have
found (Edmunds and Funch, 1969b, and unpublished) that τ appears to
vary cyclically between 30 and 40 hr (see Pavlidis, 1969a; Swade, 1969;
Eskin, this volume). This phenomenon has not yet been adequately explained.

One might reasonably expect some variation in τ in the random lighting
regimen, but not in the fixed, high-frequency cycles. Variation could be
the result of shifting subpopulations within the culture. It is also conceivable that the continually decreasing effective light intensity as the population increases during batch culture* may influence τ. Finally, variation in τ
could be due to intercellular communication.

---

*We have not yet performed these particular experiments using our semicontinuous
dilution techniques (Terry and Edmunds, 1969).

3. Implications of a Strongly Persistent Free-Running Rhythm in a Population

We have reported that the cell division rhythm in *Euglena* persists un-damped for as long as 10 days in dim LL at 25°C (Edmunds, 1966). More recently (Edmunds *et al.*, 1971) we have found that in a low-temperature (14°C) batch culture of the P_4ZUL mutant the rhythm of cell division persists for at least 21 days in DD. The rhythm also persists for extended periods of time in both the mutant and the wild-type grown on a low pH organic medium in DD at 19°C in semicontinuous culture. Similar observa-tions have been made for several rhythms in *Gonyaulax polyedra* (Sweeney and Hastings, 1958; McMurry, 1971). The rhythm of phototactic response has been observed to persist in *Euglena* in DD for 14 days (Feldman, 1967), while the motility rhythm in the same organism persists for months (Brinkmann, 1966, this volume; Terry and Edmunds, 1970b). Why, then, does cell division synchrony not appear to decay in any reasonable amount of time in a cell population under constant conditions due to the variation in individual cell doubling times arising from both extrinsic factors (fluctu-ations in the cellular environment) and of intrinsic factors (heterogeneity in karyotype and residual variations stemming from stochastic processes in the cell cycle) (Engelberg, 1964; Edmunds and Funch, 1969b)?

If we assume that no subtle geophysical factors are phasing the popula-tion, then we are left with two alternatives: either (a) the period of the free-running oscillation in a cell must be not only improbably precise, but also almost identical to that of the oscillations in other cells; or (b) some sort of intercellular communication must occur that maintains synchrony within the population of self-sustaining oscillators. Conceivably, the ob-served population synchrony is engendered solely by interactions among the cells, but this seems rather unlikely in view of the demonstrations of circadian rhythms in individual cells isolated from a population (Sweeney, 1960; Hastings and Sweeney, 1964).

The hypothesis of intercellular communication has been examined theo-retically by several workers. Wever (1965), Winfree (1967), and Pavlidis (1969b) have formally treated populations of weakly and strongly coupled (i.e., interacting) oscillators, while Goodwin (1963, 1966) has considered the biochemical mechanisms whereby control signals (e.g., metabolites, re-pressors) from one cell could phase the oscillations in neighboring cells by entering into their control circuits. Indeed, Goodwin has predicted (1967) that synchrony might be expected to develop spontaneously in the oscilla-tions of enzyme activities in cells cultured continuously in a chemostat.

Experimental tests of this hypothesis, however, have been inconclusive

Persistent Circadian Rhythm of Cell Division in *Euglena*: Some Theoretical
Considerations and the Problem of Intercellular Communication

609

or negative. Hastings and Sweeney (1958) found no evidence of cellular
interaction upon mixing two cultures of *Gonyaulax* whose free-running
rhythms of luminescent capacity were 5 hr out of phase with one another.
Similar negative evidence has been reported by Brinkmann (1966) for the
motility rhythm of *Euglena*. Finally, the experiments reported in this paper
do not suggest cellular "cross talk" in two cultures of rhythmically dividing *Euglena* that were 10 hr out of phase (but not free-running). On the
other hand, Pye (1969; this volume) reports that strong metabolic coupling exists among individual yeast cells (*Saccharomyces carlsbergensis*) in
a population; the glycolytic oscillations that were assayed, however, although stable and self-sustained, differ from circadian systems in that their
frequency *in vivo* is on the order of 1.7/min.

Despite these largely negative findings, we are proceeding on the premise
that some sort of interaction will be found. It cannot be emphasized too
strongly that the experimental conditions are critical: The proper cell concentration must be attained (threshold effects) and then maintained, pref-
erably through continuous culture; the appropriate phase angles between
the test cultures must be determined; the possible light-sensitivity and
decay of control molecules must be taken into account; statistical techniques must be developed for analysis of the data. In any case it seems
probable that cellular interaction, if it does exist, merely modulates the
circadian output of self-sustained oscillations in the individual cells of a
population and does not in some way constitute the clock mechanism itself.

ACKNOWLEDGMENTS

This work represents the concerted efforts of my laboratory; particular
acknowledgment should be made of the contributions of R. M. Jarrett and
F. M. Sulzman to the mutant studies and mixing experiments, respectively.
I thank L.-W. Chuang for her expert technical assistance. These efforts
were supported by NSF research grants #GB-4140, #GB-6892, and #GB-
12474, and SUNY/Research Foundation Grant in Aid #31-7150A to
L. Edmunds.

REFERENCES

Aschoff, J., K. Klotter, and R. Wever. 1965. Circadianer Wortschatz, p. X to XIX. *In*
    J. Aschoff [ed.], Circadian clocks. North-Holland Publ. Co., Amsterdam.
Barnett, A. 1969. Cell division: a second circadian clock system in *Paramecium multi-
    micronucleatum*. Science 164:1417–1419.

Brinkmann, K. 1966. Temperatureinflüsse auf die circadiane Rhythmik von *Euglena gracilis* bei Mixotrophie und Autotrophie. Planta 70:344–389.

Bruce, V. G. 1965. Cell division rhythms and the circadian clock, p. 125 to 138. *In* J. Aschoff [ed.], Circadian clocks. North-Holland Publ. Co., Amsterdam.

Cook, J. R., and T. W. James. 1960. Light-induced division synchrony in *Euglena gracilis* var. *bacillaris*. Exp. Cell Res. 21:583–589.

Edmunds, L. N., Jr. 1964. Replication of DNA and cell division in synchronously dividing cultures of *Euglena gracilis*. Science 145:266–268.

Edmunds, L. N., Jr. 1965a. Studies on synchronously dividing cultures of *Euglena gracilis* Klebs (strain Z). I. Attainment and characterization of rhythmic cell division. J. Cell. Comp. Physiol. 66:147–158.

Edmunds, L. N., Jr. 1965b. Studies on synchronously dividing cultures of *Euglena gracilis* Klebs (strain Z). II. Patterns of biosynthesis during the cell cycle. J. Cell. Comp. Physiol. 66:159–182.

Edmunds, L. N., Jr. 1966. Studies on synchronously dividing cultures of *Euglena gracilis* Klebs (strain Z). III. Circadian components of cell division. J. Cell Physiol. 67:35–44.

Edmunds, L. N., Jr., and R. Funch. 1969a. Circadian rhythm of cell division in *Euglena*: effects of a random illumination regimen. Science 165:500–503.

Edmunds, L. N., Jr., and R. Funch. 1969b. Effects of "skeleton" photoperiods and high frequency light-dark cycles on the rhythm of cell division in synchronized cultures of *Euglena*. Planta 87:134–163.

Edmunds, L. N., Jr., L. Chuang, R. M. Jarrett, and O. W. Terry. 1971. Long-term persistence of free-running circadian rhythms of cell division in *Euglena* and the implication of autosynchrony. J. Interdisc. Cycle Res. 2:121–132.

Ehret, C. F., and E. Trucco. 1967. Molecular models for the circadian clock. I. The chronon concept. J. Theor. Biol. 15:240–262.

Ehret, C. F., and J. J. Wille. 1970. The photobiology of circadian rhythms in protozoa and other eukaryotic organisms, p. 369–416. *In* P. Halldal [ed.], Photobiology of microorganisms. John Wiley & Sons, New York.

Engelberg, J. 1964. The decay of synchronization of cell division. Exp. Cell Res. 36:647–662.

Feldman, J. F. 1967. Biochemical and physiological studies on the circadian clock of *Euglena*. Ph.D. Thesis. Princeton University.

Goodwin, B. C. 1963. Temporal organization in cells. Academic Press, New York.

Goodwin, B. C. 1966. An entrainment model for timed enzyme syntheses in bacteria. Nature (Lond.) 209:479–481.

Goodwin, B. C. 1967. Biological control processes and time. Ann. N.Y. Acad. Sci. 138:748–758.

Hastings, J. W., and B. M. Sweeney. 1958. A persistent diurnal rhythm of luminescence in *Gonyaulax polyedra*. Biol. Bull. 115:440–458.

Hastings, J. W., and B. M. Sweeney. 1964. Phased cell division in the marine dinoflagellates, p. 307 to 321. *In* E. Zeuthen [ed.], Synchrony in cell division and growth. Interscience Division, John Wiley & Sons, New York.

Jarrett, R. M., and L. N. Edmunds, Jr. 1970. Persisting circadian rhythm of cell division in a photosynthetic mutant of *Euglena*. Science 167:1730–1733.

McMurry, L. 1971. Studies on properties and biochemistry of circadian rhythms in the bioluminescent dinoflagellate, *Gonyaulax polyedra*. Ph.D. thesis. Harvard University.

Mitchell, J. L. A. Photoinduced division synchrony in permanently bleached *Euglena gracilis*. Planta (in press).

Neal, W. K., E. A. Funkhouser, and G. A. Price. 1968. Large-scale temperature-induced synchrony of cell division in *Euglena gracilis*. J. Protozool. 15:761–763.

Persistent Circadian Rhythm of Cell Division in *Euglena*: Some Theoretical
Considerations and the Problem of Intercellular Communication

611

Ottesen, E. A. 1965. Analytical studies on a model for the entrainment of circadian systems. Senior Thesis. Princeton University.

Pavlidis, T. 1969a. An explanation of the oscillatory free-runs in circadian systems. Am. Nat. 103:31–42.

Pavlidis, T. 1969b. Populations of interacting oscillators and circadian rhythms. J. Theor. Biol. 22:418–436.

Petropulos, S. F. 1963. Physiological basis of synchronous division. In *Euglena*. Ph.D. Thesis. Princeton University.

Pittendrigh, C. S. 1965. On the mechanism of the entrainment of a circadian rhythm by light cycles, p. 277 to 297. *In* J. Aschoff [ed.], Circadian clocks. North-Holland Publ. Co., Amsterdam.

Pittendrigh, C. S. 1966. The circadian oscillation in *Drosophila pseudoobscura* pupae: a model for the photoperiodic clock. Z. Pflanzenphysiol. 54:275–307.

Pogo, A. O., and A. Arce. 1964. Synchronization of cell division in *Euglena gracilis* by heat shock. Exp. Cell Res. 36:390–397.

Pye, E. K. 1969. Biochemical mechanisms underlying the metabolic oscillations in yeast. Can. J. Bot. 47:271–285.

Russell, G. K., and H. Lyman. 1968. Isolation of mutants of *Euglena gracilis* with impaired photosynthesis. Plant Physiol. 43:1284–1290.

Schiff, J. A., H. Lyman, and G. K. Russell. The isolation of mutants of *Euglena gracilis*. *In* A. San Pietro [ed.], Methods in enzymology. Academic Press, New York. In press.

Skopik, S. D., and C. S. Pittendrigh. 1967. Circadian systems. II. The oscillation in the individual *Drosophila* pupa; its independence of developmental stage. Proc. Natl. Acad. Sci. U.S. 58:1862–1869.

Swade, R. H. 1969. Circadian systems in fluctuating light cycles: toward a new model of entrainment. J. Theor. Biol. 24:227–239.

Sweeney, B. M. 1960. The photosynthetic rhythm in single cells of *Gonyaulax polyedra*. Cold Spring Harbor Symp. Quant. Biol. 25:145–148.

Sweeney, B. M., and J. W. Hastings. 1958. Rhythmic cell division in populations of *Gonyaulax polyedra*. J. Protozool. 5:217–224.

Terry, O. W., and L. N. Edmunds, Jr. 1969. Semi-continuous culture and monitoring system for temperature-synchronized *Euglena*. Biotechnol. Bioeng. 11:745–756.

Terry, O. W., and L. N. Edmunds, Jr. 1970a. Phasing of cell division by temperature cycles in *Euglena* cultured autotrophically under continuous illumination. Planta 93:106–127.

Terry, O. W., and L. N. Edmunds, Jr. 1970b. Rhythmic settling induced by temperature cycles in continuously-stirred autotrophic cultures of *Euglena gracilis* (Z strain). Planta 93:128–142.

Volm, M. 1964. Die Tagesperiodik der Zellteilung von *Paramecium bursaria*. Z. Vgl. Physiol. 48:157–180.

Wever, R. 1965. Eizelorganismen und Populationen im circadianen Experiment. Eine methodische Analyse. Z. Vgl. Physiol. 51:1–24.

Wille, J. J., and C. F. Ehret. 1968. Light synchronization of an endogenous circadian rhythm of cell division in *Tetrahymena*. J. Protozool. 15:785–788.

Winfree, A. T. 1967. Biological rhythms and the behavior of populations of coupled oscillators. J. Theor. Biol. 16:15–42.

THÉRÈSE VANDEN DRIESSCHE

Structural and Functional Rhythms in the Chloroplasts of *Acetabularia*: Molecular Aspects of the Circadian System

The unicellular alga *Acetabularia mediterranea* displays a number of endogenous circadian rhythms; all thus far investigated are associated with the chloroplasts. In the first part of this paper, these rhythms are described and the molecular basis of the overt processes discussed in the light of particular experiments. In this connection, the possibility that two or more overt processes are various manifestations of the same rhythm is examined. All overt rhythms apparently involve proteins, and it is likely that the conformational state of the protein is significant to the rhythmic process. The second part of the paper presents experimental evidence on the basic mechanism of rhythmicity in *Acetabularia*. The discussion concludes with an analysis of the possible involvement of transcription and translation in rhythmicity in this species.

TEMPORAL MORPHOLOGY

The temporal morphology of *Acetabularia* is graphically represented in two time-maps (Figures 1 and 2). To facilitate comparison, all values are calculated as percent of the mean. The following rhythms have been described in *Acetabularia* under a light/dark cycle:

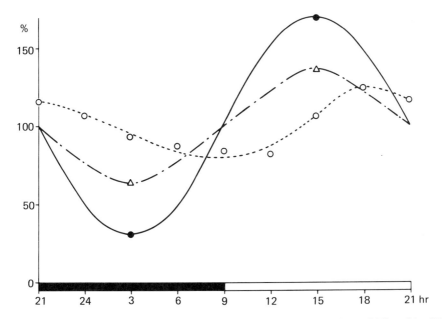

FIGURE 1   Time-map of *Acetabularia*: 1   Rhythm in photosynthetic capacity: solid line with solid circles. Rhythm in chloroplast shape: broken line with open triangles. Rhythm in RNA synthesis: dashed line and open circles. The relative value of the two first rhythms were calculated from aliquot samples in a representative experiment reported in Vanden Driessche (1966a). The relative value of the RNA-synthesis rhythm was calculated from a typical experiment reported in Vanden Driessche and Bonotto (1969).

Photosynthesis   Photosynthesis and photosynthetic capacity oscillate in parallel (Terborgh and McLeod, 1967). The maximum of both rhythms occurs at the middle of the light period (Richter, 1963). In Figure 1 a typical experiment on photosynthetic capacity is represented.

ATP content of the chloroplasts   Surprisingly, the ATP content of the sedimentable fraction of *Acetabularia* cells has been found to be significantly higher 5 hr before the midpoint of the light phase than it is at the midpoint itself (Vanden Driessche, 1970).

Choloroplast shape   The chloroplasts of *Acetabularia mediterranea* vary in shape with the time of day: They are elongated in the middle of the light period and more spherical in the middle of the dark period, as quantified by the distribution curves of the ratios between axes (Figure 1) (Vanden Driessche, 1966a).

RNA synthesis   The incorporation of ^{3}H-uridine in the acid-insoluble fraction of *Acetabularia* cells oscillates in a circadian manner, with the

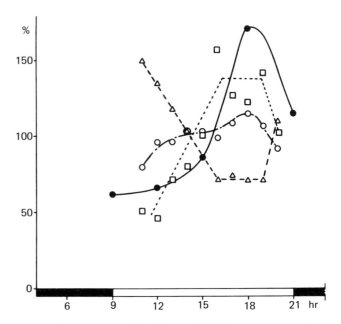

FIGURE 2   Time-map of *Acetabularia*: 2   Rhythm in the polysaccharide
content of the chloroplasts: solid line with solid circles. Rhythm in the
number of chloroplastic granules: chloroplasts with one granule—broken
line with open triangles; chloroplasts with two granules—broken line with
open circles; chloroplasts with three or more granules—dotted line with
open squares. The relative values of the rhythm in polysaccharide content
are from previously unpublished data. The relative values of the number
of granules per chloroplast are derived from the work of Puiseux-Dao and
Gilbert (1967), where the original values are the means of three experimental series.

peak occurring at about 3 hr after the middle of the light period (Figure 1)
(Vanden Driessche and Bonotto, 1969). Under the conditions of these experiments, the site of RNA synthesis is the chloroplasts.

   Chloroplastic polysaccharides   The polysaccharide content of the
chloroplasts varies with the time of the day, with the maximum about
3 hr after the middle of the light period (Figure 2) (Vanden Driessche and
Bonotto, 1968).

   Number of carbohydrate granules in the chloroplasts; plastid division
Puiseux-Dao and Gilbert (1967) have observed a diurnal variation in percentage of chloroplasts containing different numbers of carbohydrate
granules (Figure 2). These granules afford a ready estimate of the number
of "plastidial units," one of the authors having shown that some chloroplasts contain several structural and functional units (Puiseux-Dao, 1966).

Structural and Functional Rhythms in the Chloroplasts of
Acetabularia: Molecular Aspects of the Circadian System
615

The percentage of chloroplasts containing one granule decreases during
the first hours of illumination; at the same time, the percentage of chloroplasts with two and the percentage with three or more granules both increase. The three curves level off, then show a dramatic change: The percentage of chloroplasts containing one granule increases, while the converse
is observed for the other curves. This latter inflection point corresponds
to the time at which chloroplast division takes place.

All these rhythms have been observed in enucleate as well as in whole
algae except that in number of chloroplast granules, which has not yet
been examined in enucleate cells. All have been found to be endogenous
except that of chloroplast granules, which has been studied only under
LD conditions.

MOLECULAR BASIS OF THE OVERT RHYTHMIC FUNCTIONS

Terborgh and McLeod (1967) have recently shown that, in *Acetabularia
crenulata*, the variations in photosynthetic activity must result from con-
comitant variations in both light and dark reactions. However, no variation has been found in the activity of ribulose diphosphocarboxylase or
in the activity of eight other enzymes of the Calvin cycle (Hellebust,
Terborgh, and McLeod, 1967). A variation in the dark reaction, however,
could be the result of variations in activity of enzymes that have not yet
been examined or that behave differently *in vivo* and *in vitro*. Alternatively,
the rhythm could depend on a variation in the coupling between the light
and dark reactions, as suggested by Terborgh and McLeod (1967). Modification of the coupling is likely to be caused by modifications in or between
the structural subunits of the chloroplasts of which proteins are an important component. These modifications could also produce variation in the
light reaction—i.e., in the rate of energy transfer.

Whatever the underlying mechanism, it seems probable that the rhythm
in photosynthesis is expressed through proteins. This working hypothesis
is supported by experiments with inhibitors of protein synthesis both in
Sweeney's laboratory (Sweeney, Tuffli, and Rubin, 1967) and in the
author's (Vanden Driessche, 1967a): chloramphenicol, puromycin, and
cycloheximide severely reduce the amplitude of the rhythm and lower the
base line within 2 days, but affect neither the phase nor the period of the
rhythm. The basic oscillator does not seem sensitive to these inhibitors,
but the overt process, the photosynthetic rhythm, is affected and thus
must be expressed either through structural proteins or enzymes.

In connection with hypothesized variations in structural proteins, it has been found by Changeux *et al.* (1967) that biological membranes possess properties of cooperativity and that they differ from the allosteric systems in solution by a lattice constraint only. Many enzymatic proteins have been found to be also allosteric.

Rhythms in photosynthesis and in chloroplast shape are correlated (Vanden Driessche, 1966b). As in the case of mitochondria (Hackenbrock, 1968), electron transfer could result in mechanical work and possibly in a modification of the supramolecular organization of the organelle. Alternatively, discrete distortions at the molecular level might be amplified and become apparent at the substructural level. In either scheme, the amino acid sequence (which conditions the flexibility of the membrane) is of extreme importance. A small difference in sequence could explain why the circadian variation in shape appears in one species and not in another. The rhythm in shape of the chloroplasts observed in *A. mediterranea* necessarily involves proteins.

The rhythm in ATP content of the sedimentable fraction may be a result of the interaction between the rhythm in photosynthesis (generating ATP) and one or more other rhythms in other functions that use ATP to a greater extent than the increment resulting from photosynthesis, and more rapidly than the regulatory processes involving transport of ATP in the cytoplasm or transformation by adenylate kinase.

The relation of the rhythm in polysaccharide content to the rhythm in photosynthesis is still an open question. There are, however, indications suggesting that the polysaccharide rhythm could involve an autonomous oscillation. Several other systems display similar oscillations in their carbohydrate content, the most conspicuous being that in liver glycogen (Halberg *et al.*, 1961). These variations are accounted for by different enzymes, depending on the system.

The rhythm in RNA synthesis is apparently independent of the other rhythms examined; it damps much more rapidly than the others in continuous light of the same intensity as that given during the light period of a normal light–dark cycle. Rhythmicity in RNA synthesis has been previously described in several cases, including mammals (Halberg *et al.*, 1961; Mayersbach, 1967; Rückbeil, 1961), but its oscillating components remain unknown. The facts that RNA synthesis by the chloroplasts in the light is inducible (Brawerman, Pogo, and Chargaff, 1962) and that inducible enzymes are usually allosteric (Monod, 1966) support the view that allosteric proteins might be responsible for rhythmicity in RNA synthesis.

It appears, then, that all the overt rhythmic processes in *Acetabularia* involve proteins; it is possible that allostery is particularly important.

BASIC MECHANISM OF RHYTHMICITY

Actinomycin D specifically inhibits transcription by binding to the guanine residues of DNA. It strongly interferes with the rhythmicity of whole algae: At a concentration of 0.27 μg ml^{-1}, actinomycin reduces the amplitude of the rhythm in photosynthesis within about 6 days and totally abolishes the rhythm in about 2 weeks. Apparently it takes that long for the RNA previously formed to be used up, and no new RNA is synthesized (Vanden Driessche, 1966b). Rifampicin also inhibits transcription; it competitively inhibits the binding of the first nucleotide to the RNA-polymerase (Di Mauro *et al.*, 1969). But rifampicin is fairly specific to bacterial systems, affecting mammalian RNA-polymerase only slightly at much higher concentrations (Wehrli *et al.*, 1968).

Rifampicin has been found to exert a clear-cut effect on the RNA synthesis associated with the chloroplasts in *Acetabularia* (Bonotto, Janowski, and Vanden Driessche, 1968): a 90-min treatment with rifampicin at a concentration of 50 μg ml^{-1} produces 75 percent inhibition. In algae treated for much longer times—24 hr or 3 days at concentrations of rifampicin varying from 1 to 20 μg ml^{-1}—it has not been possible to detect any effect on the rhythm (Table 1). Therefore, the RNA implicated in rhythmicity must be nuclear in origin.

Further support for this suggestion comes from experiments examining the effects of RNAse (Vanden Driessche, 1966b), although the direct effects of RNAse are difficult to determine since photosynthesis itself decreases sharply. The concentration of RNAse used was 0.1 mg ml^{-1}. When RNAse-treated anucleate algae are returned to normal seawater, they do not regain rhythmicity in photosynthesis. When similarly treated whole algae are transferred to normal seawater, photosynthesis is not only resumed, but the amplitude of the rhythm is significantly higher than in controls never exposed to RNAse, provided that the recovery period has been sufficiently long.

This interesting result should be examined in the light of the following experiments: Algae were sectioned in different ways—by removing and discarding only the apex, by discarding half of the stalk, or by removing the whole stalk and leaving only the rhizoid that contains the nucleus. This fragmentation was found to stimulate the amplitude of the rhythm in photosynthetic capacity—the smaller the fragment remaining after sectioning, the greater the stimulation (Vanden Driessche, 1967b). Apparently, when the number of molecules of the specific mRNA is decreased, a compensatory mechanism stimulates its synthesis, probably by derepressing specific nuclear genes.

TABLE 1 Effect of Rifampicin on the Photosynthetic Capacity of *Acetabularia*

Experiment	Controls	1 μg ml^{-1}, 24 hr	10 μg ml^{-1}, 24 hr	10 μg ml^{-1}, 3 D	20 μg ml^{-1}, 24 hr
I	11.78	13.27	12.62		
	18.74	20.26	20.01		
	1.6	1.5	1.6		
II	24.72		29.95		
	33.58		31.78		
	1.4		1.1		
III	19.28	17.79	20.48		17.47
	28.35	31.10	26.62		26.95
	1.5	1.7	1.3		1.5
IV	15.46		14.93		
	18.81		19.09		
	1.2		1.3		
V	14.94			11.32	
	17.70			17.53	
	1.2			1.55	

Note: The three values in each group represent (top) the mean photosynthetic rate (μl of O$_2$ evolved by 25 algae in 30 min) at 5 hr before the midpoint of the light period; (middle) the mean photosynthetic rate at the midpoint of the light period; (bottom) the ratio of these two rates.

The experiments outlined above support the hypothesis that a nuclear mRNA controls rhythmicity in the chloroplasts. However, anucleate algae retain their rhythm in photosynthetic capacity. Even 40 days after enucleation, the rhythm is expressed and is undamped (Sweeney and Haxo, 1961; Schweiger, Walraff, and Schweiger, 1964; Vanden Driessche, 1966b). The difference in persistence of the rhythm between anucleate *Acetabularia* and whole *Acetabularia* treated with actinomycin D could be accounted for by a difference in the turnover rate of the specific mRNA involved in rhythmicity: the turnover could be much higher in whole cells—taking about two weeks to deplete the supply—than in anucleate cells, where the rate could be essentially zero. [Two distinct experimental approaches produced results entirely consistent with this view: test of the effect of actinomycin D on anucleate *Acetabularia*, and grafting experiments between R$^+$ and R$^-$ strains (Sweeney *et al.*, 1967; Vanden Driessche, 1966b and 1967a).]

CONCLUSIONS AND SPECULATIONS

The results discussed above are schematized in Figure 3. In the classical DNA to protein chain, the two crucial steps have been tested for their bearing on rhythmicity. In *Acetabularia*, a nuclear R N A has a central role in the primary oscillation. However, the persistence of rhythmicity in anucleate algae demonstrates that daily transcription from nuclear D N A is not necessary for the oscillation to continue. Nor is daily transcription from chloroplastic D N A a condition for rhythmicity, as the rifampicin experiments make clear. Specific R N A synthesis is regulated, in part by feedback, by the quantity of specific R N A molecules.

All available evidence suggests that the overt oscillators involve proteins. However, daily translation cannot account for the circadian organization of *Acetabularia*, since translation inhibitors such as chloramphenicol or puromycin, while they reduce the amplitude of the rhythms, do not abolish them altogether.

For the time being, a general molecular mechanism does not emerge from the wealth of experimental data obtained with various species. But when a general model does emerge, *Acetabularia* will be of particular interest since the molecules involved in the primary oscillation apparently do not have to be synthesized as frequently as in other systems. The fact that the lifetime of R N A's is greater than 24 hr suggests that the oscillating mechanism might involve transitions in populations of allosteric enzymes or allosteric proteins in the thylakoids. These transitions could be effected by pre-existing inducers or repressors that might or might not be specific for the particular protein. If this is the case, some mechanism must exist to bind and release the allosteric modifiers from the master oscillator. The oscillator itself, however, remains a "black box," since we

FIGURE 3   Bearing of transcription and translation on the photosynthetic rhythm in *Acetabularia*. From the nuclear D N A (DNA n), long-lived specific R N A(s) is (or are) transcribed; the synthesis can be stimulated by decreasing the number of specific R N A molecules. Protein(s) of the chloroplast embody the rhythm; one of them is represented (protein c). When protein synthesis is impeded, the photosynthesis is depressed, although the rhythm is still expressed.

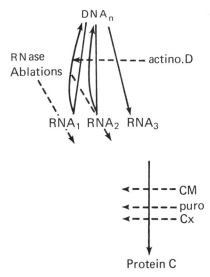

do not yet understand how RNA—probably complexed with other molecules—operates and provides circadian organization.

REFERENCES

Bonotto S., M. Janowski, and T. Vanden Driessche. 1968. Effet spécifique de la rifampicine sur la synthèse du RNA chez *Acetabularia mediterranea*. Arch. Int. Physiol. Biochim. 76:919–920.

Brawerman, G., A. O. Pogo, and E. Chargaff. 1962. Induced formation of ribonucleic acids and plastid protein in *Euglena gracilis* under the influence of light. Biochim. Biophys. Acta 55:326–334.

Changeux, J.-P., J. Thiéry, Y. Tung, and C. Kittel. 1967. On the cooperativity of biological membranes. Proc. Natl. Acad. Sci. U.S. 55:335–341.

Di Mauro, E., L. Snyder, P. Marino, A. Lamberti, A. Coppo, and G. P. Tocchini-Valentini. 1969. Rifampicin sensitivity of the components of DNA-dependent RNA polymerase. Nature (Lond.) 222:533–537.

Hackenbrock, C. R. 1968. Ultrastructural bases for metabolically linked mechanical activity in mitochondria. II. Electron transport-linked ultrastructural transformations in mitochondria. J. Cell Biol. 37:345–369.

Halberg, F., E. Halberg, C. P. Barnum, and J. J. Bittner. 1961. Physiologic 24-hour periodicity in human beings and mice, the lighting regimen and daily routine, p. 803 to 877. *In* R. R. Withrow [ed.], Photoperiodism and related phenomena in plants and in animals. 2nd ed. American Association for the Advancement of Science, Washington, D.C.

Hellebust, J. A., J. Terborgh, and G. C. McLeod. 1967. The photosynthetic rhythm of *Acetabularia crenulata*. II. Measurements of photoassimilation of carbon dioxide and the activities of enzymes of the reductive pentose cycle. Biol. Bull. 133:670–678.

Mayersbach, H. von. 1967. Seasonal influences on biological rhythms of standardized laboratory animals, p. 87 to 99. *In* H. von Mayersbach [ed.], The cellular aspects of biorhythms. Springer, Berlin.

Monod, J. 1966. From enzymatic adaptation to allosteric transitions. Science 154:475–482.

Puiseux-Dao, S. 1966. L'ultrastructure et la division des plastes chez l'*Acetabularia mediterranea*. Dasycl. Electron Microsc., Proc. 6th Int. Congr., Kyoto, p. 377.

Puiseux-Dao, S. J., and A. M. Gilbert. 1967. Rhythme de replication de l'unité plastidiale chez l'*Acetabularia mediterranea* placée dans diverses conditions d'éclairement. C. R. Acad. Sci., Ser. D. 261:870–873.

Richter, G. 1963. Die Tagesperiodik der Photosynthese bei *Acetabularia* und ihre Abhängigkeit von Kernaktivität RNS und Protein synthese. Z. Naturforsch. B. 18:1085–1089.

Ruckbeil, A. 1961. Untersuchungen zur Tagesperiodiziteit der *32 p* Einlagerung in die Verbindungen der Nucleisäurefraction. Z. Bot. 49:1–22.

Schweiger, E., H. G. Walraff, and H. G. Schweiger. 1964. Über tagesperiodische Schwankungen der Sauerstoffbilanz kernhaltiger und kernloser *Acetabularia mediterranea*. Z. Naturforsch. B. 19:499–505.

Sweeney, B. M., and F. T. Haxo. 1961. Persistence of a photosynthesis rhythm in enucleated *Acetabularia*. Science 134:1361–1363.

Sweeney, B. M., C. F. Tuffli, Jr., and R. H. Rubin. 1967. The circadian rhythm in photosynthesis in *Acetabularia* in the presence of actinomycin D, puromycin and chloramphenicol. J. Gen. Physiol. 50:647–659.

Terborgh, J., and G. C. McLeod. 1967. The photosynthetic rhythm of *Acetabularia crenulata*. I. Continuous measurement of oxygen exchange in alternating light-dark regimes and in constant light. Biol. Bull. 133:659–669.

Vanden Driessche, T. 1966a. Circadian rhythms in *Acetabularia*: photosynthetic capacity and chloroplast shape. Exp. Cell Res. 42:18–30.

Vanden Driessche, T. 1966b. The role of the nucleus in the circadian rhythms of *Acetabularia mediterranea*. Biochim. Biophys. Acta 126:456–470.

Vanden Driessche, T. 1967a. Experiments and hypothesis on the role of RNA in the circadian rhythm of photosynthetic capacity in *Acetabularia mediterranea*. Nachr. Akad. Wiss. Göttingen, Mathe.-Phys. Kl. II. 10:108–109.

Vanden Driessche, T. 1967b. The nuclear control of the chloroplasts circadian rhythms. Sci. Prog. (Lond.) 55:293–303.

Vanden Driessche, T., and S. Bonotto. 1968. Le rhythme circadian de la teneur en inuline chloroplastique de "*Acetabularia mediterranea*." Arch. Int. Physiol. Biochim. 76:205–206.

Vanden Driessche, T., and S. Bonotto. 1969. The circadian rhythm in RNA synthesis in *Acetabularia mediterranea*. Biochim. Biophys. Acta 179:58–66.

Wehrli, W., J. Nüesch, F. Knüsel, and M. Staehelin. 1968. Action of rifamycins in RNA polymerase. Biochim. Biophys. Acta 157:215–217.

DISCUSSION

EHRET: Your argument for the involvement of a stable RNA in persistent rhythmicity rests heavily on your experiments showing that preventing transcription—either chemically or by enucleating the cells—does not abolish the rhythm. I think the pitfalls of drug experiments are especially clear in this case, and you should be very careful about the inferences you draw from negative evidence. There are perhaps roughly half a million plastids in each cell, and some large percentage of these have their own DNA. Even if the actinomycin blocked transcription completely in a large number of these plastids and thus stopped some linear transcription "clocks"—the cellular rhythm would still be intact unless you stopped all the transcribing "clocks." The doses you used could hardly have accounted for more than a small fraction of these, let alone for all of them.

VANDEN DRIESSCHE: It seems to me that a great number—perhaps nearly all—the transcribing "clocks" would have to keep running in order to produce sufficient mRNA to couple without impairment the basic oscillator to the overt rhythmic processes, if indeed the mRNA were not stable and needed to be continually resynthesized.

FELDMAN: Is it possible that you have in *Acetabularia* a system where an unstable RNA is being transcribed directly off some stable RNA?

VANDEN DRIESSCHE: Our inhibition studies showed that RNA synthesis, measured by incorporation of H^3 uridine, is effectively blocked with inhibitors like actinomycin or rifampicin that act on the DNA to RNA transcription, the former in the nucleus and the chloroplasts and the latter in the chloroplasts only. We get between 90 and 100 percent inhibition of chloroplastic RNA synthesis in the presence of rifampicin; rhythmicity is nevertheless maintained. The drug competitively inhibits the binding of the first nucleotide to the RNA polymerase during the DNA-dependent RNA synthesis. However, the particular case of replicase, which is also an RNA polymerase, has not yet, to my knowledge, been tested for sensitivity to rifampicin.

E. KENDALL PYE

Periodicities in Intermediary Metabolism

While most investigations of rhythmic biological functions have been concerned primarily with descriptions of behavioral and physiological phenomena, attention has recently focused on the nature of the basic oscillations that underlie and drive these overt rhythms. Students of circadian rhythms follow two main schools of thought. The "extrinsic" school holds that the organism monitors some rhythmically changing environmental factor, possibly of cosmic origin, which drives physiological rhythmicity (Brown, 1969). The more widely accepted "intrinsic" school, on the other hand, proposes an endogenous driving oscillation, generated at the biochemical level (Pittendrigh, 1960). Considerable speculation has gone into models of biochemical oscillations, but as yet no molecular mechanism has been demonstrated for any physiological rhythm. Oscillations in intermediary metabolism provide a concrete model for the type of cyclic biochemical process that might drive circadian biorhythms.

It was only in the past decade, and particularly since the work of Chance and his colleagues (Chance, Estabrook, and Ghosh, 1964; Chance, Hess, and Betz, 1964; Ghosh and Chance, 1964; Pye and Chance, 1966), that sustained oscillations in metabolism have been generally recognized. Earlier, some doubted that metabolic control mechanisms could generate self-sustained, essentially continuous oscillations. The recognition of feedback processes in metabolism (Yates and Pardee, 1956) and the discovery of allosteric enzymes (Monod, Changeaux, and Jacob, 1963) led to a re-evaluation of the nature and behavior of metabolic control pro-

623

cesses and to the prediction, on theoretical grounds, that metabolic oscillations might generally exist (Goodwin, 1965).

Allosterism is now recognized as a fundamental feature of the mechanism of the most extensively studied oscillating metabolic system— glycolytic oscillations in yeast (Higgins, 1964; Pye, 1969). These oscillations are particularly interesting because they have properties analogous to those of many overt biorhythms; they are self-sustained and have stable frequencies and specific phase responses. Although there is no evidence so far that links these oscillations to any physiological rhythm, the system is a useful model of the general properties of biochemical oscillators.

This paper reviews some of these properties of glycolytic oscillations in yeast and proposes a mechanism for their generation that might apply, in general terms, to other metabolic pathways.

MATERIALS AND METHODS

The organism used throughout this work was an inositol-requiring strain of *Saccharomyces carlsbergensis* (ATCC 9080). The methods of growth, harvesting, and preparation of intact-cell suspensions, as well as the preparation of cell-free extracts from these cells, have been previously described (Pye, 1969).

DPNH (reduced diphosphopyridine nucleotide) in the intact cells, or in the cell-free extracts, was monitored continuously by a dual-wavelength spectrophotometer fitted with appropriate interference filters or by a specially designed temperature-controlled fluorometer (Mayer *et al.*, 1968). Output from the spectrophotometer was recorded as percentage transmission and that from the fluorometer as zero-suppressed absolute fluorescence. The two methods of DPNH measurement are mutually and internally consistent; chemical measurements have demonstrated that they do indeed reflect the relative changes in DPNH concentration. DPNH was monitored because it is the only component associated with the glycolytic pathway that can be conveniently measured both continuously and nondestructively.

RESULTS AND DISCUSSION

SELF-SUSTAINED LIMIT-CYCLE OSCILLATIONS

Glycolytic oscillations were first observed by Duysens and Amesz (1957), who reported the appearance of several highly damped oscillations in the fluorescence of yeast cells immediately after a transition from aerobic to anaerobic conditions. The fluorescence was characteristic of reduced pyridine nucleotide. Their report received little attention, however, until the phenomenon was rediscovered 7 years later (Chance, Estabrook, and Ghosh, 1964). Subsequent extensive investigation has shown that the reduced pyridine nucleotide whose concentration oscillates is cytoplasmic DPNH. Furthermore, the concentrations of all the intermediates in the glycolytic sequence, as well as the adenine nucleotides ATP (adenosine triphosphate), ADP (adenosine diphosphate) and AMP (adenosine monophosphate) oscillate with the same frequency, but not necessarily the same phase, as DPNH concentration (Ghosh and Chance, 1964). The relative amplitudes of the oscillations in ADP and AMP are the largest of all the metabolites; the concentration of each compound varies by more than 60 percent of the maximum concentration (Betz, 1966), compared with changes of approximately 20 percent in DPNH concentration. Since ADP and AMP levels strongly reflect the total energy balance, variations in their levels probably exert a controlling effect on the metabolic state.

Indications were that the oscillations arose solely within the glycolytic pathway and that they represented a pulsing, or surging, of the glycolytic flux (Pye, 1969). Recognizing that the oscillations were associated with the rather clearly defined glycolytic pathway, workers successfully produced a cell-free extract displaying similar oscillations (Chance, Hess, and Betz, 1964).

One of the major problems in this early work was that the oscillations always damped out rapidly in both the intact cell and the cell-free extract, although trains of eight or nine cycles were sometimes generated. The relatively short duration of the phenomenon made experimentation technically difficult and made it impossible to tell whether the mechanism producing these oscillations was capable of generating sustained limit-cycle oscillations. Short trains of highly damped oscillations can arise from rather simple negative feedback systems following a perturbation, but such systems are incapable of generating sustained limit-cycle oscillations except under very unusual conditions (Morales and McKay, 1967).

It became clear that oscillations in the yeast cell-free extract were damping out because of the depletion of an endogenous carbohydrate reserve that was supplying glucose to the glycolytic pathway. The compound was identified as trehalose; when this disaccharide was added to a cell-free extract, prolonged trains of oscillations, frequently in excess of 100 cycles, were produced (Pye and Chance, 1966). The oscillations would damp out as the trehalose was exhausted but could be reinitiated by further additions of trehalose. Figure 1 illustrates a portion of one such prolonged train of oscillations. Both the amplitude and the period of the oscillation

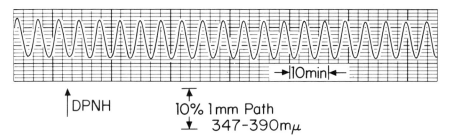

FIGURE 1   Portion of a prolonged train of DPNH oscillations in a cell-free extract of *S. carlsberg-ensis*. The DPNH concentration (vertical axis) was recorded as percentage transmission from a dual-wavelength spectrophotometer using the wavelength pair 347–390 mμ. Time runs from left to right in this and subsequent figures. The temperature of the cuvette was 22°C.

(approximately 5 min in this case) are quite stable, and it appears to be of the limit-cycle type. The waveform is sinusoidal and in this respect might be compared with the waveform of the diurnal rhythms of flashing, glow luminescence and photosynthesis observed in cultures of *G. polyedra* (Hastings and Keynan, 1965).

Highly sustained limit-cycle oscillations have been demonstrated in intact yeast cells, as well as in cell-free extracts (Pye, 1966). Figure 2 shows the relative changes in intracellular DPNH concentration in a yeast cell suspension following the addition of glucose and after the onset of anaerobiosis (due to depletion of oxygen from the suspension by the respiring cells). Following the onset of anaerobic conditions, the DPNH concentration oscillates for more than 23 cycles. As with the cell-free extract, the frequency is constant over most of the train, but the period, at approximately 35 sec, is much shorter. Furthermore, the amplitude changes slowly with time. Waveform is again sinusoidal.

Several properties of these metabolic oscillations are analogous to those of circadian rhythms, including the requirement of a stimulus to initiate rhythmicity. For example, Pittendrigh and Bruce (1957) have demonstrated the initiation of an eclosion rhythm in an arrhythmic, dark-grown culture of *Drosophila* by a single 4-hr light pulse.

WAVEFORM AND AMPLITUDE MODULATION

Although the waveform of the oscillations in intact cells is always sinusoidal, this is not the case for cell-free extracts. There, many different waveforms have been observed, ranging from a sharp single cycle occurring once every 20 min (Pye, 1969) to an approximation of a square wave (Figure 3).

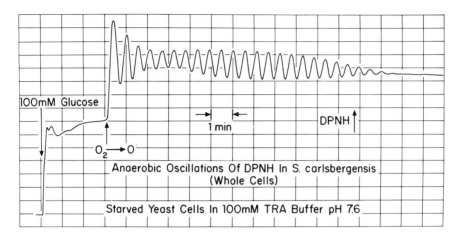

100mM Glucose

1 min

DPNH

$O_2 \rightarrow O$

Anaerobic Oscillations Of DPNH In S. carlsbergensis (Whole Cells)

Starved Yeast Cells In 100mM TRA Buffer pH 7.6

FIGURE 2   Relative changes in the intracellular DPNH concentration of a suspension of intact yeast cells, recorded by fluorescence. The DPNH concentration increases sharply on addition of glucose and again on the onset of anaerobiosis, which is followed by a long train of self-sustained oscillations. The temperature was 25°C.

Again, these features have analogs in circadian rhythms. A discrete event occurring once every 24 hr would have a waveform of the first type, while a rhythm of locomotor activity, for example, could be represented by a square wave.

Although information can be carried on a continuous waveform by either frequency or amplitude modulation, no clear case of frequency modulation has appeared in this work, while many examples of amplitude

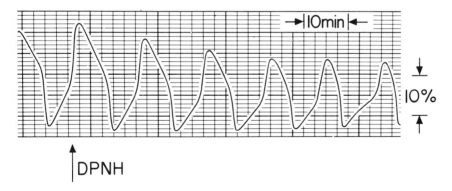

10min

10%

DPNH

FIGURE 3   Waveform of DPNH oscillations in a cell-free extract. Percentage transmission was recorded from a dual-wavelength spectrophotometer with the wavelength pair 347–390 mμ. The temperature was 22°C.

FIGURE 4   Amplitude modulation of DPNH
oscillations in a cell-free extract. Percentage
transmission was recorded from a dual-
wavelength spectrophotometer with the wavelength pair 347–390 mμ. The temperature was
22°C.

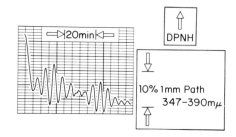

modulation have been seen in both intact cells and cell-free extracts. A typical pattern of amplitude modulation in a cell-free extract is shown in
Figure 4.

A different type of amplitude modulation seen in both systems is the
double-periodic waveform, in which a small amplitude cycle and a large
amplitude cycle alternate (Pye and Chance, 1966). Analogs to this phenomenon also occur in circadian systems.

CONTROL OF FREQUENCY

Details of the mechanism controlling frequency, amplitude, and waveform
of the glycolytic oscillations are still unknown, but some information is
available concerning the control of frequency. In intact cells, the frequency
of the oscillations was observed to decrease at low concentrations of glucose. When initial frequency was plotted against initial glucose concentration, the resulting curve had characteristics typical of saturation conditions and resembled a Michaelis-Menten plot of velocity versus substrate
concentration for an enzymatic reaction. Moreover, when other metabo-
lizable sugars were used in place of glucose, frequency decreased in the
order glucose>mannose>fructose. Figure 5 shows two typical traces obtained with identical yeast cell suspensions under the same conditions,
except that the cells received either 25 mM fructose (top curve) or 25 mM
glucose (lower curve). Frequency is clearly much lower with fructose.
These observations suggest that the frequency may be directly related to
the overall glycolytic flux rate.

In the cell-free extract, the oscillation frequency increased when apyrase (which acts as an adenosine triphosphatase) was added (Pye, 1969),
an observation that supports the interpretation suggested above. Apyrase
lowers the ATP concentration and raises the ADP concentration. Raising
ADP concentration would increase the glycolytic flux rate (since, in the
presence of excess substrate, ADP concentration controls the overall rate

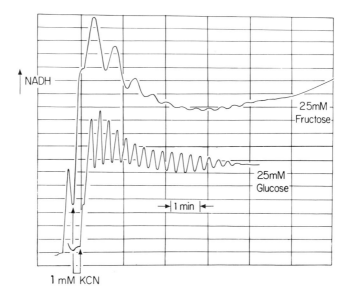

FIGURE 5   Difference in frequency of the oscillations in yeast cells metabolizing fructose (top trace) and in identical cells metabolizing glucose
(bottom trace). Traces were obtained by the fluorescence method. The
temperature was 25°C.

of glycolysis in these extracts). If the above interpretation is correct, this
should increase the frequency.

PHASE-DEPENDENT PHASE SHIFTS

Phase-dependence of the response to phase-shifting stimuli is a characteristic property of many biorhythms, especially circadian oscillations, and
is presumably the basis of the phenomenon of entrainment (DeCoursey,
1960; Pittendrigh, 1960; Pavlidis and Kauzmann, 1969). Glycolytic oscillations also show this property clearly. The oscillations in cell-free extracts
can be rapidly phase-shifted by very small pulses of ADP, but the nature
of this phase-shift is very much dependent on the phase at which the addition is made. Figure 6 illustrates an experiment in which ADP (to a final
concentration of 0.7 mM) was added at different points in the cycle. The
addition of ADP at the trough of a cycle causes a delay in the appearance
of the next peak (top, left diagram). An identical amount of ADP added on
the rising portion of the curve (top, center diagram) causes an advance in
the appearance of the next peak, while the same addition made at the peak

FIGURE 6   Phase-dependent phase-shifts produced by adding equal amounts of **ADP** to an
oscillating cell-free extract at different points in
the cycle. Traces were obtained by the fluorescence method. The temperature was 25°C.

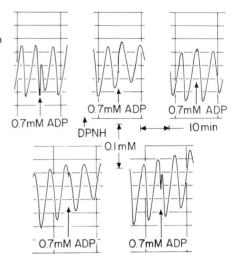

of the cycle (top, right diagram) induces no phase shift. Additions on the
falling portion of the curve (bottom two diagrams) produce phase delays,
which increase in magnitude as the trough is approached.

It is not possible to add A DP to intact cells because of the permeability
barrier presented by the cell membrane. However, phase-dependent phase-
shifts can be observed in oscillations in intact cells after addition of appropriate amounts of acetaldehyde, an intermediate in the glucose metabolism
of yeast.

Interesting similarities exist between the phase response curve of glycolytic oscillations for A DP additions and the phase response curve of eclosion in *D. pseudoobscura* for low temperature pulses (data taken from
Zimmerman, Pittendrigh, and Pavlidis, 1968) (Figure 7). The data are inadequate to show whether the A DP phase response curve resembles the
strong light effect curve or the weak light effect curve distinguished by
Pavlidis and Kauzmann (1969) for phase-shifting by light pulses in this
organism. There is evidence that these two effects may be simulated by
large and small additions of A DP (Pye, unpublished).

THE MECHANISM OF THE GLYCOLYTIC OSCILLATIONS

Analog computer studies (Higgins, 1964) have indicated three essential
components of the mechanism that generates the glycolytic oscillations.
These are (a) a product-activated reaction, (b) a constant flux into this

Phase Shift Of NADH Cycles By ADP Pulse

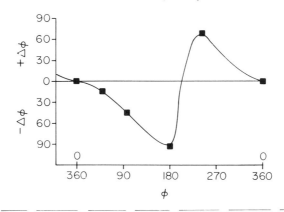

Physiological System

Phase Shift Of Eclosion In D. pseudoobscura Adults
Generated By Low Temperature Pulse

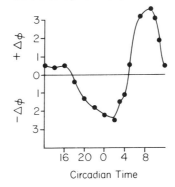

FIGURE 7 A comparison of a phase response curve for **ADP** additions
to an oscillating cell-free extract with the phase response curve of eclosion
in *D. pseudoobscura* adults generated by a low temperature pulse (data
taken from Zimmerman, Pittendrigh, and Pavlidis, 1968). In the top diagram the coordinates are in degrees and in the lower diagram they are in
hours.

FIGURE 8   A hydrodynamic analog for the
mechanism of the glycolytic oscillations in
yeast. The valve represents the activity of the
enzyme phosphofructokinase; the float system
represents the allosteric activation of phosphofructokinase by ADP; and the levels in the chambers represent the concentrations of fructose-6-
phosphate and ADP.

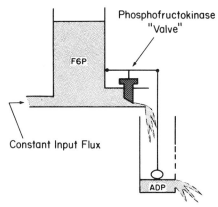

reaction of the correct order of magnitude, and (c) a reaction below first
order (i.e., enzymatic) to remove the activating product. A comprehensive
study of the theory of oscillating reactions has been published by Higgins
(1967).

Experimental studies of the glycolytic oscillations (Pye, 1969) show
that the three components predicted by Higgins (1964) do exist in this
system. The product-activated reaction is the step mediated by the enzyme
phosphofructokinase (PFK), which is allosterically activated by its product,
ADP. AMP, which changes in concert with ADP because of the equilibrium
maintained by the enzyme adenylate kinase, also activates PFK. The constant input flux results from trehalose degradation in the cell-free extract
and from the transport of glucose across the cell membrane in intact cells.
ADP is removed by the two kinase reactions in glycolysis that have ADP
as a substrate: 3-phosphoglycerate kinase and pyruvic kinase.

This mechanism has been described in detail elsewhere (Pye, 1969) but
the essential features can be represented here in terms of a hydrodynamic
analog (Figure 8). The activity of the enzyme PFK is represented by a
valve through which water flows at a rate equivalent to the reaction velocity. When lack of allosteric activation by ADP almost completely closes
the valve, the input flux rate (constant throughout) for fructose-6-phos-
phate (F 6P) formation is greater than the utilization of F 6P by the PFK
reaction. Hence the concentration of F 6P, represented by the height of
water in the top vessel, rises. With an increase in F 6P concentration, the
velocity of the PFK reaction will increase (the cooperative effect of F 6P
being represented by the sloping face of the valve), increasing the rate of
formation of the product ADP. An increase in ADP concentration will
give positive feedback through the float mechanism (representing the
allosteric activation of PFK by ADP), further increasing the velocity of

the PFK reaction (i.e., opening the valve). Eventually the ADP activation will be so great that the flow through the PFK "valve" will be greater than the input flux and the concentration of F6P will drop. When F6P is depleted, the flow through the valve will be limited to the input flux rate, ADP concentration will fall because it is being utilized at a higher rate and produced at a lower rate, and the valve will partially close. Provided the constant input flux rate is between the maximal and minimal velocities of the PFK reaction (Sel'kov, 1968), the system will oscillate continuously. Oscillations in F6P and ADP concentration will be reflected in other glycolytic intermediates, including DPNH.

CONCLUSION

Metabolic systems clearly can generate self-sustained, essentially continuous oscillations in the concentrations of associated metabolites. Furthermore, these biochemical oscillations display many features—specific waveforms, amplitude modulation, damping, initiation, and phase-dependent phase-shifts—analogous to those of rhythms observed at the behavioral and physiological levels. They do not, however, have the period length or temperature compensation characteristic of circadian rhythms (see, however, Pavlidis and Kauzmann, 1969).

The essential features of the proposed mechanism for the glycolytic oscillations could, in general form, exist in many other metabolic pathways. If the flux through these pathways was sufficiently low, very low-frequency oscillations could conceivably arise in these systems. No doubt physiological processes, provided they had suitable response times, would be forced to follow oscillations in the metabolic processes associated with them.

ACKNOWLEDGMENTS

The work reported in this paper was supported by NSF Grant GB-8270 and USPHS Grant GM-12202. I thank B. Chance for his interest and encouragement.

REFERENCES

Betz, A. 1966. Metabolic flux in yeast cells with oscillatory controlled glycolysis. Physiol. Plant. 19:1049–1054.
Brown, F. A. 1969. A hypothesis for extrinsic timing of circadian rhythms. Can. J. Bot. 47:287–298.

Chance, B., R. W. Estabrook, and A. Ghosh. 1964. Damped sinusoidal oscillations of cytoplasmic reduced pyridine nucleotide in yeast cells. Proc. Natl. Acad. Sci. U.S. 51:1244–1251.

Chance, B., B. Hess, and A. Betz. 1964. DPNH oscillations in cell-free extracts of S. carlsbergensis. Biochem. Biophys. Res. Commun. 16:182–187.

DeCoursey, P. J. 1960. Daily light sensitive rhythm in a rodent. Science 131:33–35.

Duysens, L. N. M., and J. Amesz. 1957. Fluorescence spectrophotometry of reduced phosphopyridine nucleotide in intact cells in the near ultraviolet and visible. Biochim. Biophys. Acta 24:19–26.

Ghosh, A., and B. Chance. 1964. Oscillations of glycolytic intermediates in yeast cells. Biochem. Biophys. Res. Commun. 16:174–181.

Goodwin, B. C. 1965. Oscillatory behavior in enzymatic control processes. Adv. Enzyme Regul. 3:425–438.

Hastings, J. W., and A. Keynan. 1965. Molecular aspects of circadian systems, p. 167 to 182. In J. Aschoff [ed.], Circadian clocks. North-Holland Publ. Co., Amsterdam.

Higgins, J. 1964. A chemical mechanism for oscillation of glycolytic intermediates in yeast cells. Proc. Natl. Acad. Sci. U.S. 51:989–994.

Higgins, J. 1967. The theory of oscillating reactions. Ind. Eng. Chem. 59:18–62.

Mayer, D. H., J. R. Williamson, V. Legallais, and M. Sartre. 1968. A sensitive filter fluorometer for metabolite assays. Chem. Instrum. 1:383–389.

Monod, J., J.-P. Changeaux, and F. Jacob. 1963. Allosteric proteins and cellular control processes. J. Mol. Biol. 6:306–329.

Morales, M., and D. McKay. 1967. Biochemical oscillations in controlled systems. Biophys. J. 7:621–625.

Pavlidis, T., and W. Kauzmann. 1969. Towards a quantitative biochemical model for circadian oscillators. Arch. Biochem. Biophys. 132:338–348.

Pittendrigh, C. S. 1960. Circadian rhythms and the circadian organization of living systems. Cold Spring Harbor Symp. Quant. Biol. 25:159–184.

Pittendrigh, C. S., and V. G. Bruce. 1957. An oscillator model for biological clocks, p. 75 to 109. In D. Rudnick [ed.], Rhythmic and synthetic processes in growth. Princeton Univ. Press, Princeton, New Jersey.

Pye, K. 1966. Metabolic control phenomena associated with oscillatory reactions. Stud. Biophys. 1:75–78.

Pye, E. K. 1969. Biochemical mechanisms underlying the metabolic oscillations in yeast. Can. J. Bot. 47:271–285.

Pye, K., and B. Chance. 1966. Sustained sinusoidal oscillations of reduced pyridine nucleotide in a cell-free extract of Saccharomyces carlsbergensis. Proc. Natl. Acad. Sci. U.S. 55:888–894.

Sel'kov, E. E. 1968. Self-oscillations in glycolysis. Eur. J. Biochem. 4:79–86.

Yates, P. A., and A. B. Pardee. 1956. Control of pyrimidine biosynthesis in Escherichia coli by a feed-back mechanism. J. Biol. Chem. 221:757–770.

Zimmerman, W. F., C. S. Pittendrigh, and T. Pavlidis. 1968. Temperature compensation of the circadian oscillation in Drosophila pseudoobscura and its entrainment by temperature cycles. J. Insect Physiol. 14:669–684.

DISCUSSION

EDMUNDS: Goodwin has predicted that if you take an asynchronous culture of bacteria, logarithmically growing, and let it sit through a couple of division cycles in a chemostat, cell division will become synchronized spontaneously—and will, I suppose, stay in phase, although he hasn't addressed himself to that question. Do you think you could take a random population of intact yeast cells and expect spontaneous DPNH oscillations to appear? Would you care to speculate on whether you could get persistent self-induced synchronization?

PYE: That's a difficult question to answer; I just don't know. The way we induce these oscillations, of course, is with an aerobic-to-anaerobic shock that puts all the cells in phase.

EDMUNDS: But the rhythm isn't detectable until you put the cells in phase. Or do you have a way of knowing whether or not they are rhythmic before you pulse them?

PYE: Yes, we do—with the microfluorometer. We have looked at individual cells through the microfluorometer, and they show no persistent DPNH oscillation until we give them an anaerobic shock.

EDMUNDS: If they were to behave as Goodwin predicts, would you expect them to drift out of phase again in time, or do you think they would have to stay in phase?

PYE: I think they would have to stay in phase. We can produce these very long trains of oscillations—up to about 43 cycles—and we can monitor single cells in the population. My colleagues Chance and Williamson (FEBS Abstracts, p. 237, Prague, 1968) have found that as the rhythm in the population damps out so does the rhythm in the single cell, and it damps out within a cycle or two after the damping out in the population as a whole.

EDMUNDS: That may mean that the rhythm in the population really has

nothing to do with the "clock" at all, but is merely self-induced.

PYE: Yes, the glycolytic oscillations that we see are certainly transient phenomena; you see them after the shock and they always disappear eventually. But other studies with the cell-free extract have shown that the oscillation mechanism is self-sustaining. What is apparently happening in the intact cells is that some intermediate close to the oscillation reaction mechanism is building up and eventually turning off the mechanism. Now I believe that given a situation where this intermediate doesn't accumulate, you would have a continuously oscillating system.

HOCHACHKA: In the cell-free extract, how close to saturation are the enzymes?

PYE: That really depends on which enzymes you are talking about. Aldolase is probably close to saturation because fructose diphosphate builds up to a high concentration. Other enzymes—certainly 3-phosphoglycerate kinase—are under the control of ADP concentration, and those are operating in the range of the K_m or below.

HOCHACHKA: How about PFK itself? What is the range of fructose-6-phosphate?

PYE: Once in every cycle F6P goes down nearly to zero. The maximum concentration depends on the flux rate into the system and the ATPase present. Small variations in total activity of PFK, flux rate, or ATPase will change frequency, amplitude, and almost every characteristic of the oscillations.

HOCHACHKA: If you saturate the hexokinases with glucose at the beginning, do you see the oscillations?

PYE: No; we have demonstrated in the extract a requirement for trehalose, which limits the flux into glycolysis. If you saturate hexokinase with a large dose of glucose, the oscillations disappear and don't return until the glucose has been depleted. In the intact cell, however, the flux into glycolysis is limited by the rate of glucose transport through the cell membrane, and hexokinase is never saturated.

AUDREY BARNETT
CHARLES F. EHRET
JOHN J. WILLE, JR.

Testing the Chronon Theory of Circadian Timekeeping

Proposed molecular mechanisms for circadian timekeeping can be classified by the extent to which gene action is presumed to underlie and regulate temporally distinct physiological phenomena (Figure 1). One such proposal, the chronon theory (Ehret and Trucco, 1967), lies at one extreme of such a classification; it is especially attractive to us because it is both plausible and testable. The chronon is a hypothetical linear sequence of DNA; transcription proceeds sequentially along the chronon, cistron by cistron, in such a way that it takes about a day to complete, beginning to end. Then transcription begins again at the first cistron, and a new cycle is initiated. Chronons are regarded as inevitable features of eukaryotic organization, having lengths fixed by natural selection.

In the chronon model (Figure 1, bottom section), linear sequential transcription produces labeled RNA's characteristic of certain cistrons being transcribed at the time a pulse is given; RNA-18, for instance, appears later than RNA-12 because it is synthesized at a more distal position on the chronon. Alternatively, temporally unique RNA's could arise from transcription based on various regulatory circuits, rather than linear transcription (middle section). A third possibility (top section) is that DNA

This paper is dedicated to Ernesto Trucco, who died in August 1970 at the age of 48.

template cistrons may be "on" or "off" at any time. In this case, a pulse delivered at any given time produces an equivalent charge of RNA, not necessarily unique temporally. Under the latter two general schemes, the circadian period could arise, in a still undescribed manner, from stochastic reactions among the gene products. According to the chronon model, however, circadian periodicity is directly a function of the total length of DNA (or, better, the number of cistrons) in the chronon transcription unit.

Since the last two models in Figure 1 predict temporally unique RNA's, our first task is to search for different RNA's synthesized at characteristic times of day, using the DNA–RNA molecular hybridization technique (Gillespie and Spiegelman, 1965).

MATERIALS AND METHODS

Two types of cells, both ciliated protozoans, were used. In one, *Tetrahymena*, Wille and Ehret (1968) have previously demonstrated a circadian rhythm of cell division, entrainable by light/dark cycles. *Tetrahymena pyriformis*, strain W, can be grown in defined medium and routinely reaches cell concentrations of 500,000/ml; it was with this system that the first indication of temporally unique RNA's—both in the circadian and in the heat–shock synchronized (ultradian) cell division cycles—was found (Ehret, Wille, and Trucco, 1970). The second organism used was *Paramecium multimicronucleatum*. It is less satisfactory technically— seldom reaching cell concentrations higher than 1,000/ml and growing on a bacterized plant extract—but it has a well-described circadian rhythm of mating-type reversals (Sonneborn, 1957; Barnett, 1966). An organism showing periodicity so close to the gene level was thought to be an excellent system for testing the chronon theory.

The general experimental procedure is outlined in Figure 2. The left side of the figure lists the steps in the preparation of DNA filters:

1. Cells concentrated by centrifugation were lysed in PAS (6 percent 4-amino salicylic acid, 0.02 M sodium arsenate, pH 8.5) 6×SSC (standard saline citrate buffer, 0.15 M sodium chloride and 0.015 M sodium citrate, pH 7). If all cells were not lysed, 25 percent SDS (sodium dodecyl sulfate, recrystallized) was added in 20λ portions. Phenol extraction followed (phenol saturated with 6×SSC) and the aqueous layer was separated by centrifugation. The above procedure was repeated with the heavy interface layer. Two volumes of ethanol were layered over the pooled aqueous layers and the long polymers spooled out on glass rods.

GENE ACTION AND TEMPORAL ORDER

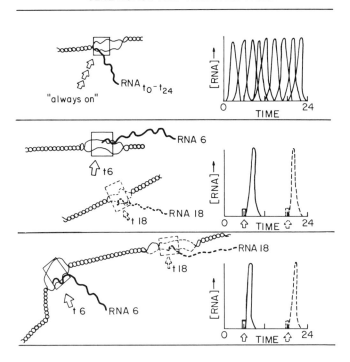

FIGURE 1   Three general classes of models: Transcription at the times
indicated at the arrows results in pulse-labeled RNA's isolated and quan-
tized as shown at the right.

2. The fibers were redissolved in 0.2×SSC and centrifuged at 50,000
rpm for 30 min to precipitate the polysaccharides. RNase at 50 µg/ml was
added for 1 hr at 37°C. The DNA/RNA ratio was determined by oricinol
and diphenylamine tests, and the nucleic acids were adjusted to a concentration of 350 µg/ml; 15 ml were loaded onto a 50-cm × 1.5-cm column
of agarose gel (CalBiochem BIO–GEL A-5m, 50 to 100 mesh, with a molecular weight exclusion limit of 5×10^6). The flow rate of 0.2×SSC was adjusted to 20 ml/hr; DNA was recovered after the first 90 ml and RNA began to appear after about 125 ml.
3. A DNA pool sufficient to make 1,200 filters was brought to a pH of
11.7 with the addition of 1 N NaOH and held there for 10 min; it was
then neutralized with 0.1 N HCl and the SSC buffer concentration adjusted to 6×SSC. The volume was adjusted so that approximately 2 ml
contained 4× the desired DNA concentration/filter. One hundred mem-

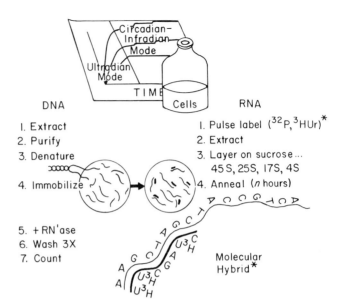

FIGURE 2   Experimental protocol. Left: preparation of DNA filters.
Right: preparation of RNA.

brane filters (.45 μ, 25 mm; obtained from the Sartorius Division,
Brinkmann Instruments) were placed on a specially designed hundred-unit
filter holder and the solution slowly (10–30 min) drawn through the filter
with a vacuum pump. Following a wash with an equal volume of 6×SSC,
the filters were removed individually and air-dried overnight, then placed
in a vacuum oven for 90 min at 80°C. Stacks of five filters were cut into
quarters and 10 "quarter filters" from 10 different filters were routinely
used for each experimental point.

Figure 2 (right side) outlines the preparation of timed RNA's.

1. Cells were grown under LD *12*:12 and concentrated at hours 6, 12,
18, 24. In labeled preparations tritiated uridine (5 mc of ICN Uridine-5-
H-3 per 3 liters of *Tetrahymena*, and 10 mc of NEN Uridine-5-H-3 per
2 liters of previously concentrated *Paramecia*) was added 1 hr before extraction. In unlabeled RNA preparations step 1 was omitted.
2. In a modification of the method of DiGirolamo *et al.* (1964), concentrated cells were lysed with a 1 percent SDS and 0.5 percent ND
(1,5 Naphthalene di-sulfonic acid disodium salt) in a medium A buffer
(0.05 M Tris·HCl, 0.003 M CaCl$_2$, 0.25 M sucrose) and extracted with

Phenol-8HQ (1 percent SDS, 0.1 percent 8-hydroxyquinoline and saturated with 0.05 M Tris). Purified Bentonite (5–10 $\mu g/ml$) was added each
time phenol was used. Following a 15 min cold extraction, the aqueous
and interface layers were recovered by centrifugation and re-extracted
two times with phenol-8HQ. The aqueous layer was saved, and the phenol
and interface layers were re-extracted twice with Med A, 1 percent SDS,
and 0.5 percent ND; the aqueous layer was saved each time. The nucleic
acids were precipitated with 0.1 M NaCl and 67 percent ethanol. The precipitate was redissolved in .01 M Tris and 0.01 M MgCl$_2$, and 50 $\mu g/ml$ of
DNase was added for 30 min. The residual protein was extracted with
1 percent SDS, 0.5 percent ND and phenol-8HQ, and the aqueous layer
separated by centrifugation. The nucleic acids were precipitated with
0.1 M NaCl and 67 percent ethanol and the precipitate, after a 5-min centrifugation, was redissolved in 0.001 M MgCl$_2$. A series of selective pre-
cipitations with 2 M potassium acetate and 25 percent ethanol (which
allows the oligionucleotides to remain in solution) was done until the DNA
content, as monitored by diphenylamine tests, was negligible. After the
last precipitation the RNA was dissolved in 0.01 M Na acetate and centrifuged to remove the polysaccharides.

3. Separation of total RNA into different classes by density gradient
ultracentrifugation was omitted in the experiments reported here.

4. The RNA's were brought to the desired concentration in a 2×SSC
buffer and 33 percent Formamide Solution (FOSC). Ten DNA filters were
added to 2.5 ml and five blank-filters to 1 ml of the RNA solution and
gently agitated for the desired time at room temperature.

5. Individual filters were transferred from the RNA solution to separate
vials, and 5 ml of RNase (5 $\mu g/ml$ in 2×SSC) were added for 30 min with
mild agitation.

6. A screen was placed over the box of vials and the solution poured off.
The filters were washed with 5 ml of 2×SSC three times, in a similar
manner.

7. Filters were transferred to new scintillation vials and 1/2 ml of water
was added to each; the vials were heated for 1 hr at 80°C. (This step is
essential to get the counts into solution immediately.) Fourteen ml of
Scintisol-Complete (Isolab Incorp.) were added and the vials were counted
in a Beckman Scintillation Counter.

Figure 3 illustrates the types of hybridization experiments that can be
used to test for the presence of distinct RNA molecules in different time
cuts. Kinetic studies provide essential preliminary information about the
hybridization system. For a given DNA and RNA concentration the counts

PARAMETERS OF MOLECULAR HYBRIDIZATION

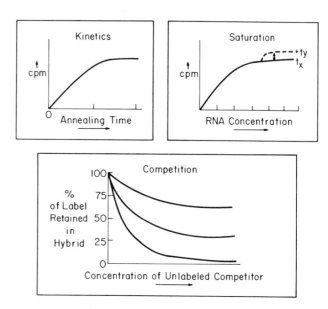

FIGURE 3   Types of hybridization experiments used to test for unique
RNA molecules present in different time cuts.

increase with annealing time until a stable plateau is reached. Saturation
studies (24 hr at room temperature in the experiments reported here)
show that as the RNA concentration (for example, of type X) is increased, the counts increase until a plateau is reached, indicating that all
available complementary DNA sequences have been saturated. If the DNA
filter is now placed in an RNA of time y, any labeled RNA's synthesized
uniquely during y-pulse should find available DNA sites for hybridization
and the counts will increase. This is the basis for the powerful saturation-
transfer equipment. Competition studies involve a third ingredient: an
unlabeled RNA competitor is included in the annealing mixture with the
DNA filter and labeled RNA preparation. As one increases the amount of
unlabeled competitor, the number of counts will decrease—since the limiting factor becomes the number of available DNA sites. The greater the
number of RNA's in common, the more the competitor will lower the
counts. (Note that in labeled preparations the bulk of the RNA's isolated
will be unlabeled either because of inefficiency of incorporation of the
label or because these RNA's were synthesized before the label was added;
thus all experiments essentially involve a strong element of competition.

As described earlier, unlabeled competitor RNA preparations were extracted from entirely separate but equivalent batch cultures of synchronized cells.)

RESULTS

RNA isolated from *Tetrahymena* grown under *LD 12*:12 (dawn at 4 am CST), pulsed with tritiated uridine at hour 17 and harvested at hour 18 (TC 18 [3]HUr RNA) was used in a kinetic study (Figure 4) and a saturation study (Figure 5). The kinetics show a rapid uptake of label in the first 3 hr with a plateau from 12 to 24 hr and a further increase during the second 24 hr. The saturation curve begins to level off around 300 μg/ml. These results suggested that a round of saturation transfer experiments would be prohibitively expensive, so we performed instead a set of competition experiments using labeled and unlabeled RNA's from hours 6, 12, 18, and 24 of the *LD 12*:12 cycle.

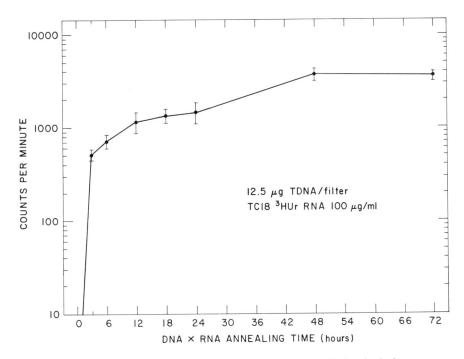

FIGURE 4   The kinetics of DNA–RNA hybridization in *Tetrahymena*. Each point is the mean ± 2SE for 10 filters. Counts on blank filters have been subtracted.

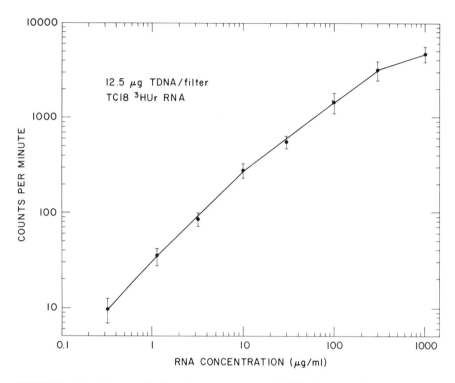

FIGURE 5 Saturation curve for *Tetrahymena* DNA–RNA hybridization. Annealing occurred for 24 hr at room temperature in FOSC. Each point represents the mean ± 2SE for 10 filters. Counts on blank filters have been subtracted.

Figure 6 shows the competition curves for the four times-RNA competitors against each of the pulse-labeled RNA's. The lower right set (TC24 ³HUr RNA) indicates that the four competing RNA's do have different degrees of homology with C24; as might have been anticipated, C24 competes best against itself, while C6 and C18 (closest to C24 in time and hence perhaps having RNA's in common) are about equally good competitors. But from the other three graphs it is clear that C24 not only competes best against itself but also is the best competitor at all four times. This suggests that C24 has the most RNA's in common with the labeled RNA's at all points. Moreover, C6, C12, and C18 are each the weakest competitors for the RNA's labeled at corresponding times. However, few of the curves have plateaued, and it is possible that the rank order at saturation would be different. In addition, these are competition curves for 24-hr annealing, and we have reason to suspect from experiments in

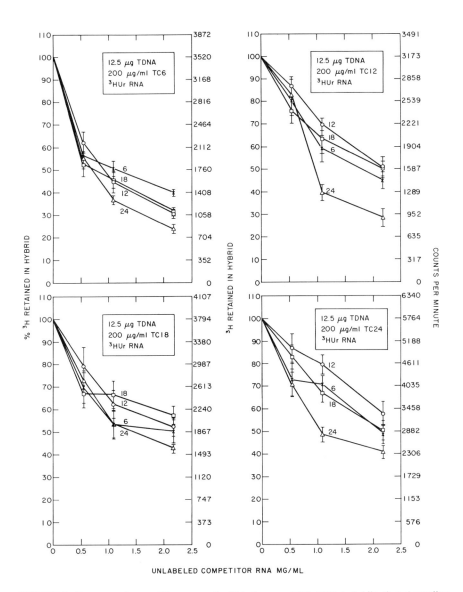

FIGURE 6   Four sets of competition curves for *Tetrahymena* DNA–RNA hybridization. Annealing occurred for 24 hr at room temperature in FOSC. Both percentages (left) and mean ± SE for 10 filters (right) are given. Counts on blank filters have been subtracted. The labeled RNA is given in the upper right corner for each plot. Unlabeled RNA isolated at hr 6 x---x; at hr 12 o---o; at hr 18 □---□ ; at hr 24 △---△.

progress that the competition curves for longer annealing times—which allow rarer RNA molecules, synthesized along only one or few copies on the DNA template, to hydridize with their complements—may differ. (See the change of slope in Figure 4.)

The results of a saturation transfer experiment with *Paramecium* are shown in Figure 7. Nucleic acids extracted from paramecia grown in bacterized medium invariably include some bacterial nucleic acids even though, by the time the mating type cycle can be assayed, the paramecia have cleared the culture of bacterial turbidity and reduced the bacterial concentration to a level too low to support further *Paramecium* cell division. To correct for bacterial nucleic acids, filters with *Aerobacter* DNA are hybridized with four labeled RNA's (C6, 12, 18, and 24) from the *Paramecium* preparations. The hatched graphs (Figure 7, upper right) show that the different timed-RNA's contain different *Aerobacter* counts and that C18 contributes about the same number of counts to all the filters regardless of the RNA with which they were annealed during the first 48 hr. C18 gains counts from the other RNA's in proportion to the number of counts originally present in that RNA (upper left), which is what is expected when there are still available DNA sites on the filter. In the bottom right graphs (hatched) the counts bound to filters prepared from *Paramecium* (which may have traces of *Aerobacter* DNA) differ in rank order (C6 > C18 > C12 > C24) from those bound to *Aerobacter* filters. The lower left graphs show that C18 adds counts even to filters that had already been exposed to C18 for 48 hr; apparently "true" saturation had not been reached. However, the number of counts added by the RNA's of the other time points is greater, even though only C6 contributed more counts during the first 48-hr annealing, and the rank order for the additional counts (C12 > 6 > 24 > 18) differs from that seen in the first 48-hr annealing. Equally important, reciprocal additions are seen with *Paramecium* DNA filters; i.e., C12 adds the most counts after C18 annealing and C18 adds the most counts after C12 annealing, even though C6 had twice as many initial counts as C12. Thus, despite the presence of *Aerobacter* nucleic acids and lack of full saturation, there is some indication that labeled RNA species are present at time 12 that are not present at time 18, and *vice versa*.

DISCUSSION

The molecular hybridization studies presented are just the beginning of a matrix of saturation transfer and competition experiments (including interspecific crosses) planned.

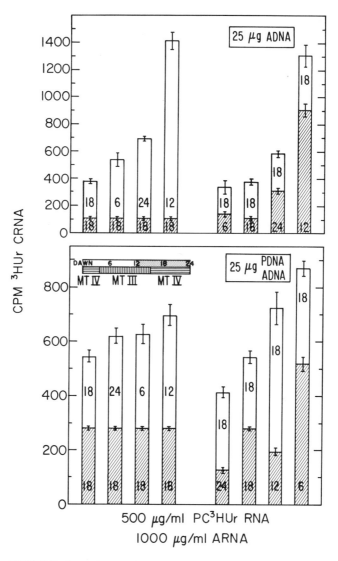

FIGURE 7 Saturation-transfer experiment with *Paramecium*. Labeled RNA's isolated from different phase points in the entrained mating type cycle (see insert) were annealed with DNA filters prepared from *Aerobacter* cultures (top) and *Paramecium* cultures (bottom). After 48 hr the filters were rinsed in the annealing buffer (FOSC) and transferred to a second labeled RNA for an additional 48 hr. (Unlabeled *Aerobacter* RNA was added to all RNA mixtures.) Bar graphs show the total counts at the end of 96 hr; hatched portions represent the number of counts picked up in the first 48 hr and retained after RNase treatment and a second 48-hr annealing period in FOSC. (Blank filter counts have been subtracted; the mean ± SE for 10 filters is shown.)

TESTING CHRONON THEORY:

Topology of Temporally Unique RNA's

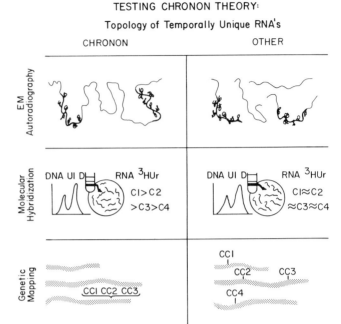

FIGURE 8   Three general classes of experiments that distinguish the chronon theory from other models of gene action accounting for temporally unique RNA.

In the introduction, we briefly described alternative mechanisms by which ordered genetic transcription could produce temporally unique RNA's (Figure 1; see also Ehret and Trucco, 1967). Figure 8 diagrams some conceptually simple, but technically difficult, experiments that would distinguish between these general hypotheses. The genetic approach (Figure 8, bottom) involves accumulating "clockless" mutants—organisms in which circadian rhythmicity is abolished in two or more normally rhythmic processes. Mapping the "arrhythmicity" mutants would distinguish between at least one version of the chronon theory and other hypotheses: If there is only one chronon per haploid genome, such mutations should map as alleles or pseudoalleles, rather than appearing throughout the chromosome. However, redundancy of chronons, that is, two or more identical chronons, would prevent such mutants from being detected. Moreover, the genetic approach would be very difficult experimentally; both producing and assaying "clockless" mutants and the subsequent breeding analysis might be prohibitively time-consuming.

The techniques of molecular hybridization and electron microscopic autoradiography are more promising. In the former method (middle section, Figure 8). DNA is pulse labeled with deuterium (D) during ultradian growth at a particular time (say, U–1) in a synchronously dividing population of cells. The polycistronic segments of DNA labeled at this time can be separated from "earlier" and "later" regions by equilibrium density gradient ultracentrifugation; this temporally defined class of DNA can then be denatured and applied to filters. Chronon theory predicts a conspicuous rank ordering in hybridization efficiencies of temporally unique circadian RNA's (*viz.*, that C1 > C2 > C3 > C4 >, etc.); for other regulatory mechanisms there is no compelling reason to expect any such rank-ordering or association with DNA ordinality.

The most direct test of the chronon theory would be visualization of molecular hybrids by electron microscopic autoradiography. With this method, one could see the consequences of *in vitro* molecular hybridization of unsheared stretches of DNA with pulse-labeled RNA taken from different phases of the circadian cycle. Chronon theory (upper diagram, Figure 8) predicts that temporally sequential RNA will hybridize at spatially sequential points along the DNA strand; other proposed mechanisms do not demand such correspondence. In order to be meaningful, however, experiments of this type would have to be performed on carefully prepared unsheared DNA templates of native length.

ACKNOWLEDGMENTS

This work was performed under the auspices of the U.S. Atomic Energy Commission. We thank A. Zadylak and R. Januszyk for their invaluable help in all phases, and student trainees Kenneth Dobra for his contributions to the DNA purification methods, L. M. Lewis for her work with *Aerobacter*, and B. Rasmussen and C. Duffy for their work with scintillation-counting efficiency.

REFERENCES

Barnett, A. 1966. A circadian rhythm of mating type reversals in *Paramecium multimicronucleatum*, syngen 2, and its genetic control. J. Cell. Physiol. 67:239–270.
Bonner, J., G. Kung, and I. Bekhor. 1967. A method for the hybridization of nucleic acid molecules at low temperature. Biochemistry 6:3650–3653.
DiGirolamo, A., E. C. Henshaw, and H. H. Hiatt. 1964. Messenger ribonucleic acid in rat liver nuclei and cytoplasm. J. Mol. Biol. 8:479–488.

Ehret, C. F., and E. Trucco. 1967. Molecular models for the circadian clock. I. The chronon concept. J. Theor. Biol. 15:240–262.

Ehret, C. F., J. J. Wille, and E. Trucco. The circadian oscillation: An integral and undissociable property of eukaryotic gene-action systems. *In* B. Chance [ed.], Biological and Biochemical Oscillations. Academic Press, New York. (In press.)

Gillespie, D., and S. Spiegelman. 1965. A quantitative assay for DNA-RNA hybrids with DNA immobilized on a membrane. J. Mol. Biol. 12:829–842.

Sonneborn, T. M. 1957. Diurnal change of mating type in paramecia. Anat. Rec. 128: 626. (Abstr.)

Wille, J. J., Jr., and C. F. Ehret. 1968. Light synchronization of an endogenous circadian rhythm of cell division in *Tetrahymena*. J. Protozool. 15:785–789.

DISCUSSION

ZIMMERMAN: With regard to the genetic test of the chronon theory, I think you are overlooking an important possibility. That is, it is conceivable that two species of mRNA, controlling two physiological events (hands of the clock), are synthesized at the same time, yet "turned on" at different times later. In that case, the fact that you had a mutant with both hands of the clock knocked out at once wouldn't tell you anything.

BARNETT: Yes, it would, if the mutations were independently derived. If they were chronon mutations, they should map at the same site on the same chromosome.

EDMUNDS: Would you care to speculate on the comparison between mRNA synthesized at the same time in the circadian cycle in two different species of organisms? Have you tried any cross-hybridization experiments?

BARNETT: We are very interested in trying cross-hybridization. If the chronon is, as I suspect, a very primitive segment of the genome, showing great evolutionary homology, then you might expect to be able to cross-hybridize. We do have some evidence—

although it really isn't much at this point—that you do get higher competition of these temporally labeled RNA's between *Paramecium* and *Tetrahymena* than between *Paramecium* and *Aerobacter*.

RENSING: Did I understand correctly that the time constant, or frequency of the rhythm, is determined by the transcription process? Or is there feedback from the end to the beginning?

BARNETT: The frequency is determined by the total length of the chronon. Of course, transcription may not go on continuously—that is, there may be pauses, or intercistronic events—but the time constant would still be determined by the total length of the transcribing segment.

RENSING: Do you have any idea what happens between the different templates of the chronon, or what determines the duration of the pauses?

BARNETT: Until we have some better idea of how the chronon works, we really can't say.

RENSING: Do you think that activity of RNA polymerase, or perhaps hormones, could influence the time constant in your model?

BARNETT: I have no idea.

HILLMAN: One of the most puzzling things about clocks is their temperature-compensated nature. Does your model make any suggestions about this? What is unique about this transcription that it should be temperature-compensated? It seems to me no better than a simple chemical model in that respect.

BARNETT: The chronon model is a chemical model—but the thing that is stressed is not the transcription process itself but the length of tape that must be read sequentially. As I recall, Ehret and Trucco considered the temperature-compensation problem in their original paper.

JERRY F. FELDMAN
NICKOLAS M. WASER

New Mutations Affecting Circadian Rhythmicity in Neurospora

Genetic analysis has been a powerful tool for unraveling a number of complex cellular functions. For example, the use of single-gene mutants, in which one element of the system is altered at a time, helped to elucidate the nature of metabolic pathways (Beadle and Ephrussi, 1937; Srb and Horowitz, 1944; Ames, 1957) and their control (Ames and Garry, 1959; Jacob and Monod, 1961). More recently phenomena at higher levels of organization, such as bacteriophage assembly (Wood *et al.*, 1968) and even behavioral responses (Benzer, 1967), are yielding to this approach.

On the other hand, genetic approaches to the study of biological clocks have been few. The experiments that have been done, however, suggest that a concerted effort in this direction could provide striking insights into the cellular and molecular basis of circadian rhythmicity. Research suggesting genic control of the period length or the phase of the rhythm includes the work of Bünning (1935) on plants and Pittendrigh (1967) on *Drosophila*. Polygenic inheritance has also been suggested as the basis for determination of critical day length in the photoperiodic induction of diapause in certain insects (Danilyevsky, 1965, pp. 171–184). None of these studies, however, has identified a single-gene effect on the clock.

Another class of experiments has identified single genes that control the ability of the organism to express a rhythm. Such genes are known in *Paramecium* (Barnett, 1966, 1969), *Neurospora* (Stadler, 1959; Sussman, Lowry, and Durkee, 1964; Sargent, Briggs, and Woodward, 1966), and

Podospora (Van Huong, 1967). In all cases, however, these genes probably do not affect the clock mechanism but only one of the particular processes controlled by it.

The patch strain of *Neurospora crassa* exhibits a circadian rhythm of conidia (asexual spore) formation, which gives it a "patched" appearance in a growth tube. The rhythm is entrained by light–dark cycles and is damped in constant light; it persists in continuous darkness with a temperature-compensated period length of about 24 hr (Pittendrigh *et al.*, 1959; Stadler, 1959). The vast backlog of genetic and biochemical data on *Neurospora* gives the organism enormous potential as a tool for genetic analysis of the circadian clock. The experiments reported here represent the results of a pilot study to determine the feasibility of obtaining new single-gene mutants in *patch* that affect some easily observed parameter of the rhythm. Such mutants, which would include loss of rhythmicity or alteration of period length, phase relationships, light-sensitivity, and temperature-coefficients, might represent mutations in the clock mechanism itself. Hopefully, the genetic and biochemical analysis of such mutants could result in identification of specific gene functions in the operation of the clock.

METHODS AND MATERIALS

Strain The strain used for mutagenesis was a *patch* reisolate (C11a) obtained by crossing *patch* (FGSC #981) (obtained from the Fungal Genetics Stock Center, Dartmouth College) with 74A (FGSC #262) (obtained from Dr. N. H. Horowitz, California Institute of Technology).

Mutagenesis Mutations were induced in strain C11a by incubating conidia in $0.1N$ ethylmethane sulfonate (EMS) (Malling and deSerres, 1967). Conidia were plated on a sorbose-complete medium; single colonies were picked and tested for their rhythmic properties.

Assay of Circadian Rhythm Conidia from the single colony isolates were inoculated into growth tubes containing Gray's medium (Brandt, 1953). The rhythmic parameters of each isolate were measured in four different growth conditions: continuous light (350 ft-c, cool white fluorescent, 25°C) and continuous darkness at 20, 25, and 30°C.

Crosses Crosses were carried out on Westergaard and Mitchell's (1947) crossing medium by standard techniques.

RESULTS

Approximately 400 colonies from conidia exposed to EMS were isolated and tested for rhythmic properties. Of these nearly 50 percent showed some difference in the expression of the conidiation rhythm from the original C11a *patch*. In most isolates, the rhythm was simply less marked, with poorly defined bands and variable amounts of damping. However, three isolates appeared to warrant further study (see Table 1):

1. CM89a. This strain shows no rhythmicity. Results of crosses to 74A suggest that CM89a carries a second mutation, unlinked to *patch*, which prevents the expression of rhythm.
2. CM211a. The rhythm in this strain does not damp out in constant light as does the original *patch*. Crosses to wild-type were nearly sterile.
3. CM113a. The period length of this strain at 30°C is 16 hr, significantly shorter than the original *patch*, although at 20 and 25°C, the period length is the same as that of *patch*. Again, crosses to wild-type were nearly sterile (Table 2).
4. CM70a. The periodicity of this mutant is normal even though its growth rate is significantly slower than the original *patch* (Table 2).

DISCUSSION

The new mutants obtained in this study suggest that isolating single-gene mutants of *Neurospora crassa* in which some parameter of the circadian rhythm is altered is feasible. CM89a carries a second mutation, unlinked to *patch*, that abolishes the rhythm. This mutation could simply be a suppressor of *patch*, having nothing to do with the clock, or it may represent a mutation in some part of the clock mechanism itself. In order to distinguish these alternatives, it will be necessary to determine whether the 89 mutation abolishes circadian rhythmicity in some rhythm unrelated to conidia formation, such as the recently-discovered rhythm of CO_2 production (Sargent, personal communication).

TABLE 1   Mutants with Altered Rhythmicity

Mutant	Phenotype	Type of Inheritance
CM 89a	No rhythm	Single gene
CM 211a	Rhythm in constant light	Not yet known
CM 113a	Short period length at 30°C	Not yet known

TABLE 2   Temperature Effect on Growth Rate and Period of the Rhythm

Strain	Period Length (hr)				Growth Rate (cm/day)			
	20°	25°	30°	Q_{10}	20°	25°	30°	Q_{10}
C 11a	24.0	21.9	19.6	1.2	3.2	4.8	6.6	2.06
CM 70a	24.0	22.2	19.5	1.2	2.4	3.5	4.7	1.96
CM 113a	24.0	22.0	16.0	1.7	3.1	4.4	5.5	1.8

Standard deviations ranged from 0.7 to 1.5 hr for period length and 0.1 to 0.7 cm/day for growth rate.

In CM211a the sensitivity of the rhythm to visible light has either been lost entirely or the threshold intensity increased. Such a mutant could arise from a qualitative or quantitative change in the receptor pigment or in any function necessary to couple the light to the oscillation mechanism.

CM113a, with a shortened period length in DD at 30°C, but with normal periods at 20 and 25°C, might represent a mutation affecting the temperature-compensation mechanism of the oscillation.

Of these three mutants, genetic data on the segregation pattern of the observed alterations exist only for CM89a, because of the apparent sterility of CM211a and CM113a. The usefulness of such mutants is, of course, severely limited if the mutation cannot be transferred to a more fertile strain.

That mutants affecting different parameters of the rhythm can be isolated suggests that the circadian oscillation as we observe it represents a conglomerate of components—e.g., the oscillation itself, the light sensitivity system, the temperature-compensation system—whose coupling has been effected by natural selection. The frequent occurrence in fungi of rhythms lacking one or more of these components and the apparent flexibility of fungal rhythms (Jerebzoff, 1965; Sussman, Durkee, and Lowry, 1965; Berliner and Neurath, 1966) is consistent with this speculation and suggests that these "noncircadian" rhythms may be more closely allied to typical circadian rhythms than appears at first glance.

ACKNOWLEDGMENTS

The research reported in this paper was supported by a postdoctoral fellowship of the U.S. Public Health Service, #5 F02 GM38017. We wish to thank N. H. Horowitz, in whose laboratory these experiments were carried out, for his interest and helpful suggestions, and S. Benzer, for several stimulating discussions.

REFERENCES

Ames, B. N. 1957. The biosynthesis of histidine. D-erythro-imidazole glycerol phosphate dehydrase. J. Biol. Chem. 228:131–143.

Ames, B. N., and B. Garry. 1959. Coordinate repression of the synthesis of four histidine biosynthetic enzymes by histidine. Proc. Natl. Acad. Sci. U.S. 45:1453–1461.

Barnett, A. 1966. A circadian rhythm of mating type reversals in *Paramecium multimicronucleatum*, syngen 2, and its genetic control. J. Cell. Physiol. 67:239–270.

Barnett, A. 1969. Cell division: a second circadian clock system in *Paramecium multimicronucleatum*. Science 164:1417–1418.

Beadle, G. W., and B. Ephrussi. 1937. Development of eye colors in Drosophila: Diffusible substances and their interrelations. Genetics 22:76–86.

Benzer, S. 1967. Behavioral mutants of *Drosophila* isolated by countercurrent distribution. Proc. Natl. Acad. Sci. U.S. 58:1112–1119.

Berliner, M., and P. Neurath. 1966. Control of rhythmic growth of a *Neurospora* "clock" mutant by sugars. Can. J. Microbiol. 12:1068–1070.

Brandt, W. H. 1953. Zonation in a prolineless strain of *Neurospora*. Mycologia 45:194–208.

Bünning, E. 1935. Zur Kenntnis der erblichen Tagesperiodizität bei den Primarblättern von *Phaseolus multiflorus*. Jahrb. Wiss. Bot. 81:411–418.

Danilyevsky, A. S. 1965. Photoperiodism and seasonal development of insects. Oliver & Boyd, Edinburgh.

Jacob, F., and J. Monod. 1961. Genetic regulatory mechanisms in the synthesis of proteins. J. Mol. Biol. 3:318–356.

Jerebzoff, S. 1965. Manipulation of some oscillating systems in fungi by chemicals, p. 183 to 189. *In* J. Aschoff [ed.], Circadian clocks. North-Holland Publ. Co., Amsterdam.

Malling, H. V., and F. J. deSerres. 1967. N-methyl-N-nitro-N-nitrosoguanidine (MNNG) as a mutagenic agent for *Neurospora crassa*. Genetics 56:575.

Pittendrigh, C. S. 1967. Circadian systems. I. The driving oscillation and its assay in *Drosophila pseudoobscura*. Proc. Natl. Acad. Sci. U.S. 58:1762–1767.

Pittendrigh, C. S., V. G. Bruce, N. S. Rosensweig, and M. L. Rubin. 1959. A biological clock in *Neurospora*. Nature (Lond.) 184:169–170.

Sargent, M. L., W. R. Briggs, and D. O. Woodward. 1966. Circadian nature of a rhythm expressed by an invertaseless strain of *Neurospora crassa*. Plant Physiol. 41:1343–1349.

Srb, A. M., and N. H. Horowitz. 1944. The ornithine cycle in *Neurospora* and its genetic control. J. Biol. Chem. 154:129–139.

Stadler, D. R. 1959. Genetic control of a cyclic growth pattern in *Neurospora*. Nature (Lond.) 184:170–171.

Sussman, A., T. L. Durkee, and R. J. Lowry. 1965. A model for rhythmic and temperature-independent growth in "clock" mutants of *Neurospora*. Mycopathol. Mycol. Appl. 25:381–396.

Sussman, A. S., R. J. Lowry, and T. Durkee. 1964. Morphology and genetics of a periodic colonial mutant of *Neurospora crassa*. Amer. J. Bot. 51:243–252.

Van Huong, N. 1967. Ph.D. Thesis. Université de Toulouse.

Westergaard, M., and H. K. Mitchell. 1947. Neurospora. V. A synthetic medium favouring sexual reproduction. Am. J. Bot. 34:573–577.

Wood, W. B., R. S. Edgar, J. King, I. Lielausis, and M. Henninger. 1968. Bacteriophage assembly. Fed. Proc. Fed. Amer. Soc. Exp. Biol. 27:1160–1166.

Participants

DR. P. L. ADKISSON, Department of Entomology, Texas A&M University, College of Agriculture, College Station, Texas 77843

DR. KRAIG ADLER, Department of Biology, University of Notre Dame, Notre Dame, Indiana 46556

PROF. DR. JÜRGEN ASCHOFF, Max-Planck Institut für Verhaltensphysiologie, 8131 Erling-Andechs, West Germany

DR. AUDREY BARNETT, Department of Zoology, University of Maryland, College Park, Maryland 20742

DR. RICHARD E. BELLEVILLE, Bioscience Programs, National Aeronautics and Space Administration, Washington, D.C.

DR. MIRIAM F. BENNETT, Department of Biology, Sweet Briar College, Sweet Briar, Virginia 24595

MISS SUE ANN BINKLEY, Patterson Laboratory, University of Texas, Austin, Texas 78712

DR. JOHN BRADY, Department of Zoology, Imperial College Field Station, Silwood Park, Ascot, Berks., England

DR. K. BRINKMANN, Institut für Molekularbiologie, 3301 Stöckheim, West Germany

G. A. BROWN, Division of Biological and Medical Research, Argonne National Laboratory, Argonne, Illinois 60439

DR. VICTOR G. BRUCE, Department of Biology, Princeton University, Princeton, New Jersey 08540

PROF. DR. ERWIN BÜNNING, Institut für Biologie, Der Universität Tübingen, 7400 Tübingen, Auf der Morgenstelle, West Germany

DR. THOMAS J. CROWLEY, Department of Psychiatry, University of Colorado Medical Center, 4200 East 9th Street, Denver, Colorado 80220

DR. BRUCE G. CUMMING, Department of Biology, University of New Brunswick, Fredericton, N. B., Canada.

DR. DAVID E. DAVIS, Department of Zoology, North Carolina State University, Raleigh, North Carolina 27607

DR. PATRICIA DeCOURSEY, Department of Biology, University of South Carolina, Columbia, South Carolina 29208

DR. ALICE DENNEY, Department of Plant Science, Utah State University, Logan, Utah 84321

MR. RICHARD DRISKILL, Department of Biological Sciences, 117 Wolf Hall, University of Delaware, Newark, Delaware 19711

DR. LELAND N. EDMUNDS, Department of Biological Sciences, State University of New York, Stony Brook, New York 11790

DR. C. F. EHRET, Division of Biological and Medical Research, Argonne National Laboratory, Argonne, Illinois 60439

DR. J. T. ENRIGHT, Scripps Institution of Oceanography, La Jolla, California 92037

DR. ARNOLD ESKIN, Division of Biology, Calif. Institute of Technology, Pasadena, California 91109

PROF. DONALD S. FARNER, Department of Zoology, University of Washington, Seattle, Washington 98105

DR. JERRY F. FELDMAN, Department of Biological Sciences, State University of New York at Albany, Albany, New York 12203

DR. SILVIA FROSCH, Department of Botany, The University of Western Ontario, London, Ontario, Canada

J. J. GAMBINO, Veterans Administration Center, Los Angeles, California

MISS SUZANNE GASTON, Patterson Laboratory 319, University of Texas, Austin, Texas 78712

URSULA GERECKE, Max-Planck Institut für Verhaltensphysiologie, 8131 Erling-Andechs, West Germany

DR. S. A. GORDON, Division of Biological and Medical Research, Argonne National Laboratory, Argonne, Illinois 60439

DR. EBERHARD GWINNER, Max-Planck Institut für Verhaltenphysiologie, 8131 Erling-Andechs, West Germany

DR. R. HALABAN, Department of Biology, Brookhaven National Laboratory, Upton, L.I., New York 11973

PROF. FRANZ HALBERG, Department of Pathology, University of Minnesota Medical School, Minneapolis, Minnesota 55455

PROF. KARL HAMNER, Department of Botanical Sciences, University of California, Los Angeles, California 90024

DR. W. M. HAMNER, Department of Zoology, University of California, Davis, California 95616

DR. J. W. HASTINGS, Biological Laboratories, Harvard University, 16 Divinity Avenue, Cambridge, Massachusetts 02138

P. HAYDEN, Northrop Corporate Laboratories, Hawthorne, California

DR. D. K. HAYES, USDA-ARS, Entomology Res. Division, Pesticide Chemicals Research Branch, Chemical and Physical Methods Investigations, 110 Center Bldg, ARC, Beltsville, Maryland 20705

FRANK H. HEPPNER, Department of Zoology, University of Rhode Island, Kingston, Rhode Island 02881

DR. WILLIAM S. HILLMAN, Department of Biology, Brookhaven National Laboratory, Upton, New York 11973

PROF. HOCHACHKA, Department of Zoology, University of British Columbia, Vancouver, B.C., Canada

DR. KLAUS HOFFMANN, Max-Planck Institut für Verhaltensphysiologie, 8131 Erling-Andechs, West Germany

DR. K. HOMMA, Department of Veterinary Medicine and Animal Sciences, Faculty of Agriculture, University of Tokyo, Bunkyo-ku, Tokyo, Japan

DR. HANS-WILLI HONEGGER, Zoologisches Institut der Universität zu Köln, 5 Köln-Lindenthal, Weyertal 119, West Germany

DR. T. HOSHIZAKI, Space Biology Laboratory, Brain Research Institute, Center for Health Sciences, University of California, Los Angeles, California 90024

DONALD IVES, Department of Biology, Amherst College, Amherst, Massachusetts 01002

PROF. JON W. JACKLET, Department of Biological Sciences, State University of New York at Albany, Albany, New York 12203

DR. K. E. JUSTICE, School of Biological Sciences, Department of Population and Environment Biology, University of California, Irvine, California 92664

MR. FRANK KARP, Department of Biology, New York University, University Avenue and 181st St., Bronx, New York

MR. RONALD J. KONOPKA, Division of Biology, California Institute of Technology, Pasadena, California 91109

MR. KENNETH KRAMM, Department of Population and Environmental Biology, University of California, Irvine, California 92664

DANIEL F. KRIPKE, Department of Psychiatry, Albert Einstein College of Medicine, New York City

ARMIN KURECK, Max-Planck Institut für Verhaltensphysiologie, 8131 Erling-Andechs, West Germany

PROF. ANTHONY D. LEES, Imperial College Field Station, Ashurst Lodge, Sunninghill, Ascot, Berks., England

PROF. MARVIN LICKEY, Department of Psychology, University of Oregon, Eugene, Oregon 97403

DR. R. G. LINDBERG, Life Sciences Laboratory, Northrop Corporate Laboratories, 3401 West Broadway, Hawthorne, California 90250

DR. MARY LOBBAN, National Institute for Medical Research, Hamstead Laboratories, Holly Hill, Hamstead, London N.W. 3, England

DR. LAURA McMURRY, Biological Laboratories, Harvard University, 16 Divinity Avenue, Cambridge, Massachusetts 02138

DR. ALBERT H. MEIER, Zoology Department, Louisiana State University, Baton Rouge, Louisiana 70803

DR. MICHAEL MENAKER. Patterson Laboratory 319, University of Texas, Austin, Texas 78712

DOROTHEA H. MINIS, Department of Biological Sciences, Stanford University, Stanford, California 94301

MR. JOHN MITCHELL, Department of Biology, Princeton University, Princeton, New Jersey 08540

MR. TADASHI OISHI, Department of Zoology, University of Alberta, Edmonton 7, Canada

DR. THEO. PAVLIDIS, Department of Electrical Engineering, Princeton University, Princeton, New Jersey 08540

G. VERNON PEGRAM, Holloman Air Force Base, New Mexico

PROF. COLIN S. PITTENDRIGH, Department of Biological Sciences, Stanford University, Stanford, California 94301

DR. R. J. PLANCK, Department of Zoology, University of Western Ontario, London, Ontario, Canada

DR. HERMANN POHL, Max Planck Institut für Verhaltensphysiologie, 8131 Erling-Andechs, West Germany

DR. THOMAS POULSON, Department of Biology, Yale University, New Haven, Connecticut 06520

DR. KENDALL PYE, School of Medicine, Department of Biochemistry, University of Pennsylvania, Philadelphia, Pennsylvania 19104

DR. LUDGER RENSING, I. Zoologisches Institut, Universität Göttingen, 34 Göttingen, West Germany

PETER RIEGER, Max-Planck Institut für Verhaltensphysiologie, 8131 Erling-Andechs, West Germany

S. H. ROACH, Cotton Insects Laboratory, Entomology Research Division, U.S. Department of Agriculture, Florence, South Carolina

DR. SHEPHERD K. ROBERTS, Department of Biology, Temple University, Philadelphia, Pennsylvania 19122

MR. BRUCE ROUSE, Patterson Laboratory 317, University of Texas, Austin, Texas 78712

YOSHIKAZU SAKAKIBARA, Department of Veterinary Medicine and Animal Sciences, Faculty of Agriculture, University of Tokyo, Bunkyo-ku, Tokyo, Japan 113

DR. FRANK SALISBURY, Plant Science Department, Utah State University, Logan, Utah 84321

DR. ROBERT G. SCHWAB, Department of Animal Physiology and the Institute of Ecology, University of California, Davis, California 95616

DR. HORST O. SCHWASSMANN, Scripps Institution of Oceanography, P.O. Box 109, La Jolla, California 92037

DR. STEVEN D. SKOPIK, Department of Biological Sciences, 117 Wolf Hall, University of Delaware, Newark, Delaware 19711

PROF. A. SOLLBERGER, Department of Psychiatry, Yale University Medical School, 34 Park Street, New Haven, Connecticut 06508

DR. KONRAD F. SPRINGER, Springer-Verlag, Neuenheimer Landstrasse 28–30, 69 Heidelberg, West Germany

URSULA von ST. PAUL, Max-Planck Institut für Verhaltensphysiologie, 8131 Erling-Andechs, West Germany

DR. F. STRUMWASSER, Division of Biology, California Institute of Technology, Pasadena, California 91109

DR. RICHARD H. SWADE, Biology Department, San Fernando Valley State College, Northridge, California 91324

DR. BEATRICE M. SWEENEY, Department of Biological Sciences, University of California, Santa Barbara, California 93106

DR. ERNESTO TRUCCO, Division of Biological and Medical Research, Argonne National Laboratory, Argonne, Illinois 60439

MR. JAMES TRUMAN, Biological Laboratories, Harvard University, Cambridge, Massachusetts 02138

MR. HERBERT UNDERWOOD, Patterson Laboratory 319, University of Texas, Austin, Texas 78712

DR. THÉRÈSE VANDEN DRIESSCHE, Université Libre de Bruxelles, Faculté des Sciences, 67, rue des Chevaux, Rhodes-St-Genèse, Brussels, Belgium

DR. EDGAR WAGNER, Department of Botany, Technical University, 51 Aachen, Hainbuchenstr. 20, West Germany

DR. GÖRAN WAHLSTRÖM, Farmakologiska Institutionen, Uppsala Universitet, Dag Hammarskjöldsvag 19, Uppsala, Sweden

NICKOLAS M. WASER, Department of Biological Sciences, State University of New York at Albany, Albany, New York 12203

DR. H. MARGUERITE WEBB, Marine Biological Laboratory, Woods Hole, Massachusetts; Goucher College, Towson, Maryland

DR. RÜTGER WEVER, Max Planck Institut für Verhaltensphysiologie, 8131 Erling-Andechs, West Germany

PROF. MALCOLM B. WILKINS, Department of Botany, The University, Glasgow W2, Scotland

DR. J. J. WILLE, Department of Biological Sciences, University of Cincinnati, Cincinnati, Ohio

PROF. A. O. D. WILLOWS, Dept. of Zoology, University of Washington, Seattle, Washington 98105

DR. W. O. WILSON, Department of Poultry Husbandry, University of California, Davis, California 95616

DR. ARTHUR T. WINFREE, University of Chicago, Department of Theoretical Biology, 939 E. 57th Street, Chicago, Illinois 60637

DR. C. M. WINGET, Physiology Branch, Ames Research Center, NASA, Moffett Field, California 94035

MR. KATSUHIKO YOKOYAMA, Department of Zoology, University of Washington, Seattle, Washington 98105

MR. SHELDON ZACK, Department of Psychology, University of Oregon, Eugene, Oregon 97403

DR. WILLIAM F. ZIMMERMAN, Department of Biology, Amherst College, Amherst, Massachusetts 01002